Farm and Ranch Financial Records

Farm and Ranch Financial Records

James D. Libbin *New Mexico State University*
Lowell B. Catlett *New Mexico State University*

Macmillan Publishing Company
NEW YORK
Collier Macmillan Publishers
LONDON

Printed in the United States of America

Macmillan Publishing Company
866 Third Avenue, New York, New York 10022

Collier Macmillan Canada, Inc.

Library of Congress Cataloging-in-Publication Data

Libbin, James D.
Farm and ranch financial records.

Includes index.
1. Agriculture—United States—Accounting.
I. Catlett, Lowell B. II. Title.
HF5686.A36L53 1987 657'.863'0973 86-18235
ISBN 0-02-370570-1

Printing: 1 2 3 4 5 6 7 8 Year: 7 8 9 0 1 2 3 4 5 6

ISBN 0-02-370570-1

Preface

The image of American agriculture is changing rapidly, both from the perspective of consumers and non-agriculturally oriented business and from the perspective of farmers, ranchers, and other agricultural industry participants. Farm and ranch booms and busts, foreclosures, news reports, and industry analyses all point out clearly that financial concerns now outweigh production concerns as the primary problems facing American agriculture. Simply stated, the financial-information revolution that changed American industry also pervades American agriculture.

Business and financial management (generally defined to include marketing, forward planning, financial control, cost cutting, use of credit, farm and ranch organization, tax planning, and risk control) is a main concern of modern commercial farmers and ranchers and the individuals who sell to, buy from, or advise them. Farmers and ranchers are bombarded with questions relating to their financial circumstances. Each one of these questions requires financial information about the individual's business, information that is usually difficult to provide. As agricultural economists concerned with the financial health of all of American agriculture, we sense the farmer's and rancher's frustration with obtaining the information necessary to make the very difficult, make or break, business and financial management decisions that now seem to arise frequently.

It is in this vein that we have written this book. We attempt to lay the groundwork for making the decisions by taking the first step in the information-decision-implementation process. That first step requires an adequate, accurate, complete farm or ranch business records system. Although this book concentrates on the form and development of farm and ranch business and financial records, we do not intend to imply that those records are an end in themselves. Business and financial records are solely a means to the end of enhanced, intelligent business and financial management decisions. These records provide merely the information so necessary to make informed decisions about each farm or ranch business.

Whereas this book will likely be used most extensively in the college classroom, we have attempted to write it for a broader audience. We believe it will be a useful guide and reference for farmers and ranchers, in fact we will use it in general extension and farm and ranch management association workshops. It should also find applications in Vocational Agriculture and Future Farmers of America programs. Our primary goal is to help farmers and ranchers and those high school and college students who soon will enter production agriculture. A secondary goal is to train future agricultural lenders and other agribusiness professionals in the quirks and necessities of farm and ranch record-keeping.

Many colleges and universities now offer a farm and ranch records course as a prelude or prerequisite to farm and ranch management and agricultural finance courses. We applaud this effort and have written this book to fit that niche. As stated previously we believe the first step is to develop a records system to provide the information needed to make marketing, business, and financial decisions.

This book has been divided into five major parts to introduce farmers, ranchers, and students to farm and ranch record-keeping issues and problems. The first part (Chapters 1–4) attempts to describe the need for and purpose of records and reviews the basics of accounting. It does not, however, presume to condense the entire accounting profession into one or two chapters. Readers are encouraged to study accounting in addition to this material. Part II (Chapters 5–12) presents, in detail, the components of a records system and includes many blank forms and examples that can be used for the development of a personalized system. Chapter 7 also includes a detailed discussion of depreciation methods, one of the most difficult topics and parts of a records system. Chapter 8 discusses another very difficult topic—scheduled loan repayment and interest analysis. Part III (Chapters 13–16) described how to summarize daily and annual records to prepare financial statements for financial analysis of the farm or ranch business and for credit application, whereas Part IV (Chapters 17–21) presents the steps of the farm and ranch income tax reporting process. Part V (Chapters 22–24) very briefly introduces the various types of ratios, measurements, and methods of a complete business management analysis. The last chapter discusses the design of a personalized records system. Chapter summaries draw out the most important points of each chapter into a condensed form for quick recall and review.

This book stresses *practical* application in that it draws heavily on forms and record books easily available to farmers and ranchers. It also draws heavily on the work of many dedicated records specialists around the country who have long strived to assist farmers and ranchers to *use* and *understand* their records.

The book also draws heavily on the comments and suggestions of several reviewers as well as many students, farmers, and ranchers who have survived our classes and workshops. We especially wish to acknowledge the thorough reviews, comments, advice, and guidance of Gary Condra, Texas Tech University; Larry Held, University of Wyoming; Mike Jones, Eastern New Mexico Farm and Ranch Business Analysis Association; Bruce Taylor, Purdue University; Tom Frey, University of Illinois; Gordon Bethards, Century Communications; and George Dawson, New Mexico State University. Undoubtedly there will be many comments, disagreements, arguments, and concerns with a presentation that is biased clearly by our preferences. Comments and suggestions for future revisions are welcome, just do not blame our reviewers and counselors, they did all that they could to educate us.

J.D.L.
L.B.C.

Contents

PART I

Records in Modern Agriculture

American agriculture is changing. Revolution after revolution has confronted agriculture and has forced the individual farmer or rancher to either change with or leave the industry. Past revolutions have caused agriculture to move from hand labor to horses and mules, to adopt new machinery, and eventually to replace horses with tractors. The direction of agriculture has also been changed by the introduction of nonnative crops such as soybeans, the development of hybrid seed, and the use of chemical fertilizers and pesticides.

Another revolution permanently changed the face of agriculture in the 1970s. Although less visible than the previous technological revolutions, the financial revolution in agriculture has had an effect just as dramatic as the introduction of the tractor. The financial revolution has included an explosion of debt financing, a tremendous increase in price volatility, high interest rates, and high inflation rates. It has also set the stage for the next major revolution: the information age. The advent of microcomputers, word processors, communications satellites, and computer networks will permanently change the business environment in which every farmer and rancher must compete.

The growing sophistication of financial management and the data processing environment has forced individual farmers and ranchers to make the choice once again between adapting to the change or leaving the industry. Only if an individual is debt-free, financially independent, and satisfied with less-than-market rates of return on labor and investment can one of those two choices be avoided. For the young farmer or rancher or one who is not debt-free, the choice is no longer between prosperity and mediocrity; it is a matter of survival.

There are two major tools that are now critical to survival and eventual success: financial records and economic analysis of business activities. Without in-depth analysis and planning, a farm or ranch business will not be able to grow. It may not survive or succeed even if the analysis has been accurately and completely done, but without analysis and planning its survival chances are very small.

A popular story can serve as a comparison with current conditions in agriculture. The World Series of Poker, held annually in Las Vegas, Nevada, draws the world's best players. Only those who have extensive training, discipline, and experience and have seriously studied the game are able to enter the Series. Almost by definition, the players in the final rounds are all fine poker players. But not all will win. Did the losers fail to win because they were not highly skilled or well prepared? Definitely not—the luck or the cards just did not fall for them. In fact, the 1984 Series champion won with a final hand of only a pair of tens! The analogy for agriculture is that although planning will not necessarily guarantee success, only those farmers and ranchers who seriously study and prepare for the challenge facing them will be able to survive.

It is often stated that farming or ranching on paper is much more successful than farming or ranching in the real world. Penciling a profit on paper will not guarantee a profit in fact, but if a decision cannot pencil in a profit, it almost assuredly will not lead to a real profit.

The four chapters that make up Part I of this text introduce the basic necessities of a modern farm and ranch records system and detail why records must be maintained. Additionally, Chapter 2 reviews the highlights of accounting theory applicable to a family-size sole proprietorship. This chapter is not a complete study of accounting theory. Students, farmers, and ranchers are encouraged to further study the accounting field, some knowledge of which is critical to business survival. Chapters 3 and 4 will briefly discuss the types of records needed and generally what they are used for, as well as the types of records systems commonly used on farms and ranches today. Detailed analysis of the necessary components will be deferred to Part II.

CHAPTER 1

Farm and Ranch Records Systems

Decision making, economic analysis, and financial management are all necessary for future prosperity and survival. But, before decisions can be made or analyzed, the information necessary for the decisions must be available. In other words, all financial management, analysis, and decisions must have a source of accurate information pertaining to the particular farm or ranch business. It is the task of the farm and ranch records system to provide the required information in an orderly, accurate, and complete fashion.

Purposes of Records

Business management analysis and control should be the primary goal of any farm or ranch records system. In other words, the records system should be geared toward the farm or ranch manager—if the records system is not used, it is for all practical purposes worthless. Because many of the uses of the records system relate to individuals other than the manager, its most important reason or purpose is to communicate financial information.

As indicated in the organizational chart shown in Figure 1–1, the records system can and should be a vital link in the information flow between a manager and others concerned with his business. In fact, the records system ties all major management functions together by providing the information necessary for decision making. The financial information of a business firm must flow between the bookkeeper and the manager (although on small or medium-sized farms or ranches these two roles may be played by one individual), between the manager and his financial support staff, between the manager and tax authorities, and between the manager and his ownership/management team. The manager's support staff may include such individuals as his lender(s), accountant, management and tax advisors, banker (for purposes in addition to lending), trust officer, attorney, and investment advisors. Tax authorities include federal, state, and local

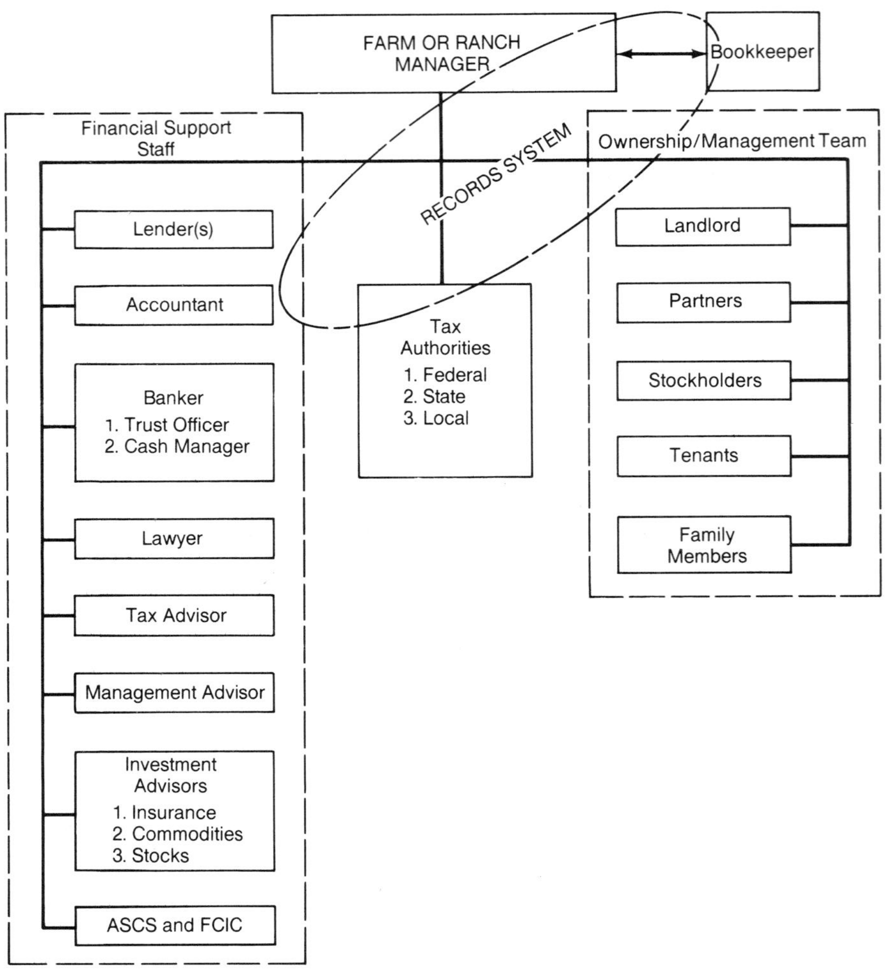

FIGURE 1–1. The farm or ranch manager and the management team.

agency representatives for income, social security, estate and inheritance, and property tax purposes. The ownership/management team may or may not include individuals other than the manager and the spouse, depending upon the business's particular legal form of organization.

To summarize the major purposes of a farm or ranch records system in outline form, the most important uses are listed below. The order in which these purposes is listed is not necessarily the order of greatest importance.

1. Income tax reporting and tax management
2. Credit application
3. Cash flow and financial control
4. Business management analysis and decision making

Although these four purposes may be the most important for the vast majority of farm and ranch managers, there are other reasons for keeping records, which can be extremely important in specific situations on any given farm or ranch. They include:

5. Corporate distributions
6. Partnership distributions
7. Share lease distributions
8. Gift, inheritance, and estate taxes
9. Off-farm or off-ranch investment opportunities

Tax Requirements

The Internal Revenue Service (IRS) and most state income tax authorities require that enough business records be kept to justify all income and expense claims reported on an income tax return. There are no specific legal requirements for records on unincorporated farms and ranches. However, the IRS can require that all items of income, deductions and credits be verified by sales slips, invoices, cancelled checks, or other documents. IRS comments on farm (and ranch) record requirements, taken from Chapter 1 of the annual Farmer's Tax Guide, are displayed in Box 1–1.

The lack of standardized requirements for a minimum acceptable set of records has led some farmers and ranchers to store their cash register receipts, invoices, bank statements, and cancelled checks in a box or file drawer and to do very little more. Legally, such records are sufficient. This system can become an extremely expensive one, however, during an IRS examination, if, the manager is called upon to substantiate claims made on an old tax return. Memories can fade, short explanatory notes can lose significance, and receipts can be lost. If all entries on the tax return cannot be supported, additional tax liability and penalties can be assessed. A complete, accurate records system can help farmers and ranchers avoid memory problems and present a businesslike, comfortable, assured attitude in negotiations with tax authorities; furthermore, the records system can provide verifiable sources of information to save additional tax dollars.

Credit Application

During the 1970s, an explosion in the number of dollars changing hands took place. Much of this explosion was due only to inflated prices, but high interest rates and increased amounts of debt capital also increased cash flow needs. Many lenders financed and refinanced farmers and ranchers solely on the basis of inflating land values; as land values increased and dollar-denominated mortgages remained constant, net equities increased and provided collateral surety to lenders. Many farmers and ranchers, however, refinanced to such an extent that their equities were stretched very thinly or were even lost altogether. After a fairly long period of rising land values, the bubble burst in the early 1980s. Land prices no longer increased, and in many areas they fell. Leveling and declining land values caused havoc in a lending market dominated by collateral interests rather than the capacity of the business to repay the loan.

In more recent years, lenders have come to stress repayment capacity of loans as well as collateral. In other words, most borrowers need to show that the

BOX 1–1

1. Importance of Good Records

A farmer, like other taxpayers, must keep records to prepare an accurate income tax return, and be sure only the tax owed is paid. Items such as cancelled checks and paid bills that support entries in your books, should be filed in an orderly manner and stored in a safe place.

Advantages of good records. Good records can save you money. They are needed for efficient farm management and are useful in preparing credit applications. Here are some ways in which they may help.

Identify source of receipts. The money or property you receive can come from many sources. Your records can identify the source of your receipts. For example, you can prove that they are from nonbusiness sources, or that they are not taxable.

Keep track of deductible expenses. Expenses may be forgotten when you prepare your tax return unless you record them when they occur. An overlooked item of $25 could cost you $3.00 or more in taxes.

Figure depreciation allowance. You should record the farm assets you can depreciate in a permanent record. Without a record of the cost and other information on your assets, you cannot figure your depreciation deductions. If the assets are sold, become fully depreciated, or capital improvements are made to them, only a permanent record will show how much of their cost you have not recovered. A sample depreciation record is in Chapter 19. The information there was used to prepare the Form 4562. *Depreciation and Amortization,* in Chapter 19.

Take advantage of capital gain and loss laws. If your records are good and complete, they show the date an asset was acquired, what it was used for, and whether it was sold, traded, destroyed, or otherwise disposed of. With this information you may be able to use the capital gain or loss laws. You may be allowed to put off paying tax on certain gains, or to deduct 100% of certain losses that would otherwise not be deductible, or would be only partly deductible.

Figure your earnings for self-employment social security tax. The self-employment tax is part of the system for providing social security coverage for people who work for themselves. The social security benefits that you can get when you retire, or are disabled, or that your family can get when you die, depend on how much you earn. Your records should show how much of your earnings you pay self-employment tax on.

Support items reported on your income tax return. If your income tax return is examined by the Internal Revenue Service, you may be asked to explain the items reported. Adequate and complete records are always supported by sales slips, invoices, receipts, bank deposit slips, cancelled checks, and other documents. They can explain the entries on your tax return.

Recordkeeping. No special kind of record books are required. However, your books should contain all the information you need to figure your income, deductions, credits, and other items that you have to show on your income tax return.

The source of receipts and expenditures for the year should be shown in your records. Only by keeping good records can you be sure that you will pay only the tax you owe.

Retaining your records. You must keep your records for as long as they may be needed in the administration of any Internal Revenue law.

Keep records that support an item of income or expense on your tax return until the statute of limitations for the return runs out. Usually this is 3 years after the date the return is due or filed, or 2 years from the date the tax was paid, whichever occurs later.

Keep records that verify your basis in property for as long as they are needed to figure the basis of the original or replacement property.

A farm income and expense record is illustrated in Chapter 19. It is part of a typical farm account book that is used to record farm business receipts and expenses for the tax year.

Source: Internal Revenue Service, Publication 225, *Farmer's Tax Guide* investment for which the loan is intended will be able to generate enough income to pay back the interest and principal owed within the specified time period.

To determine their repayment capacities, lenders need cash flow projections and other financial documentation. Both of these needs have put pressure on the farm and ranch records system to provide the necessary information. Not only will a complete, accurate records system provide the information, but it will also project a businesslike image.

Cash Flow and Financial Control

Just as lenders are concerned with cash flows and repayment capacity, financial advisors and astute business managers have also become greatly concerned with cash flow management. How much to borrow either in long-term investments or in operating credit is only half of the story. When and how much to pay back borrowed funds is just as important as tight control of cash reserves. Paying operating money back after a sale may not be the wisest option if it later puts the business in a cash flow bind. Interest charges must be analyzed in addition to liquidity needs of the business and of the family.

Preparation of a realistic cash flow budget is one of the first, vital steps in the annual record-keeping process. It is a difficult process because it involves all the steps required in the whole farm or ranch planning process: marketing (including price projections for inputs as well as outputs), yield goals and projections, and enterprise combinations. Despite the difficulty of preparation, the cash flow budget helps document managerial abilities and loan repayment capacities. Furthermore, the cash flow budgeting process can be extended one more step to provide an extremely effective financial control device. If monitored monthly, or even quarterly, the cash flow budget can indicate potential problems before they arise. This timing or ability to foresee problems will allow the manager to adjust before the fact rather than react afterwards. Finally, the cash flow record of the past year, if maintained in the records systems on a regular basis (usually a monthly basis), can become an extremely useful guide in developing a cash flow budget for the next year.

Business Management Analysis

If a farmer or rancher will discipline himself sufficiently to develop and maintain a records system to meet income tax reporting and credit application needs, he will have virtually all the information assembled and effort expended to meet the fourth and probably most important goal of a farm or ranch records system: business management analysis.

Is a good farm or ranch manager indicated by the ease with which he can negotiate an equitable solution with the Internal Revenue Service? Is a good farm or ranch manager indicated by his ability to obtain every loan applied for? The answer is partly yes, but there is more to it than that.

A good farm or ranch business manager knows exactly what his variable and total costs of production are; he knows whether he is meeting the goals of his marketing plan or his cash flow budget. He has analyzed his strengths and his weaknesses, both in physical terms and in financial terms. In short, he knows where his business has been, where it is now, and where it is going.

It is the comprehensive farm or ranch records system that provides the background information necessary to know costs, achievements, strengths and weaknesses, and business directions. As aptly stated by Guenthner and Wittman, "Some of the most common applications for good records are:

1. Measuring operating and financial performance
2. Supporting requests for operating and capital credit
3. Establishing proper insurance coverages for farm principals and capital assets
4. Providing a basis for estate valuation and estate planning
5. Serving as a basis for analyzing investments in farm expansions or depreciable assets
6. Measuring the profitability of individual enterprises where multiple enterprises exist in an operation
7. Keeping track of production inventories and
8. Developing a sound marketing plan."[1]

The Businesslike Attitude

In each preceding section we have alluded to a businesslike posture or attitude. Each farmer and rancher has many goals for his life, his business and his family. Maximization of profit or income and equity growth may be important goals, but they are probably not the only or even the most important goals. Several studies have been undertaken to determine the goals which farmers and ranchers view as most important to them. Seldom is the maximization of profit or equity listed as the most important goal. Rather, such goals as staying in business, providing a good life and environment for their family, reducing debt and providing a base for continued education of their children rate the highest among farmers and ranchers. Despite the low ranking of profit maximization, it should be clear that all of these goals are interrelated. A businesslike attitude about financial affairs and making the most of the business can be extremely effective forces in helping each farmer or rancher achieve other goals.

Corporations and Partnerships

Multiple-owner or legal forms of business organization other than the sole proprietorship require more detailed records because tax reporting requirements

[1]J. F. Guenthner and R. L. Wittman, "Selection and Implementation of a Farm Records System" (Moscow, Idaho: University of Idaho, Cooperative Extension Service, Bulletin No. 633, June 1984).

are more intricate, because state corporation laws may require the organization to file certain documents, and because lenders need more documentation when the chain of responsibility and liability is complicated by legal form and the larger number of individuals involved. But perhaps the most important need for more detailed records in partnerships and corporations stems from the greatly expanded likelihood of problems and potential conflicts among the individuals involved.

It is commonly felt among farm and ranch business management specialists, accountants, lawyers, tax advisors, and consultants that partnerships and corporations are easier to form than to dissolve. Many of the problems that arise during the dissolution of these business organizations result from the joint ownership of assets which cannot be broken into pieces in such a way that each party receives a fair share. Confusion also results from questions about the past contribution of each party.

Although on a somewhat less critical scale, there is always the potential for similar confusion about the distribution of contributions and profits in multiple-owner businesses. A detailed record-keeping system can easily pay for itself in terms of the savings it generates during disagreements among family and friends. Often, a strictly kept detailed records system can be cited as an important factor in keeping a multiple-owner business together.

Lease and Family Distributions

As with multiple-owner forms of business organization, individuals involved in informal family partnerships or joint ventures need to rely upon a detailed records system to ensure fairness in distribution of profits and contributions. Similarly, lessees and lessors will benefit if all financial transactions are recorded.

The bottom-line rule for all multiple-individual businesses is to write down everything possible in a simple, easily understood but detailed records system. The work involved is easily justified through the avoidance of conflict between owners when it comes time to make sure that everyone involved received a fair share of the benefits in proportion to the burden of contribution.

Other Taxes and Investments

Estate tax planning is similar in concept to all other forms of tax, marketing, and management planning. Its major purpose is to deal with and possibly avert a problem before it arises rather than to pass it on to someone else at some future date. Estate, gift, and property tax problems can be minimized if complete and accurate records are maintained.

Furthermore, the ability to participate in investments outside of normal farm or ranch business activities can be enhanced by having the information readily available to determine whether a particular investment opportunity is financially feasible at a particular point in time.

Developing a Records System

Professional accountants often remind themselves and others that the critical element in all financial records is substance, not form. In other words, the most important part of a records system is that it must be complete and it must be used. It must be flexible enough to allow the bookkeeper and the manager (who are often the same person) to record the essential facts in a logical fashion that permits retrieval. The least important part of a records system is the "right" paper or a neat look. The only reason for using the best laid-out forms is that typically they have drawn upon the experience of others to achieve a simple but logical system. However, a system that achieves completeness, accuracy, and simplicity may be the best.

The individual who wishes to study his current records system or who wishes to develop a new one must first be concerned with what needs to be included in the system. The following chapters in this book describe the records needed as well as how to record necessary information. Despite the emphasis on form, the purposes of the records system described earlier are not forgotten. The main goal, once again, is to develop a system that allows easy retrieval of information for tax reporting and management, credit application, management analysis, and cash flow control.

Summary

The financial environment in which modern farmers and ranchers must compete has changed dramatically since the 1960s. No longer can farmers and ranchers rely solely upon hard work and dedication to producing crops and/or livestock. To survive in agriculture today and in the future, farmers and ranchers must become astute business managers as well as efficient producers. The first step to becoming a business manager is to develop a thorough, complete and accurate records system.

This chapter has pointed out that:

1. The farm or ranch records system is a vital link in providing financial information for decisions.
2. The major purposes of a records system are
   - **a.** income tax reporting and tax management
   - **b.** credit application
   - **c.** cash flow and financial control
   - **d.** business management analysis and decision making.
3. Tax records are often insufficient for management analysis.
4. Most lenders now stress repayment capacity and require more records and projections before approving a loan.
5. Cash flow is a critical element in business survival.
6. Business management analysis is vital to success.
7. Corporations, partnerships and leases often require additional records due to the potential for conflict among participating individuals.

CHAPTER 2

The Basics of Accounting

As in every other discipline, accounting has its own terminology—terminology that is used to convey precise meaning to specific concepts. As is true with every other discipline, learning that terminology is a major stumbling block for many students and business people alike. A brief introduction to several of the major accounting concepts should lay the groundwork for the organization of a comprehensive farm and ranch records system.[1]

The Fundamental Accounting Equation

Virtually the entire discipline of modern financial accounting revolves around one basic concept, the *fundamental accounting equation.* That equation is

$$\text{Assets} = \text{Liabilities} + \text{Net Worth}$$

or simply

$$A = L + NW$$

An *asset* is any item owned or controlled by the business that has value, or any item that is owed to the business by other individuals or businesses. For example, cash on hand, machinery, market and breeding livestock, buildings, and land are all assets. Each of these is also a tangible asset, but there are other assets that may not be so obvious. A note receivable is an asset; it is a contract that you hold which requires another person to repay money that you lent to him. Similarly, cash invested in a growing crop (such as winter wheat) is an asset, even though that crop cannot be sold in the form of grain on a particular day (say January 1).

To be an asset, the item does not necessarily need to be owned free and clear.

[1]This book does not presume to replace an accounting text. Every reader of this book is encouraged to study accounting at a college or university or through a correspondence course.

In other words, there could be a lien, a mortgage, or some other loan outstanding (called a note payable) against that item. Although you may feel that the bank owns more of the item than you do, you would classify the entire value of the item as an asset and the amount of money owed on it as a liability.[2]

A *liability,* then, is any amount *owed* to some other person or business entity. Liabilities are normally financial items rather than physical ones. They may be contractual obligations such as loans, mortgages, or other notes payable; or accrued obligations such as wages earned by employees but not yet paid; or potential obligations (called *contingent liabilities*) such as the income taxes due only if an item or the entire business was sold.

Net worth is defined as the difference between the sum of the asset values and the sum of the liabilities. In other words,

$$NW = A - L$$

Net worth, often called *owner's equity,* thus represents what the business owner is worth, or the sum of his claims on the collective set of assets controlled by the business. To understand the concept, define *net worth* to be what the owner would be worth (how much money he would have in his hands) if all assets were sold or collected and all liabilities were paid. Thus, assets are what the whole business is worth and net worth is the owner's share (or equity) in the business. Liabilities represent other individual's shares of the business. In other words, liabilities and net worth are claims against the assets owned by a person or business. Liabilities represent claims of individuals other than the owner, whereas the owner's net worth represents his claim on those assets. Claims by individuals other than the owner almost always have priority or first claim on the assets.

The Fundamental Equation and Transactions

The concept of the fundamental accounting equation can be carried far beyond the balance sheet listing of assets, liabilities, and net worth. Use of the equation can, in fact, be extended to analyze every business transaction by recognizing that every transaction affects the equation in a balanced manner. To show this we will introduce the T-account and provide a few examples. But first, let's return to terminology to define *debit* and *credit. Debit* simply means the left-hand side of an account and *credit* means the right-hand side. Accounts are set up for three major categories: assets, liabilities and net worth. Under net worth there are three major subdivisions: equity, income, and expense. The structure of the overall system and the effect of changes in each of the accounts are depicted in Figure 2–1.

In the diagram, each T-account will likely have both increases and decreases, thus both debits and credits, but the effect of an increase or decrease will vary depending upon the type of account.

Consider the numbered transactions listed following, their effects on the T-accounts, and their effects on the fundamental accounting equation.

[2]Conversely, the bank would classify the loan as a note receivable and thus an asset.

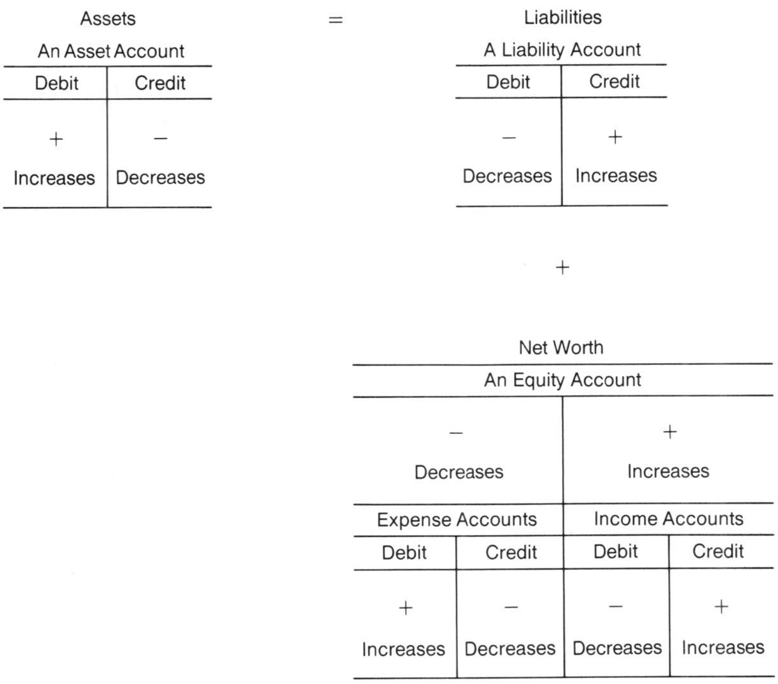

FIGURE 2-1. Relationship of types of accounts to the fundamental equation.

1. A tractor is purchased for $75,000. The entire amount is financed through a local bank.
2. A tractor is purchased for $75,000; $25,000 is paid in cash and $50,000 is financed through a local bank.
3. Fertilizer is purchased for use in the current year. Cash is paid.

Transaction 1: A tractor is purchased for $75,000. The entire amount is financed through a local bank.

***Analysis*:** The addition of the tractor increases the machinery account (an asset) by $75,000—a debit. The new loan increases the notes payable account (a liability) by $75,000, a credit.

***T-accounts*:**

Machinery			*Notes Payable*	
+ $75,000				+ $75,000

***Equation*:**

$$A = L + NW$$
$$75{,}000 = 75{,}000 + 0$$

Transaction 2: A tractor is purchased for $75,000; $25,000 is paid in cash and $50,000 is financed through a local bank.

***Analysis*:** The addition of the tractor increases the machinery account by $75,000—a debit. The down payment reduces cash (another asset) by $25,000—a credit. The new loan increases the notes payable account by $50,000—a credit.

T-accounts:

Machinery	
+	
$75,000	

Notes Payable	
	+
	$50,000

Cash	
	–
	$25,000

Equation:

$$A = L + NW$$
$$\$75{,}000 - 25{,}000 = 50{,}000 + 0$$

TRANSACTION 3: Fertilizer is purchased for $5,000 for use in the current year. Cash is paid.

Analysis: Cash is reduced by $5,000—a credit. Fertilizer expense is increased by $5,000—a debit—which reduces net income and thus reduces owner's equity when expense and revenue accounts are summarized into owner's equity.

T-accounts:

Cash	
	–
	$5,000

Fertilizer Expense	
+	
$5,000	

Equation:

$$A = L + NW$$
$$-5{,}000 = 0 - 5{,}000$$

We will return to the T-account concepts and the balancing of the fundamental equation in later chapters. But for now three major points should be highlighted:

1. For every transaction there is at least one debit and at least one credit.
2. The fundamental accounting equation must *always* remain in balance. This means that debits must always equal credits.
3. Debit means left and credit means right, *not* increase or decrease. Table 2–1 summarizes the relationships between increases/decreases and debits/credits.

TABLE 2–1 Relationships Between Debits/ Credits and Increases/Decreases

Debits	Credits
1. Increase assets	1. Decrease assets
2. Increase expenses	2. Decrease expenses
3. Decrease income	3. Increase income
4. Decrease equity	4. Increase equity
5. Decrease liabilities	5. Increase liabilities

Accrual versus Cash Accounting

There are two major methods of maintaining accounting records and reporting summarized accounting information, the accrual accounting method and the cash accounting method. The difference between the two lies primarily in the time when a transaction is recorded and in the interpretation of summary statements.

Cash Accounting Method

Under the cash accounting method, every transaction is recorded at the time that payment is made or received. The critical element is that cash (including transfer of cash by check) must change hands. For example, if fertilizer was purchased on a specific date and payment was made that day (either in cash or by check), the transaction would be reported on that day. But if the fertilizer was charged to an account, the transaction would not be recorded until the charge account bill was paid. Similarly, if a longer-term asset, such as a tractor, was purchased on a specific date and payment was made, the tractor would be reported on that day. If the tractor was financed by a loan, however, no transaction would be reported until the first loan payment.

Accrual Accounting Method

Under the accrual accounting method, transactions are recorded whenever they actually take place regardless of when payment is made—in other words, whenever assets and/or liabilities are transferred. Using the same fertilizer example of Transaction 3, we note that the transaction is recorded on the day the agreement is made. If cash changed hands, the farmer transferred an asset (cash) to the fertilizer dealer and the fertilizer dealer transferred ownership of an asset (fertilizer) to the farmer. If the price of the fertilizer has been placed on a charge account, the farmer has a new liability (account payable), and the fertilizer dealer has a new asset (accounts receivable) other than cash. The effect on the earnings of both individuals is the same whatever the form of payment. In the tractor example of Transaction 2, the transaction is recorded on the day the agreement is made. The tractor is recorded as a new asset on that day, regardless of the form of payment. If it is purchased with cash, the farmer transfers cash to the equipment dealer and the dealer transfers the tractor. If the purchase is financed, the farmer accepts both the tractor and a new liability.

Receipts versus Income

To clarify these differences and to distinguish forms of payment further, we can define two terms on the revenue side of a transaction. Consider a transaction that brings earnings into a business, for example the sale of market livestock. If payment accompanies that sale, call the payment a *receipt*. If the market livestock has been raised within the current accounting period and should be considered earnings in that period, call the transaction *income* regardless of the timing of payment.

Record a receipt when cash is received (cash method) and record income when cash is realized or earned (accrual method).

A receipt is a payment in cash; income is value (either in the form of cash or any other asset) generated by a factor of production such as land, labor, or any other investment. Any given transaction may be both a receipt and income, or it may be one but not the other in any given time period. Consequently, as shown diagramatically in Figure 2–2, there are three possibilities: a transaction may be a receipt but not income, income but not a receipt, or it may be both.

Consider the following transactions and decide when or whether they would be classified as a receipt or as income.

4. Corn produced in 19X1 and sold for cash in November 19X1.
5. Corn produced in 19X1, delivered to an elevator in November 19X1 and sold for cash in January 19X2.
6. Loan proceeds received from a bank in December 19X1 to be used to purchase land.
7. Cash received in November 19X1 when a used tractor is sold to a neighbor.

Transaction 4: Corn produced in 19X1 and sold for cash in November 19X1.

Analysis:
- **a.** Classify as a *receipt* in November 19X1 since cash is received (cash method).
- **b.** Classify as *income* in November 19X1 since the crop was produced and value was earned in 19X1 (accrual method).
- **c.** The transaction is *both* a receipt and income in 19X1.

Transaction 5: Corn produced in 19X1, delivered to an elevator and sold for cash in January 19X2.

Analysis:
- **a.** Classify as a *receipt* in January 19X2 (cash method).
- **b.** Classify as *income* in November 19X1 since the crop was produced, the value was earned, and the crop was available for sale in 19X1 (accrual method). In 19X2, an entry in the accounting books would reflect that inventory was transferred for cash received, but no income was earned in 19X2. Only the form of the asset changed.
- **c.** The transaction is income in 19X1 but not a receipt.
- **d.** The transaction is a receipt in 19X2 but not income.

Transaction 6: Loan proceeds received in December 19X1 for purchase of land.

Analysis:
- **a.** Classify as a *receipt* in 19X1 (cash method).
- **b.** Loan proceeds are never classified as income. Under the accrual method, loan proceeds would require an account-

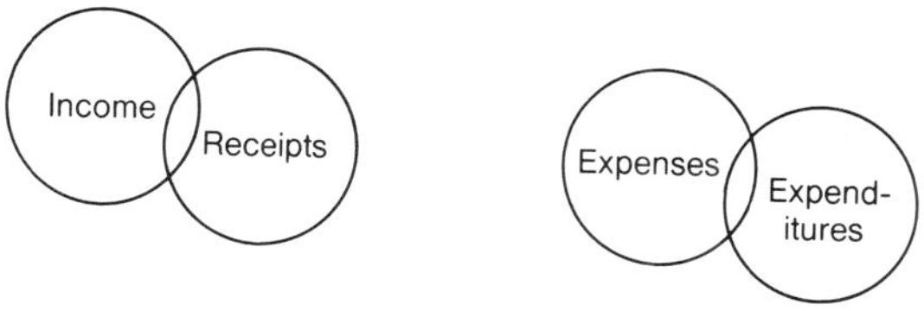

FIGURE 2–2. Receipts versus income and expenditures versus expenses.

ing entry that increases cash and increases notes payable (a liability).

TRANSACTION 7: Cash received when a used tractor is sold.

Analysis:

a. Classify as a *receipt* in November 19X1 since cash is received (cash method).

b. Cash received from the sale of an intermediate or long-term asset is usually not classified as income. Under the accrual method, sale of an asset would require an entry that reduces (that is, credits) the tractor or machinery account and increases (that is, debits) cash. A portion of this sale *might* be classified as income if the amount of cash received is different from the value of the asset as listed in the tractor account. Only the difference would be classified as income, since that difference is a gain or loss.

Expenditure versus Expense

As there are two terms and concepts for revenues coming into the business, there are also two terms to classify payments going out of the business. Consider a transaction such as Transaction 3, the purchase of fertilizer (page 14). If payment accompanies the purchase, call the payment an *expenditure.* But if the fertilizer was to be applied within the period of the purchase, call the transaction an *expense.*

Record an expenditure when cash is paid (cash method), and record an expense when it is realized or incurred (accrual method).

As with revenues, a given transaction may be both an expenditure and an expense; or it may be one but not the other in any given time period. Consequently, as shown in Figure 2–2, there are again three possibilities: a transaction may be an expenditure but not an expense, an expense but not an expenditure, or both.

Consider the following transactions and decide when or whether each one would be classified as an expenditure or an expense.

8. Fertilizer purchased for $5,000 for use in 19X1 and paid for in cash in February 19X1.

9. Fertilizer purchased for $5,000 in December 19X1 for use in 19X2 and paid for in cash in December 19X1.

10. Fertilizer purchased for $5,000 on account in March 19X1 for use in 19X1. Charge account paid in January 19X2.
11. Fertilizer purchased for $5,000 on account in December 19X1 for use in 19X2. Charge account paid in April 19X2.
12. Loan payment made in November 19X1 ($4,000 principal and $3,000 interest).
13. Depreciation on owned machinery realized in 19X1.
14. Tractor purchased for $80,000 in 19X1.

TRANSACTION 8: Fertilizer purchased in 19X1 for use in 19X1 and paid for in cash.

Analysis:
a. Classify as an *expenditure* in 19X1 since cash is paid (cash method).
b. Classify as an *expense* in 19X1 since the fertilizer is used in the 19X1 production period (accrual method).
c. The transaction is *both* an expenditure and an expense in 19X1.

TRANSACTION 9: Fertilizer purchased in 19X1 for use in 19X2 and paid for in 19X1.

Analysis:
a. Classify as an *expenditure* in 19X1 since cash is paid (cash method).
b. Classify as an *expense* in 19X2 since the fertilizer was used in 19X2 (accrual method). In 19X1, an entry in the accounting books would reflect that cash was transferred for fertilizer received, but no expense occurred in 19X1. Only the form of the asset changed.
c. The transaction is an expenditure in 19X1 but not an expense.
d. The transaction is an expense in 19X2 but not an expenditure.

TRANSACTION 10: Fertilizer purchased on account in 19X1 for use in 19X1. Cash paid in 19X2.

Analysis:
a. Classify as an *expenditure* in 19X2 since cash is paid (cash method).
b. Classify as an *expense* in 19X1 since the fertilizer was used in 19X1 (accrual method).
c. The transaction is an expenditure in 19X2 but not an expense.
d. The transaction is an expense in 19X1 but not an expenditure.

TRANSACTION 11: Fertilizer purchased on account in 19X1 for use in 19X2. Cash paid in 19X2.

Analysis:
a. Classify as an *expenditure* in 19X2 (cash method).

b. Classify as an *expense* in 19X2 (accrual method).
c. The transaction is both an expenditure and an expense in 19X2.

TRANSACTION 12: Loan payment made in 19X1.

***Analysis*:**
a. Classify as an *expenditure* in 19X1 (cash method).
b. Classify interest portion as an *expense* in 19X1 (accrual method).
c. Principal portion is never classified as an expense.

TRANSACTION 13: Depreciation on machinery for 19X1.

***Analysis*:**
a. Classify as an expense in 19X1 (accrual method).
b. Depreciation is never classified as an expenditure.

TRANSACTION 14: Tractor purchased in 19X1.

***Analysis*:**
a. Classify as an *expenditure* in 19X1 (cash method).
b. The purchase price of the tractor will be prorated over its economic useful life. That proration is called depreciation and will be classified as an expense through the years as recognized.

Which Method Should Be Used?

Generally accepted accounting principles (GAAP), established by accountants, indicate that the accrual method of accounting is always preferred. GAAP are often given the force of law through the Securities and Exchange Commission (SEC) and the Internal Revenue Service (IRS). All publicly held corporations are required by the SEC to maintain accounting records on the accrual basis and are further required by the IRS to report income tax liability based on the accrual method. Many small businesses, however, are allowed by the IRS to report income tax liability on a slightly modified cash method. No other government agency requires small businesses to maintain accounting records.

Under the current tax law structure, virtually all farmers and ranchers should report their tax liability on the cash basis, but to make the information reported useful, financial statements should be prepared on the accrual basis. Because of these two conflicting goals and because one of the primary (although unstated) goals of the records system is to make record keeping as easy as possible, a modified mixed cash/accrual system will fit many farmers and ranchers.

The major advantages of the accrual accounting method are:

1. It more clearly leaves an *audit trail* (see Chapter 5, page 54).
2. It allows nearly instantaneous retrieval of a current balance sheet and income statement.
3. It more accurately reflects net income in the current period.

In other words, the accrual accounting method provides a financial track that can be more easily retraced, while the two most basic financial statements can be generated quickly so that on any given day a current balance sheet could be written and for any given period of time (such as one week, one month, or one quarter) a current income statement could be developed.

In defense of the cash method (or rather a modified cash method), a farmer or rancher using it seldom needs a formal audit and he very rarely needs to prepare financial statements more often than once each year.[3] All business firms that work with farms and ranches realize that income from these is received in lumps at harvesttime or when livestock or livestock products are sold and that cash-short periods always occur near planting time or when the farmer or rancher is purchasing livestock or feed. Consequently, creditors typically need the farmer's or rancher's statements just once each year.

There is no doubt that the accrual method provides a much more accurate income picture than does the cash method, but with certain modifications that will be discussed in the next chapter, the cash method can be modified slightly to provide all the information necessary to prepare accrual financial statements.

For most farm and ranch businesses, a modified cash/accrual accounting system will be sufficient. The purpose of the modification is to capture all the advantages of the accrual method in the context of the valuable information provided. Furthermore, the advantage of the modified method is that it captures the simplicity of the cash method. But even if a cash or modified cash accounting system is chosen, an understanding of accrual accounting will help the producer analyze the effects of any one business action, and will provide him with a more solid grasp of the action's impact on his net worth.

Double-Entry versus Single-Entry Bookkeeping

There are two major parts of a double-entry bookkeeping system: the *journal* and the *ledger*. The journal is the book of original entry. In other words, each transaction of the business is recorded first in the journal (sometimes called the *general journal* or *transactions journal* or, under the cash accounting system, the *cash journal*). The date is recorded along with the amount of money involved. Furthermore, the accounts, or items affected by the transaction, are recorded with a very short explanation.

The ledger is the second stage of the recording process. Very simply, the ledger is the book that contains all the separate accounts (sometimes called the *chart of accounts*). In other words, the ledger is the collection of all the various T-accounts, although in everyday business practice the diagram of the T-account has been superseded by a different visual format.

Because each transaction affects two accounts and each amount is recorded twice, the system is called *double entry*. Double-entry bookkeeping systems are often mistakenly equated with accrual accounting. While all accrual accounting systems use double-entry books, double-entry bookkeeping does not necessarily imply that the underlying accounting system is accrual based.

[3]The clarity of an audit trail is more a function of the completeness and clarity of the record book than of whether the book is accrual-based or cash-based.

Under the very simplest, pure cash accounting methods, only one book is used: the cash journal. Again, the date, an explanation, and an amount are recorded. Generally, a running cash balance is calculated at the same time. Since the transaction is written only once and in one book, the system is called *single entry*. An example of a single-entry system is a checkbook register.

With a modified system, a transaction is usually recorded on only one line of a book, but the amount is entered twice: once under a total column and once under an account column. Some systems, however, use two separate pages, sections, or locations to record each transaction. Both types thus use two separate books: a journal and a ledger.

Accounting Periods

Each business must select the accounting period over which all its records, and especially its financial statements and income tax returns, will be based. GAAP and income tax laws generally require that one consecutive 12-month period be selected as the accounting year. The year can begin on January 1 (calendar year) or any other date (fiscal year). Statements may be prepared for a shorter time frame, such as a month or a quarter, but those statements are considered to be just an update or a part of the overall reporting system, secondary to the annual statements.

When starting a business, the choice of the accounting or fiscal year is open. And if an existing business can "show that there is a substantial business purpose for the change,"[4] the accounting period then being used can be changed. An application must be filed with the IRS, however, and approval granted before the change is made for tax purposes.

Generally speaking, the calendar year will very well suit most cash-based farmers and ranchers. If income taxes are reported on an accrual basis, a fiscal year other than the calendar year may be more logical, as it may fit the annual crop and livestock production patterns more closely. For example, a winter wheat producer's production season may be better suited to a fiscal year that begins August 1.

Summary

Most farmers and ranchers are not highly trained accountants, nor do they need to be. However, an understanding of some of the basic accounting principles can be very helpful in knowing the effect of business activities on financial position.

The most basic accounting principle is that *Assets = Liabilities + Net Worth*. Assets are all those items that are of value to a business and liabilities are all those

[4]Internal Revenue Service Publication 225, *Farmer's Tax Guide*.

items owed to someone outside the business. Net worth is the remainder, or the claim of the owner if all assets were sold and all liabilities paid.

The fundamental accounting equation can be used to analyze every transaction of the business and must remain in balance at all times. Debits and credits are the two possible entries in a T-account. They have specific effects on each account:

Debits	Credits
1. Increase assets	1. Decrease assets
2. Increase expenses	2. Decrease expenses
3. Decrease income	3. Increase income
4. Decrease equity	4. Increase equity
5. Decrease liabilities	5. Increase liabilities

Two major accounting methods are available: accrual and cash. The primary difference between the two is the timing of cash receipt or payment. The timing of payment has no effect on accrual accounting since this method recognizes income when earned and expenses when incurred. The timing of payment is of utmost importance to cash accounting since this method stresses receipts when received and expenditures when paid.

The choice of which method to use depends upon the complexity and needs of a business. Accrual methods (all are double entry) offer an audit trail, instantaneous retrieval of balance sheets and income statements, and a more accurate reflection of net income. Cash methods are simpler, more closely match the tax reporting procedures of most farms and ranches, and can be modified relatively easily to provide most of the information of an accrual system.[5]

[5]The rest of this text will solve this problem by discussing a modified cash system, which should meet the management information needs of a large majority of commercial farms and ranches.

CHAPTER 3

Records Needed

A basic farm and ranch records system begins with the simplest, most obvious and most time-consuming component—the cash transactions records. All cash inflow and outflow items must be recorded and together they form the basis for the rest of the records system. They also provide nearly all the records necessary for a farmer or rancher to file an adequate income tax return. (The major exception is the depreciation records covered in Chapter 7). Further, they provide most of the information necessary to meet lender requirements for loan application documentation. (Major exceptions include outstanding loan balance and debt service records and inventory records necessary to complete a balance sheet.)

Cash Transactions

As noted above and in the outline provided in Box 3–1, cash transactions begin the record-keeping task. The other items in Box 3–1 follow, but their ordering or ranking is not important. In cash accounting terminology, a *receipt* is an item of cash flow into the business, regardless of whether the money is earned or comes from other sources such as a gift, money borrowed, or the sale of assets. In the same terminology, an *expenditure* is any item of cash flow out of the business, regardless of whether the money is spent for a production expense or for non-business requirements such as transfers for family living expenses, for repayment of principal borrowed, or for purchase of assets.

Since the checkbook and bank balances are the most important day-to-day financial concerns of the businessman, the farm or ranch records system should be built around the cash transactions. These transactions records can form the basis around which the rest of the records system can be built. They will be discussed in detail in Chapter 5.

Box 3–1

Outline of Records Needed

I. Cash Transactions
  A. Receipts
  B. Expenditures

II. Physical Inventories (Amounts and Values)
  A. Land and water resources
  B. Business assets
    1. Buildings
    2. Machinery and equipment
    3. Breeding, dairy and work livestock
    4. Orchards and vineyards
    5. Land improvements
  C. Crops, livestock and feed
  D. Supplies
    1. Fertilizer, chemicals, seed, and fuels
    2. Other materials

III. Financial Inventories
  A. Cash accounts
    1. Business checking
    2. Personal/family checking
    3. Savings
  B. Loan balances
  C. Accounts payable
  D. Accounts receivable
  E. Futures markets and other securities

IV. Financial Statements
  A. Balance sheet
  B. Income statement
  C. Cash flow statement

V. Production Records (For Each Enterprise)
  A. Crops
    1. Yields and prices
    2. Fertilizer requirements
    3. Seeding rates
    4. Approximate labor requirements
    5. Chemical applications
    6. Irrigation applications
    7. Disease and pest control
    8. All other purchased inputs
  B. Livestock
    1. Weights and prices bought and sold
    2. Feed fed
    3. Approximate labor requirements
    4. Veterinary and medical costs
    5. Pest control
    6. All other purchased inputs

VI. Labor and Machinery Records
  A. Labor use
  B. Social security and income tax withholdings
  C. Machinery and equipment
    1. Hours
    2. Fuel and lubricants
    3. Repairs and maintenance
  D. Depreciation schedule

VII. Family Accounts
  A. All tax deductible expenditures
  B. Nonfarm or nonranch income
  C. Normal periodic family living expenditures
  D. Major financial commitments
    1. Loans outstanding
    2. Payment schedule
    3. Other securities

VIII. Records Summaries
  A. Financial statements
  B. Enterprise records
  C. Whole-farm or whole-ranch plans and budgets
  D. Financial ratios and accuracy checks

Physical Inventories

The assembling of physical inventory information begins the process of converting our way of thinking from a cash-based records system to a comprehensive management accounting system. Physical inventories may be maintained through a full accrual accounting system (see Chapter 2) plus an annual inventory check, or they may be completed on an annual appraisal basis. The choice partially depends on whether a cash accounting system, an accrual accounting system, or a modified cash/accrual system is selected. The major purpose of physical inventory records is to provide a basis for the proper valuation of the overall business for analysis and for purposes of credit application. The inventories also form the starting point for the whole-farm or whole-ranch planning process. Consequently, the physical inventories are usually brought up to date or recorded at the beginning of each calendar (or fiscal or production) year.

Land and Water Resources

The logical, though not necessary, starting point for physical inventory records is to appraise land and water resources. These resources need to be recorded in amount, value, and availability, and the quality should be determined. Consequently, soil tests may be needed and soil types should be determined from USDA Soil Conservation Service maps, aerial photographs, and soil surveys. Water quality may need to be tested and such items as depth to water, irrigation ditches, and aquifer thickness and porosity should be recorded.

Depreciable Business Assets

After land and water resources (the main fixed assets) have been recorded, attention should be directed to depreciable business asset inventories. Buildings, machinery and equipment, long-lived livestock, orchards and vineyards, and land improvements such as irrigation ditches, tiling systems, or terraces should be evaluated and transferred to a depreciation schedule. Unlike land and water resources, the cost of these assets are recognized (or deducted from income) through time by using the depreciation methods allowed by the Internal Revenue Service (see Chapter 7).[1] Like land and water resources, these inventories are used to value the business and to begin the annual planning process.

Other Inventories

Amounts and values of all other *saleable crops or livestock* and items to be used in some stage of the production process of another commodity also need to be recorded. These inventories need to be taken, whether the item has been purchased or raised. The list should include such things as grain stored, silage or hay on hand, market livestock being fed, and inputs that have previously been purchased.

Once again, these inventories are used to value the overall farm or ranch business, but they are also used to calculate net income. The value of crops or livestock raised but not sold is income in the production period, despite the fact that cash has not changed hands. Similarly, production inputs purchased but not used artificially reduce income in the period purchased and artificially raise income in the period used.

[1]Certain assets, such as raised breeding livestock, may not be depreciable. The reason for this is explained further in Chapter 7.

Financial Inventories and Statements

In addition to inventories of physical assets on hand, records of all financial accounts are needed for business valuation purposes, for financial analysis purposes, and for an accurate determination of net farm or ranch income. Because of the importance of these accounts, inventories, and statements, several later chapters (Chapters 8 and 13 through 16) are devoted to them.

Financial Accounts

The records system should provide places to record openings, closings, and any changes in cash accounts (checkbooks for business and for personal/family accounts and bank statements will provide this information if they are maintained and stored in an accessible location), loans and mortgages outstanding (again, bank or lending agency statements should provide the necessary information), and other financial accounts. In addition to dollar balances and changes, each account should be identified with the names authorized to sign on the account, the financial institution, the beneficiaries, the account number, and other relevant information.

Most farmers and ranchers have some accounts payable but few have accounts receivable; nonetheless, the records system should provide for both. An *account payable* is simply an amount of money owed to someone else but not technically due in cash yet. Accounts payable are generally charge accounts at input supply firms or credit card companies. A very loose definition of accounts payable may include accrued liabilities, which are items such as employee wages or loan interest that others have earned but that are not yet due.

An *account receivable* is simply an amount of money due from someone else but technically not yet due in cash. One example of an account receivable that may be common on some farms and ranches is loans made to employees or children. Other examples might be a forward pricing contract under which grain is transferred to an elevator for payment at a later date, or a sale of grain or cotton to a cooperative if payment is expected after the cooperative has finished its crop-year sales.

Futures market accounts, whole-life insurance policies, corporate stock and bond accounts, and any other financial securities held should also be recorded. As with bank accounts, all information necessary to locate and track these accounts, in addition to dollar amounts, should be recorded.

Financial Statements

The basic financial statements are truly the purest form of financial communication. They have become the language of business. At a bare minimum, every business needs a balance sheet and an income statement for each year, and the cash flow statement is becoming a necessity.

The *balance sheet* (also called the *net worth statement* or sometimes simply the *financial statement*) is completed as of a specified date and lists all the assets of the business (all items owned) and all its liabilities (items owed) in order to calculate the difference, which is the owner's equity or net worth.

The *income statement* (also called the *profit-and-loss* or *P&L statement*) lists all business income and expenses over a given period of time—usually one year. To be useful, the income statement must be calculated on an accrual basis. In other words, it must reflect changes in physical and financial inventories in order to represent the amount of income actually earned during the period of time under consideration, rather than merely representing cash inflows and outflows.

In addition to the balance sheet and the income statement, the *cash flow summary* should be included in the records system, because it can become one of the most important business management tools available to modern farmers and ranchers. While not truly a record of historical activities, a projected *cash flow budget* should be filed in or near the record-keeping package. If well thought out, the cash flow budget will provide a comprehensive short-term plan for financing the business. It will lay out a plan for the short-term borrowing needs and repayment capabilities of the business (in terms both of timing and of amount) by noting when cash payments are scheduled or most likely to be required, and when cash receipts are most likely to occur. Once the cash flow budget has been projected and short-term credit secured, the budget should be *monitored*. In other words, actual cash inflows and outflows should be compared to those budgeted. The cash flow budget is simply a plan; it should not totally dictate business activity. But differences between the budgeted and actual cash flows can indicate potential needs for additional short-term credit, or indicate where borrowings can be repaid earlier than planned to save interest costs.

The three major financial statements taken together comprise a comprehensive body of financial information in a form that lends itself to analysis. Any one of the three statements is useful, but all three are really necessary, as the real reasons for business problems or failure are often hidden and can seem to have several solutions which, however, will not stand up to in-depth scrutiny.

Production and Input Records

Most of the records listed to this point have been financially oriented, but obviously the farm or ranch must do something physically as well as manage its business and financial affairs. Agricultural economists often imply that the production of commodities and the choice of inputs to produce those commodities are not as important as managing business affairs, financing the business, and marketing the output. The implication is usually not intentional, but stems from the fact that more farms and ranches have been sold as a result of poor business management than from poor technical expertise in production. Nevertheless, the purpose of this book is to present methods to record necessary information and to indicate how that information can be analyzed. Because of poor (or at times convenient) memories, both financial and production data must be promptly reported in the farm and ranch records system.

The production records and labor and machinery records indicate the amounts, qualities, varieties, brand names, breeds, and values of the inputs used in the production process. They also indicate amounts, qualities, and values of the intermediate products and final outputs produced. Production and input records will be discussed in greater detail in Chapters 9, 10, and 11, but attention can also be directed to specialized records systems such as those developed by the Dairy Herd Improvement Associations and breed associations.

Crop and Livestock

For the entire business, quantities of specific inputs and outputs should be recorded. But to be most useful, these inputs and outputs should be recorded separately for each crop or livestock enterprise produced. For example, for each crop species (enterprise) grown, production records should include: yield; price received; fertilizer, chemical, seed and irrigation applications (date applied, quantity, variety, quality, and analysis); and other purchased inputs such as baling wire, crew labor, harvest packaging, or any item specific to that crop. Similarly for each livestock enterprise, production records should include weights and prices at which market livestock or products are bought and sold, feed fed (amount, analysis, and price), feed supplements, and any other purchased inputs.

Labor and Machinery

It is also useful to know labor and machinery requirements for each crop and livestock enterprise. But keeping labor and machinery records for each enterprise can be very time consuming and expensive. Cost and return estimates (enterprise budgets) prepared and released by universities and the USDA may provide sufficient detail to approximate labor and machinery use on a crop or livestock enterprise, especially if total labor and machinery records are kept to check for applicability and comparability of the published reports.

At the very minimum, labor records should include the total hours worked per pay period for each nonsalaried employee, the wage rate paid, the gross wages per pay period, and any withholdings. In many circumstances, farmers and ranchers are required to withhold Social Security taxes from employees' paychecks. They may also withhold federal and state income taxes if the employee chooses.

Beyond the minimum, the amount and precision of labor records needed is questionable. The optimum will depend on the type of employees hired (full-time general, full-time assigned to a particular job, part-time, seasonal, or task-oriented) and the duties performed. Employees may or may not be asked to keep detailed records or even estimates as to the job performed or enterprise on a daily, weekly, or monthly basis.

Machinery use records are similar to labor records in that a particular item of machinery may be used for different jobs and different enterprises on any given day. Recording detailed use or even estimates of such use may prove to be very burdensome. If a particular item is used for only one task, for example a manure spreader on a one-species livestock farm, recording use by hours and enterprise may not be difficult. But, for most items, recording detailed use can be very costly.

For each machine, there are several items that must be recorded: annual hours of use, fuel and lubricant schedules, repairs and maintenance schedules. In addition to use and service records, a detailed depreciation schedule covering all machinery and equipment must be maintained. Depreciation will be discussed in Chapter 7.

Family Accounts

In most family-sized farms and ranches, the family will designate one person as the record keeper for their business and financial affairs. And, in most cases, that responsibility will be extended to include keeping records on the personal financial affairs of the owner and his family, since personal and business concerns are so often closely intertwined in agriculture. Furthermore, most lenders and all tax preparers will require family financial information as well as business information, even though an original loan or tax return question was framed as applying only to the business.

Family records may be maintained in as great detail as the business records but usually do not need to be. At a minimum, the family records must include all tax-deductible expenditures such as contributions to churches and other charitable organizations, medical expenses, personal taxes, interest on nonbusiness credit, and personal property casualty losses. The family records must also contain all amounts and sources of off-farm or off-ranch income and all major financial inventories such as checking and savings accounts, loans and other debts outstanding, payment schedules, corporate stocks, and all other financial securities.

Family records need not be kept by the same person, in the same format, or even in the same location as the business records. But, since the two so often go hand in hand, it is generally most convenient to delegate authority to one person to keep all such records together.

Other Financial Records

Just as family records must be maintained separately from farm or ranch business records, financial records dealing with other investment or nonfarm business activities should be separated as well. For analysis and tax purposes, it is usually wise to split activities into as many components as practical and to maintain accounting records on the components (or *accounting entities*) separately. This division into entity components helps to identify the sources and uses of funds, net income, and wealth.

Separate record books should be kept, for example, for a farm business, family accounts, a seed corn sales sideline, and a welding shop, even if all four entities are physically headquartered at the same home (office) and are all headed by the same individual. This records separation can then be used to see whether the farm is subsidizing the welding business or the other way around.

Records Summaries

The final summary component of a complete farm or ranch records system will not be discussed in detail now. Rather, that discussion will be delayed to Chapters 13 through 16 and subsequent chapters. At this point, it is sufficient to note that almost all of the information recorded so far is in raw form and must be summarized in some fashion in order to accomplish the four main tasks of a records system—income tax reporting and management, credit application, cash flow and financial control, and business management analysis and decision making—in addition to the other purposes discussed in Chapter 1.

The first and foremost summaries have actually already been discussed; the financial statements are also summaries. Other summaries include enterprise budgets that detail costs and returns of producing each commodity, whole-business plans and budgets that are primary decision tools, and financial ratios that begin to provide keys to discovering business management strengths and weaknesses.

Summary

A complete farm and ranch records system is composed of many parts. Those parts include:

1. Cash transactions
2. Physical inventories
3. Financial inventories
4. Financial statements
5. Production records
6. Labor and machinery records
7. Family accounts and non-farm business records
8. Records summaries

The detail required in each part depends upon the needs of the business and the time available.

Each records system must start with a complete listing of all cash receipts and expenditures, as these records will form the basis for the income tax return and the projected cash flow budget. Physical and financial inventories are necessary to prepare accrual-based financial statements, especially the income statement. They also provide most of the information required for the balance sheet. Production records are necessary for later analysis. Physical inventories and production records should include physical amounts as well as dollar values. Labor and machinery records involve use records, payroll amounts and withholdings, and the depreciation schedule. Family accounts are a necessary component of a farm or ranch records system, because the family and the business are usually intertwined, because personal cash withdrawals for the family are a priority item on the cash flow statement, and because personal tax-deductible items must be recorded.

Records summaries are the most usable and readable documents in the records system, because they pull together the most important parts of all of the raw information available in the first seven components.

CHAPTER 4

Records Systems

Many different types of farm and ranch records systems have been used in the past and are currently being used. It would stretch the truth only slightly to say that there are almost as many different farm and ranch records systems in use today as there are farmers and ranchers.

Figure 4–1 lists most of the various general types of systems being used. They are listed in order according to this writer's preference and each one will be discussed briefly below. Not one of these systems is necessarily bad or necessarily good. Good and bad depend upon how the individual business uses the system. However, some systems are likely to be better or more efficient than others.

Simple Systems

Probably the two most commonly used records systems are facetiously referred to as the "barn door" (or "pickup door") and "shoe box" systems. Essentially, the farmer or rancher writes all receipts and expenditures on the barn door and refers to those notes only at tax time, or he may keep all relevant scraps of paper in a shoe box and turn the box over to the tax preparer.

If used diligently, either one of these systems can be good and sufficient. But the likelihood of finding everything that is needed when it is needed to complete financial statements, loan applications, or tax returns is fairly low. Memories never serve us well. Often information is lost. Furthermore, organizing the information into a form that can be used for statements, loans, or taxes is a difficult, frequently frustrating, time-consuming process.

Journals

As the book of original entry, the journal is the first segment in any system more advanced than the barn door or shoe box. The *general journal* generally lists

1. Barn Door
2. Shoe Box
3. Cash Journal
4. Check Processing
5. Accountant
6. System of Your Own Design
7. Prepared System
8. Microcomputer System
9. Supervised System
   a. Record Book
   b. Microcomputer

FIGURE 4–1. Records system.

all cash transactions but may also include charge account transactions on the day of the charge, changes in inventory accounts, and non-cash recognition, as well as all other possible transactions.

A *cash journal,* an example of which is shown in Figure 4–2, will mirror the checkbook as it will list all cash transactions. A double-entry journal will look significantly different. However, in either case, a cash journal will generally list receipts, expenditures, and a cash balance. It may also include additional notes.

Cash journals or transactions journals are sometimes computerized and offered as a service by local banks, lenders, and some private companies. These systems, lumped together in Figure 4–1 as Check Processing, operate from a business checking account. In using this type of service, the farmer or rancher codes checks and deposit slips according to the purpose of the transaction. After the check has been cleared through normal banking channels, the cancelled checks are sent to a third party (which may be inside the bank) and processed to develop a cash journal report. Examples of checks, deposit slips, and reports are shown in Figure 4–3. The system may or may not allow noncash transactions to be entered periodically; this would transform the cash journal into a general journal.

These computerized journal systems can be excellent records systems, especially for income tax filing purposes, if maintained diligently. To avoid month-end correction entries, all transactions must be handled by check or deposit slip and must be coded at the time the check is written. Because of the detailed problems in using, processing, and correcting information within these systems, their popularity has never been very widespread. However, several companies (such as Agro Systems Corporation) offer good systems that are much improved over earlier systems and that can be used quite efficiently by farmers and ranchers.

Accountants

Many farmers and ranchers have simply given up on trying to maintain their own records system and have turned over the task to a professional bookkeeper or accountant. There are several advantages and disadvantages to turning over the system to someone else. The major advantages are:

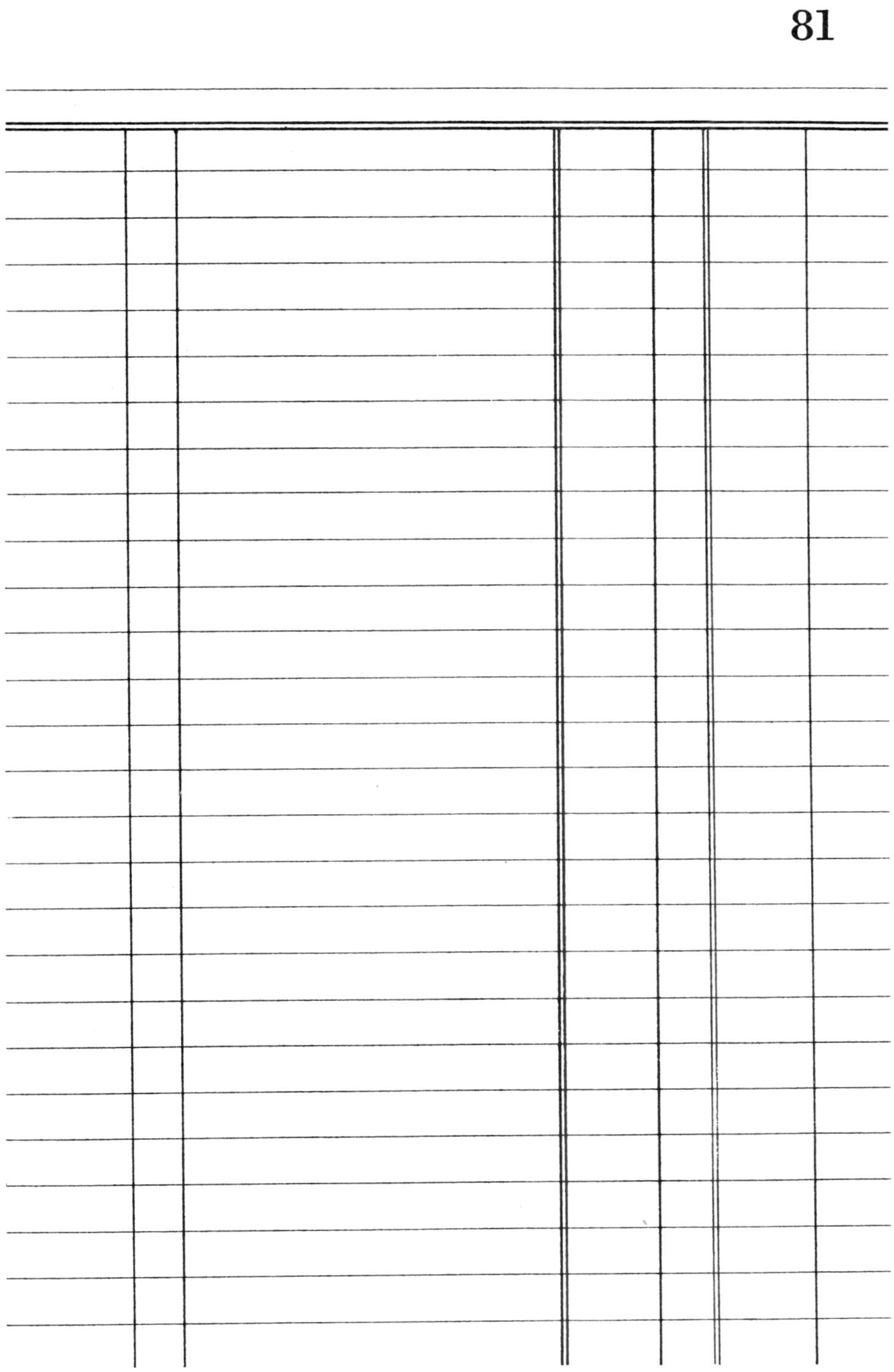

FIGURE 4–2. Single-entry cash journal.

AGRICULTURAL CHECK (family)

POCKET

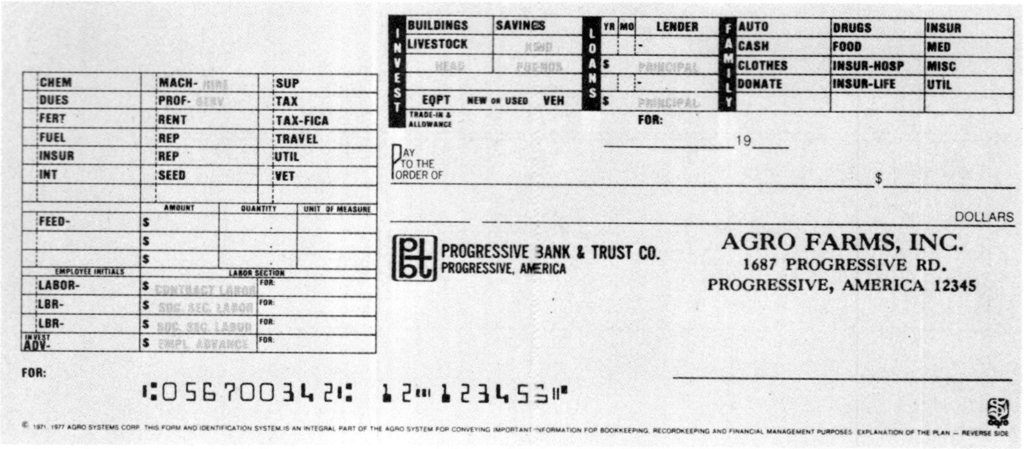

POCKET WITH END STUB and SMALL 3-PER PAGE

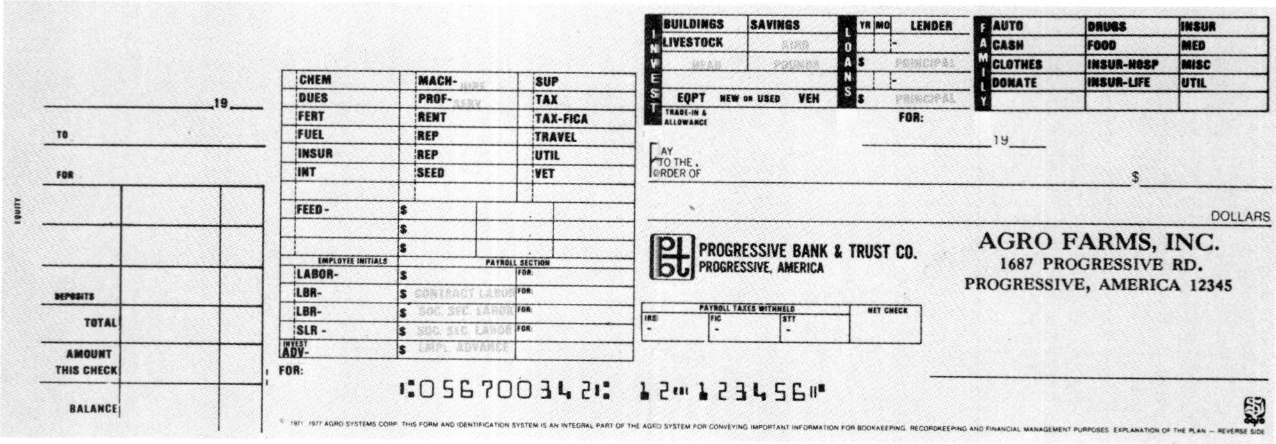

LARGE 3-PER PAGE

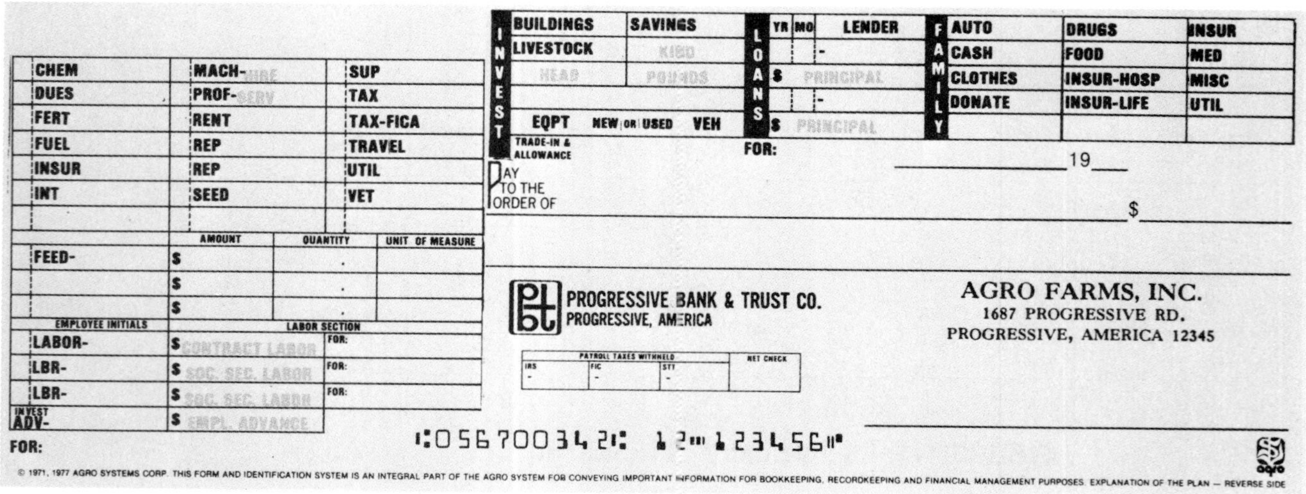

FIGURE 4–3. Check processing example.

Source: Agro Systems Corporation

1162

19

TO

FOR

EQUITY 7618

DEPOSITS

TOTAL

AMOUNT THIS CHECK

BALANCE

CHEM	MACH-HIRE	SUP
DUES	PROF-SERV	TAX
FERT	RENT	TAX-FICA
FUEL	REP	TRAVEL
INSUR	REP	UTIL
INT	SEED	VET
FEED-	$ QUANTITY	UNIT OF MEASURE
	$ QUANTITY	UNIT OF MEASURE
	$ QUANTITY	UNIT OF MEASURE

EMPLOYEE INITIALS	PAYROLL SECTION	
LABOR-	$ CONTRACT LABOR	FOR:
LBR-	$ SOC. SEC. LABOR	FOR:
LBR-	$ SOC. SEC. LABOR	FOR:
SLR -	$ OFFICER SALARIES	FOR: (CORP. ONLY)
INVEST ADV-	$ EMPL. ADVANCE	

FOR:

INVEST		LOANS	YR	MO	LENDER	EQUITY	CORPORATION	PARTNERSHIP
BUILDINGS	SAVINGS							
LIVESTOCK	KIND		-				DIV- NAME	DRAW- NAME
HEAD	POUNDS		$ PRINCIPAL				$	$
			-				DIV- NAME	DRAW- NAME
EQPT NEW OR USED	VEH		$ PRINCIPAL				$	$

TRADE-IN & ALLOWANCE

FOR:

19 1162

PAY TO THE ORDER OF $

DOLLARS

PROGRESSIVE BANK & TRUST CO.
PROGRESSIVE, AMERICA

AGRO FARMS, INC.
1687 PROGRESSIVE RD.
PROGRESSIVE, AMERICA 12345

⑆056700342⑆ 12⑈123456⑈

© 1971, 1977 AGRO SYSTEMS CORP. THIS FORM AND IDENTIFICATION SYSTEM IS AN INTEGRAL PART OF THE AGRO SYSTEM FOR CONVEYING IMPORTANT INFORMATION FOR BOOKKEEPING, RECORDKEEPING AND FINANCIAL MANAGEMENT PURPOSES. EXPLANATION OF THE PLAN — REVERSE SIDE

980

19

TO

FOR

FAMILY 76188

DEPOSITS

TOTAL

AMOUNT THIS CHECK

BALANCE

CHEM	MACH-HIRE	SUP
DUES	PROF-SERV	TAX
FERT	RENT	TAX-FICA
FUEL	REP	TRAVEL
INSUR	REP	UTIL
INT	SEED	VET

	AMOUNT	QUANTITY	UNIT OF MEASURE
FEED-	$	.	
	$	.	
	$	.	

EMPLOYEE INITIALS	LABOR SECTION	
LABOR-	$ CONTRACT LABOR	FOR:
LBR-	$ SOC. SEC. LABOR	FOR:
LBR-	$ SOC. SEC. LABOR	FOR:
INVEST ADV-	$ EMPL. ADVANCE	

FOR:

INVEST		LOANS	YR	MO	LENDER	FAMILY		
BUILDINGS	SAVINGS					AUTO	DRUGS	INSUR
LIVESTOCK	KIND		-			CASH	FOOD	MED
HEAD	POUNDS		$ PRINCIPAL			CLOTHES	INSUR-HOSP	MISC
			-			DONATE	INSUR-LIFE	UTIL
EQPT NEW OR USED	VEH		$ PRINCIPAL					

TRADE-IN & ALLOWANCE

FOR:

19 980

PAY TO THE ORDER OF $

PROGRESSIVE BANK & TRUST CO.
PROGRESSIVE, AMERICA

AGRO FARMS, INC.
1687 PROGRESSIVE RD.
PROGRESSIVE, AMERICA 12345

⑆056700342⑆ 12⑈123456⑈

© 1971, 1977 AGRO SYSTEMS CORP. THIS FORM AND IDENTIFICATION SYSTEM IS AN INTEGRAL PART OF THE AGRO SYSTEM FOR CONVEYING IMPORTANT INFORMATION FOR BOOKKEEPING, RECORDKEEPING AND FINANCIAL MANAGEMENT PURPOSES. EXPLANATION OF THE PLAN — REVERSE SIDE

1. The drudgery work is delegated to someone else, thus freeing the manager's time for other tasks.
2. A professional bookkeeper or accountant knows how to handle the perplexing questions that arise with some transactions.
3. The individual who controls the records is very likely also qualified to complete income tax returns and is familiar with the records necessary to file the return.
4. The records are generally maintained in a very orderly fashion and can be summarized into financial statements.

Despite the advantages, there are some serious drawbacks to using an outside accountant to do all of the business bookkeeping:

1. The manager can quickly lose control of his financial affairs, especially day-to-day cash flow.
2. The manager will face the tendency to turn over financial management, as well as bookkeeping, to the accountant.
3. The service can become costly.
4. It is often difficult to find an accountant familiar with agriculture.

On balance, the advantages and disadvantages may cancel each other out, as long as the manager does not lose control of day-to-day and long-run financial management of the business. However, using a qualified accountant as a backup to review and audit records maintained by the manager and to prepare tax returns and financial statements is probably the best system for most farmers and ranchers. Most of the bookkeeping and some of the analysis can be done on the farm or ranch but the accountant should be consulted about the things that are most difficult for the farmer or rancher.

Expanded Systems

Beyond the extremely simple systems and purchasing the direct service lie a group of record systems maintained by the manager without outside assistance. These systems generally take the form of an expanded cash journal/ledger combination bound in a book. Most frequently, multiple-column accounting analysis forms are converted into a modified cash/accrual accounting system. Cash and noncash transactions are entered in two places: as they affect the total balance and in a second column that shows the type of expenditure or receipt.

These systems can be designed from the ground up using columnar pads (Figure 4–4), or they can be obtained from business schools, colleges, and universities or from governmental agencies (Figures 4–5 and 4–6)[1] or purchased from private companies or office supply stores (Figure 4–7).

[1]County Extension Agents and the Departments of Agricultural Information at land grant universities are excellent sources of information regarding the availability of prepared systems.

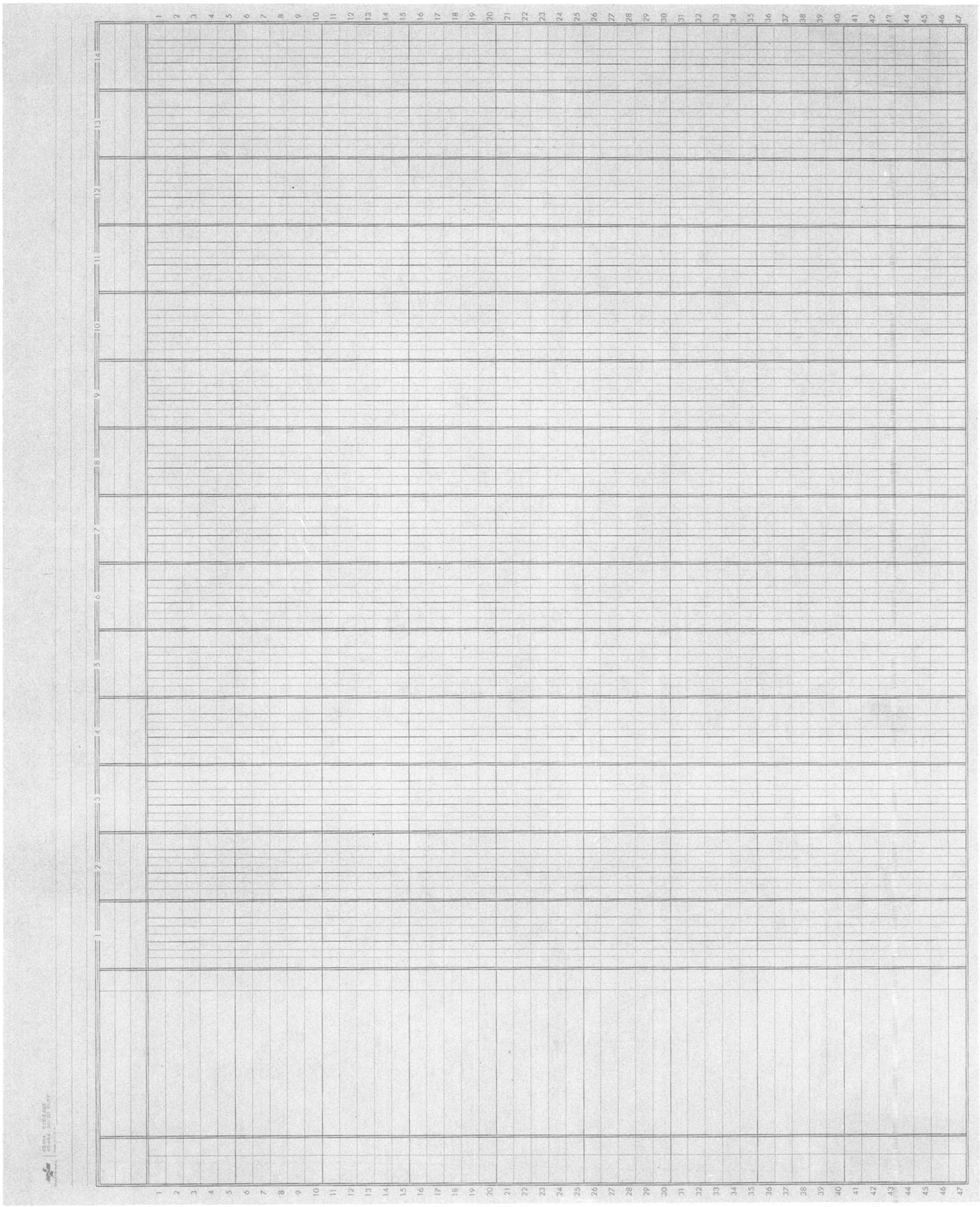

FIGURE 4–4. Accountant's analysis pad.

RANCH EXPENDITURES (Operator's share)

Date	Description		Total Amount Paid	Use these columns for share lease			Purchases of Capital Items		Purchases of Breeding Stock			Purchases of Stocker Livestock								
												Cattle			Sheep			______		
	Paid to	Item		Personal Share	Operator Share	Landlord Share (Carry to pages 50 and 51.)	Buildings and improvements	Machinery and equipment	No.	Weight	Value	No.	Weight	Value	No.	Weight	Value	No.	Weight	Value

Note: Add all columns by months. After the last entry for each month, skip a line and add columns. Then skip a line or go to the next page and begin entries for the next month. Circle all monthly totals. Carry monthly totals to pages 48 and 49.

(Operator's share) RANCH EXPENDITURES

Feed Purchased		Livestock Expense	Hired Labor	Machinery Repairs	Gas, Fuel, and Oil for Truck and Tractor	Irrigation Fuel, Oil and Electricity	Machine Hire	Auto Expense	Electricity and Telephone			Crop Supplies	Mis-cellaneous Supplies	Repairs of Buildings, Fence, Wells and Irrigation Equipment	Taxes	Cash Rent	Insurance	Hedging	Miscel-laneous	Interest Paid	Principal Payments on Notes and Mortgages
Bushels, tons, pounds	Value																				

Note: Add all columns by months. After the last entry for each month, skip a line and add columns. Then skip a line or go to the next page and begin entries for the next month. Circle all monthly totals. Carry monthly totals to pages 48 and 49.

FIGURE 4–5. New Mexico expenditures records.

NEW MEXICO
FRBR

Form 1A

TRANSACTION JOURNAL

Account No. ______________

Ln No	Day	Dep or Chk No	Description To/From Whom	Item	Ref No Pg-Ln	Ent. No.	Quantity	Unit	Receipts	Payments	Balance
			Balance Forward								
1											
2											
3											
4											
5											
6											
7											
8											
9											
10											
11											
12											
13											
14											
15											
16											
17											
18											
19											
20											
					TOTALS/Balance Forward						

Page ______ **Month** ______ **Year** ______

NEW MEXICO
FRBR

Form 5A

CROP EXPENDITURES

Enterprise _______
Month _______
Page _______

Ln No	TJ Ref Pg-Ln	Seed	Fertilizer	Chemicals	Crop Supplies	Drying and Storage	Crop Insurance	Rents and Leases	Hedge Acct. Deposits	________
1										
2										
3										
4										
5										
6										
7										
8										
9										
10										
11										
12										
13										
14										
15										
16										
17										
18										
19										
20										
Total										
		5A1	5A2	5A3	5A4	5A5	5A6	5A7	5A8	5A9

Page Total ________

FIGURE 4–6. New Mexico transactions journal and crop expense accounts.

Source: Michael L. Jones and James D. Libbin, *New Mexico Farm and Ranch Business Record Book,* New Mexico Cooperative Extension Service, New Mexico State University, 1986.

Ideal™ **CASH DISBURSEMENTS FARM AND RANCH** FORM 33R

	DATE 19__	CHECK NUMBER	1 TO WHOM PAID	2 AMOUNT	3 FEED PURCHASED	4 SEEDS & PLANTS PURCHASED	5 FERTILIZER LIME & CHEMICALS	6 MACHINE HIRE	7 SUPPLIES & SMALL TOOLS	8 BREEDING FEES	9 VETERINARY FEES	10 SALARY & WAGES	11 SALARY & WAGE DEDUCTIONS	
1														1
2														2
3														3
4														4
5														5
6														6
7														7
8														8
9														9
10														10
11														11
12														12
13														13
14														14
15														15
16														16
17														17
18														18
19														19
20														20
21														21
22														22
23														23
24														24
25														25
26														26
27														27
28														28
29														29
30														30
31														31
32														32
33														33

THE IDEAL SYSTEM, REG. U.S. PAT. OFFICE. MADE IN U.S.A.

FIGURE 4–7. Ideal expenditures records.

Source: Ideal Bookkeeping and Tax Record: Farm and Ranch, Esselte Pendaflex Corporation, Garden City, NY, 1982.

Finally, many records systems are now available for farms or ranches for use with personal computers. Great advances have been made over the past few years in computer hardware (the physical computer) and software (the written programs necessary to run the computer). Many excellent (and a few very poor) personal computer accounting packages have been written recently. Two examples of microcomputer output are shown in Figures 4–8 and 4–9. Packages range from simple cash journal systems through modified single-entry systems to complete double-entry systems.

Most personal computer packages are relatively inexpensive and can be readily adapted to individual farms and ranches, but should be evaluated thoroughly before purchase to ensure that the particular package will meet expectations and needs. Chapter 24 will describe how to select and evaluate a computer system or any prepared system.

DUSK TO DAWN FARMS

09:40:44 06-24-1986 TRIAL BALANCE 06-24-86 GLTBAL PAGE 1

ACCOUNT	DESCRIPTION	MONTHLY AMOUNT	----YEAR TO DATE---- DEBITS	CREDITS
1000 0	NON FARM ASSETS	0.00		
1010 0	CASH	0.00	1,680.00	
1020 0	SAVINGS	0.00	2,300.00	
1030 0	CHECKING ACCOUNT	1,063.94-	11,978.83	
1040 0	CASH VALUE OF LIFE INS.	0.00	875.00	
1200 0	INVENTORIES	0.00		
1210 1	DAIRY COWS	0.00	23,000.00	
1310 2	CORN-EAR	0.00	6,750.00	
1312 2	CORN-SHELLED	0.00	17,100.00	
1410 1	FEED-PURCHASED	0.00	1,050.00	
1420 2	SEED CORN	0.00	2,800.00	
1430 2	FERTILIZER	0.00	1,600.00	
1440 2	CHEMICALS	0.00	250.00	
1510 0	MISC. INVENTORY	0.00	640.00	
1600 0	LONG TERM ASSETS	0.00		
1610 0	HOUSE	0.00	36,500.00	
1620 0	LAND	0.00	326,400.00	
1630 0	FARM BUILDINGS	0.00	56,500.00	
1631 0	FARM BLDG (ACCUM DEPR)	0.00		23,200.00-
1650 0	TRUCK & AUTO	0.00	7,200.00	
1651 0	TRUCK & AUTO (ACCUM DEPR)	0.00		3,100.00-
1660 0	MACHINERY	0.00	85,502.00	
1661 0	MACHINERY (ACCUM DEPR)	0.00		41,050.00-
1670 1	LIVESTOCK EQUIPMENT	0.00	35,300.00	
1671 1	LIVESTK EQUIP (ACCUM DEPR)	0.00		17,650.00-
2000 0	CURRENT LIABILITIES	0.00		
2010 0	ACCOUNTS PAYABLE	0.00		5,536.00-
2110 0	DISCOUNTS TAKEN	0.00		
2210 0	NOTES PAYABLE	0.00		20,122.00-
2310 0	CHATTEL MORTGAGE	1,420.00		30,701.10-
2500 0	LONG TERM LIABILITY	0.00		
2510 0	REAL ESTATE MORTGAGE	0.00		25,710.00-
2610 1	MILKING PARLOR EQP. MORT.	0.00		8,270.00-
299099	DRAWING ACCOUNT	0.00		1,047.82-
2992 0	OWNER CONTRIBUTION	0.00		400,381.79-
2998 0	INCOME	0.00	80.40	
2998 1	INCOME	0.00		570.79-
2998 2	INCOME	0.00		507.22-
299899	INCOME	0.00	641.55	
2999 0	RETAINED EARNINGS	0.00		
2999 1	RETAINED EARNINGS	0.00		31,460.00-
2999 2	RETAINED EARNINGS	0.00		27,245.00-
299999	RETAINED EARNINGS	0.00	18,760.00	
3000 0	FARM INCOME	0.00		
3110 1	PURCHASED COWS-SOLD	0.00		
3112 1	RAISED COWS-SOLD	0.00		
3130 1	MILK SALES	3,752.55-		7,505.10-

FIGURE 4–8. FMS record output (a modified single-entry microcomputer bookkeeping system).

Source: FMS Financial Manager, Farm Management System, New Lenox, IL.

Despite the tremendous growth in computer accounting packages and the wide interest in personal computers, this text will not address computers directly. Most of the problems encountered in using computer packages are the same as those for noncomputer systems, namely determining what is needed in the records and how to handle difficult questions. The computer will not answer these questions, it will simply make the format easy to use and perform the necessary calculations quickly and easily.

Crop Expenses

DATE	DESCRIPTION	Pesticides	Starter Fertilizer	Potash Fertilizer	Anhydrous Fertilizer	Seed	Machine Hire	Diesel
4/ 2/84	Agro Fertilizer				62760.00			
4/10/84	Agro Fertilizer		2615.51					
4/19/84	FS Petroleum Corporation							2450.00
4/ 5/84	Pioneer Seed Company					3458.00		
4/ 5/84	Pioneer Seed Company					6392.00		
4/25/84	Tyler Grain Company					3620.00		

The Farm Management System

TRANSACTION SUMMARY
for
DEMONSTRATION VERSION
All operators

Month 4/84 Page No. 1

Crop Expenses

DATE	DESCRIPTION	Pesticides	Starter Fertilizer	Potash Fertilizer	Anhydrous Fertilizer	Seed	Machine Hire	Diesel
	Total	0.00	2615.51	0.00	62760.00	13470.00	0.00	2450.00
	Year to date	22195.00	2615.51	0.00	63060.00	23169.23	0.00	2450.00

Crop Income

		Corn Sales			Soybean Sales		
DATE	DESCRIPTION	Bushels	Price	Dollars	Bushels	Price	Dollars
	Total	0.00	0.00	0.00	0.00	0.00	0.00
	Year to date	6000.00	3.15	18900.00	1273.09	9.85	12540.00

FIGURE 4–9. Red Wing records output (a double-entry bookkeeping system).

Source: Red Wing Financial Management Software, Red Wing Business Systems, Red Wing, MN.

Supervised Systems

The primary difficulty with each one of the self-maintained expanded systems, and the simple systems as well, has already been mentioned: almost all of the decisions on how to handle difficult transactions are left to a manager who is probably not very well trained in accounting. No matter how well planned the records system is and no matter how well written the instructions are (and also no matter whether the system is pencil-and-paper based or computer based), there will always be a few transactions or accounting problems that will require professional assistance. We can safely say that expert assistance should be sought because those difficult cases almost always involve large amounts of money and/or have significant tax consequences.

Accountants and Tax Consultants

Virtually all accountants and tax consultants will be willing to assist in decisions of how to handle certain items. They will likely charge a fee for those services, but generally those fees are low relative to the large potential consequences of mishandling the decision.

Records Associations

In many states, farm and ranch records and business analysis associations have been organized to assist producers to maintain their records and to analyze business management decisions, strengths, and weaknesses. These associations are generally nonprofit corporations or cooperatives owned by members, built around the common goal of improving business management skills. A professional farm and ranch business analyst is hired by the association to provide the day-to-day and long-term expertise required to maintain and analyze necessary financial information.

The major advantage of these associations is that the common goal allows member producers access to a consultant trained in farm and ranch business management. Furthermore, the association will summarize all of its members' records annually to provide averages and comparisons with other similar farms and ranches.

Some associations offer only a pencil-and-paper record book, but most are expanding to offer personal computer systems as well. The primary advantage of computers to the association is that less time is required to transmit data at the end of each year to be summarized, and thus less time is required to process and summarize the annual records.

Summary

Several different types of records systems are available to and are used by farmers and ranchers. These systems range from very simple "barn door" or "pickup door" systems all the way to sophisticated, supervised computer-based systems. These systems may use no formalized paper or books at all, they may use blank accounting pads or purchased books or computer programs, or they might be kept by a public accountant. The cash journal, ledger, or modified journal might form the basis for any system more sophisticated than the shoe box.

Each system has its own set of advantages and disadvantages. The most important element in the selection of any records system is that the system must be maintained on a regular basis and must be usable. It must have the flexibility to provide the information necessary to meet the primary purposes of the records system and to support business management analysis needs.

PART II

Components of a Complete Records System

The discussion so far has presented an overview of a complete farm or ranch records system as well as some accounting conceptual background. The next eight chapters discuss in detail all the components of a financial records system up to the point where financial statements should be prepared.

The progression from Chapter 5 through Chapter 12 will generally follow the outline of the types of necessary records presented in Chapter 3. The progression of the chapters is really a list of priorities.

First and foremost, cash transactions records must be maintained. Then all the inventories must be recorded. Those inventories are split into three groups: short-term inventories, long-lived asset inventories, and financial inventories, primarily because the three differ widely as to means of recording and calculation needs. For instance, only long-lived assets are depreciable; consequently, much of Chapter 7 concentrates on depreciation methods and basis in property. Financial inventories must involve loan amortization methods, whereas short-term inventories are characterized by valuation problems.

The remaining chapters in Part II briefly cover physical, production, labor, machinery, and family records. Each of these types of records can be important to business analysis.

CHAPTER 5

Cash Transactions Records

The first priority for all farmers and ranchers in a financial records system is to control and record cash flow. If that priority is taken here as well, we must first decide what form the cash records will take. Earlier discussions of cash versus accrual accounting methods and single- versus double-entry bookkeeping systems in Chapter 2 have pointed out at least two of the decisions that must be made. Further, the choice must be made whether a cash journal or general journal (or neither) is to be used. Some sophisticated accrual systems essentially bypass the journal and proceed directly to the ledger. On the other hand, some cash-based systems use the checkbook to perform the functions of the journal.

Recording Transactions

To illustrate the different methods of recording transactions, we will refer to the transactions presented in Chapter 2 and enter each 19X1 entry in both cash and general journals. Those transactions have been renumbered and are listed following.

1. Corn produced in 19X1 and sold for cash in November 19X1. Amount: $50,000.
2. Corn produced in 19X1, delivered to an elevator in November 19X1 and sold for cash in January 19X2. Amount: $50,000.
3. Loan proceeds received in December 19X1 for purchase of land. Amount: $100,000.
4. Cash received in November 19X1 when a used tractor is sold to a neighbor. Amount: $10,000.
5. Fertilizer purchased for $5,000 for use in 19X1 and paid for in cash in February 19X1.
6. Fertilizer purchased for $5,000 in December 19X1 for use in 19X2 and paid for in cash in December 19X1.

7. Fertilizer purchased for $5,000 on account in March 19X1 for use in 19X1. Charge account paid in January 19X2.
8. Fertilizer purchased for $5,000 on account in December 19X1 for use in 19X2. Charge account paid in April 19X2.
9. Loan payment made in November 19X1 ($4,000 principal and $3,000 interest).
10. Depreciation ($8,000 total) on owned machinery realized in 19X1.
11. Tractor purchased for $30,000 in November 19X1.

Journal Entries—Double-entry System

First, we will enter each of the transactions in the general journal shown in Figure 5–1, assuming that we have selected an accrual-based double-entry system.

TRANSACTION 1: Corn produced in 19X1 and sold for cash November 19X1.

Analysis: Transaction 1 is a fairly simple cash transaction. When the raised corn is sold for cash, both the cash account and corn income account increase. The cash account is debited (increased since it is an asset) and the corn income account is credited (also increased since it is an income account).

TRANSACTION 2: Corn produced in 19X1, delivered to an elevator in November 19X1 and sold for cash due in January 19X2.

Analysis: Transaction 2 is not a cash transaction, but that makes no difference to a general journal in an accrual accounting system. Having the corn ready for sale is enough to count this item as a financial transaction despite the fact that no cash has changed hands. Given that a transaction has occurred, there must be at least one debit and at least one credit. In this case, the harvested corn was placed in inventory at a local elevator. Thus, corn inventory was debited (asset increased) and corn income was credited (income increased).

TRANSACTION 3: Loan proceeds received in December 19X1 for purchase of land.

Analysis: Transaction 3 is a receipt, that is, cash has flowed into the business, but it is not income. A new liability was created (mortgage payable) and that account was credited. The cash inflow is refleced by a debit to the cash account.

TRANSACTION 4: Cash received when used tractor is sold.

Analysis: Like the loan receipt of Transaction 3, Transaction 4 is a receipt since cash flowed into the business but it is not income. Cash inflow debits the cash account, whereas the sale of the tractor reduces (credits) the value of machinery owned. Proper handling of this transaction can become very tricky if depreciation had been taken on the machine and/or if a gain or loss was recognized on the sale.

TRANSACTION 5: Fertilizer purchased for use in 19X1 and paid for in cash in February 19X1.

TRANSACTION 6: Fertilizer purchased in December 19X1 for use in 19X2 and paid for in cash in December 19X1.

TRANSACTION 7: Fertilizer purchased on account in March 19X1 for use in 19X1. Charge account paid in January 19X2.

TRANSACTION 8: Fertilizer purchased on account in December 19X1 for use in 19X2. Charge account paid in April 19X2.

***Analysis*:** Transactions 5 through 8 are slightly different from one another, based on the timing of the payment and the time period in which the fertilizer is to be used. But because in each case the fertilizer was contracted for and payment was committed, each is a transaction requiring a journal entry. Transaction 5 is a straight cash transaction so that fertilizer expense was debited (assuming the fertilizer was applied immediately) and cash was credited (decreased). The fertilizer purchased in Transaction 6 was intended for application in the following year so that an asset account, fertilizer inventory, was debited or increased and cash was credited or decreased. When payment is delayed until the following year, as in Transactions 7 and 8, a liability account, accounts payable, is credited. In 19X2, when the account is paid, the account payable will be debited (decreased) and cash will be credited (also decreased).

TRANSACTION 9: Loan payment made in November 19X1.

***Analysis*:** Loan payments generally will carry not two but three items to record in the journal, as depicted under Transaction 9 in Figure 5–1. Most often, the loan payment will carry both a component of interest and one of principal. Consequently, two accounts are debited, interest payable and mortgage payable (both liabilities, thus both accounts are decreased), and therefore cash is credited.

TRANSACTION 10: Depreciation on owned machinery realized in 19X1.

***Analysis*:** As in Transactions 2, 7, and 8, no cash changed hands in Transaction 10. However, depreciation is an expense, thus depreciation expense is debited. Most accountants prefer to separate the original cost of machinery from the depreciation taken on that machinery by using a *contra account*. A contra account, as the name implies, acts in an opposite manner to the parent account. The easiest way to understand the concept of a contra account is through an example. In the case of Transaction 10, the contra account called accumulated depreciation is credited when depreciation expense is recognized. The machinery assets account is *not* credited with the depreciation. When a machine is bought or sold, the machinery assets account is affected, but when depreciation expense is realized, only the accumulated depreciation account is affected. To find the adjusted basis (or remaining value) of the machinery, you must subtract the accumulated depreciation account balance from the machinery assets account balance. The net result is that the machinery assets account contains only the total *original* basis of machinery while the accumulated depreciation account shows all of the depreciation ever recognized on the machinery owned. Contra accounts are used sparingly but are included when it is useful to know the original information (such as original basis) as well as the net information (such as adjusted basis or remaining value).

TRANSACTION 11: Tractor purchased in 19X1.

***Analysis*:** Transaction 11 is a simple transaction in that one asset (cash) is exchanged for another (machinery). The action is exactly opposite that of Transaction 4.

Ledger Entries—Double-Entry System

As noted earlier, the journal is the book of original entry, but to complete the process, the journal entries must be *posted* to a ledger. In other words, an entry is listed in the journal by specifying the account to be debited and the account to be credited, as well as the amount or value involved. Secondly, each debit and each credit is carried to the *ledger* where all debits and credits affecting any single account are summarized on one page. The process of transferring the journal entries to the ledger is called *posting*. An example of the ledger entries for the cash account is shown in Figure 5–2.

The visual form of ledgers may vary significantly, as shown by an additional example in Figure 5–3. Two-column ledgers will list only debits and credits, without a balance. The choice of form should be a matter of personal preference, as the premise of all accounting is function over form.

Note the use of account numbers and page numbers in Figures 5–1 and 5–2. Journal pages are usually numbered consecutively like the pages in any book. It is also normal practice to assign reference numbers to each account in the ledger. Those numbers may be assigned consecutively or in some other pattern. These

General Journal
Mayfarms, Inc.
19X1

Page
1

19X1		Acct	Debit	Credit
Nov 1	Cash	101	50000 00	
	Corn Income	501		50000 00
	Raised corn sold for cash			
Nov 2	Corn Inventory - County Coop Elevator	108	50000 00	
	Corn income	501		50000 00
	Raised corn delivered to elevator			
Dec 3	Cash	101	100000 00	
	Mortgage Payable - 1st National Bank	232		100000 00
	Loan proceeds received			
Nov 10	Cash	101	10000 00	
	Machinery Inventory	120		10000 00
	Used tractor sold			
Feb 4	Fertilizer Expense	405	5000 00	
	Cash	101		5000 00
	Fertilizer purchased for cash			
Dec 5	Fertilizer Inventory	118	5000 00	
	Cash	101		5000 00
	Fertilizer purchased for use in 19X2			
Mar 6	Fertilizer Expense	405	5000 00	
	Accounts payable - County Service Company	214		5000 00
	Fertilizer purchased on account			
Dec 7	Fertilizer Inventory	118	5000 00	
	Accounts Payable - County Service Company	214		5000 00
	Fertilizer purchased on account for use in 19X2			
Nov 8	Interest Payable - Farmers State Bank	218	3000 00	
	Mortgage Payable - Farmers State Bank	218	4000 00	
	Cash	101		7000 00
	Loan payment			
Dec 9	Depreciation Expenses			
	Accumulated Depreciation - Machinery	416	8000 00	
	Depreciation on machinery	127		8000 00

Nov 15	Machinery Inventory			
	Cash	120	30000 00	
	Tractor purchased	101		30000 00

FIGURE 5–1. Sample general journal entries.

General Ledger
Mayfarms, Inc.
19X1

		Initials	Date
Prepared By			
Approved By			

Acct 101	Cash	Page	Debit	Credit	Balance
	Previous Balance				42000 00
Nov 1	Raised corn sold	1	50000 00		92000 00
Dec 3	Loan Proceeds received	1	100000 00		192000 00
Nov 10	Tractor sold	1	10000 00		202000 00
Feb 4	Fertilizer purchased	1		5000 00	197000 00
Dec 5	Fertilizer purchased	1		5000 00	192000 00
Nov 8	Loan payment	1		7000 00	185000 00
Nov 15	Tractor purchased	1		80000 00	105000 00

4803 (84803) — Buff
8803 (88803) — Green

Made in U.S.A.

FIGURE 5–2. Sample general ledger entries.

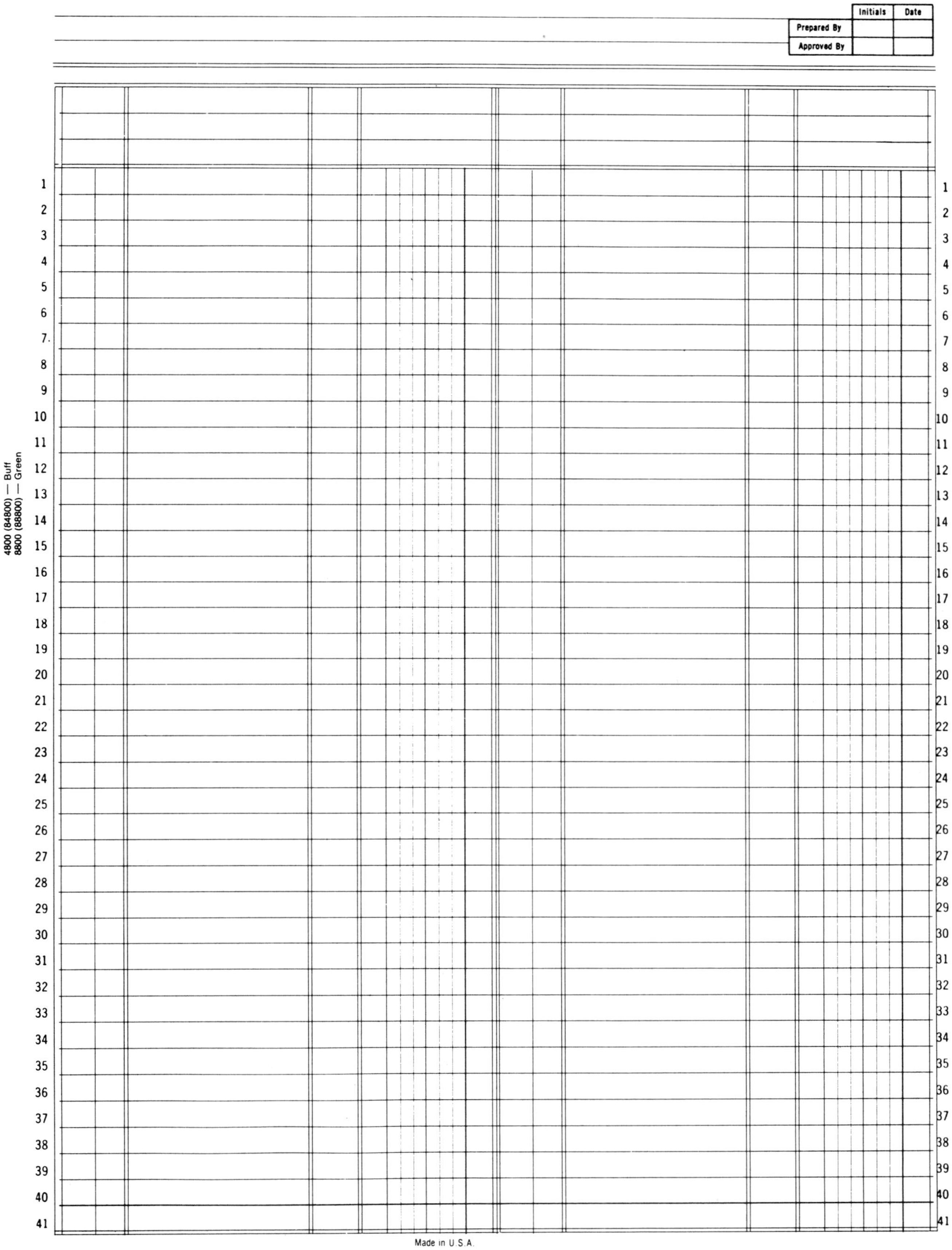

FIGURE 5–3. Additional example of a double-entry ledger.

account numbers are used to organize the ledger and also appear on the journal, as can be seen under the ACCT column in Figure 5–1. The pattern used here, which should be modified to fit individual needs, is shown in Figure 5–4. A relatively simple modification, which would allow a greatly expanded potential chart (or list) of accounts, would be to use 10,000, 20,000, 30,000 and so on instead of 1000, 2000, 3000 and so on as shown in Figure 5–4.

The two main purposes of any page and numbering system are to organize both the journal and the ledger in a logical, orderly fashion and to allow transactions to be *cross referenced.* Cross referencing simply means that for each journal entry, a ledger account number is listed so that the bookkeeper knows which ledger page to turn to when posting that entry. And further, the journal page is listed on the ledger so that the original journal may be located easily for each ledger entry.

Cross referencing (as well as maintaining all original paper receipts, checks, invoices, purchase orders, and similar vouchers) lays the groundwork for what is called an *audit trail.* An audit trail is simply a set of orderly paths for an independent accountant to follow when a verification of records is required.

The primary difficulty involved in cross referencing is that the bookkeeper must remember or locate account numbers for each transaction. In other words, the bookkeeper must know and be able to easily use the *chart of accounts,* which is the complete listing of the firm's account titles and reference numbers. The chart of accounts can also be thought of as the table of contents of the ledger.

Cash Journal Entries—Single-entry System

The single-entry cash accounting system could be considered an extremely simplified subset of the double-entry accrual method. Only one book is used; the ledger is not a part of this system. Furthermore, only the accrual system's cash account is maintained. Consequently, the cash journal will appear to be very similar to the cash account in the double-entry ledger, as can be seen in Figure 5–5 for the transactions listed at the beginning of this chapter. For all practical pur-

Account Number		
1000		Assets
	1100	Current Assets
	1200	Intermediate Assets
	1300	Long-Term Assets
2000		Liabilities
	2100	Current Liabilities
	2200	Intermediate Liabilities
	2300	Long-Term Liabilities
3000		Net Worth
4000		Income
5000		Expense

FIGURE 5–4. Example of a ledger account numbering system.

Cash Journal
Mayfarms, Inc.
19X1

Page 1

	Initials	Date
Prepared By		
Approved By		

19X1			Check/ Deposit No.	Payments (Expenditures)	Deposits (Receipts)	Balance
		Previous Balance				42000 00
Nov	1	Raised corn sold	092		50000 00	92000 00
Dec	2	Loan proceeds received	104		100000 00	192000 00
Nov	10	Tractor sale	095		10000 00	202000 00
Feb	9	Fertilizer purchased	651	5000 00		197000 00
Dec	5	Fertilizer purchased	923	5000 00		192000 00
Nov	8	Loan payment	889	7000 00		185000 00
Nov	15	Tractor purchase	895	30000 00		155000 00

4803 (84803) — Buff
8803 (88803) — Green

Made in U.S.A.

FIGURE 5–5. Sample cash journal entries.

poses, the cash journal is a duplicate of a business checkbook register with the addition of purchases and receipts made with currency rather than check.

Journal Entries—Modified Single-entry System

Most modified single-entry systems are cash accounting systems, but they go beyond the pure single-entry system to allow the business manager to extract nearly as much information from the records as does the accrual method. The primary concept underlying modified single-entry systems is that each cash transaction is described once, but the dollar amount is listed twice. For each entry, the dollar amount is listed once in a total column, and once in a column set aside for particular activities. In essence, these activities are conceptually the same as the expense and income accounts of the accrual method, except that they are cash only. One other distinctive feature of most modified single-entry systems is that receipts are usually recorded on pages separate from expenditures.

Entries for the example transactions listed on pages 47 and 48 are shown in Figure 5–6 (receipts) and Figure 5–7 (expenditures) under a homemade modified single-entry system. The names and accounts used in Figures 5–6 and 5–7 can be selected by the individual if a homemade journal is used. The major alternative to this type of journal is the preprinted books discussed in Chapter 4 and depicted in Figures 4–6 and 4–7. Although the preprinted record books decrease naming flexibility, the column names used most often are those used in federal farm income tax returns. Consequently, most homemade systems will use those names as well.

Enterprise and Joint Ownership Accounting

Most farm and ranch businesses produce more than one commodity (enterprise) and many involve more than one owner. As a result, many managers wish to record receipts, income, expenditures, and/or expenses in such a way as to associate the particular transaction with the enterprise or individual to which it belongs. Consider the numbered transactions listed below to help clarify this choice.

9. Loan payment made in November 19X1 ($4,000 principal and $3,000 interest).

12. Payment of $9,500 made to farm supply store for the following items in November 19X1 for use in 19X2:

a. Corn fertilizer	$5,000
b. Soybean fertilizer	3,000
c. Alfalfa baling wire	1,000
d. Machinery oil and grease	500

13. Stocker calves purchased for $25,000 for backgrounding in December 19X1. Partner 1 purchased 40 percent and partner 2 purchased 60 percent.

Receipts
Mayfarms, Inc.
19X1

Date	Item	Total	Market Cattle	Breeding Cattle	Corn	Soybeans	Hay	Machine Work	Other Income	Capital Sales	Money Borrowed
Nov 1	Corn sales	5000000			50000 00						
Dec 3	1st National Bank Mortgage	10000000									100000 00
Nov 10	Tractor sale	1000000								10000 00	

FIGURE 5–6. Sample receipts entries in modified single-entry system.

Expenditures
Mayfarms, Inc.
19X1

Date	Item	Total	Hired Labor	Repairs	Fertilizer	Seed	Fuel & Oil	Supplies	Livestock	Interest	Principal Payments	Capital Purchases
Feb 4	Fertilizer	5000 00			5000 00							
Dec 5	Fertilizer	5000 00			5000 00							
Nov 8	Loan payment	7000 00								3000 00	4000 00	
Nov 15	Tractor purchase	30000 00										30000 00

FIGURE 5–7. Sample expenditures entries in modified single-entry system.

There are many options for handling the recording of these transactions, but perhaps the easiest is to assign a letter or number code to each enterprise or commodity produced by the farm or ranch. For partnerships, corporations, and even landlord–tenant or father–son operating agreements, it is usually desirable to maintain one set of books for the operation in order to first view the business as a whole, although the results must be split eventually among cooperating individuals. Consequently, another number or letter coding system may be needed.

TRANSACTION 9: Loan payment made in November 19X1.

Analysis: Transaction 9 was analyzed earlier and handled sufficiently by using two debits and one credit in the general journal (Figure 5–1), with two separate accounts for interest and principal, under the double-entry system. If a modified single-entry system is used, two columns are again sufficient to record this transaction (Figure 5–7). If a single-entry cash system is used, the transaction may need to be written on two lines of the cash journal (see Figure 5–8).

TRANSACTION 12: Payment of $9,500 made to farm supply store for several items in November 19X1 for use in 19X2.

Analysis: To record Transaction 12, consider assigning codes such as:

Corn	01	or	C
Soybeans	02		S
Alfalfa	03		A
General (or Unallocated)	49		G

Then, recording the transaction in a general journal would require five lines (Figure 5–8), as it would in the cash journal (Figure 5–9) and the modified single-entry journal (Figure 5–10). Each of these systems allows entries to be added to calculate a total figure, a total figure by category (such as interest or fertilizer) and a total figure by category and by enterprise or owner. A sample coding scheme developed for use in Illinois is shown in Figure 5–11, but this scheme must be modified to fit different agricultural producing regions and different types of farm or ranch operations.

TRANSACTION 13: Stocker calves purchased for $25,000 for backgrounding in December 19X1. Partner 1 purchased 40 percent and Partner 2 purchased 60 percent.

Analysis: Unequal splits such as the stocker purchase in Transaction 13 can be handled quite easily as shown in Figures 5–9, 5–10 and 5–11. One of the easiest methods of showing the separate contributions (from bank accounts that are not commingled) is to record an initial or a name together with the amount contributed by each participant. Although this

Cash Journal
Mayfarms, Inc.
19X1

Page 2

	Initials	Date
Prepared By		
Approved By		

	19X1		Check/deposit no.	Expenditures (1)	Receipts (2)	Balance (3)	
1						182,000.00	1
2	Nov 8	Loan payment - Interest $7000.	889	3,000.00			2
3		- Principal		4,000.00		175,000.00	3
4	Nov 10	Casa Frijole Supply - Corn Fert.		5,000.00			4
5		- Soybean Fert. $9500.	894	3,000.00			5
6		- Alfalfa Wire		1,000.00			6
7		- Grease		500.00		165,500.00	7
8	Dec 11	Stocker calf purchase - Donald $25,000.	936	15,000.00			8
9		- Jim		10,000.00		140,500.00	9
10							10
11							11
12							12
13							13
14							14
15							15
16							16
17							17
18							18
19							19
20							20
21							21
22							22
23							23
24							24
25							25
26							26

FIGURE 5–8. Sample cash journal entries.

General Journal
Mayfarms, Inc.
19X1

Page 2

19X1		Acct	Debit	Credit
Nov 8	Interest Payable – Farmers State Bank	218	3000 00	
	Mortgage Payable – Farmers State Bank	222	4000 00	
	Cash	101		7000 00
	Loan Payment			
Nov 10	Fertilizer Inventory	118-01	5000 00	
	Fertilizer Inventory	118-02	3000 00	
	Baling wire Inventory	119-03	1000 00	
	Lubricant Inventory	120-49	500 00	
	Cash	101		9500 00
	Supplies purchased for use in 19X2			
Dec 11	Stocker calf Inventory – Donald	110-D	15000 00	
	Stocker calf Inventory – Jim	110-J	10000 00	
	Cash	101		25000 00
	Stocker calves purchased			

© WILSON JONES COMPANY G7202 GREEN 7202 BUFF 682 MADE IN U.S.A.

FIGURE 5–9. Sample general journal entries.

Expenditures
Mayfarms, Inc.
19X1

Date	Item	Code		Repairs	Fertilizer	Seed	Fuel & Oil	Supplies	Purchased Livestock	Interest	Principal Payments
Nov 8	Loan Payment - Farmers State Bank		7000 00							3000 00	4000 00
Nov 10	Supplies: Corn fertilizer	01	5000 00		5000 00						
	Soybean fertilizer	02	3000 00		3000 00						
	Baling wire	03	1000 00					1000 00			
	Grease	49	500 00				500 00				
Dec 11	Stocker calves - Donald	D	15000 00						15000 00		
	- Jim	J	10000 00						10000 00		

FIGURE 5–10. Sample modified single-entry journal entries.

DEPOSIT CODE (Receipts)
*01 — Crop or Feed Sale
*02 — Livestock, Milk & Egg Sales
*03 — Livestock Sale (Capital)
*Enterprise No. Necessary
11 — Custom Work
12 — Federal Gas Tax Refund
13 — State Gas Tax Refund
14 — Patronage Refund
15 — Building Receipts
16 — Cash Settlements
17 — ASC Diverted Acres
18 — ASC Wool
19 — ASC Others
20 — Misc. Receipts & Auto
25 — Machinery, Building Sales (Capital Sales)
61 — Non-Farm Income
62 — Non-Taxable Receipts
63 — Rental Property
81 — Money Borrowed

FBFM — FARM DATA SYSTEM — Management / Analysis / Program

ENTERPRISE NOS. (Use with *01, 02, 03)
(Circle the ones you use as reminder)

CROPS
02 — Corn
03 — Soybeans
04 — Wheat
05 — Oats
06 — Rye
07 — Barley
08 — Corn Silage
09 — Hay Silage
10 — Legume — Grasses
11, 13, 14 — Misc. Crops
12 — Protein

LIVESTOCK
28, 29, 30, 31 — Hogs
25, 26, 27 — Dairy Cattle
20, 21, 22, 23, 24 — Beef Cattle
32, 33, 34 — Sheep
35, 36, 37, 38 — Poultry
39 — Misc. LS
15 — Milk
16 — Eggs

EXPENSE CODE (Expenditures)
*31 Crop or Feed Purchase
*32 Livestock Purchase
*ENTERPRISE NO. MANDATORY
55 Mach., Bldg Purchases (Capital Purch)
82 Principal Payments

FARM
33 Interest
34 Hired Labor
35 Fertilizer
36 Seed & Crop
37 Livestock Expense
38 Utilities
39 Insurance
40 Misc.
41 Taxes
42 Cash Rent
43 Machinery Repairs
44 Machinery Hire
45 Gas & Oil
46 Auto
47 Building & Fence
48 Soil & Water Conservation

PERSONAL
70 Contributions
71 Drugs
72 Medical
73 Life Insurance Cost
**74 Family Living Expense
75 Family Living (Capital)
76 Social Security Income Tax
77 New Savings & Investments
78 Non-Farm Expense
79 Rental Property Expense

FBFM — FARM DATA SYSTEM — Management / Analysis / Program

ENTERPRISE NOS.
(Circle the ones you use as reminder)

CROPS
02 — Corn
03 — Soybeans
04 — Wheat
05 — Oats
06 — Rye
07 — Barley
08 — Corn Silage
09 — Hay Silage
10 — Legumes — Grasses
11, 13, 14 — Misc. Crops
12 — Protein

LIVESTOCK
28, 29, 30, 31 — Hogs
25, 26, 27 — Dairy Cattle
20, 21, 22, 23, 24 — Beef Cattle
32, 33, 34 — Sheep
35, 36, 37, 38 — Poultry
39 — Misc. LS
15 — Milk
16 — Eggs

FAMILY LIVING NOS.
**(Optional)

85 Food
86 Household Operating
87 Household Equipment
88 House Repairs, Rent
89 Clothing
90 Personal Items
91 Entertainment, Recreation
92 Education, Reading
93 Gifts
94 Transportation
95 Personal Interest Principal
(Optional with **74)

FIGURE 5–11. Illinois FBFM enterprise coding scheme.

method may be easy for this transaction, it may lead to difficulties later. It would be much easier to maintain a commingled or joint business account.

Business and Personal Checking Accounts

A major problem in accounting and records systems on many farms and ranches is that business and personal expenditures and assets are often mixed. Many producers, most of whom do not like to keep records to begin with, do not want to keep as detailed a set of records for their personal or family affairs as they do for business purposes. In addition, more individuals are generally involved in making family purchases and more purchases are made with currency. This mixture of people, currency, checks, and purposes can cause much confusion. But there is a very easy way to avoid this confusion: establish two separate checking accounts, one for the business and one for the family. Periodically, or whenever necessary, a transfer can be made from the business to the family account, thus leaving an accurate, easily followed track of total family living expenditures.

Most financial institutions now offer a wide variety of checks and checkbook covers to help people visually distinguish among different accounts and even to carry more than one checkbook at a time.

One final tip will help make bookkeeping a lot easier: never pay for any business purchase in currency. Always use checks to provide a written record. Investigate, as well, the use of carbonless copy checks for an immediate written copy.

Summary

Cash transactions records form the basis for many of the remaining types of records and are usually the most important and time-consuming part of the recordkeeping task. Before the form of the cash transactions records takes shape, the accounting and bookkeeping methods to be used must be chosen.

The textbook double-entry bookkeeping system uses two books. The general journal is the book of original entry. The information recorded in the journal is then posted to the general ledger, which maintains all of the various accounts.

Pure cash or modified cash accounting systems will use just one book. A cash journal replicates the checkbook register with the addition of noncheck purchases. Most modified systems also use just one book, which is actually a hybrid between a journal and a ledger since it performs the functions of both.

Many typical farm and ranch cash transaction recording problems can be avoided if separate personal and business checking accounts are maintained.

CHAPTER 6

Physical Inventories

Maintaining physical inventory records is often a perplexing issue to the farm or ranch financial manager and bookkeeper. Inventory control, also called *asset management,* is an important component of cash flow management, income tax management, and the preparation of financial statements. Inventory control involves several major groups of farm and ranch assets: financial assets (which will be discussed in detail in Chapter 8), supplies, commodities produced primarily for sale, depreciable assets, and other assets held for producing other commodities but which are not depreciable. (Intermediate and long-term inventories will be covered in Chapter 7.)

There are two main problem areas associated with inventory control and record keeping: (1) valuation of the assets and (2) how often and in how much detail inventory records should be maintained. The answer to both depends in part upon the choice of the cash or the accrual accounting method and upon the goals of the business manager—what business measures he will need and how often he will need financial statements.

Accrual Accounting

If a pure accrual accounting method is selected, inventory records are automatically maintained within the general ledger—at least values are, but the inventory ledger accounts can be modified to include physical units as well. However, many bookkeepers will shortcut the journal/ledger posting system to declare a purchase of supplies, for example, as an expense immediately rather than as an inventory of supplies that will later be transferred to an expense account as they are used. The process of debiting an inventory account upon purchase, and later crediting the inventory account while debiting the expense account, can quickly become cumbersome and impractical. Accounting for each sheet of paper, each gallon of gasoline, as well as spillage and other losses would be both a tedious and a very expensive process. The pure process is followed neither on farms and ranches nor by industries and merchandisers. Instead, estimates of use will be made and that use will be expensed. Then a manual inventory count will be taken periodically, using a form such as the example shown in Figure 6–1, and both inventory and expense accounts will be adjusted.

Modified Cash Accounting

The periodic inventory count and adjustment process used in most accrual accounting systems closely resembles the process used by modified cash accounting systems. Because the timing of payment rather than the timing of actual use is the critical distinction for cash-based accounting systems, inventories are not maintained. But to accurately reflect actual earnings within any accounting period inventory changes must be taken into consideration, and to accurately prepare a balance sheet inventories must be included.

To meet these needs, most agricultural producers will undertake an annual inventory count to establish their level of inventories for a balance sheet. The income statement's need for data on inventory changes can be met by comparing two successive balance sheets.

Inventory Valuation

Beyond the need to maintain physical inventories lies the question of how to properly value those inventories. A major controversial issue within the accounting profession is the proper valuation of inventories. The issue has significant impact upon reporting of income, the determination of total asset value, and thus upon stockholders' equity and upon income taxes. Many journal articles, books, and other publications have been written about the relative merits of the various recognized inventory valuation methods. Much of the argument about the methods has arisen because of the inflation in the United States since the early 1960s.

Very briefly, the two most common and widely accepted inventory valuation methods are First-In-First-Out (FIFO) and Last-In-First-Out (LIFO). Under FIFO the inventory quantities purchased first are assumed to be used for production or to be sold first, thus the inventory accounts are argued to be the most current since they reflect the cost of the most recently purchased items. On the other hand, under LIFO the inventory quantities purchased last are assumed to be sold or used first, thus the current cost of production is argued to be reflected most accurately within the income statement, and net income consequently best shows current operating conditions.

Much of the significance of this argument is lost on businesses that do not deal in extensive inventories of purchased supplies and finished goods and that do not face income tax liabilities measured in millions of dollars. Moreover, the significance of the argument is certainly lost for cash-based taxpayers.

The most common inventory valuation method used by farmers and ranchers as well by as other small businesses is to value inventories at the time the physical

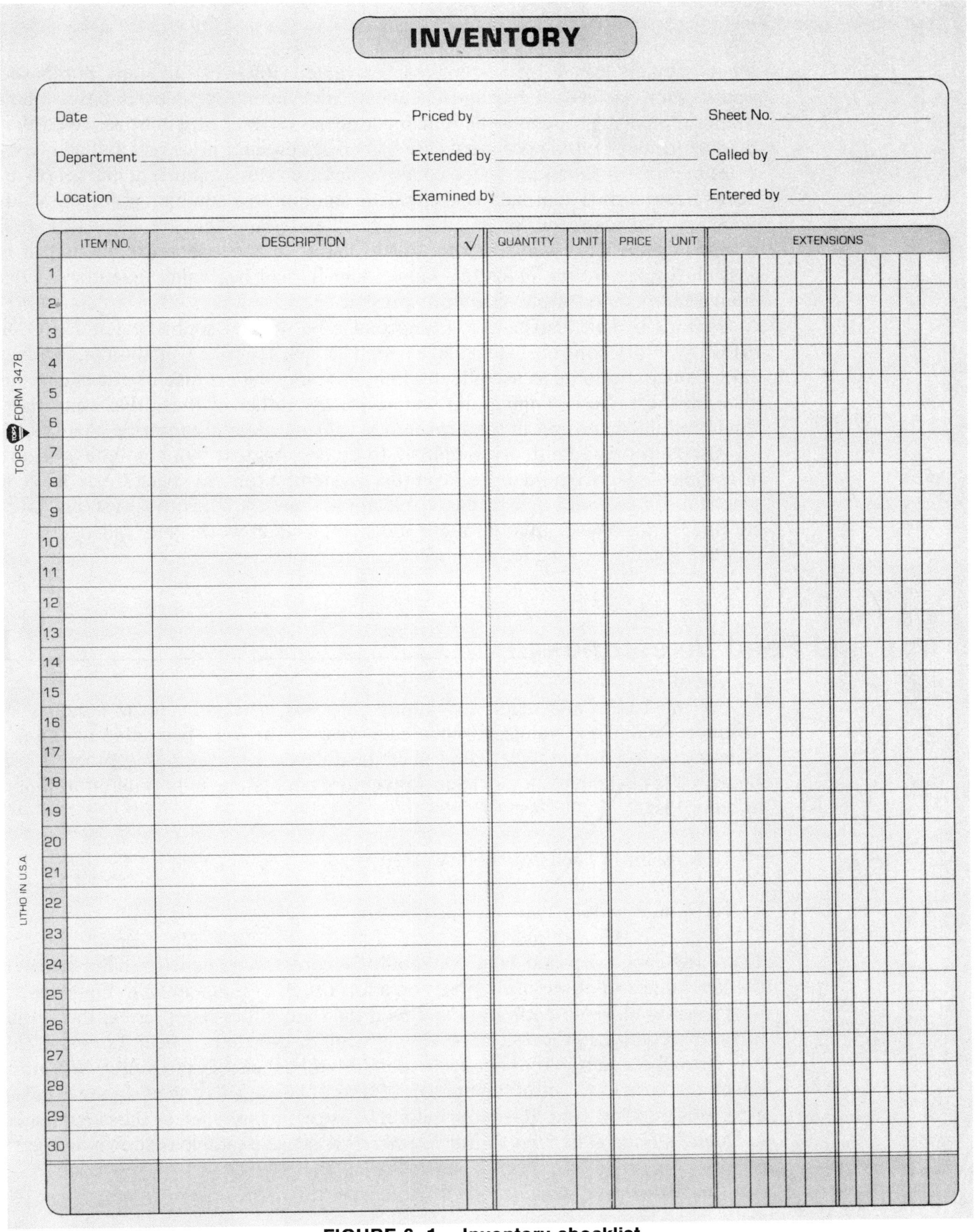

INVENTORY

Date ______ Priced by ______ Sheet No. ______

Department ______ Extended by ______ Called by ______

Location ______ Examined by ______ Entered by ______

	ITEM NO.	DESCRIPTION	√	QUANTITY	UNIT	PRICE	UNIT	EXTENSIONS
1								
2								
3								
4								
5								
6								
7								
8								
9								
10								
11								
12								
13								
14								
15								
16								
17								
18								
19								
20								
21								
22								
23								
24								
25								
26								
27								
28								
29								
30								

TOPS FORM 3478

LITHO IN U.S.A.

FIGURE 6–1. Inventory checklist.

count is taken, based on current market value. However for many purchased inputs, such as feed, fuels, and chemicals, the inventory value is based upon purchase price. Throughout the remainder of this chapter, it will be assumed that all inventories of purchased inputs are valued at purchase price and that all inventories of farm- or ranch-produced commodities are valued at current market price. Recognize that this approach is a simplistic solution to a complex problem. Valuing assets such as feeder or breeder livestock and grains that are subject to wide swings in price can be misleading. For example, if prices increase enough, it is possible to see ending inventory values higher than beginning inventory, even though the number of physical units present decreased.

Perhaps a more satisfactory compromise between a sophisticated LIFO or FIFO inventory control system and a simplistic market price valuation on the date of inventory would be to use either a long-term moving average of prices or even fixed on the same beginning and ending prices. Either of these two companies would reduce distortion in total inventory values due solely to price changes.

Although it is admittedly simplistic to do so, this text will continue with the price-times-quantity-on-date-of-inventory system, primarily because it is very practical and provides a very reasonable approximation. The other systems, journal and compromise, involve more judgment and/or work, and will likely not produce significantly better or more accurate results.

Crop and Feed Inventories

For most farm and ranch accounting purposes, it is sufficient to conduct an inventory count and valuation once each year at the very beginning of the accounting period (or at the very end since the difference between December 31 and January 1 is negligible). As with most inventory counts, the important information to record is:

1. Amount of each commodity
2. Value per unit
3. Date inventory is taken

Thus, for each crop and feed commodity record the amount on hand and the market value and/or purchase price, on a form such as that shown in Figure 6–2.

Probably more so with grain and feed than any other inventories, the actual inventory counting process can be very difficult or even impractical. For example, it is virtually impossible to count the number of bushels of corn stored in a grain bin on the farm or to count the number of tons of hay in a hay shed. In these cases estimates must be used. It is often helpful to use estimates such as those published in Doane's *Facts and Figures For Farmers.*[1] A silage example is shown in Figure 6–3.

[1]Published by Doane Agricultural Service, Inc., St. Louis, Missouri, 1972.

INVENTORY

Line		Beginning of year					End of year				
		Operator			Landlord		Operator			Landlord	
		Quantity	Price per unit	Value	Quantity	Value	Quantity	Price per unit	Value	Quantity	Value
401	Corn	bu.			bu.		bu.			bu.	
402											
403											
404	Soybeans	bu.			bu.		bu.			bu.	
405											
406	Oats										
407											
408	Wheat	bu.			bu.		bu.			bu.	
409											
410	Hay	ton			ton		ton			ton	
411											
412	Silage	ton			ton		ton			ton	
413	Protein and mineral feeds										
414	Total										

QUANTITY OF FEED AND GRAIN

Line		Corn (bushels)	Soybeans (bushels)			Hay (tons)	Silage (tons)
451	End-of-year inventory (401 thru 412)						
452	Sales (p. 12)						
453	Fed to livestock (473)						
454	Used as seed						
455	Total (451 thru 454)						
456	Beginning inventory (401 thru 412)						
457	Purchased for feed (p. 50)						
458	Produced (p. 46)						
459							
460	Total (456 thru 459)						
461	Shrinkage or overrun (460 — 455)						

VALUE OF FEED AND GRAIN

Line		Source	Operator share of value	Landlord share of value	Total farm value
415	End-of-year inventory	414			
416	Sales of grain, hay, etc.	118			
417	Fed to livestock	475			
418	Used as seed	454			
419					
420	Total (415 thru 419)				
421	Beginning inventory	414			
422	Purchased for feed	157			
423					
424	Total (421 thru 423)				
425	Value of crop production (420 — 424)				
426	Seed and crop expenses	167 168			
427	Value above seed and crop expense (425 — 426)				
428	Fertilizer cost	166			
429	Value of crops above seed and fertilizer (427 — 428)				

SUMMARY OF FEED FED TO LIVESTOCK

(Source: Livestock feed record, pages 54 through 57.)

Line		Protein, minerals, mixed feed (pounds)	Corn (bushels)	______ (bushels)	Oats (bushels)		Hay (tons)	Silage (tons)	Pasture
462	Dairy cattle								
463	Beef cattle								
464	Feeder cattle								
465									
466	Hogs								
467									
468	Poultry								
469									
470	Sheep								
471									
472									
473	Total fed								
474	Average price								
475	Total value								

FIGURE 6–2. Crop and feed inventory record.

Source: Robert B. Schwart, *Illinois Farm Record Book*, Illinois Cooperative Extension Service, University of Illinois.

Depth of settled silage	Inside diameter of silo in feet								Mean weight of settled silage per cubic foot
	10	12	14	16	18	20	24	30	
(Feet)					(Tons)				(Pounds)
10	10	15	20	26	33	41	59	93	26.3
11	12	17	23	30	38	47	68	106	27.3
12	13	19	26	34	43	53	77	120	28.2
13	15	21	29	38	48	59	86	134	29.1
14	16	24	32	42	53	66	95	148	30.0
15	18	26	36	47	59	73	105	163	30.9
16	20	29	39	51	65	80	115	179	31.7
17	22	31	43	56	70	87	125	195	32.5
18	24	34	46	60	76	94	136	212	33.3
19	25	37	50	65	82	102	147	229	34.1
20	27	39	54	70	89	110	158	247	34.9
21	29	42	58	75	95	117	169	264	35.6
22	31	45	62	80	102	125	181	282	36.3
23	33	48	66	86	108	134	193	301	37.0
24	36	51	70	91	115	142	205	320	37.7
25	38	54	74	97	122	151	217	339	38.4
26	40	57	78	102	129	159	229	358	39.0
27	42	60	82	107	136	168	242	378	39.6
28	44	64	87	113	143	177	255	398	40.2
29	46	67	91	119	151	186	268	418	40.8
30	49	70	96	125	158	195	281	439	41.4
31	51	74	100	131	166	205	295	460	42.0
32	54	77	105	137	173	214	308	482	42.6
33	56	81	110	143	181	224	322	504	43.2
34	58	84	115	150	189	234	337	526	43.8
35	61	88	120	156	198	244	352	549	44.4
36	64	92	125	163	206	254	366	572	45.0
37	66	95	130	169	214	264	381	595	45.5
38	69	99	135	176	222	275	395	618	46.0
39	71	103	140	182	231	285	410	641	46.5
40	74	106	145	189	239	295	425	664	47.0
41			150	196	248	306	441	688	47.5
42			155	203	257	317	456	713	48.0
43			161	210	265	328	472	737	48.5
44			166	217	274	339	488	762	49.0
45			171	224	283	350	504	787	49.5
46			177	231	293	361	520	813	50.0
47			183	239	302	373	537	839	50.5
48			188	246	311	385	554	865	51.0
49			194	254	321	396	571	892	51.5
50			200	261	331	408	588	919	52.0
52					349	431	621	970	52.8
54					368	455	655	1023	53.6
56					388	479	689	1077	54.4
58					407	503	724	1132	55.2
60					428	528	760	1188	56.0

Data adapted from the National Silo Association

FIGURE 6–3. Silage amount approximations.

Source: Doane's Fact and Figure For Farmers, Doane Publishing Company, St. Louis, MO, 1972.

Livestock Inventories

Like crop and feed inventories, amounts and market values of each class of livestock should be recorded at least once each year. Livestock classes obviously should be designated by species but also by age groups or weight classes. It is especially important when valuing livestock inventories to also classify livestock according to purpose (that is, whether the livestock in being held primarily for market, breeding, dairy, or work) and according to whether it was originally purchased or raised on the farm or ranch.

These classification distinctions are made for two primary reasons: for easier valuation and for income tax purposes. If animals are classified by age/weight groups, better and more accurate average weights and market price estimates may be obtained. An example of a typical livestock inventory record is shown in Figure 6–4.

Raised versus Purchased Market Livestock

The only real reason for distinguishing between raised and purchased market livestock (livestock held primarily for sale) is that current federal income tax laws make that distinction and require that the sales be reported in different places on the farm income tax return (see Figure 6–5 and Chapter 17). If income taxes are filed on the cash basis, income from sales of livestock is always reported in the year of the sale. Similarly, purchases of feed and other inputs are reported in the year of the purchase, with one major exception: purchases of livestock to be fed or otherwise held for resale are deducted in the year of the *sale,* not the year of the *purchase*. Consequently, a record of livestock purchases must be maintained in addition to other inventory records. An example form for recording market livestock purchases is shown in Figure 6–6.

Raised versus Purchased Breeding, Dairy, and Work Livestock

Although it is important for valuation and balance sheet classification purposes, the main reason for separating market livestock from other breeding, dairy, and work livestock is once again for income tax purposes. Livestock held for purposes other than resale often qualifies for capital gains tax exclusions, a special set of provisions of the tax law that favors investments if the asset is held for a specified period of time (see Chapter 18). Furthermore, livestock held for breeding, dairy, or work purposes which have been *purchased* will qualify for depreciation and investment tax credit (see Chapter 7), while livestock that have been raised will not qualify. Record forms for breeding, dairy, and work stock are very similar to those for market livestock with the major exception of showing additional categories (see Figure 6–7).

Other Purchased Inputs

Once again, when compiling inventories of other purchased inputs, the important information is the *amount* (weight, number of gallons, number of pieces) and the *value*. Because current market value may not be very easy to determine for each item, because an item was not really purchased for resale, and because the purchase price is not likely to have changed significantly, it is generally sufficient

56 **FEEDER CATTLE OR FEEDER HOG ENTERPRISE**

Inventory / **Death Loss**

Line	Description (list by type, sex, and weight class)	Beginning of year			End of year			Death Loss	
		No.	Weight	Value	No.	Weight	Value	No.	Weight
501	Feeder stock								
502									
503									
504									
505									
506									
507									
508									
509									
510									
511									
512									
513									
514	Total								
515	Operator share								

Breeding Record

Month	Females bred[1]	Off-spring born	Off-spring weaned
Jan.			
Feb.			
Mar.			
Apr.			
May			
June			
July			
Aug.			
Sept.			
Oct.			
Nov.			
Dec.			
Total			
Offspring per animal bred			

Feed Record

Month	Farm grains raised and purchased: Corn (bushels)	(bushels)	(bushels)	Hay (tons)	Silage (tons)		Pasture[2]: Date or days	Units
Jan.								
Feb.								
Mar.								
Apr.								
May								
June								
July								
Aug.								
Sept.								
Oct.								
Nov.								
Dec.								
Total								
Unit pr.								
Tot. val.								

	Livestock Summary	Source	No.	Total weight	Operator share of value	Landlord share of value	Total farm value
516	End-of-year inventory	514, 515					
517	Livestock sales, net value	109					
518	Home consumption	back cover					
519	Death losses	514, 515					
520	Total (add 516 thru 519)						
521	Beginning-of-year inventory	514, 515					
522	Purchases	154, 155, or 156					
523							
524	Total (add 521 thru 523)						
525	Net increase (520-524) / decrease (524-520)		0				

	Feed Summary	Total for year: Pounds	Total for year: Value	Per unit of product[5]: Pounds	Per unit of product[5]: Value
526	Total farm grains[3]				
527	Protein, minerals, mixed feeds[4]				
528	Total concentrates				
529	Hay, silage				
530	Pasture allowance[2]				
531	Total all feeds				

FOOTNOTES

[1] Females bred to produce offspring during this year.

[2] Pasture charge may be cash rent or estimated costs on a per head basis (see page 1).

[3] Convert bushels to pounds for all farm grains and enter totals in feed summary. In calculating value of feeds, use average yearly price.

[4] Enter the total pounds and value of protein, minerals, and mixed feeds for the year from feed inventory and purchase record (page 49).

[5] Divide the totals for the year by the total units of product produced.
- 1 dairy unit = 100 pounds of meat or 1,000 pounds of milk
- 1 beef unit = 100 pounds of beef
- 1 hog unit = 100 pounds of pork
- 1 poultry unit = 1½ pounds of meat or 1 dozen eggs
- 1 sheep unit = 100 pounds of lamb or mutton

[6] Does not apply for feeder livestock enterprise.

BREEDING RECORDS

Livestock Enterprise Analysis

551	Livestock returns (525)	$
552	Product returns (milk, eggs, or wool) (p. 53)	$
553	Total enterprise returns (551 and 552)	$
554	Total value of feed fed (531)	$
555	Total returns above feed cost (553 − 554)	$
556	Total value of cash costs (EH, p. 39)	$
557	Total returns above feed and cash costs (554 − 555)	$
558	Returns per $100 feed fed (553 ÷ 554 × 100)	$
559	Returns above feed per cow, litter, hen, ewe (555 ÷ 564)[6]	$
560	Feed costs per cow, litter, hen, ewe (554 ÷ 564)[6]	$
561	Feed costs per unit of product (531)	$
562	Pounds of feed per unit of product (531)	
563	Weight produced (525), lb.	
564	Females bred (513 breeding record)[6]	
565	Pounds of livestock per female bred (563 ÷ 564)[6]	
566	Product produced (milk, eggs, or wool, p. 53)[6], lb. or doz.	
567	Number of livestock (hens, cows, or ewes, p. 53)[6]	
568	Product per unit of livestock (566 ÷ 567),[6] lb. or doz.	

FIGURE 6–4. Feeder livestock inventory record.
Source: Robert B. Schwart, *Illinois Farm Record Book*, Illinois Cooperative Extension Service, University of Illinois.

SCHEDULE F (Form 1040) Department of the Treasury Internal Revenue Service (X)	**Farm Income and Expenses** ▶ Attach to Form 1040, Form 1041, or Form 1065. ▶ See Instructions for Schedule F (Form 1040).	OMB No. 1545-0074 **1985** 14

Name of proprietor(s)	**Social security number (SSN)**
A Agricultural Activity Code. (Write in the code that best describes your principal income-producing activity. The codes are listed on page 2 of this schedule.) ▶	**B Principal Product.** (Describe in one or two words your principal crop or output for the current tax year.) ▶
If you disposed of commodities received under the payments-in-kind (PIK) program, check the box(es) that apply: ☐ Feed for livestock ☐ Sold and reported in income	**Employer ID number (NOT SSN)**

Part I Farm Income—Cash Method—Complete Parts I and II

(Accrual method taxpayers complete Parts II and III, and line 12 of Part I.)

Do not include sales of livestock held for draft, breeding, sport, or dairy purposes; report these sales on Form 4797.

1 Sales of livestock and other items you bought for resale	1	
2 Cost or other basis of livestock and other items you bought for resale	2	
3 Subtract line 2 from line 1	3	
4 Sales of livestock, produce, grains, and other products you raised	4	
5 a Total distributions received from cooperatives (from Form 1099-PATR) 5a		
b **Less:** Nonincome items 5b		
6 Net distributions. Subtract line 5b from line 5a	6	
7 Agricultural program payments:		
a Cash	7a	
b Materials and services	7b	
8 Commodity credit loans under election (or forfeited)	8	
9 Crop insurance proceeds	9	
10 Machine work	10	
11 Other income, including Federal and state gasoline tax credit or refund (see instructions)	11	
12 **Gross income**. Add amounts on lines 3, 4, 6, and 7a through 11. If accrual method taxpayer, enter the amount from Part III, line 52 ▶	12	

Part II Farm Deductions—Cash and Accrual Method

Do not include personal or living expenses (such as taxes, insurance, repairs, etc. on your home), which do not produce farm income. Reduce the amount of your farm deductions by any reimbursements before entering the deduction below.

13 Breeding fees		26 Mortgage interest paid to financial institutions (see instructions)	
14 Chemicals		27 Other interest	
15 Conservation expenses		28 Pension and profit-sharing plans	
16 Depreciation, and section 179 expense deduction (from Form 4562)		29 Rent of farm, pasture	
17 Employee benefit programs other than on line 28		30 Repairs, maintenance	
		31 Seeds, plants purchased	
18 Feed purchased		32 Storage, warehousing	
19 Fertilizers and lime		33 Supplies purchased	
20 Freight, trucking		34 Taxes	
21 Gasoline, fuel, oil		35 Utilities	
22 Insurance		36 Veterinary fees, medicine	
23 a Labor hired		37 Other expenses (specify):	
b Jobs credit		a	
c Balance (subtract line 23b from line 23a)		b	
		c	
24 Land clearing (see instructions)		d	
25 Machine hire		e	

38 **Total deductions from Part II**. Add amounts in columns for lines 13 through 37e ▶	38	
39 **Net farm profit or (loss)** (subtract line 38 from line 12). If a profit, enter on Form 1040, line 19, and on Schedule SE, Part I, line 1. If a loss, you **MUST** go on to line 40. (Fiduciaries and partnerships, see the instructions.)	39	

40 If you have a loss, you **MUST** answer this question:
"Do you have amounts for which you are not at risk in this farm (see instructions)?" ☐ Yes ☐ No
If "Yes," you **MUST** attach **Form 6198**. If "No," enter the loss on Form 1040, line 19, and on Schedule SE, Part I, line 1.

For Paperwork Reduction Act Notice, see Form 1040 Instructions. **Schedule F (Form 1040) 1985**

FIGURE 6–5. Livestock sales reporting for income tax purposes.

SUMMARY OF LIVESTOCK PURCHASED FOR RESALE (Use only for reference for filing income tax on cash basis)

Kind	Purchases					Disposition				Number on hand end of year
							Sold			
	Date	Number	Weight	Total cost	Cost per head	Number died	Number	Weight	Value	

Note: *At the end of the year, carry information in last column along with date of purchase and cost per head to same page of new book.*

FIGURE 6–6. Market livestock purchase record.

Source: Robert B. Schwart, *Illinois Farm Record Book*, Illinois Cooperative Extension Service, University of Illinois.

54 **DAIRY OR BEEF COW ENTERPRISE**

Inventory

Line	Description	Beginning of year No.	Beginning of year Weight	Beginning of year Value	End of year No.	End of year Weight	End of year Value
501	Cows						
502							
503							
504							
505	Bulls (over 1 yr.)						
506							
507	Bred heifers						
508							
509							
510	Open heifers						
511							
512	Calves (under 1 yr.)						
513							
514	Total						
515	Operator share						

Death Loss

No.	Weight

Breeding Record

Month	Females bred[1]	Off-spring born	Off-spring weaned
Jan.			
Feb.			
Mar.			
Apr.			
May			
June			
July			
Aug.			
Sept.			
Oct.			
Nov.			
Dec.			
Total			
Offspring per animal bred			

Feed Record

Month	Farm grains raised and purchased: Corn (bushels)	(bushels)	(bushels)	Hay (tons)	Silage (tons)		Pasture[2]: Date or days	Units
Jan.								
Feb.								
Mar.								
Apr.								
May								
June								
July								
Aug.								
Sept.								
Oct.								
Nov.								
Dec.								
Total								
Unit pr.								
Tot. val.								

Line	Livestock Summary	Source	No.	Total weight	Operator share of value	Landlord share of value	Total farm value
516	End-of-year inventory	514, 515					
517	Livestock sales, net value	103					
518	Home consumption	back cover					
519	Death losses	514, 515					
520	Total (add 516 thru 519)						
521	Beginning-of-year inventory	514, 515					
522	Purchases	151, 154					
523	Young stock born	513					
524	Total (add 521 thru 523)						
525	Net increase (520-524) decrease (524-520)		0				

Line	Feed Summary	Total for year Pounds	Total for year Value	Per unit of product[5] Pounds	Per unit of product[5] Value
526	Total farm grains[3]				
527	Protein, minerals, mixed feeds[4]				
528	Total concentrates				
529	Hay, silage				
530	Pasture allowance[2]				
531	Total all feeds				

FOOTNOTES

[1] Females bred to produce offspring during this year.

[2] Pasture charge may be cash rent or estimated costs on a per head basis (see page 1).

[3] Convert bushels to pounds for all farm grains and enter totals in feed summary. In calculating value of feeds, use average yearly price.

[4] Enter the total pounds and value of protein, minerals, and mixed feeds for the year from feed inventory and purchase record (page 49).

[5] Divide the totals for the year by the total units of product produced.
- 1 dairy unit = 100 pounds of meat or 1,000 pounds of milk
- 1 beef unit = 100 pounds of beef
- 1 hog unit = 100 pounds of pork
- 1 poultry unit = 1½ pounds of meat or 1 dozen eggs
- 1 sheep unit = 100 pounds of lamb or mutton

[6] Does not apply for feeder livestock enterprise.

BREEDING RECORDS

Livestock Enterprise Analysis

551 Livestock returns (525) . . . . $____
552 Product returns (milk, eggs, or wool) (p. 53) . . . . . $____
553 Total enterprise returns (551 and 552) . . . . . $____
554 Total value of feed fed (531) . . . . . $____
555 Total returns above feed cost (553 − 554) . . . . . $____
556 Total value of cash costs (EH, p. 39) . . . . . $____
557 Total returns above feed and cash costs (554 − 555) $____
558 Returns per $100 feed fed (553 ÷ 554 × 100) . . . . . $____
559 Returns above feed per cow, litter, hen, ewe (555 ÷ 564)[6] . . . . . $____
560 Feed costs per cow, litter, hen, ewe (554 ÷ 564)[6] . . . $____
561 Feed costs per unit of product (531) . . . . . $____
562 Pounds of feed per unit of product (531) . . . . . ____
563 Weight produced (525), lb. . . . . . ____
564 Females bred (513 breeding record)[6] . . . . . ____
565 Pounds of livestock per female bred (563 ÷ 564)[6] . . . ____
566 Product produced (milk, eggs, or wool, p. 53)[6], lb. or doz. . . . . . ____
567 Number of livestock (hens, cows, or ewes, p. 53)[6] . . . . ____
568 Product per unit of livestock (566 ÷ 567),[6] lb. or doz. ____

FIGURE 6–7. Dairy stock inventory record.

Source: Robert B. Schwart, *Illinois Farm Record Book*, Illinois Cooperative Extension Service, University of Illinois.

Inventory of Fertilizer, Lime, Spray Materials, and Supplies

Line	Description	Beginning of year		End of year	
		Weight	Value	Weight	Value
601					
602					
603					
604					
605					
606					
607					
608					
609	Total				
610	Operator share				

FIGURE 6–8. Supplies inventory record.

to use the original purchase price as the measure of unit value. An example record form is shown in Figure 6–8. Include such items as welding rods and gas, fuels, fertilizer, chemical pesticides, baling wire, and repair parts. Once again, estimates of amounts may be necessary.

Summary

Physical inventories are an important part of complete farm and ranch records systems. These inventories are especially important for the preparation of balance sheets and income statements, as will be seen in later discussions of financial statements. In the search for an accurate yet practical inventory control system, the manager must decide whether a formal LIFO or FIFO control system is desired, or whether a simplistic system of counting (or estimating) amounts on hand is sufficient. The choice will depend on time available, patience of the bookkeeper and manager, size of operation, and management information needs.

Regardless of the system chosen, it is vital to record and verify amounts and values of all short-term assets on hand on the dates of preparation of major financial statements (usually once each year). These major short-term assets include crop and feed inventories, livestock inventories (broken down by raised versus purchased and market versus breeding), and other purchased inputs.

CHAPTER 7

Long-lived Asset Inventories

Long-lived assets are those with a useful life of over one year. The inventory records of these long-lived assets are usually prepared separately from the current asset inventories discussed in Chapter 6. The distinction is made for several major reasons, including these:

1. Long-lived assets typically turn over (are bought, sold, traded, or exchanged) much less frequently than do current assets.
2. The cost of each long-lived asset is relatively large and cannot be charged as a business expense in any one year.
3. The cost of many of these assets is recovered over a period of years through *depreciation*.
4. The income tax treatment on the purchase, sale, or trade of long-lived assets is significantly different from that of current assets.
5. Two values, *current market value* and *book value,* must be recorded for each asset item.

Because of the major difference in the manner in which the values of depreciable and nondepreciable assets will be recorded, it is first necessary to distinguish between the two—that is, to define what depreciation is and which assets are depreciable.

Depreciation

Depreciation is an economic concept, an accounting concept, and a tax convention that describes how the cost of a long-lived asset should be charged as an expense to the business. The basic rule is that since the asset is long-lived and its usefulness to the business will extend over a period of years, the cost of the asset should be spread over that period. The cost may be spread out in order to reflect the fact that the asset is a necessary input in the production process and the cost of any input should be charged when used, not when cash is paid. Thus, since outputs are produced with the use of the asset over a period of years, the cost of the asset should be allocated, expensed, deducted, or depreciated over that period of years.

In addition to the concept of multiple-year use, depreciation has also been justified as a means to recognize that most assets will eventually wear out, will grow old so that productivity declines, and will become obsolete. It has also been justified as a means to reflect the fact that as many assets age their market resale value drops.

Problems with Accepted Depreciation Methods

Practically speaking, it is difficult at the beginning of the life of any long-lived asset purchase to determine how many years or how much output can be produced with it. Consequently, conventions have been developed to estimate the amount of original cost to be charged in any one year. These conventions have been given distinct acceptable forms by the accounting profession and by the Internal Revenue Service, and they have been modified several times to reflect changing economic conditions.

Many of the conventions still in use today were developed in noninflationary periods and consequently several appear to be archaic in modern business times. The conventions have been based on: mathematical methods, how fast depreciation will be taken in the early years of life relative to the later years, how many years the asset will be used, and how much the asset can be sold for when its usefulness to the business is over.

Unfortunately, the depreciation rules, methods, and conventions that we are allowed to use for income tax purposes bear little resemblance to the original goals of both economists and accountants. Economists and accountants both recognized that assets decline in value over time and developed the concept of depreciation to approximate ownership costs when measuring the income actually earned during a specific year or production period. Tax conventions, however, overrule those goals. This conflict leaves an unresolved problem of two realities: the tax reality and the actual cost reality.

In practice, it may be necessary to consider depreciation in both ways depending upon the particular need. For tax reporting purposes we must use IRS-approved methods and we should select the method that provides the most tax benefit. But for such needs as estimating long-term profitability we should rid ourselves of the tax conventions and estimate actual useful lives and ownership costs of purchased long-lived assets. For example, agricultural economists will usually select a simple straight-line depreciation method (which will be discussed later) over expected useful life when estimating long-term costs and returns of an agricultural enterprise, recognizing however that for tax purposes farmers and ranchers must use the IRS methods.

Nondepreciable Assets

Despite the fact that they have long lives, are requirements of production and have fairly high purchase prices, certain long-lived assets *cannot* be depreciated for income tax purposes and should not be depreciated for business analysis purposes. Nondepreciable long-lived assets will generally fall into one of three major categories:

1. Land
2. Personal or nonbusiness assets
3. Assets that have no basis (see the next section)

In turn, the reason why each of these categories is considered nondepreciable should become clear.

Depreciation, for tax purposes and for analysis purposes, is reserved for business assets with a determinable or finite useful life. Thus, land is not depreciated because it is generally assumed to have an infinitely long useful life. When we spread a finite cost over an infinite number of years, the amount to be allocated to any one year becomes zero. And nonbusiness assets such as personal automobiles, personal residences, and furniture are not depreciable even though their resale value also declines through time.

Basis

Assets that have no basis comprise the third category of nondepreciable assets, probably the most difficult to explain and understand. The first problem is to understand the concept of *basis*. The basis of any piece of property is equal to the amount of money and/or the value of any other goods and services paid or given in exchange for that property. The *adjusted basis* is the original basis minus any *accumulated depreciation* taken. For a piece of property to have no basis, then, means that nothing was directly paid for the property. Two examples of property with no basis are gifts and raised livestock. Gifts are fairly easy to understand but raised livestock may not be. Technically, no raised livestock has a basis because nothing was actually paid for the newborn animal. Certainly there were costs attributable to the production of that animal, such as the feed costs, veterinary bills and other costs which went into maintaining its mother. But for a cash-basis taxpayer those costs were written off as expenses in the year in which they were incurred. Consequently, no "purchase price" is left to be attached to the newborn animal. Since there is no purchase price, there really is nothing to depreciate. If the costs were accumulated on an accrual basis and not expensed in the year paid, then the amount of those costs could be depreciated. But for a cash-basis taxpayer depreciation would not be taken. Although this may seem harsh at the moment, the process really benefits the cash-basis taxpayer in the long run because the costs are expensed immediately rather than spread over a period of years.

Depreciable Assets

All long-lived assets other than those listed above (land, personal, and zero-basis assets) qualify for depreciation. These assets include grain storage facilities, machinery, equipment, purchased breeding or dairy livestock, buildings, fences, shop tools, trucks, and business automobiles, among others that have a useful life of over one year and that are purchased and used for business purposes.

Calculating Depreciation

Three major depreciation methods were recognized by IRS for purchases of depreciable assets prior to 1981: *straight-line, sum-of-years digits,* and *declining-balance methods*. To calculate depreciation under any of the three methods, four data items were needed: *date of purchase, purchase price, salvage value,* and *useful life*.

Although only the straight-line method remains available as an option to the *Accelerated Cost Recovery System* (ACRS), which became effective on January 1, 1981 (see pages 82 and following), items purchased before 1981 cannot be changed to ACRS. Therefore, it is likely that each established farm and ranch will continue to depreciate some items under the old methods until all of those assets are completely depreciated or sold. It is no longer necessary, though, to discuss the advantages and disadvantages of each method or to discuss criteria for determining appropriate useful life or salvage value.

Under each method, be careful to not depreciate below the salvage value originally specified.

Straight-Line Method

To calculate depreciation under the straight line method, simply divide the number of years of useful life into the depreciable balance (purchase price minus salvage value).

$$\text{Depreciation} = \frac{\text{Purchase Price} - \text{Salvage Value}}{\text{Years of Useful Life}}$$

As implied by the name, the amount of depreciation taken is the same in each year of the selected useful life.

Sum-of-Years Digits Method

Depreciation calculations under the sum-of-years digits (SYD) method are only slightly more complicated. First, the SYD factor shown below must be calculated by adding the digits of the years in the useful life. For example, if a five-year useful life was specified, the factor would be $1 + 2 + 3 + 4 + 5 = 15$. Then, in the first year, depreciation would be calculated using 5 for *YR*, and 4 for the second year, 3 for the third and so on.

$$\text{Depreciation} = \frac{\text{YR (Purchase Price} - \text{Salvage Value)}}{\text{SYD Factor}}$$

Declining-Balance Methods

To calculate depreciation under the declining balance method, first determine the rate at which depreciation was to take place. Those rates were specified by the IRS as a maximum allowed for various classes of assets and included 1, 1.25, 1.5, 2, or any rate less than those. The rate chosen at time of purchase must stand for the remaining years of useful life for each asset. Typically the rate chosen for machinery and many other assets was 2, thus the name *double declining balance*.

In year 1, depreciation was calculated as:

$$\text{Depreciation} = \frac{\text{Rate (Purchase Price)}}{\text{Useful Life}}$$

At the end of Year One, the remaining value was calculated:

$$\text{Remaining Value} = \text{Purchase Price} - \text{Depreciation}$$

In year 2, depreciation was calculated as:

$$\text{Depreciation} = \frac{\text{Rate (Remaining Value)}}{\text{Useful Life}}$$

and a new remaining value was calculated by subtracting current depreciation from the last remaining value. All succeeding years are calculated in the same manner.

Notice that salvage value is *not* used in the equations anywhere, but the remaining value can never fall below the specified salvage value. Furthermore, the rate chosen and used in the equations will not change throughout the useful life of the item.

Depreciation Examples

To illustrate the mechanics of the three depreciation methods, consider example asset purchases 1 and 2 of the following list. These two depreciation problems are solved in Tables 7–1 and 7–2.

1. Tractor purchased for $50,000 on January 1, 19X0. Useful life selected to be 5 years, salvage value selected to be $5,000.
2. Truck purchased for $20,000 on July 1, 19X0. Useful life selected to be 4 years, salvage value selected to be $4,000.
3. Tractor purchased for $50,000 on February 1, 19X1.
4. Grain combine purchased for $65,000 on November 15, 19X1.
5. Half-ton pickup purchased for $9,000 on June 23, 19X1.

TABLE 7–1 Depreciation Analysis of Asset Purchase 1

Asset 1—Straight Line

Purchase Price: $50,000; Salvage Value: $5,000; Useful Life: 5 years

Year	Remaining Value, Beginning of Year	Depreciation	Remaining Value, End of Year
19X0	$50,000	($50,000 − $5,000) ÷ 5 years = $9,000	$41,000
19X1	41,000	(50,000 − 5,000) ÷ 5 years = 9,000	32,000
19X2	32,000	(50,000 − 5,000) ÷ 5 years = 9,000	23,000
19X3	23,000	(50,000 − 5,000) ÷ 5 years = 9,000	14,000
19X4	14,000	(50,000 − 5,000) ÷ 5 years = 9,000	5,000
19X5	5,000	0	5,000

Asset 1—Sum-of-Years Digits

Purchase Price: $50,000; Salvage Value: $5,000; Useful Life: 5 years

Year	Remaining Value, Beginning of Year	Depreciation	Remaining Value, End of Year
19X0	$50,000	5/15* × ($50,000 − 5,000) = $15,000	$35,000
19X1	35,000	4/15 × (50,000 − 5,000) = 12,000	23,000
19X2	23,000	3/15 × (50,000 − 5,000) = 9,000	14,000
19X3	14,000	2/15 × (50,000 − 5,000) = 6,000	8,000
19X4	8,000	1/15 × (50,000 − 5,000) = 3,000	5,000
19X5	5,000	0	5,000

*SYD Factor: 1 + 2 + 3 + 4 + 5 = 15

Asset 1—Double Declining Balance

Purchase Price: $50,000; Salvage Value: $5,000; Useful Life: 5 years

Year	Remaining Value, Beginning of Year	Depreciation	Remaining Value, End of Year
19X0	$50,000	2/5** × $50,000 = $20,000	$30,000
19X1	30,000	2/5 × 30,000 = 12,000	18,000
19X2	18,000	2/5 × 18,000 = 7,200	10,800
19X3	10,800	2/5 × 10,800 = 4,320	6,480
19X4	6,480	2/5 × 6,480 = 2,592 or 1,480***	5,000
19X5	5,000	0	5,000

**Double declining balance implies *rate* = 2.
***Cannot depreciate below salvage value.

ACRS

Since January 1, 1981, all purchases of depreciable assets fall under the Accelerated Cost Recovery System (ACRS). ACRS nullifies all depreciation methods except straight-line, *for purchases made after December 31, 1980*. Calculations under ACRS are easier than under any of the other previously described depreciation methods. Salvage value was eliminated by the legislation that created

TABLE 7–2 Depreciation Analysis of Asset Purchase 2

Asset 2—Straight Line

Purchase Price: $20,000; Salvage Value: $5,000; Useful Life: 3 years

Year	Remaining Value, Beginning of Year	Depreciation	Remaining Value, End of Year
19X0	($20,000)	($20,000 − $5,000) ÷ 3 years × ½* = $2,500	$17,500
19X1	17,500	(20,000 − 5,000) ÷ 3 years = 5,000	12,500
19X2	12,500	(20,000 − 5,000) ÷ 3 years = 5,000	7,500
19X3	7,500	(20,000 − 5,000) ÷ 3 years × ½* = 2,500	5,000
19X4	5,000	0	5,000

*½ year allowed.

Asset 2—Sum-of-Years Digits

Purchase Price: $20,000; Salvage Value: $5,000; Useful Life: 3 years

Year	Remaining Value, Beginning of Year	Depreciation	Remaining Value, End of Year
19X0	($20,000)	3/6** × ($20,000 − $5,000) × ½ = $3,750	$16,250
19X1	16,250	3/6 × (20,000 − 5,000) × ½ + 2/6 (20,000 − 5,000) × ½ = 3,750 + 2,500 = 6,250	10,000
19X2	10,000	2/6 × (20,000 − 5,000) × ½ + 1/6 (20,000 − 5,000) × ½ = 2,500 + 1,250 = 3,750	6,250
19X3	6,250	1/6 × (20,000 − 5,000) × ½ = 1,250	5,000
19X4	5,000	0	5,000

**SYD Factor: 1 + 2 + 3 = 6

Asset 2—1.5 Declining Balance

Purchase Price: $20,000; Salvage Value: $5,000; Useful Life: 3 years

Year	Remaining Value, Beginning of Year	Depreciation	Remaining Value, End of Year
19X0	($20,000)	1.5/3 × $20,000 × ½ = $5,000	$15,000
19X1	15,000	1.5/3 × 15,000 = 7,500	7,500
19X2	7,500	1.5/3 × 7,500 = 3,750 *or* 2,500	5,000
19X3	5,000	0	5,000

ACRS, but most important, the useful life of any long-lived asset purchase is predetermined. Each depreciable asset now falls into one of four categories: 3-year, 5-year, 10-year or 19-year property).[1] Box 7–1 presents a list of the types of property that fit into each category. Generally, buildings will fall into the 19-year

[1]The Economic Recovery Tax Act of 1981 originally specified a 15-year class of property, but the Tax Reform Act of 1984 redefined that category as an 18-year class for property placed in service after March 15, 1984. Since then, the class of property has been restructured to be 19-year property (for property placed in service after May 8, 1985).

BOX 7–1

ACRS Property Classes

3-year property:	Tangible personal property with a present guideline life of four years or less, or used in connection with research and experimentation: cars, light-duty trucks (under 13,000 pounds), special tools, purchased breeding swine, racehorses over 2 years old, and other horses over 12 years old when placed in service.
5-year property:	All property not included as 3-, 10-, or 19-year property: heavy-duty trucks; most office furniture; machinery and equipment; single-purpose agricultural and horticultural structures such as swine facilities, milking parlors, and greenhouses; purchased breeding sheep, beef, and dairy animals; fence and tile. *Nearly all* depreciation for most farmers and ranchers will be in this case.
10-year property:	Property with a present guideline life of 12.5 years or less. One possible application to agriculture is "manufactured housing" (for example, a mobile home for an employee).
19-year property:	Property with present guideline lives of over 12.5 years; depreciable real estate will fit here. General-purpose or multipurpose farm builngs are about all that will fit in this class. There are no differences in the treatment of new and used property.

class; mobile homes are 10-year property; breeding swine, cars, and pickups will be classified as 3-year property. Virtually everything else, including all machinery, equipment, special use buildings, breeding livestock, fences, and grain storage facilities will be 5-year property.

To depreciate capital assets using the ACRS system, simply multiply the applicable depreciation percentage rate shown in Tables 7–3 or 7–4 by the original purchase price in each year of the tax-specified useful life. Note that the numbers in each column of both Tables 7–3 and 7–4 add up to 100 percent. Consequently, the original purchase price is multiplied each year by the percentage listed.

ACRS Alternatives

Although the percentage system described in the previous section is now the dominant depreciation calculation method for new purchases, the Economic Recovery Tax Act of 1981 (ERTA) allows business owners to elect to use the straight-line method. However, three restrictions are placed on this election:

1. No salvage value may be used in the calculations.
2. Only three different periods may be used for the useful life.
3. All assets purchased in any one year that fall within the same property class must be treated in an identical manner (except for 19-year property).

To explain these three concepts in order, consider the following examples of Assets 3 through 5, as listed on page 81, which were purchased in 19X1. Before any decisions can be made, the ACRS property class to which each of these assets

TABLE 7–3 Percentage Depreciation Rates for ACRS Property

	ACRS Property Class		
Recovery Year	3-Year %	5-Year %	10-Year %
1	25	15	8
2	38	22	14
3	37	21	12
4		21	10
5		21	10
6			10
7			9
8			9
9			9
10			9

belongs must be identified. Assets 3 and 4 fall into the "remainder" category, the 5-year class, which includes all machinery and equipment according to Box 7–1. Asset 5, a light-duty truck, is 3-year property.

After the property class has been identified, the recovery method (percentage or straight-line) must be chosen. Suppose that the farmer wishes to depreciate the tractor using the straight-line method.[2] If he uses the straight-line method for the tractor, he must also use it for the combine, since both belong to the same property class (Restriction 3 above), but he may select either method for the pickup since it belongs to a different class.

If the straight-line method is selected, then a useful life must be selected (Restriction 2). Only under the straight-line alternative may the useful life be

TABLE 7–4 Percentage ACRS Depreciation Rates for Real Estate

	19-year Real Property											
	Month Placed in Service											
Year	1	2	3	4	5	6	7	8	9	10	11	12
1st	8.8	8.1	7.3	6.5	5.8	5.0	4.2	3.5	2.7	1.9	1.1	0.4
2nd	8.4	8.5	8.5	8.6	8.7	8.8	8.8	8.9	9.0	9.0	9.1	9.2
3rd	7.6	7.7	7.7	7.8	7.9	7.9	8.0	8.1	8.1	8.2	8.3	8.3
4th	6.9	7.0	7.0	7.1	7.1	7.2	7.3	7.3	7.4	7.4	7.5	7.6
5th	6.3	6.3	6.4	6.4	6.5	6.5	6.6	6.6	6.7	6.8	6.8	6.9
6th	5.7	5.7	5.8	5.9	5.9	5.9	6.0	6.0	6.1	6.1	6.2	6.2
7th	5.2	5.2	5.3	5.3	5.3	5.4	5.4	5.5	5.5	5.6	5.6	5.6
8th	4.7	4.7	4.8	4.8	4.8	4.9	4.9	5.0	5.0	5.1	5.1	5.1
9th	4.2	4.3	4.3	4.4	4.4	4.5	4.5	4.5	4.5	4.6	4.6	4.7
10–19th	4.2	4.2	4.2	4.2	4.2	4.2	4.2	4.2	4.2	4.2	4.2	4.2
20th	0.2	0.5	0.9	1.2	1.6	1.9	2.3	2.6	3.0	3.3	3.7	4.0

[2]The choice of methods depends upon the farmer's or rancher's income tax management strategy. The percentage system provides a faster write-off—more depreciation in early years—than does the straight-line method. The faster write-off benefits taxpayers relatively more when they are in a higher income tax bracket.

TABLE 7–5 Depreciation or Recovery Alternatives Under ACRS

	Years to Recover or Depreciate			
Asset Class	Accelerated Rates	Straight Line		
3-year property	3	3	5	12
5-year property	5	5	12	25
10-year property	10	10	25	35
19-year property	19	19	35	45

changed from the number of years specified in the property class designations of Box 7–1. These alternatives are listed in Table 7–5. For example, the tractor and the combine may be depreciated over five years, twelve years, or twenty-five years if the straight-line method is selected, but they must be handled in an identical manner.

To calculate depreciation in the year of purchase, a concept called the *half-year convention* must be used. The half-year convention was optional prior to ERTA but has been mandatory for all items purchased since January 1, 1981. It simply requires that one half of the allowable annual depreciation be taken in the first year of ownership for 3-, 5-, and 10-year property regardless of how long that asset was owned during the year. The half-year convention is assumed by and built into the percentage depreciation rates shown in Table 7–4, but must be calculated when using the straight-line option. The half-year convention does not apply to 19-year property; instead, first-year depreciation is prorated according to months of service.

ACRS Examples

To show how depreciation is calculated under ACRS, consider again Asset Purchases 3, 4, and 5, assuming Assets 3 and 4 will be depreciated using the straight-line method over five years and Asset 5 will be depreciated using the accelerated percentage rates over three years. The analysis of these asset depreciation problems is presented in Tables 7–6 and 7–7.

Basis in Property

Although *basis in property* was defined in an earlier section of this chapter (see page 79), it might be useful to return to the definition and provide an example. To review, basis is equal to the amount of money and/or value of any other goods and services paid or given in exchange. It is the basis value that will be depreciated, not just the amount of cash paid.

Consider the following examples:

ASSET PURCHASE 6: Tractor purchased on September 1, 19X1.
Original list price was $58,000. The seller agreed to accept

TABLE 7–6 ACRS Depreciation Analysis of Asset Purchases 3 and 4

Assets 3 and 4—ACRS Straight-Line, 5 Years			
Purchase Price: $115,000			
Year	**Remaining Value, Begining of Year**	**Depreciation**	**Remaining Value, End of Year**
19X1	($115,000)	$115,000 ÷ 5 years × ½ = $11,500	$103,500
19X2	103,500	115,000 ÷ 5 years = 23,000	80,500
19X3	80,500	115,000 ÷ 5 years = 23,000	57,500
19X4	57,500	115,000 ÷ 5 years = 23,000	34,500
19X5	34,500	115,000 ÷ 5 years = 23,000	11,500
19X6	11,500	115,000 ÷ 5 years × ½ = 11,500	0

TABLE 7–7 ACRS Depreciation Analysis of Asset Purchase 5

Asset 5—ACRS Accelerated Rates, 3 Years			
Purchase Price: $9,000			
Year	**Remaining Value, Beginning of Year**	**Depreciation**	**Remaining Value, End of Year**
19X1	($9,000)	$9,000 × .25 = $2,250	$6,750
19X2	6,750	9,000 × .38 = 3,420	3,330
19X3	3,330	9,000 × .37 = 3,330	0

$50,000 in cash, $4,000 in grain and $2,000 in labor services for the tractor.

Basis: $50,000 cash[3] + $4,000 goods + $2,000 services = $56,000

ASSET PURCHASE 7: Combine purchased on October 1, 19X1. Original list price was $75,000. The seller agreed to accept $60,000 in cash and a used combine which was determined to have a fair market value of $12,000. The remaining value of the combine in the owner's depreciation records was $4,000.

Basis: $60,000 cash + $4,000 remaining value = $64,000.

This basis calculation for Asset Purchase 7 seems at first glance to be very strange, but the reason for doing it this way is to avoid paying tax on the *gain* on the old combine. The owner's depreciation records indicate that the remaining value of the combine is $4,000 but the owner could sell it for $12,000. There is consequently an $8,000 gain, which would be taxable income if the combine was

[3]The amount of cash offered in this type of arrangement is often called the cash "boot."

sold. To avoid taxation of that gain, the farmer could *roll over* that gain into the new combine. In essence, he will recognize the gain at a later time.

The basis calculated can be viewed in an alternative way:

Negotiated Selling Price	=	$72,000
Price Paid in Cash	=	$60,000
+ Price Paid in Property	=	$12,000
= Total Price Paid	=	$72,000

So the seller receives $72,000 worth of cash and machinery for the new machine. But, to avoid taxation of the gain, the buyer subtracts that gain:

Total price paid	=	$72,000
Minus gain	=	–$8,000
Basis	=	$64,000

The farmer would consequently place the new combine on his depreciation schedule at a basis of $64,000 and calculate depreciation on that amount. However, he would list the *market value* of the machine at $72,000.

This process of not recognizing gain on items traded in is called a *nontaxable exchange*. In most instances, this process works to the advantage of the taxpayer. But it is not a matter of choice. That is, the owner of the property traded in *must* handle the transaction in this manner; he cannot choose or elect to recognize and pay tax on the $8,000 gain and depreciate a $72,000 machine. He *must* calculate the basis to be $64,000.

Suppose the farmer sees this trade coming and he decides that he would like to recognize the gain and depreciate the full amount of the negotiated purchase price. So he arranges to sell the old combine for $12,000 either to the dealer or to a third person. Then he later pays the dealer $72,000 in cash.

Even under these circumstances, his basis in the new combine is $64,000 because the Internal Revenue Service sees this sale and purchase as being in fact a trade. Generally, a sale and purchase of "like-kind" property (see below) occurring within a 60-day period is viewed as a trade for tax purposes.

ASSET PURCHASE 8: Plow purchased from a neighbor. The negotiated price was $5,000. The buyer paid $3,000 in cash and traded a bull having a fair market value of $2,000 and a book value of $1,000.

***Basis*:** $3,000 cash + $2,000 bull = $5,000

The purchase of Asset 8 illustrates a limitation to the nontaxable exchange concept. Since the bull and the plow are not "like-kind" property, the transaction is viewed as a sale and purchase and a taxable exchange. Therefore, the plow buyer would depreciate a $5,000 plow and pay tax on a $1,000 gain on the sale of the bull.

The key to determining whether the trade was a taxable or a nontaxable exchange lies in the definition of *like-kind property*. Exchanges or trades of like-kind property are nontaxable, trades of non-like-kind property are taxable. The IRS has defined *like-kind* to refer to the nature or character or use or purpose of the property. Any truck or any other machine is like-kind property as to any other machinery. A trade of farmland for apartment buildings or timberland is also a like-kind exchange, since all of these forms of property were held to generate income. However, a trade of machinery for land, livestock of one sex for those of the other sex, and an old bull for a young bull, and other such trades are not like-kind. The IRS rules are very strict and should be checked before handling the transaction either as a taxable or a nontaxable exchange.

Investment Tax Credit

The *Investment tax credit* (or simply *investment credit* or ITC) is a provision of the Internal Revenue Code that allows businesses to reduce tax liability if qualified long-lived assets are purchased. Since ITC was first enacted in 1962 for the purpose of encouraging businesses to invest in new plant and equipment (in order to encourage national economic growth and development), Congress has changed its provisions many times.[4] But the basic mechanism has generally remained the same: an amount equal to a stated percentage of the purchase price of the newly acquired asset is subtracted from the amount of taxes owed. Note especially that the investment tax credit is a *credit,* not a *deduction.*

A deduction reduces taxable income before tax liability is calculated. A credit reduces taxes. Consequently, a dollar of tax credit is much more valuable to the taxpayer than is a dollar of deduction. For example, assume a taxpayer is in the 35 percent marginal tax bracket (that is, on the next dollar earned, an additional 35 cents is owed in taxes). If the taxpayer finds an additional $100 in deductions, his tax liability would fall by $35. On the other hand, if the taxpayer found $100 in tax credits rather than deductions, his tax liability would fall by the full $100. Consequently, tax credits are very valuable income tax management tools for all businessmen.

What Qualifies for Investment Credit?

The purchases of property that qualify for investment credit are, firstly, all new or used depreciable personal property placed in service in the business during the tax year. There are a few exceptions, however. The property must have a useful life of at least three years.[5] In general, buildings have not qualified, except for *special-purpose facilities,* including grain handling and storage facilities, milking parlors, and some horticultural and confined livestock-raising facilities. The 1978 tax legislation specifically extended investment credit to *single-purpose livestock and horticultural structures.* Livestock structures must be specifically de-

[4]After this text was written, Congress passed the Tax Reform Act of 1986 and eliminated ITC retroactively to January 1, 1986.

[5]Under certain conditions property that normally would qualify for ITC is specifically excluded. For example, if you purchase property from a related person, you may not claim ITC on that item. Or, if you previously owned, that item, you may not claim ITC if you later repurchase it. Check with your tax advisor for more specific details.

signed, constructed, and used for housing, raising, and feeding livestock (and poultry) and their produce. Likewise, horticultural structures, such as greenhouses, will qualify if they are designed, constructed, and used for commercial production of plants and do not have a second use, such as selling plants.

Breeding livestock qualify unless they are replacing *substantially identical* animals. Young breeding animals, replacing others culled because of age or performance, are not substantially identical and so would qualify.

Investment credit should be taken in the earlier of (1) the year the item is placed in service or (2) the year in which depreciation is first taken. Normally these would be the same year. Investment credit cannot be claimed in a later year unless an amended return is filed for the year the property was acquired (limit of 3 years later). When an item is disposed of earlier than intended, and investment credit has been claimed, part or all of the credit must be *recaptured* or repaid.

Investment credit is *not* considered optional by the IRS. So early disposal of an asset that was eligible for investment credit could necessitate payment of *recaptured investment credit* even though the credit was not claimed when the item was purchased. Early disposal involves any sale, trade, or loss before the end of the item's originally estimated useful life.

How Much Credit Can Be Taken?

Under the rules of the Economic Recovery Tax Act of 1981 and the Tax Equity and Fiscal Responsibility Act of 1982, there are three rates of investment credit for items purchased after 1980. For items in the 3-year class the rate is either 4 or 6 percent, and for items in the 5-year class, the rate is either 8 or 10 percent. There would not be investment credit on items in the 19-year class, normally. These rates would be applied to the qualifying investment, discussed below.

Qualifying Investment

The amount of the investment used to figure the credit—the qualifying investment—is not necessarily what the farmer or rancher would consider the purchase price. This amount actually depends on several things: Is it new or used property? Does it replace an old item? Was it a trade or a true, separate sale and purchase? Was there recapture of unused investment credit on property being replaced because it was disposed of before its original expected useful life was over? The possible combinations are these:

1. *New* property, with or without trade-in sale of an item replaced with or without recapture or investment credit on a previous item; qualifying investment is *always the basis,* that is, cost if purchased outright (regardless of whether purchase is before or after sale, or how long before or after) or cash paid (boot) plus adjusted cost basis (remaining depreciable value) of the item traded in.
2. *Used* property, purchased outright (no item being disposed of by sale or

trade, therefore no investment credit recapture); qualifying investment will be the *full purchase price*.

3. *Used* property with a trade-in or a sale of property being replaced within 60 days before or after acquisition of the replacement property, no recapture on the item disposed of: The qualifying investment is only the *boot in a trade*, or the *difference in the basis* of the item acquired and the one disposed of in a separate purchase and sale.
4. *Used* property with a trade or sale of old property within 60 days of acquiring the replacement, with recapture on the item disposed of: Qualifying investment is *boot plus basis in a trade*, or *full purchase price in a purchase-sale*.

These combinations are summarized in Table 7–8.

Trade or Buy and Sell?

While the preceding discussion suggests there may be investment credit advantages to a purchase and sale rather than a trade, there are also potential income

TABLE 7–8 Amount Eligible for Investment Credit

Was the Item Acquired New or Used?	Did the New Item Replace an Old One?	Was it a Trade or Purchase & Sale?	Was Investment Credit Recaptured on Item Replaced?	Amount Qualifying for Investment Credit
New	No	—	—	Full purchase price (cost)
New	Yes	Trade	Yes	Boot plus basis on old
New	Yes	Trade	No	Boot plus basis on old
New	Yes	P & S	Yes	Full purchase price (cost)
New	Yes	P & S	No	Full purchase price (cost)
Used	No	—	—	Full purchase price (cost)
Used	Yes	Trade	Yes	Boot plus basis on old
Used	Yes	Trade	No	Boot only
Used	Yes	P & S[1]	Yes	Full purchase price (cost)
Used	Yes	P & S[1]	No	Difference between expenditure for replacement item and basis for item replaced
Used	Yes	P & S[2]	Yes	Full purchase price (cost)
Used	Yes	P & S[2]	No	Full purchase price (cost)

[1] "Replaced" item sold within 60 days before or after purchase of "replacement."
[2] "Replaced" item *not* sold within 60 days before or after purchase of "replacement."

tax disadvantages because any gain on the item sold is taxable as ordinary income, to the extent of the depreciation taken.

The decision to trade or to make a separate purchase and sale can significantly affect the amount of investment credit that can be claimed. It would also affect depreciation and possibly result in taxable gain or loss. Whether acquiring new or used property, the relationships among (1) purchase cost or boot for the replacement, (2) remaining basis of the old item, (3) trade-in or sale value of the old item, and (4) tax bracket are critical. They may cause a trade to be preferable in one case or a separate sale and purchase in another.

Limitations on Credit

Any amount of new property purchased can qualify for investment credit. However, no more than $125,000 of the cost of qualifying used property can be so counted in any one year.

The total amount of investment credit that can be used in a given year is limited to the income tax liability on the return, or $25,000 plus a portion of the tax liability in excess of $25,000, whichever is *less*. For most farmers and ranchers, this means that the amount of investment credit that can be used in a given year is limited by the amount of tax owed. The portion of tax liability beyond $25,000 is 85 percent for tax years starting after 1981. Credit in excess of tax liability can be used to offset tax in other years.

Recording Capital Asset Inventories

Inventory records of long-lived assets should be divided into four categories:

1. Real estate, including land and buildings
2. Personal property
3. Depreciable assets
4. Nondepreciable assets

For each item, it is vital to record all of the following information:

1. Date acquired
2. Amount or number
3. Cost, purchase price or basis
4. Current book value (also called remaining value or adjusted basis)
5. Estimated current market value
6. Depreciation criteria (if applicable)

An example form to record inventories of real estate is shown in Figure 7–1. Personal real property would be recorded in a similar manner.

SCHEDULE 5 — FARM REAL ESTATE

OWNED:	Legal Description	Title Held in Name of	Percent Owned	Date Acquired	Acres	Cost or Basis	Accumulated Depreciation	Book Value	Estimated Mkt. Value
Tract 1 Land						$	$	$	$
Residence									
Service bldgs.									
Improvements									
Tract 2:									
				TOTAL		$	$	$	$

RENTED: Legal Description	Landlord	Type of Lease	Expires	Acres	Annual Cash Rent
					$
			TOTAL		$

SCHEDULE 6 — NON-FARM REAL ESTATE

Description	Title Held in Name of	Percent Owned	Date Acquired	Cost or Basis	Accumulated Depreciation	Book Value	Estimated Mkt. Value
				$	$	$	$
			TOTAL	$	$	$	$

FIGURE 7–1. Real estate inventory record form.

Source: Thomas L. Frey and Danny A. Klinefelter, *Coordinated Financial Statements for Agriculture,* 2nd ed. (Skokie, IL: AgriFinance, 1980).

Depreciable assets can be summarized in a record such as the example form shown in Figure 7–2, but it is also very useful to maintain a separate longer-term depreciation schedule such as the one shown in Figure 7–3. Generally, the depreciation schedule will have separate sections for machinery and equipment, orchards and vineyards or other soil/land improvements, dairy or breeding livestock, and buildings. Examples of a filled-in depreciation schedule are presented in Figure 7–4 and a computer-maintained depreciation schedule is shown in Figure 7–5.

Nondepreciable raised livestock inventories are usually maintained in the same fashion as market livestock on a form like that shown in Figure 6–6.

Determining Market Value

Determining the current fair market value of capital assets is often a difficult process, primarily because each piece of property is different in location, quality, condition, age, resale potential, size, and characteristics. For many assets, though, current values can be found easily by consulting such sources as local livestock auctions for breeding livestock or the *Tractor Blue Book*[6] for major pieces of machinery. Furthermore, many equipment dealers will assist in this valuation process if called upon.

Valuing real estate is trickier because of the large values involved. A formal appraisal is the best way to establish value as of a specific date. Although a professionally prepared independent appraisal can be very valuable, it can be expensive and is probably not necessary on an annual basis for inventory records and balance sheet purposes. Estimates based on past appraisals and current trends are probably sufficient. Further, because of their need to establish collateral values when extending a loan, most agricultural lenders are also appraisers or have access to an appraisal department within their institution. Since they will also benefit from accurate records and financial statements, most agricultural lenders should be willing to update a past appraisal or provide a reasonable estimate of value.

Summary

Depreciation is a concept central to maintenance of long-lived asset inventory records. First, depreciation is just an economic concept that recognizes that as many assets age their value falls and many assets will eventually wear out or become obsolete. Second, depreciation is an accounting concept in that it provides a mechanism to distribute the cost of an asset over its productive life, thus matching costs and associated benefits. Third, and possibly most importantly,

[6]National Market Reports, Inc., *National Farm Tractor and Implement Blue Book* (Chicago: Maclean-Hunter Publishing Corporation).

SCHEDULE 9 — MACHINERY, EQUIPMENT AND TRUCKS

Item	Year, Make & Model	Date Acquired	Percent Owned	Cost or Basis	Accumulated Depreciation	Book Value	Estimated Mkt. Value
				$	$	$	$
					$	$	$

FIGURE 7–2. Machinery and equipment inventory record form.

Source: Thomas L. Frey and Danny A. Klinefelter, *Coordinated Financial Statements for Agriculture,* 2nd ed. (Skokie, IL: AgriFinance, 1980).

OPERATOR'S DEPRECIATION SCHEDULE FOR

Kind of machine or equipment	Date acquired	Cost or other basis		Invest-ment credit	Salvage value	Depre-ciable balance	Depreci-ation, life & method		19___ Remaining costs at beginning of year	
Total										

DO NOT DEPRECIATE BELOW SALVAGE VALUE

FIGURE 7–3. Depreciation schedule.

19__	19__		19__		19__		19__	
Depreciation for this year	Remaining costs at beginning of year	Depreciation for this year	Remaining costs at beginning of year	Depreciation for this year	Remaining costs at beginning of year	Depreciation for this year	Remaining costs at beginning of year	Depreciation for this year

DO NOT DEPRECIATE BELOW SALVAGE VALUE

Source: Robert B. Schwart, *Illinois Farm Record Book,* Illinois Cooperative Extension Service, University of Illinois.

2

INSTRUCTIONS FOR CONTINUOUS DEPECIATION SCHEDULES

The depreciation schedules are set up on a continuing basis for a period of five years. This eliminates the need to copy such records into a new book each year and also avoids errors in copying.

Depreciable property

The deduction for depreciation is allowed only for property used in your trade or business or held for the production of income. The purpose of the depreciation deduction is to let you recover your investment in depreciable property over the period of useful life of the property. For additional information about depreciation, refer to the New Mexico Farm and Ranch Business Management Handbook (NMFRBMH) pages 10-109.

Basis for depreciation

The basis for determining depreciation is usually the cost of property. If you materially improve the property, the additional costs are added to the basis. If casualty losses are sustained, the basis must be reduced. The depreciation deduction is never determined by reference to replacement cost. For additional information on basis in property, see NMFRBMH see pages 10-111.

Useful life and salvage value

Prior to 1981, useful life and salvage value were specified by the taxpayer. Since January 1, 1981, all purchases of depreciable assets fall into one of four classes of assets so that useful life is prespecified.

Also, salvage value was specified by the taxpayer prior to 1981, but since then, there is no salvage value used on any purchases.

Methods of Figuring Depreciation

There are four major methods of computing depreciation in general use: ACRS, Straight-line, Declining balance, and Sum-of-Years digits. Only the first two are legal for purchases made since January 1, 1981. The last three were used for purchases made prior to that date and must continue to be used until those purchases are fully depreciated. See NMFRBMH pages 10-109 and 10-112 for more information.

ACRS

To use the Accelerated Cost Recovery System (ACRS), simply define the class of property that the purchase belongs to, locate the correct percentage figures (see Appendix tables 1 and 2), and

Illustration of Five-Year

Kind of machine or equipment	Date acquired	Cost or other basis		Investment credit	Salvage value	Depreciable balance	Depreciation, life & method		19 81 Remaining costs at beginning of year	
Tractor #2	7-1-83	50,000	00	5,000	0	47,500	ACRS	5		
Combine	1-1-81	40,000	00	4,000	0	40,000	ACRS	5	40,000	
Beef Cow	1-1-80	2,000	00	200	500	1,500	SL	10	1,850	
Tractor #1	1-1-80	40,000	00	4,000	2,000	38,000	DDB	8	30,000	
Dairy Cow	1-1-78	2,000	00	67	500	1,500	SL	4	875	

FIGURE 7–4. Example depreciation schedule.

multiply the applicable rate times the purchase price in each year.

The straight-line method

Under this method, the cost or other basis of the property less its estimated salvage value is deducted in equal annual amounts over the period of its estimated useful life. The allowance for depreciation for each taxable year is determined by dividing the adjusted basis of the property at the beginning of the taxable year, less salvage value, by the remaining useful life of the property at such time.

Declining balance method

Under this method, the largest depreciation deduction was taken in the first year, and a gradually smaller allowance in each of the following years. The amount of depreciation taken each year is subtracted from the basis of the property before figuring the next year's depreciation, so that the same depreciation rate is applied to a smaller or declining balance each year.

You may change from the declining balance method to the straight-line method (for property acquired after December 31, 1953) at any time during the life of the property without permission.

Sum of the year's digits method

Under this method, you apply a different fraction each year to the basis of the property less its estimated salvage value. The denominator or bottom of the fraction is the total of the numbers representing the years of useful life of the property. For example, if the useful life is 5 years the denominator is 15 (1 + 2 + 3 + 4 + 5 = 15). The numerator or top of the fraction is the number of years of life remaining at the beginning of the year for which the computation is made.

SPECIAL NOTE: The information in the depreciation schedules between the heavy lines (salvage value, depreciable balance, etc.) is used to determine annual depreciation. The "cost or other basis" less depreciation equals "remaining costs at the beginning of the year." Salvage value is included in remaining costs; therefore, **do not depreciate below salvage value.**

Depreciation Record

19 81	19 82		19 83		19 84		19 __	
Depreciation for this year	Remaining costs at beginning of year	Depreciation for this year	Remaining costs at beginning of year	Depreciation for this year	Remaining costs at beginning of year	Depreciation for this year	Remaining costs at beginning of year	Depreciation for this year
				7,125	40,375			
6,000	34,000	8,800	25,200	8,400	16,800			
150	1,700	150	1,550	150	1,400			
7,500	22,500	5,625	16,875	4,219	12,656			
375	500	0	500	0	Sold 7-15-83			

FARM management analysis program

1982 **DEPRECIATION SCHEDULE**

Name: SAMPLE FARM
Address:

OPERATOR NO 1

County: 201
Farm: 127
Fieldman: 199

Category (3)	Detail (4)	Enterprise (5)	Item Number (6)	New-1 Used-2 (9)	Description	Date Bought Mo. (10)	Date Bought Yr. (10)	Initial Yr. Full-1 None-0 (11)	Total Basis Cash Plus Remaining Cost of Items Traded (12)	Remaining Cost of Items Traded (13)	Cash Difference (14)	Salvage Value (15)	Investment Credit Thirds 0,1,2,3,4,5,6 (16)	Investment Credit Amount Taken (17)	Additional 20% & Expense Election Yes-1 No-2 (18)	Additional 20% Amount Taken (19)	Depr. SL 1 SL 2 DB 3 SD 4 AC 5 SL 6 (20)	Method & Yrs. Amount Years (21)	Remaining Cost Beginning of Year (22)	Depreciation (23)	Remaining Cost End of Year (24)
** BUILDINGS				&	IMPROVEMENTS **																
02	06		01	1	MACHINE SHED 80X120	12	76		29974				0		2		2	2242	18764	2242	16522
** MACHINERY				**																	
03	01		81	1	LIGHT VEHICLE 81 SKYLARK	03	81		4525	525	4000		3	272	2		5	3YR	3393	1720	1673
03	02		01	1	COMBINE JD 6620	09	79		36250	6250	30000		3	5625	2		1	7YR	24166	5179	18987
03	04		81	1	PLANTER, ROW JD 8R-30"	04	81		16237	237	16000		3	1624	2		5	5YR	13801	3572	10229
03	05		01	1	CULTIV.-ROW 8-ROW 30"	04	73		4800				3	336	2		1	8YR			
03	06		01	1	DISC KEWANE 24'	04	78		7000	1000	6000		3	700	2		1	7YR	3775	1000	2775
03	07		01	2	CULTIV-FIELD	01	76		5300				2	353	2		1	5YR			
03	09		01	2	HARROW GATOR	04	75		2850				3	285	2		1	7YR	103	103	
03	12		02	1	PLOW 5-16	11	77		4200				3	420	2		1	7YR	1700	600	1100
03	12		81	2	PLOW 7-16	11	80	0	5850				3	585	2		1	7YR	5014	836	4178
03	13		01	1	ROTARY HOE JD 20FT	04	70		2200			100	0		2		1	8YR	100		100
03	14		01	1	TRACTOR JD 4430	03	74	1	14000	4000	10000		3	980	1	2000	1	8YR			
03	14		02	1	TRACTOR JD 4640	11	77		27500				3	2750	2		1	7YR	11130	3929	7201
03	14		81	1	TRACTOR STEIGER	04	80		45540	3540	42000		3	4554	2		1	7YR	34154	6506	27648
03	15		01	1	TRUCK IHC 2-TON	07	71		8400				3	588	2		1	8YR			
03	15		02	1	TRUCK GMC PICKUP	02	78		6500	500	6000		2	433	2		1	5YR	1408	1300	108
03	16		01	1	WAGON, GRAIN 2	08	76		5100				3	510	2		1	7YR	1213	729	484
03	19		81	1	PLOW, CHISEL GLENCO 12'	10	81		4243	243	4000		3	424	2		5	5YR	3606	933	2673
03	26		01	1	COMBINE HEAD JD 6ROW	09	79		10000	1500	8500		3	1500	1	1700	3	7YR	3832	1095	2737
03	26		02	1	COMBINE HEAD PLATFORM	09	79		6800	800	6000		3	680	1	1200	3	7YR	2653	758	1895
03	31		81	1	ELECTRIC MOTR 10 HP	09	81		500				3	50	2		5	5YR	425	110	315
03	57		81	1	OFFICE EQUIP CALCULATOR	11	81		300				3	30	2		5	5YR	255	66	189
03	71		81	1	MOWER, ROTARY WOODS	06	81		2400				3	240	2		5	5YR	2040	528	1512
** LIVESTOCK				**																	
09	31		01	1	COW, BEEF 3 HEREFORD	12	76	0	1500			750	2	100	2		1	5YR	750		750
09	54		01	1	BOAR 1 YORK	07	79		365			200	0		2		3	3YR	200		200
** BASIS - REFERENCE **																					
10	01		02	1	* BASIS LAND 55A #$1675	03	75		92125				0		2		1	YR	92125		92125
10	01		03	1	* BASIS LAND 40A #$2400	03	78		96000				0		2		1	YR	96000		96000

SUMMARY OF DEPRECIATION

	03 MACHINES	02 BUILDINGS	04 FERT.	SUB-TOTAL	09 LIVESTK	TOTALS GRAND
PURCHASES						
REM COST BEG OF YEAR	112768	18764		131532	950	132482
DEPRECIATION-TOTAL	28964	2242		31206		31206
REM COST END OF YEAR	83804	16522		100326	950	101276
R.C. SALES-CASUALTY						
ADDITNL 1ST-YR DEPR						
TOTAL COST OR BASIS	220495	29974		250469	1865	252334

*02+*03+*04 REF.

BASIS QUALIFIED

RECOVERY PROPERTY
NEW 3-YEAR
NEW OTHER
USED 3-YEAR
USED OTHER
NONRECOVERY PROPERTY
NEW 3-4 YRS
NEW 5-6 YRS
NEW 7+ YRS
USED 3-4 YR
USED 5-6 YR
USED 7+ YR
TOTAL QUALIFIED INV.
TOTAL TENTATIVE CREDIT
TOTAL CREDIT RECAPTURE

1281-45M
SEPTEMBER 02, 1982

Farm Management Ass'ns and Farm Analysis groups cooperating with University of Kentucky and University of Illinois, Departments of Agricultural Economics.

FIGURE 7–5. Computer-maintained depreciation schedule.

Source: Illinois Farm Business Farm Management Association, Farm Management Analysis Program.

depreciation is an income tax concept. Taxes overrule economic and accounting considerations in that the IRS has established the accepted rules and procedures.

Only nonland business assets that were purchased are depreciable. Land is not depreciable since it has an infinite useful life. Depreciation allowances (or deductions) are extended only to businesses, therefore personal assets are not depreciable. And finally, if an asset was not purchased for an amount greater than zero, there is no basis to depreciate and thus no depreciation.

The basis is generally the remaining value of all assets given up to gain the services of the new asset. Thus basis is the cash boot paid plus the remaining value of any asset traded in.

Depreciation is calculated under one of four methods: straight-line, sum-of-years digits, declining-balance or ACRS. Only the first three were allowed prior to 1981 and only straight line and ACRS are allowed for assets purchased after 1980. ACRS rules are somewhat complex and very particular, thus the *Farmers Tax Guide* should always be consulted for up-to-date details.

Investment tax credit, prior to 1981, did not explicitly affect the amount of depreciation allowed, but since ERTA was enacted there is a partial tradeoff between the amount of ITC allowed and the amount of depreciation that can be claimed.

Depreciation schedules usually provide forms for recording remaining values and depreciation deductions for several years so that information transfers (and thus errors) are minimized. In addition to book value, market values should be estimated since our currently used tax depreciation methods bear very little if any relationship to current market conditions.

CHAPTER 8

Financial Inventories

For most purely financial assets, such as cash, checking accounts, savings accounts, stocks and bonds, commodity futures contracts, and loans, inventory records generally require a different type of recording procedure than do physical assets. The calculations that must be performed are significantly different, but the market values are generally much easier to determine than are those for physical assets. Furthermore, the funds held in checking or savings accounts or invested in life insurance, stocks, bonds, certificates of deposit (CDs), and commodity futures contracts are usually deposited with a financial institution such as a bank or brokerage business. Those institutions regularly notify their customers of the status of their accounts through periodic statements. Those statements provide a written record of transactions and value changes that can supplement the investor's own records and can be used to verify those records.

Financial Asset Records

The most common form of checking account records are the *check register* (Figure 8–1) or *check stubs,* although many financial institutions now offer carbonless copy checks (Figure 8–2). These records are used to complete the expenditures and receipts records as discussed in Chapter 5 and are verified, or "the account is balanced," by comparing with the checking *account statement* such as the example represented in Figure 8–3.

Savings and Checking Records

Usually the financial institution will include the transactions in and the status of savings accounts on the same statement as the checking account. Hand-kept records of savings accounts can be maintained in record books such as Figure 8–4

RECORD ALL TRANSACTIONS THAT AFFECT YOUR ACCOUNT

NUMBER	DATE	DESCRIPTION OF PAYMENTS—DEBITS—FEES OR DEPOSITS—CREDITS	AMOUNT OF PAYMENT/DEBIT OR FEE (−)	T ✓	AMOUNT OF DEPOSIT OR CREDIT (+)	BALANCE FORWARD	
		To				Pay't or Dep.	
		For				Bal.	
		To				Pay't or Dep.	
		For				Bal.	
		To				Pay't or Dep.	
		For				Bal.	
		To				Pay't or Dep.	
		For				Bal.	
		To				Pay't or Dep.	
		For				Bal.	
		To				Pay't or Dep.	
		For				Bal.	
		To				Pay't or Dep.	
		For				Bal.	
		To				Pay't or Dep.	
		For				Bal.	
		To				Pay't or Dep.	
		For				Bal.	
		To				Pay't or Dep.	
		For				Bal.	
		To				Pay't or Dep.	
		For				Bal.	
		To				Pay't or Dep.	
		For				Bal.	

REMINDER: RECORD AUTOMATIC PAYMENTS OR DEPOSITS ON DATE OF TRANSACTION

FIGURE 8–1. Check register.

to provide verification. A vital element in recording information concerning all bank accounts, whether checking or savings, is to keep a separate list of all accounts, the name of the financial institution, and the account number assigned by the institution.

Life Insurance Records

Similar to the checking and savings account records, all *life insurance* policies should be recorded on forms like the example in Figure 8–5. Typically, when a life insurance policy is purchased, the *face value* is quoted, such as "a $100,000 policy." The company that sold the policy, the person(s) named as *beneficiary* and the face value should be listed as well as whether the policy is whole-life or term insurance. Term life policies are insurance that continues in force as long as premium payments are maintained. An analogy to term life insurance is the com-

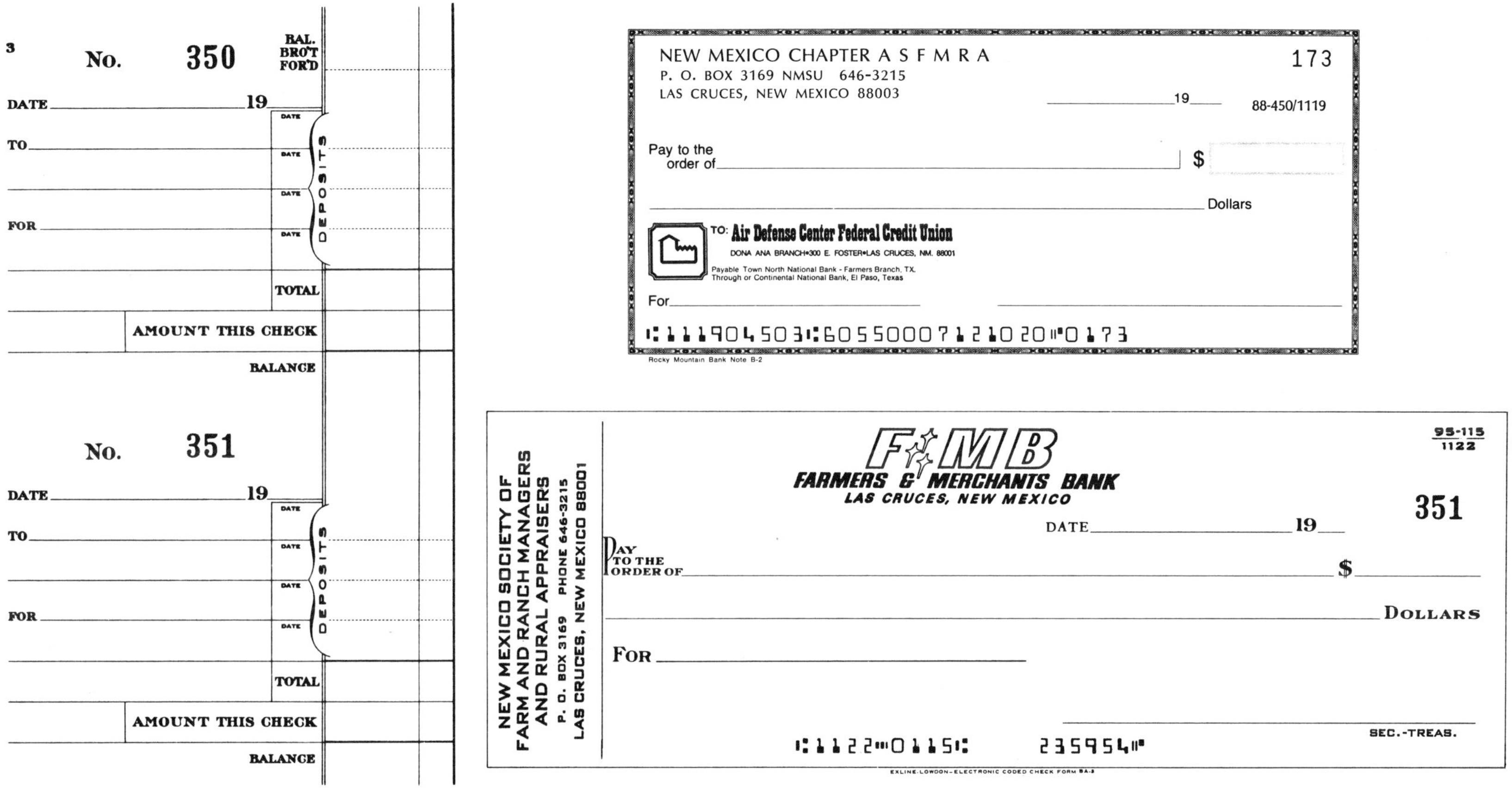

FIGURE 8–2. Check stubs and carbonless copy checks.

FIRST NATIONAL BANK
OF DONA ANA COUNTY

PHONE 505-526-7000 • P.O. BOX FNB, LAS CRUCES, NEW MEXICO 88004
LORETTO MALL • WATER STREET • NORTH MAIN BRANCH • SOUTH MAIN BRANCH • SOLANO BRANCH
SOLANO MOTOR BANK • TELSHOR BRANCH • WHITE SANDS MISSLE RANGE • HATCH, N.M. • ANTHONY, NEW MEXICO
IN ACCOUNT WITH

FULL SERVICE STATEMENT

OléCARD™ MEMBER FDIC A FULL SERVICE BANK EQUAL HOUSING LENDER

CUSTOMER NO.	PAGE NO.
[illegible]	2
7-02-82 TO 8-03-82	
STATEMENT PERIOD	

IF NO ERROR IS REPORTED IN TEN DAYS THE ACCOUNT WILL BE CONSIDERED CORRECT.

01 CHECKING

DATE	DESCRIPTION		AMOUNT	DAILY BALANCE	
7-27	TRANSFER FROM SAVINGS 107 - 24-HOUR OLE'CARD MAIN OFFICE		200.00	7-30	1,893.73
7-27	CASH WITHDRAWAL 107 - 24-HOUR OLE'CARD MAIN OFFICE		50.00	8-02	1,562.13
7-30	MISCELLANEOUS CREDIT	MI	1,383.52	8-03	1,557.13
8-02	AUTOMATIC TRANSFER DEBIT		100.00		
8-02	AUTOMATIC TRANSFER DEBIT		231.60		
8-03	SERVICE CHARGE		5.00		

CHECK NO.	DATE	AMOUNT	CHECK NO.	DATE	AMOUNT
	7-16	6.00		7-16	60.00
	7-16	380.00		7-29	50.00

01 G/M CREDIT

DATE	DESCRIPTION	AMOUNT	DAILY BALANCE
7-16	ADVANCE	300.00	300.00
8-03	** FINANCE CHARGE **	3.26	303.26

20 REGULAR SAVINGS

DATE	DESCRIPTION	AMOUNT	DAILY BALANCE
7-27	TRANSFER TO CHECKING 107 - 24-HOUR OLE'CARD MAIN OFFICE	200.00	43.46
8-02	AUTOMATIC TRANSFER DEPOSIT	100.00	143.46

50 INSTALLMENT LOAN

DATE	DESCRIPTION	AMOUNT	DAILY BALANCE
8-02	AUTOMATIC PAYMENT	231.60	
	PRINCIPAL	231.60	3,010.80
8-03	ENDING LOAN BALANCE		3,010.80

ENCLOSURES 7

FIGURE 8–3. Checking and savings account statement.

SAVINGS AND INVESTMENTS

	Enter name of account or investment below								Total for month	
Balance Jan. 1										
	Additions	Withdrawals	Additions	Withdrawals	Additions	Withdrawals	Additions	Withdrawals	Additions	Withdrawals
Jan.										
Feb.										
Mar.										
Apr.										
May										
June										
July										
Aug.										
Sept.										
Oct.										
Nov.										
Dec.										
Total										
Balance Dec. 31										

INCOME AND SOCIAL SECURITY TAXES

	Taxes on self-employment income		Taxes withheld from wages and salaries	
	Income tax	Social Security	Income tax	Social Security
January				
February				
March				
April				
May				
June				
July				
August				
September				
October				
November				
December				
Total				

FIGURE 8–4. Savings and investments record.

Source: Marilyn Dunsing and A. G. Mueller, *Illinois Family Account Book,* Illinois Cooperative Extension Service, University of Illinois.

B3

SCHEDULE 7 — LIFE INSURANCE

Insurance Company	Beneficiary	Type of Policy	Face Value	Cash Value
			$	$
		TOTAL	$	$

SCHEDULE 8 — INTERMEDIATE AND LONG TERM LIABILITIES

	To Whom	Purpose and/or Security	Due Date	Interest Rate	Payment Dates	Accrued Interest	Current Principal Balance	Portion of Principal Due within 12 months	Portion of Principal Due beyond 12 months
INTERMEDIATE									
Notes						$	$	$	$
						$	$	$	$
Sales Contracts						$	$	$	$
					TOTAL	$	$	$	$
Life Insurance Policy Loans						$	$	$	$
					TOTAL	$	$	$	$
Other						$	$	$	$
					TOTAL	$	$	$	$
LONG TERM									
Farm Real Estate Mortgages						$	$	$	$
					TOTAL	$	$	$	$
Land Contracts						$	$	$	$
					TOTAL	$	$	$	$
Non-Farm Real Est. Mortgages						$	$	$	$
					TOTAL	$	$	$	$
Other						$	$	$	$
					TOTAL	$	$	$	$

FIGURE 8–5. Life insurance record.

Source: Thomas L. Frey and Danny A. Klinefelter, *Coordinated Financial Statements for Agriculture,* 2nd ed. (Skokie, IL: AgriFinance, 1980).

mon automobile insurance policy: if payments lapse, the insurance lapses. No value is generated through the term policy other than the face value, which will be paid on the death of the holder. On the other hand, whole-life insurance policies act partly as insurance and partly as an annuity. That is to say, they build a *cash value* that can be borrowed on or that reverts to the policyholder if he chooses to stop paying the premiums. The face value, again, is the amount quoted upon purchase and payable at death. Cash value is the actual amount that can be withdrawn *before* death. That amount increases with each payment and can be determined by contacting the insurance company or agent.

Securities

Records for securities, such as stocks, bonds, CDs, and commodity futures are also similar to checking and savings records with one major exception: while it is also necessary to record the type of security (such as IBM common stock, 20-year Treasury bond due in 1995, or December cattle futures), the seller or broker, date of transaction, purchase price, and current market value must be recorded. As discussed previously, increases in value for intermediate and long-term assets are taxable as capital gains upon sale. Similarly, increases in value of securities are taxable capital gains when the security is sold. An example of a form to record securities values is shown in Figure 8–6. Market values of most securities are relatively easy to find by reading the market quotations published by the major securities exchanges in local newspapers and financial newspapers such as *Barron's* or *The Wall Street Journal.*

Financial Liability Records

In addition to the records maintained to show how much is owned, records of how much is currently owed to others should be kept. *Accounts payable* (short-term credit due generally within 30 or 60 days) records and *operating loan* (short-term credit due generally within the same calendar year) records can usually be maintained in a format similar to investment records such as the example shown in Figure 8–4. For each account payable, the company that extended the credit, the amount owed, and all transactions should be recorded on a form such as the example in Figure 8–7. All accounts payable should be summarized in a form like that shown in Figure 8–8.

Similar forms (or the same form) should be sufficient for short-term operating loans.

Intermediate and long-term records are significantly different, however, primarily because these loans are usually not repaid in one payment but rather in a series of scheduled payments. Part of each payment is interest and part is principal. Before records of longer-term liabilities are discussed, it is important to understand the background behind the way loans are repaid and interest calculated.

SCHEDULE 1 — SECURITIES

No. of Shares or Face Value	Description	Date Acquired	Percent Owned	Cost or Basis		Current Market Value	
				Per Unit	Total	Per Unit	Total
		MARKETABLE					
				$	$	$	$
				TOTAL	$		$
		NOT READILY MARKETABLE					
				$	$	$	$
				TOTAL	$		$

SCHEDULE 2 — SUPPLIES

Item	Quantity	[illegible]
	TOTAL	$

SCHEDULE 3 — PERSONAL & RECREATIONAL VEHICLES

Make & Model	Year	Value
		$
	TOTAL	$

SCHEDULE 4 — BREEDING STOCK

[illegible]	Description	Age	Wt.	$/Unit	Value
				$	$
				TOTAL	$

Developed by Thomas L. Frey and Danny A. Klinefelter, Dept. of Agricultural Economics, Cooperative Ext. Service, University of Illinois. Available from Agri Finance, 5520-G Touhy Ave., Skokie, IL 60077

FIGURE 8–6. Securities record.

Source: Thomas L. Frey and Danny A. Klinefelter, *Coordinated Financial Statements for Agriculture,* 2nd ed. (Skokie, IL: AgriFinance, 1980).

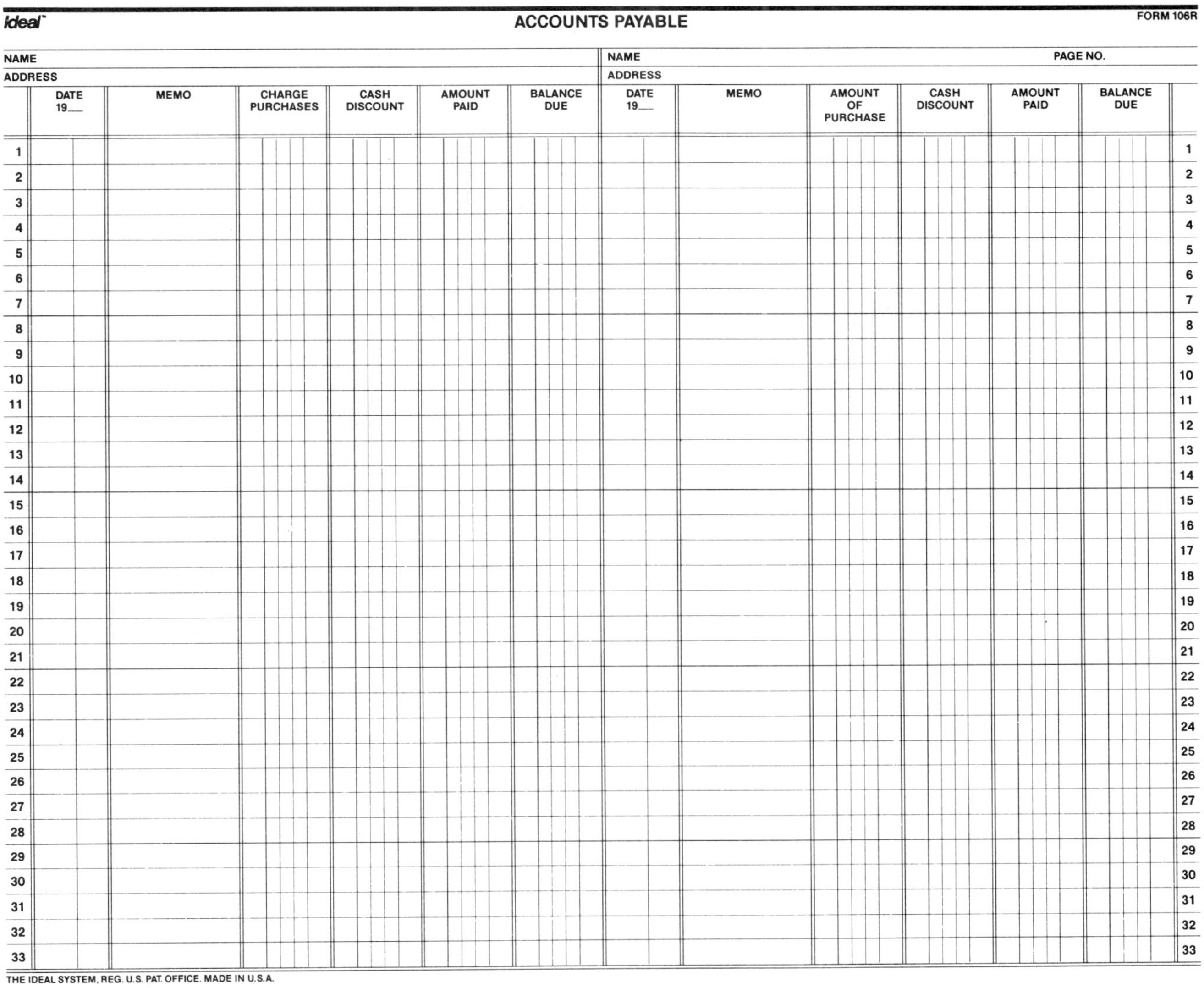
Ideal ACCOUNTS PAYABLE FORM 106R

NAME | NAME PAGE NO.

ADDRESS | ADDRESS

	DATE 19__	MEMO	CHARGE PURCHASES	CASH DISCOUNT	AMOUNT PAID	BALANCE DUE	DATE 19__	MEMO	AMOUNT OF PURCHASE	CASH DISCOUNT	AMOUNT PAID	BALANCE DUE	
1													1
2													2
3													3
4													4
5													5
6													6
7													7
8													8
9													9
10													10
11													11
12													12
13													13
14													14
15													15
16													16
17													17
18													18
19													19
20													20
21													21
22													22
23													23
24													24
25													25
26													26
27													27
28													28
29													29
30													30
31													31
32													32
33													33

THE IDEAL SYSTEM, REG. U.S. PAT. OFFICE. MADE IN U.S.A.

FIGURE 8–7. Ideal charge account record form.

Source: Ideal Bookkeeping and Tax Record: Farm and Ranch, Esselte Pendaflex Corporation, Garden City, NY, 1982.

Long-Term Loan Repayment Methods

When money is borrowed to make long-term investments, it is generally repaid in a series of annual, semiannual or monthly payments (called *amortization*). There are several different ways by which the amount of these payments can be calculated. For example, loans may be repaid in a series of:

1. Equal total payments per time period;
2. Equal principal payments per time period; or
3. Equal payments over a specified time period, with a balloon payment due at the end to repay the remaining balance.

When the equal-total-payment plan method of amortizing a loan is used, each payment includes the *accrued interest* on the unpaid balance, plus some principal.

CHARGE ACCOUNTS

Date	Amount	Account Name & Number	Date	Amount	Account Name & Number

SUMMARY OF CHARGE ACCOUNTS

	Unpaid balance, first of month		Purchases charged this month, Col. B		Paid on charge accounts this month, Col. A		Unpaid balance, end of month
January		+		−		=	
February		+		−		=	
March		+		−		=	
April		+		−		=	
May		+		−		=	
June		+		−		=	
July		+		−		=	
August		+		−		=	
September		+		−		=	
October		+		−		=	
November		+		−		=	
December		+		−		=	

FIGURE 8–8. Charge account summary.

TABLE 8–1 Example of Loan Amortization—Equal Total Payment Plan Loan Amount $10,000, Annual Interest Rate 12%, 8 Annual Payments

Year	Annual Payment	Principal Payment	Interest	Unpaid Balance
				$10,000.00
1	$ 2,013.03	$ 813.03	$1,200.00	9,186.97
2	2,013.03	910.59	1,102.44	8,276.38
3	2,013.03	1,019.86	993.17	7,256.52
4	2,013.03	1,142.25	870.78	6,114.27
5	2,013.03	1,279.32	733.71	4,834.95
6	2,013.03	1,432.83	580.20	3,402.12
7	2,013.03	1,604.77	408.26	1,797.35
8	2,013.03	1,797.35	215.68	0.00
Total	$16,104.24	$10,000.00	$6,104.24	—

The amount of the interest paid each period will decrease while the amount of the annual payment applied toward the principal will increase with each payment. The example in Table 8–1 illustrates how the amounts of interest and principal in an equal total payment change over the loan period. The equal-principal-payment plan also provides for the payment of accrued interest on the *unpaid balance* plus an equal amount of principal. The total payment will decline through time because as the remaining principal balance declines the amount of interest accrued declines. The example in Table 8–2 illustrates interest and principal payments over the loan period.

These two payment plans are the most common methods used by financial institutions to compute loan payments on long-term investments. Individual lenders will use both of these, but may also use a balloon system. The *balloon-payment* method is often used to reduce the size of periodic payments, but also to shorten the total time period until the loan is fully repaid. To accomplish this, a portion of the principal will not be amortized (paid off in a series of payments) but

TABLE 8–2 Example of Loan Schedule—Equal Principal Plan Loan Amount $10,000, Annual Interest Rate 12%, 8 Annual Payments

Year	Annual Payment	Principal Payment	Interest	Unpaid Balance
				$10,000
1	$ 2,450	$ 1,250	$1,200	8,750
2	2,300	1,250	1,050	7,500
3	2,150	1,250	900	6,250
4	2,000	1,250	750	5,000
5	1,850	1,250	600	3,750
6	1,700	1,250	450	2,500
7	1,550	1,250	300	1,250
8	1,400	1,250	150	0
Total	$15,400	$10,000	$5,400	—

will be due in a lump sum (or balloon) at the end of the loan period. For many borrowers, this means that the amount to be repaid in the lump sum must be refinanced, which may be difficult.

Borrower's Use of Loan Repayment Principles

To calculate the amount of the loan payments all of the terms of the loan must first be agreed upon. These terms include the interest rate, the timing of payments (for example, monthly, quarterly, annually), the length of the loan (time period over which the loan is to be repaid), and the original amount of the loan. The process of calculating the amount of the payments is fairly simple. Commercial lenders are able to rapidly compute the loan repayment schedule and process the loan contract for the borrower. They can do it accurately and they will periodically advise borrower of the current status of the loan.

But it is also necessary for the borrower to understand how loans are amortized, how to calculate loan payment amounts and remaining principal balances as of a particular date, and how to calculate the principal and interest portions of the next payment. This information is valuable for planning purposes before an investment is actually made, for tax management and planning purposes before the loan statement is received, and for preparation of financial statements. Since many farmers and ranchers now have electronic calculators or personal computers, the calculations can be carried out relatively easy and quickly. The use of printed tables is still fairly common, but these provide less flexibility because of the limited number of interest rates and time periods for which the numbers in the tables have been calculated.

Regardless of whether a table or a calculator is used, working through an example will be helpful in applying the concepts and formulas to a specific case.

Lenders' Use of Different Methods

Different lenders will use different methods for calculating loan repayment schedules depending upon their specific needs, the needs of the borrower, the institution's interest rate policy (whether rates are fixed or variable), the time length of the loan, and the purpose of the borrowed money. Typically, home mortgage loans, automobile and truck loans, and consumer installment loans through banks, savings and loan institutions, finance companies, or dealerships will be amortized using the equal-total-payment method.

The Farmer's Home Administration (FMHA) also generally requires equal total payments for its intermediate and long-term loans.

The Farm Credit System (FCS) will use the equal total payment method for many of their loans, but under certain conditions they may require that more of the principal be repaid earlier in the life of the loan, so they will use the equal-principal-payment method. For example, in marginal farming areas or for ranches with a high percentage of grazing land in nondeeded permits, the FCS may require equal principal payments.

Choice of Methods

Oftentimes, lenders will try to accommodate the desires and needs of their borrowers and let the borrower choose which loan payment method will be used. A quick comparison of Tables 8–1 and 8–2 indicates that there are advantages and disadvantages to each plan. The equal-principal-payment plan incurs less total interest over the life of the loan because the principal is repaid more rapidly, but it requires higher annual payments in the earlier years when money to repay the loan is typically shortest. Furthermore, since the principal is repaid more rapidly, interest deductions for tax purposes are slightly lower. Remember that principal payments are not tax deductible and that the choice of repayment plans has no effect on cost recovery (depreciation).

The reason for the difference in amounts of interest due in any time period is a simple one: interest is calculated and paid on the amount of money that has been loaned but not repaid (the remaining unpaid principal balance of the loan). In other words, interest is almost always calculated as a percentage of the unpaid (or remaining) balance[1]:

$I = i \times R$ I = interest payment
i = interest rate
R = unpaid balance

Using Amortization Tables

An *amortization table* can be used to determine the annual payment when the amount of money borrowed, the interest rate, and the length of the loan are known. For example, an eight-year loan of $10,000 made at an annual interest rate of 12 percent would require a $2,013 payment each year. By referring to Table 8–3 under the 12% column and across from 8 annual payments, we find the factor 0.20130. This indicates that for each dollar borrowed, the payment for interest and principal to retire the loan in eight years will require 0.20130 cents per year. Thus, the annual loan payment is $10,000 × 0.2013 = $2,013. Table 8–3 can be used to determine the annual payments for loans with the interest rates from 8 to 15 percent financed for the period shown in Column 1.

Using the Formulas

As noted above, many farmers, ranchers, and students have access to electronic calculators or personal computers, which can handle the loan payment calculations quickly and easily. These tools have the capability to calculate loan payments for an infinite number of interest rate and time period combinations, far more than can feasibly be printed in a table, especially when fractional interest

[1]The remaining-balance method of interest calculation, which will be discussed and compared with other methods in the following section.

TABLE 8–3 Amortization Table: Annual Principal and Interest Paid Per $1 Borrowed By Length of Loan and Interest Rates

No. of Annual Payments	Annual Interest Rate											
	8%	9%	10%	10½%	11%	11½%	12%	12½%	13%	13½%	14%	15%
3	.38803	.39505	.40211	.49566	.40921	.41278	.41635	.41993	.42352	.42712	.43073	.43798
4	.30192	.30867	.31547	.31889	.32233	.32577	.32923	.33271	.33619	.33969	.34320	.35027
5	.25046	.25709	.26380	.26718	.27057	.27398	.27741	.28085	.28431	.28779	.29128	.29832
6	.21632	.22292	.22961	.23298	.23638	.23979	.24323	.24668	.25015	.25365	.25716	.26424
7	.19207	.19869	.20541	.20880	.21222	.21566	.21912	.22260	.22611	.22964	.23319	.24036
8	.17401	.18067	.18744	.19087	.19432	.19780	.20130	.20483	.20839	.21197	.21557	.22285
9	.16008	.16680	.17364	.17711	.18060	.18413	.18768	.19126	.19487	.19851	.20217	.20957
10	.14903	.15582	.16275	.16626	.16980	.17338	.17698	.18062	.18429	.18799	.19171	.19925
11	.14008	.14695	.15396	.15752	.16112	.16475	.16842	.17211	.17584	.17960	.18339	.19107
12	.13270	.13965	.14676	.15038	.15403	.15771	.16144	.16519	.16899	.17281	.17667	.18448
13	.12652	.13357	.14078	.14445	.14815	.15190	.15568	.15950	.16335	.16724	.17116	.17911
14	.12130	.12843	.13575	.13947	.14323	.14703	.15087	.15475	.15867	.16262	.16661	.17469
15	.11683	.12406	.13147	.13525	.13907	.14292	.14682	.15076	.15474	.15876	.16281	.17102
20	.10185	.10955	.11746	.12149	.12558	.12970	.13388	.13810	.14235	.14665	.15099	.15976
25	.09368	.10181	.11017	.11443	.11874	.12310	.12750	.13194	.13643	.14095	.14550	.15470
30	.08883	.09734	.10608	.11053	.11502	.11956	.12414	.12876	.13341	.13809	.14280	.15230
35	.08580	.09464	.10369	.10829	.11293	.17760	.12232	.12706	.13183	.13662	.14144	.15113
40	.08386	.09296	.10226	.10697	.11172	.11650	.12130	.12613	.13099	.13586	.14075	.15056

rates are charged or when the length of the loan is not standard. Furthermore, the now common practice of variable interest rates and the use of interest rates carried out to two or three decimal places increases the difficulty of using tables.

Equal Total Payments

For equal total payment loans, the total amount of the periodic payment can be calculated using the following formula:

$$B = \frac{i \times A}{1 - (1 + i)^{-N}}$$

The principal portion due in period n is:

$$C_n = B \times (1 + i)^{-(1+N-n)}$$

the interest due in period n is:

$$I_n = B - C_n$$

and the remaining principal balance due after period n is:

$$R_n = \frac{I_n}{i} - C_n$$

A = amount of loan
B = periodic total payment
C_n = principal portion due in period n
I = interest portion due in period n
R_n = remaining principal balance at the end of period n
i = periodic interest rate
n = period under consideration
N = total number of periods in loan

Equal Principal Payments

For equal principal payment loans, the principal portion of the total payment is calculated as:

$$C = \frac{A}{N}$$

the interest due in period n is:

$$I_n = (A - C \times (n - 1)) \times i$$

and the remaining principal balance due after period n is:

$$R_n = \frac{I_n}{i} - C$$

Calculating Payments with Variable Interest Rates

Many lenders (especially the Farm Credit System) now use *variable interest rates,* which greatly complicates the calculation of the amount of payment due. Most commonly, loans are amortized under a variable interest rate by calculating the amount of principal due based upon the interest rate in effect on the date the payment is due. The interest payment is then calculated in the normal fashion.

To illustrate, assume the loan terms used in Tables 8–1 and 8–2: A \$10,000 loan at 12 percent interest and an 8-year repayment schedule using the equal-total-payment method. But also assume that the interest rate is variable; it remains at 12 percent for the first 6 months of the year but then changes to 13 percent for the last 6 months. Instead of calculating the principal due at the end of the first year on the basis of 12 percent, it is calculated using 13 percent. Applying the formulas of the previous section, we get:

$$C_1 = \frac{i \times A}{1 - (1 + i)^{-N}} \times (1 + i)^{-(1+N-n)} = \$783.87$$

using $i = .13$. Consequently, the principal payment is \$783.87 instead of \$813.03. The interest payment is calculated at 12 percent for 6 months and at 13 percent for 6 months:

$$I_1 = \$10{,}000 \times .12 \times \frac{6}{12} + \$10{,}000 \times .13 \times \frac{6}{12} = \$1{,}250$$

Thus the total payment for the first year is:

$$B_1 = \$783.87 + \$1{,}250 = \$2{,}033.87$$

and $R_1 = \$10{,}000 - \$783.87 = \$9{,}216.13$

Carrying this example one step further, assume that the interest rate in Year 2 of the note remains at 13 percent for 2 months and then moves to 14 percent and stays there for 10 months. The same formula is used, but now C is calculated using $i = .14$ and $n = 2$. Thus,

$$C_2 = \$861.50$$

and interest is:

$$I_2 = \$9{,}216.13 \times .13 \times \frac{2}{12} + \$9{,}216.13 \times .14 \times \frac{10}{12} =$$

$$\$199.68 + \$1{,}075.22 = \$1{,}274.90$$

$$R_2 = \$9{,}216.13 - \$861.50 = \$8{,}354.63$$

$$B_2 = \$861.50 + \$1{,}274.90 = \$2{,}136.40$$

This method of computing the amount of principal and total payments is used only for *equal-total-payment* loans. If the loan schedule was originally specified as the equal-principal-payment plan, the calculations are much easier since C (principal payment) remains the same for each period. Interest is calculated in the same manner as in the example above.

Calculating Interest Charges on Loans

Any time that money is borrowed, interest will obviously be charged. The method by which interest is computed is often confusing, as is the manner in which the interest rate is stated. The federal Truth in Lending Act of 1968 (and its successor, the Truth in Lending Simplification and Reform Act of 1980)

> was enacted in order to assure a meaningful disclosure of credit terms and to protect the consumer against inaccurate and unfair credit billing and credit card practices. The act imposes detailed reporting requirements on lenders. However, agricultural transactions are fully excluded from its application.[2]

Despite the exclusion from the act, most agricultural lenders follow its spirit and intent.

One of the main provisions of the Truth in Lending Act was that lenders must calculate and show borrowers total finance charges over the life of the loan as well as the *annual percentage rate* (APR). The APR provides a common basis for comparing interest charges among lenders and it includes both interest and noninterest charges associated with the loan. The APR may or may not be the same as the *contractural rate* (the rate actually stated on the loan contract). The difference between the two rates comes about because of the different methods of computing the total interest charge.

Methods of Computation

Lenders use three major methods to calculate interest charges: (1) the add-on method, (2) the discount method and (3) the remaining balance method. The first two methods are not commonly used by the major agricultural lenders, but they are used by finance companies, which make some agricultural loans.

[2]D. L. Uchtmann, J. W. Looney, N. G. P. Krausz, and H. W. Hannah. *Agricultural Law: Principles and Cases* (New York: McGraw Hill, Inc., 1981), p. 362.

Add-on Method. Under the add-on method, the lender calculates the total interest charge by multiplying the entire amount of the loan by the contractual interest rate and then by the number of periods covered by the loan. The interest charge is then added on to the principal to determine the total amount to be repaid.

The total interest charge is thus:

$$I = A \times i_c \times N$$

The periodic payment is:

$$B_n = \frac{A + I}{N}$$

I = total interest charge over the life of the loan

I_n = interest payment in period n

N = number of periods covered by the loan

B_n = total payment in period n

i_c = contractual interest rate per time period

For an example of add-on interest, assume a $3,000 loan to be repaid in two annual installments. The annual contractual interest rate is 6 percent. The total interest charge is then:

$$I = \$3{,}000 \times .06 \times 2 = \$360$$

and the annual payments will be:

$$B_n = \frac{\$3{,}000 + 360}{2} = \$1{,}680$$

Discount Method. The discount method calculates total interest in the same way as the add-on method, with one exception: the interest is subtracted from the loan amount and the borrower receives the balance. The total interest charge is:

$$I = A \times i_c \times N$$

The amount of money that the borrower receives is:

L = Loan Proceeds

$$L = A - I$$

and the periodic payment is:

$$B_n = \frac{A}{N}$$

Using the sample example as before ($3,000 loan amount, 6 percent annual interest rate, over 2 years), the total interest charge is again $360:

$$I = \$3{,}000 \times .06 \times 2 = \$360$$

The borrower would receive $2,640:

$$L = \$3{,}000 - \$360 = \$2{,}640$$

and would repay two installments of $1,500 each:

$$B_n = \frac{\$3{,}000}{2} = \$1{,}500$$

Remaining-balance Method. When the remaining-balance method is used, the interest charge is computed in each period by multiplying the contractual interest rate by the principal balance remaining at the beginning of the period (the *unpaid* balance). The major difference between this method and the previous two, beyond the complexity of the mathematical calculations, is that here interest is *not* charged on principal that has been repaid.

The total interest charge, the periodic interest payment, and the periodic principal payment all depend upon the method selected for repayment.

To illustrate computation of interest by the remaining-balance method for the two most common types of repayment methods, equal total payment and equal principal payment, assume a \$10,000 loan at a 12 percent annual contractual interest rate to be repaid in 8 annual payments.

Equal Total Payments. Under the equal-total-payment method, the annual payment for this loan would be \$2,013.03 for each of the 8 years.

The portion of each payment that is interest and the portion that is principal will vary with each payment. At the end of Year 1, interest is charged on the full \$10,000 principal outstanding:

$$I_1 = \$10{,}000 \times .12 = \$1{,}200$$

Thus, the principal payment is the difference:

$$B_1 = \$2{,}013.03 - \$1{,}200 = \$813.03$$

The remaining principal balance after the first payment is:

$$R_1 = \$10{,}000 - \$813.03 = \$9{,}186.97$$

Interest in Year 2 is charged on the remaining balance:

$$I_2 = \$9{,}186.87 \times .12 = \$1{,}102.44$$

Which yields:

$$B_2 = \$2{,}013.03 - \$1{,}102.44 = \$910.59$$

and

$$R_2 = \$9{,}186.97 - \$910.59 = \$8{,}276.38$$

A similar set of steps is followed for Year 3 and thereafter.

Equal Principal Payments. Under the equal-principal-payment plan, interest charges are calculated in a similar manner.

$$I_1 = \$10{,}000 \times .12 = \$1{,}200$$

But the principal payment is:

$$B_1 = \frac{\$10{,}000}{8} = \$1{,}250$$

Thus, the total payment for the first year is:

$$C_1 = \$1{,}250 + 1{,}200 = \$2{,}450$$

and the remaining principal balance is:

$$R_1 = \$10{,}000 - \$1{,}250 = \$8{,}750$$

In Year 2,

$$I_2 = \$8{,}750 \times .12 = \$1{,}050$$

$$C_2 = \$1{,}250$$

$$B_2 = \$1{,}250 + \$1{,}050 = \$2{,}300$$

$$R_2 = \$8{,}750 - \$1{,}250 = \$7{,}500$$

Comparison of Interest Charges

Given a contractual interest rate and other loan terms, the total interest charges will vary significantly. To illustrate, assume that a $10,000 loan is taken out at an annual contractual interest rate of 12 percent to be repaid in annual payments over 8 years. Table 8–4 shows how the different methods of computing interest charges affect the total interest cost.

The add-on and discount methods result in significantly higher interest charges and APRs. Thus the contractual interest rates substantially understate the true or annual percentage rate. In fact, the discount method produces ridiculous results for a loan even as short as 8 years. But this method is usually used only for very short-term loans such as 30-, 60-, or 180-day loans by finance companies. When the add-on or discount methods are used, lenders usually quote an interest rate substantially lower than the rates quoted by other lenders. Thus for short-term loans the extremely high APR figures indicated by Table 8–4 will not be correct. Very few agricultural loans are written using these methods, but some people may have consumer loans that use either add-on or discount interest.

Effects of Service Charges and/or Stock Purchase Requirements

The APR stated for loans should also reflect all loan service costs (such as loan origination fees, closing costs, or points) and stock purchase requirements (such as those of the Farm Credit System). Table 8–5 shows an example of the effect of service charges on the APR for different loan maturities. Note that as the maturity is lengthened, the effect of the service charge is diminished.

Variable Interest Rates and Partial Year Loans

Generally, interest rates are stated on an annual basis, but sometimes they are quoted on a monthly basis or even on a weekly or daily basis. In fact, most department store or bankcard charge accounts are stated on a monthly basis. Consequently, it is important to determine the time period for which the stated interest rate applies. Furthermore, even if an annual interest rate is stated, the length of most agricultural operating loans is less than one full year, and most credit card charges are paid off in much less time than one year.

TABLE 8–4 Example of Interest Charge Methods (Principal $10,000, Interest Rate 12%, 8 Annual Payments)

	Add-On	Discount	Remaining Balance	Remaining Balance[1]
Amount received	$10,000	$ 400	$10,000	$10,000
Total amount repaid	19,600	10,000	16,104	15,400
Total interest paid	9,600	9,600	6,104	5,400
Equal annual payments	2,450	1,500	2,013	1,925[2]
APR (percent)[3]	17.97	375.00	12.00	10.74

[1] Remaining balance method with equal *principal* payment.
[2] Average annual payment.
[3] APR is defined by the remaining balance method of interest calculation assuming loan amortization is based on *equal total* payments. To find APR for other methods, a direct comparison of total interest paid must be made with APR-generated interest in order to solve for the applicable APR rate.

Partial Years

If interest is stated on an annual basis and is calculated using the remaining-balance method, but the loan is paid off in less than one year, the amount of interest due can be calculated as:

$$I = A \times i \times \frac{\text{(Number of Months Money Is Used)}}{12}$$

or

$$I = A \times i \times \frac{\text{(Number of Days Money Is Used)}}{365}$$

If interest is stated on a monthly basis (i_m), calculate interest as:

$$I = A \times i_m \times \text{(Number of Months Money Is Used)}$$

Variable Interest

In recent years, it has become common for lending institutions to adopt a variable-interest-rate policy. Typically, the interest rate will be stated on an annual basis and will not change more often than once each month. If interest is

TABLE 8–5 APR Under Different Maturity and Service Charge Assumptions ($10,000 Loan, 10% Interest Rate)

Service Charge % of Loan Amount	APR% by Length of Loan Repayment			
	3 yrs.	5 yrs.	10 yrs.	20 yrs.
0	10.00%	10.00%	10.00%	10.00%
1	10.58	10.40	10.24	10.15
2	11.16	10.80	10.48	10.30
3	11.74	11.20	10.72	10.45
4	12.32	11.60	10.96	10.60
5	12.90	12.00	11.20	10.75

calculated on the remaining balance method, it can be calculated in a way similar to the process for partial year loans.

For example, if the interest rate was 12 percent for 2 months, 13 percent for 3 months and 14 percent for 7 months, the annual interest is:

$$I = \left(A \times .12 \times \frac{2}{12}\right) + \left(A \times .13 \times \frac{3}{12}\right) + \left(A \times .14 \times \frac{7}{12}\right)$$

Longer-Term Liability Records

Given the methods of calculating loan repayments and interest, the next step is to record that information for specific loans. For each loan, the lender, purpose or collateral, due date, interest rate, payment dates, and loan repayment/interest calculation method should be recorded in a form similar to the example shown in Figure 8–9. Furthermore, for each loan interest and principal payments and the current principal balance should be recorded. But the dollar amounts should be split into at least four categories:

1. Accrued interest
2. Principal balance
3. Principal due within 12 months
4. Principal due beyond 12 months

The purpose of this split concerns the different uses of the information in the balance sheet, which will be discussed in Chapter 13.

Accrued Interest

The required principal calculations were discussed earlier as were interest calculations, but accrued interest remains. *Accrued interest,* or in general *accrued liabilities,* is an accrual accounting concept that relates the use of a resource to the cost incurred even if that cost is not paid in the current time period. Assume for a moment that money is borrowed on January 15, 19X1. Assume further that $100,000 is borrowed at 15 percent interest, interest is calculated by the remaining balance method, and the loan is to be repaid in annual payments.

The formulas of previous sections indicate that $100,000 × 0.15 = $15,000 in interest will be due on January 15, 19X2. How much is due on December 31, 19X1? Technically, no interest is *due* on December 31 since the loan contract specifies that the first payment is due January 15, 19X2. But that does not mean that an interest expense has not been incurred in 19X1 or that a liability does not exist on December 31. Although not due on December 31, an expense has been incurred or accrued. The amount has accrued for 350 days. Thus, the accrued interest is:

$$AI = \$100{,}000 \times .15 \times \frac{350}{365} = \$14{,}383.56$$

SCHEDULE 7 — LIFE INSURANCE

Insurance Company	Beneficiary	Type of Policy	Face Value	Cash Value
			$	$
		TOTAL	$	$

SCHEDULE 8 — INTERMEDIATE AND LONG TERM LIABILITIES

	To Whom	Purpose and/or Security	Due Date	Interest Rate	Payment Dates	Accrued Interest	Current Principal Balance	Portion of Principal: Due within 12 months	Portion of Principal: Due beyond 12 months
INTERMEDIATE									
Notes						$	$	$	$
					TOTAL	$	$	$	$
Sales Contracts						$	$	$	$
					TOTAL	$	$	$	$
Life Insurance Policy Loans						$	$	$	$
					TOTAL	$	$	$	$
Other						$	$	$	$
					TOTAL	$	$	$	$
LONG TERM									
Farm Real Estate Mortgages						$	$	$	$
					TOTAL	$	$	$	$
Land Contracts						$	$	$	$
					TOTAL	$	$	$	$
Non-Farm Real Est. Mortgages						$	$	$	$
					TOTAL	$	$	$	$
Other						$	$	$	$
					TOTAL	$	$	$	$

FIGURE 8–9. Longer-term liability record.

Source: Thomas L. Frey and Danny A. Klinefelter, *Coordinated Financial Statements for Agriculture,* 2nd ed. (Skokie, IL: AgriFinance, 1980).

Loan Example

To provide an example of how a loan should be recorded, assume the loan conditions used in the example of Table 8–1: a $10,000 loan at 12 percent interest to be repaid in eight-equal annual payments. Assume further that the loan was entered into on January 15, 19X1.

Then the total payment due January 15, 19X2 is:

$$B = \frac{i \times A}{1 - (1 + i)^{-N}} = \frac{.12 \times 10{,}000}{1 - (1 + .12)^{-8}} = \$2{,}013.03$$

the principal due on January 15, 19X2 is:

$$C_1 = \mathrm{B} \times (1 + i) - (1 + N - n) = \$2{,}013.03 \times (1 + .12)^{-(1+8-1)}$$
$$= \$813.03$$

and the interest due January 15, 19X2 is:

$$I_1 = B - C_n = \$2{,}013.03 - \$813.03 = \$1{,}200$$

or

$$I_1 = i \times R = .12 \times \$10{,}000 = \$1{,}200$$

As a result of these calculations, accrued interest on December 31, 19X1 can be calculated as:

$$AI = I_1 \times \frac{\text{days}}{365} = \$1{,}200 \times \frac{350}{365} = \$1{,}150.68$$

The principal due within 12 months of December 31, 19X1 is the January 15, 19X2 payment: $813.03. The principal due beyond 12 months is the remaining or unpaid balance: $10,000 − $813.03 = $9,186.97.

This information can be recorded in the December 31, 19X1 loan inventory as shown in Figure 8–10.

Summary

Financial asset and liability records generally follow a different pattern from physical or long-lived asset records mainly because of the different calculation and valuation methods that must be used. Financial asset record forms, however, are often very simple because the financial institutions (banks, S&Ls, brokers, insurance companies) usually provide periodic, easy-to-read statements that provide the necessary backup documentation.

Financial liability records, on the other hand, often involve several lengthy and sometimes difficult calculations. To properly record financial liabilities a business manager must have an appreciation and understanding of the language and methods used by lenders. An astute manager will be aware that three major repayment methods (equal total payment, equal principal payment, and balloon) are used for amortized loans and will know how to use them. That astute manager will also be familiar with the three major methods of interest computation (add-on, discount, and remaining balance). The cash-flow effect of the choice of repayment

INTERMEDIATE AND LONG TERM LIABILITIES

	To Whom	Purpose and/or Security	Due Date	Interest Rate	Payment Dates	Accrued Interest	Current Principal Balance	Portion of Principal: Due within 12 months	Portion of Principal: Due beyond 12 months
INTERMEDIATE									
Notes						$	$	$	$
	1st National Bank	Combine	1-15-x9	12	1-15	1,150.68	10,000	813.03	9,186.97
					TOTAL	$	$	$	$
Sales Contracts						$	$	$	$
					TOTAL	$	$	$	$
Life Insurance Policy Loans						$	$	$	$
					TOTAL	$	$	$	$
Other						$	$	$	$
					TOTAL	$	$	$	$
LONG TERM									
Farm Real Estate Mortgages						$	$	$	$
					TOTAL	$	$	$	$
Land Contracts						$	$	$	$
					TOTAL	$	$	$	$
Non-Farm Real Est. Mortgages						$	$	$	$
					TOTAL	$	$	$	$
Other						$	$	$	$
					TOTAL	$	$	$	$

FIGURE 8–10. Example loan inventory analysis.

Source: Thomas L. Frey and Danny A. Klinefelter, *Coordinated Financial Statements for Agriculture,* 2nd ed. (Skokie, IL: AgriFinance, 1980).

method is often significant, but is usually much less damaging than a misunderstanding of how interest is computed. All borrowers should be very sure that they understand all the terms of a loan and that they know the APR before the loan document is signed.

CHAPTER 9

Production Records

Throughout the first eight chapters of this text financial records have been stressed, primarily because those are the records most neglected but most needed by farmers and ranchers in general. We must be careful, however, to avoid underemphasizing the importance of crop and livestock production practices; without adequate production, the best set of financial records and the best laid financial management plans will leave a producer in a disastrous situation. On the other hand, while financial management (including recordkeeping) will not ensure financial success, the typical problem of American farmers and ranchers is *not* the lack of technical production skills or the ability to produce.

Production records can take many forms, but once again it is important to find a records system that is both simple and complete. As with financial records, the primary purpose of production records is to supplement memories and to provide a written track of practices used in order to analyze current problems.

Oftentimes written records can point out the cause of current problems. For example, many crop disease, insect, weed, and other pathological problems can be traced to chemical application patterns used in the past.

Production records are also very useful to forward planning as they can provide the information necessary to determine yield expectations for particular soil types, crop varieties, and livestock breeds under local conditions. These records can also provide background information concerning crop yield and livestock response to chemical applications, specific amounts and formulations of fertilizer, specific rations, and other inputs. Probably the best use of this information is to adapt crop and livestock research results generated by university experiment stations and other research agencies to particular local conditions.

Production records should provide for yields, varieties planted, livestock output, all inputs applied, disease and pest control methods employed, breeding records, and feed use. Production records *must* also include the year-end inventories discussed in Chapter 6. Because of the major biological process differences, crop and livestock records must take greatly different forms. But both should start with a farm or ranch map.

The Farm or Ranch Map

Because land and buildings are the basis and primary constraints of virtually all farm and ranch operations, the best place to start production records is with a sketch (or aerial photograph) of the layout of the entire farm or ranch unit. The map does not necessarily have to be detailed, but should delineate fields, pastures, fences, roads, drainage or irrigation ditches, windmills, irrigation wells, streams, ponds, tanks, and any other major physical characteristics of the land base. Furthermore the map should identify the location of service buildings, storage facilities, domestic or livestock wells, residences, and other permanent structures. These physical characteristics set a large number of constraints that the farmer or rancher must live within. Examples of draw-your-own maps are provided in Figures 9–1 through 9–4; however, an aerial photograph can serve the purpose just as well for nonartists. These aerial photographs are usually available from the local USDA Soil Conservation Service or Agricultural Stabilization and Conservation Service.

The map can provide a convenient location to mark fertilizer applications, crop and variety, and chemical applications on a field-by-field basis for cropland. It can also be used to record forage varieties, range or pasture improvement practices, and grazing systems on a pasture-by-pasture basis for grazing land. At the very least, fields and pastures should be numbered so that later records can refer to a particular pasture or field.

Crop Records

As noted in the outline shown in Figure 3–1, crop production records should include yields and prices received, fertilizer applications, seeding rates, approximate labor requirements (which will be discussed in Chapter 10), chemical applications, irrigation requirements, disease and pest control measures, and all other purchased inputs. Typical crop production record forms are shown in Figures 9–5 and 9–6. Generally the crop production records will require much less record-keeping time and effort than the financial records. Consequently, they should not be a large imposition on a producer's time. The more detail provided in crop production records, the better and more useful the records will be in the long run.

Livestock Records

Livestock records can take several different formats, depending upon the type of livestock produced. Registered beef cattle, hog, or sheep producers will normally keep much more detailed records than will commercial producers. Similarly, dairy producers, especially if they belong to a Dairy Herd Improvement Association (DHIA), will typically maintain much more detailed records on an individual animal basis than will commercial meat livestock producers or egg producers.

Legend
SCALE: 1 sq. = ________Acres

FIELD MAP

Field No. ________________
Acres ________________
Soil Type ________________

Soil Test Record

Date tested	lbs/acre			Soil pH
	NO_3	P	K	

PROPERTY BOUNDARY
DIVISION LINE (NO FENCE)
FENCE-WOVEN WIRE
BARBED WIRE
HEDGE
SECTION CENTER
PUBLIC ROADS (SURFACED ROADS IN RED)
PRIVATE ROAD
BRIDGE
RAILROAD
PIPE LINE

HOUSE
BARN
OTHER BUILDINGS
SCHOOL
CHURCH
CEMETERY
PERMANENT STREAM (CROSSABLE)
INTERMITTENT STREAM OR DRAINAGEWAY (CROSSABLE)

LAKE, POND OR SLOUGH
DRAINAGE DITCH
TILE (SIZE IN INCHES)
WELL
WINDMILL
SPRING, FLOWING WELL
RESERVOIR
FRUIT
ROCK
EROSION

FIGURE 9–1. Farm map form.

Source: Field and Equipment Record Book (St. Louis: Doane, 1983).

Year 1

Past	AUM	Date	Date	Weight	Weight	Notes
No.		In	Out	In	Out	

Year 2

Past	AUM	Date	Date	Weight	Weight	Notes
No.		In	Out	In	Out	

66

FIGURE 9–2. Ranch map form.

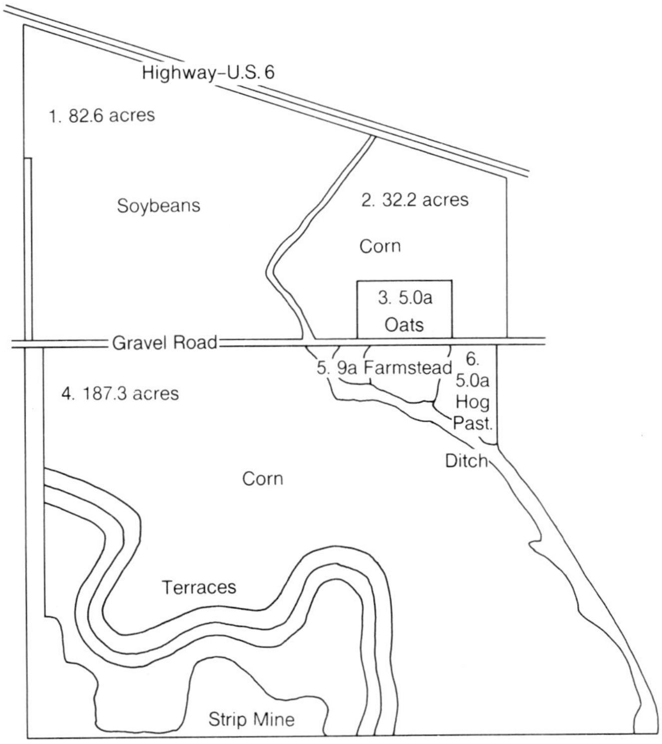

Cropland	313		
Strip Mine Area	25	Corn	219.5
Ditch	12	Beans	82.6
Farmstead	4	Oats	5.0
Roads, etc.	6	Hog Pasture	5.9
Farm Total	360 acres	Cropland	313.0 acres

FIGURE 9–3. Example crop plant.

Source: ASFMRA, *Farm and Ranch Management Manual.* American Society of Farm Managers and Rural Appraisers (Denver: ASFMRA, 1980), p. 74.

For commercial meat, egg, and wool producers, the decision to be made first is whether to maintain records on a per-animal basis, a group basis, or a whole-herd basis. It is very likely that most producers will opt for group or whole-herd records. That decision should be based upon the value per head and the number of animals required to make up an economic-sized herd. If the group option is chosen, a combined inventory-production record such as the examples shown in Figures 6–6 and 9–7 should provide the necessary information. Another example of a form for livestock group records is shown in Figure 9–8. Data such as pregnancy rates, weaning weights, weaning rates, purchase and sale weights and dates, purchase and sale prices, veterinary and medical costs, feed fed, pest control measures, and all other purchased inputs should be recorded.

Similar data should be recorded for cow-calf, ewe-lamb and sow-litter producers, but records should be kept on a per-head basis for each female breeding animal, in addition to group records. These records can be extremely valuable in determining which animals to cull each year. Culling is an economic decision as well as a biological/technical decision as there are depreciation, capital gains, and cost considerations. But in any culling decision, the weakest and least reliable producers should be culled first.

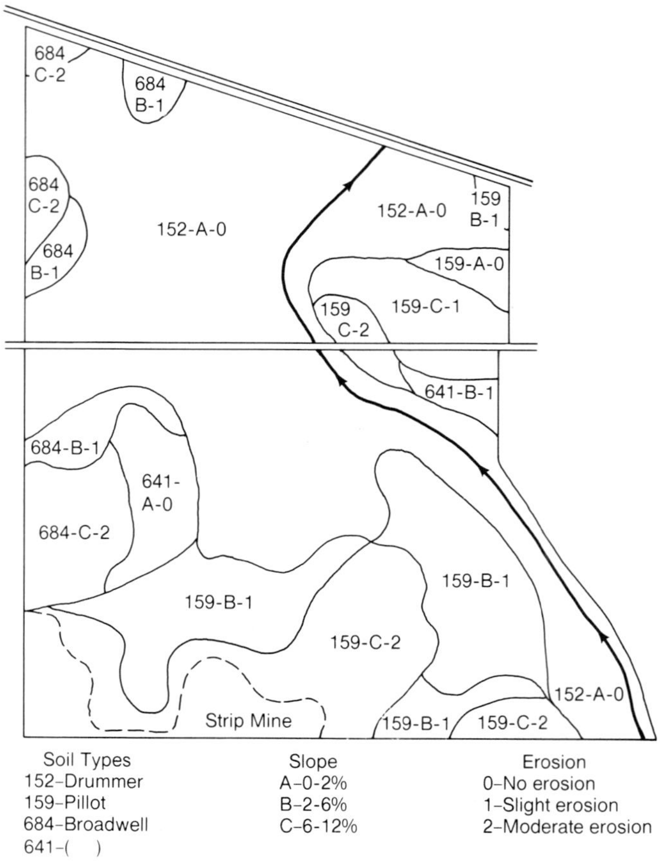

FIGURE 9–4. Example soil map.

Source: ASFMRA, *Farm and Ranch Management Manual.* American Society of Farm Managers and Rural Appraisers (Denver: ASFMRA, 1980), p. 74.

Because individual breeding animals may turn over (be replaced, sold, purchased) more rapidly than pieces of machinery, parcels of land, or the herd as a whole, it is usually advantageous to maintain individual records in a loose-leaf or card file system. An excellent example of a cow-calf record card system is shown in Figure 9–9. For each mother cow, vital statistics on her pregnancy are recorded. To cross reference a group of calves to their individual mothers, ear tags are recorded both on the cow record card and on the calf record sheet of Figure 9–8.

Producers should be careful enough with breeding records to record the sire if possible and to note females that do not conceive. If weaning rates are calculated *after* the females that do not conceive are excluded from the herd records, paper efficiency rates will look much better than actual efficiency. A system such as the Sow-Month Breeding Herd record form of Figure 9–10 avoids some of these fictitious efficiency boosters by counting every month that a breeding female remains in the herd. The total months are used to calculate an efficiency rate because the barren female requires feed and care just as the pregnant female does.

A very simple example can point out how records can be "fudged" to portray a story. Assume Ron Rancher runs a 700-head beef cow herd in Nevada. Ron likes

CROP RECORD

Crop	Field No.	Acres			Yield per acre	Production			Irrigation		Fertilizer			Seed & Chemicals		
		Owned	Rented	Total		Total (bushels)	Operator share (bushels)	Landlord share (bushels)	Amount per Acre	Total	Analysis	Amount per Acre	Total	Kind	Amount per Acre	Total
Total																
Total																
Total																
Diverted acres																
Total Crops																

72

FIGURE 9–5. Crop production record form.

Crop	Field No.	Acres			Yield per acre	Production		
		Owned	Rented	Total		Total (tons)	Operator share (tons)	Landlord share (tons)
Silage								
Total								

						Production		
						Total (tons)	Operator share (tons)	Landlord share (tons)
Hay and rotation pasture					Hay			
					1st cutting			
					2nd cutting			
					3rd cutting			
Total hay and rotation pasture								
Total grain crops								
Total __________ silage					Grass silage			
Total other crops					1st cutting			
Total hay and rotation pasture					2nd cutting			
Idle tillable land					3rd cutting			
Total tillable land								
Nontillable pasture								
Woodland								
Farmstead, roads, etc.								
Total farm								

Land Inventory	Acres			Value per acre	Total value		
	Owned	Rented	Total		Owned	Rented	Total
Tillable cropland							
Nontillable pasture - Range							
Other land							
Total							

FIGURE 9–6. Forage crop production record form.

Inventory / **Death Loss**

Line	Description	Beginning of year: No.	Beginning of year: Weight	Beginning of year: Value	End of year: No.	End of year: Weight	End of year: Value	Death Loss: No.	Death Loss: Weight
501	Cows								
502									
503									
504									
505	Bulls (over 1 yr.)								
506									
507	Bred heifers								
508									
509									
510	Open heifers								
511									
512	Calves (under 1 yr.)								
513									
514	Total								
515	Operator share								

Breeding Record

Month	Females bred[1]	Off-spring born	Off-spring weaned
Jan.			
Feb.			
Mar.			
Apr.			
May			
June			
July			
Aug.			
Sept.			
Oct.			
Nov.			
Dec.			
Total			
Offspring per animal bred			

Feed Record

Month	Farm grains raised and purchased: Corn (bushels)	(bushels)	(bushels)	Hay (tons)	Silage (tons)		Pasture[2]: Date or days	Pasture[2]: Units
Jan.								
Feb.								
Mar.								
Apr.								
May								
June								
July								
Aug.								
Sept.								
Oct.								
Nov.								
Dec.								
Total								
Unit pr.								
Tot. val.								

	Livestock Summary	Source	No.	Total weight	Operator share of value	Landlord share of value	Total farm value
516	End-of-year inventory	514, 515					
517	Livestock sales, net value	61-103					
518	Home consumption	p.60					
519	Death losses	514, 515					
520	Total (add 516 thru 519)						
521	Beginning-of-year inventory	514, 515					
522	Purchases	61-201 61-204					
523	Young stock born	513					
524	Total (add 521 thru 523)						
525	Net increase (520-524) decrease (524-520)		0				

	Feed Summary	Total for year: Pounds	Total for year: Value	Per unit of product[5]: Pounds	Per unit of product[5]: Value
526	Total farm grains[3]				
527	Protein, minerals, mixed feeds[4]				
528	Total concentrates				
529	Hay, silage				
530	Pasture allowance[2]				
531	Total all feeds				

FOOTNOTES

[1] Females bred to produce offspring during this year.

[2] Pasture charge may be cash rent or estimated costs on a per head basis.

[3] Convert bushels to pounds for all farm grains and enter totals in feed summary. In calculating value of feeds, use average yearly price.

[4] Enter the total pounds and value of protein, minerals, and mixed feeds for the year from feed inventory and purchase record.

[5] Divide the totals for the year by the total units of product produced.
1 dairy unit = 100 pounds of meat or 1,000 pounds of milk
1 beef unit = 100 pounds of beef
1 hog unit = 100 pounds of pork
1 poultry unit = 1½ pounds of meat or 1 dozen eggs
1 sheep unit = 100 pounds of lamb or mutton

BREEDING RECORDS

Livestock Enterprise Analysis

551 Livestock returns (525) .... $____
552 Product returns .... $____
553 Total enterprise returns (551 and 552) .... $____
554 Total value of feed fed (531) .... $____
555 Total returns above feed cost (553 − 554) .... $____
556 Total value of cash costs .... $____
557 Total returns above feed and cash costs (554 − 555) $____
558 Returns per $100 feed fed (553 ÷ 554 × 100) .... $____
559 Returns above feed per cow, (555 ÷ 564) .... $____

560 Feed costs per cow, (554 ÷ 564) .... $____
561 Feed costs per unit of product (531) .... $____

562 Pounds of feed per unit of product (531) .... ____
563 Weight produced (525), lb. .... ____
564 Females bred (513 breeding record) .... ____
565 Pounds of livestock per female bred (563 ÷ 564) .... ____
566 Product produced lb. .... ____

567 Number of livestock (cows) .... ____
568 Product per unit of livestock (566 ÷ 567), lb. ____

75

FIGURE 9–7. Livestock production record form.

Source: Robert B. Schwart, *Illinois Farm Record Book,* Illinois Cooperative Extension Service, University of Illinois.

CALF RECORD SHEET NMSU

TAG	DAM	SEX	BIRTH DATE	REMARKS

FIGURE 9–8. NMSU calf record sheet.

to brag at the local coffee shop that he always achieves a 95 percent calf crop. But did Ron actually mean that he weaned 665 calves (95 percent or 700)? Or did Ron discount the 75 cows that did not get pregnant and were culled several months before calving started? If he did forget about those cows (and maybe a few others), he may well be weaning less than 600 calves and still claiming his 95 percent calf crop. Livestock producers are certainly not the only businessmen with convenient memories—we have all heard stories of the grain farmer who divides his total production by the yield he wants to find out how many acres he planted. These fibs might work in the coffee shop, but they can be very dangerous for intensive business analysis.

Feed Records

Inventory records of feed on hand at the beginning and at the end of the year are a vital first step in accounting for one of the most difficult areas to maintain accurate records—feed. In addition to inventory records it is also necessary to maintain feed purchase records and records of sales of farm-grown feeds so that good estimates of actual feed use can be maintained for cost analysis.

One of the very best ways to maintain feed use records is to adapt a budgeting worksheet from the *Coordinated Financial Statements* package's cash flow forms.[1] This set of financial statements will be discussed in detail in Part III of this text, but a look at the budget sheet for crop usage and feed requirements will be

[1]Thomas L. Frey and Danny A. Klinefelter, *Coordinated Financial Statements for Agriculture*, 2nd ed. (Skokie, IL: AgriFinance, 1980).

1 2 3 4 5 6 7 8 9 10 11 12

NMSU cow record card

Ear Tag	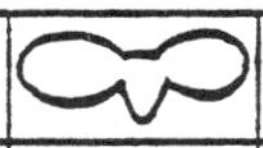	Born	Wean Age	Vaccinations
Dam	Brand	Wean Weight	Adj W Wt	Disposal
Sire		ADG	W/Wratio	

Year	Calved	Calf No.	Sex	Wean Age	Wean Wt	Wean Adj Wt	Adg	W/W ratio	Disposi-tion	FEEDLOT Gain	FEEDLOT Effic	GRADE CARC	GRADE Yld

DATE	CLINICAL DATA	CALF VACCINATION RECORD Date	No.	Vaccination	Vaccination	Vaccination

FIGURE 9–9. NMSU cow–calf record card.

useful now. This form (shown in Figure 9–11) follows a flow of feed concepts, that is, it forces the sources of feed to equal the uses of feed. Sources of feed include the beginning inventory (from the crop production records of Figures 9–5 and 9–6) and the amount purchased (taken usually from expenditure records). Uses of feed include amount fed, any amount held back for seed or other uses, sales (from receipt records) and ending inventory. This sources = uses format forces a proper

☐ All Sows
☐ All Gilts
☐ Combined

FEMALE MONTH BREEDING HERD RECORD

name ______________________ county ______________

Changes during the Month of	Number of Females: Added 1/	Number of Females: Sold, Died & Transferred	Total Females in Breeding Herd at End of Month (Female Months) 2/		Number of Pigs: Farrowed during month	Number of Pigs: Weaned during month
			(Jan. 1)	( )		
January			Jan. 31			
February			Feb. 28			
March			Mar. 31			
April			Apr. 30			
May			May 31			
June			June 30			
July			July 31			
August			Aug. 31			
September			Sept. 30			
October			Oct. 31			
November			Nov. 30			
December			Dec. 31			
TOTALS						3/
FEMALE YEARS (Divide Total Female Months by 12)						
PIGS FARROWED PER FEMALE YEAR						
PIGS WEANED PER FEMALE YEAR						

1/ Gilts are counted as added to the breeding herd in the month they reach market weight.

2/ Previous month's total plus number added minus number sold, died, and transferred equals total at end of current month.

Number of unweaned pigs on inventory, January 1. ______________ 3/
Number of unweaned pigs on inventory, December 31. ______________ 3/

3/ The total number of pigs weaned for the year should be adjusted by the difference in number of unweaned pigs on inventory, January 1 and December 31.

A copy of the female month breeding herd record can be forwarded to the Urbana office for summarization at check-in time by the fieldman.

(AE-4333)

Adopted for use with the Illinois Farm Business Record in cooperation with the Illinois FBFM Association and the Department of Agricultural Economics, University of Illinois at Urbana-Champaign.

FIGURE 9–10. Illinois sow-month breeding herd record.

C2 **BUDGET SHEET FOR CROP PRODUCTION** C2

Crop To Be Grown	Field		Yield Per Acre	Total Production	Seed			Fertilizer			Chemicals		
	No.	Acres			Variety	Rate Per Acre	Total	Analysis	Rate Per Acre	Total	Kind	Rate Per Acre	Total

SAMPLE

Developed by Thomas L. Frey and Danny A. Klinefelter, Dept. of Agricultural Economics, Cooperative Ext. Service, University of Illinois ©1980 AGRI FINANCE Available from Agri Finance, 5520-G Touhy Ave., Skokie, IL 60077

FIGURE 9–11. Budget sheet for crop usage and feed requirements.

Source: Thomas L. Frey and Danny A. Klinefelter, *Coordinated Financial Statements for Agriculture,* 2nd ed. (Skokie, IL: AgriFinance, 1980).

accounting. Feed consumption by livestock can be recorded by pen or by group in systems similar to the bottom half of Figure 9–11. Some large hog and cattle feeding operations will use a form similar to this, except that it will be printed on a card or heavy paper and maintained daily for each pen.

Summary

Production records are an extremely important component of a complete farm and ranch records system. The physical units of input and output are necessary to force consistency. If 95 animals are fed, the financial records cannot show 100 sold, unless the inventories match. Farm and ranch business analysts across the country spend a great deal of time in accounting for movements of supplies into and out of the business as well as movements of products into and out of the business.

A production records system should begin with a farm or ranch map with fields or pastures identified as well as other physical characteristics. Production records should also indicate animals born, grazing system used, soil treatments, chemical applications, and seed varieties in addition to amounts of products raised or fed.

CHAPTER 10

Labor Records

Two major records should be kept on labor use in a farm or ranch operation: (1) labor use by purpose and (2) payroll. The primary value of labor use records is that they can provide estimates of how labor resources are allocated and they can help determine which commodities use more than their fair share of labor. Thus, labor use records can be a major input into determining which crop or livestock commodities should be raised and which should not. Payroll records, on the other hand, identify total hours worked, wages paid, and taxes withheld. Although they may seem related, these two records are quite different: labor use records apply to both hired and operator labor, but payroll records only apply to hired labor. Furthermore, labor use records are a management tool not required by any outside agency whereas payroll records are necessary for financial statements and are required by the IRS and the Social Security Administration.

Labor Use

As noted above, labor use records are not required but can be a very valuable management tool. Since most farm and ranch operators stretch their time very thinly as they try to do the best job possible and since hired labor is very expensive and often difficult to obtain, the manager with a good idea of how his available labor resources are being used will have a major advantage in allocating that labor efficiently in the future.

The format of the labor use records does not need to be complicated and these records may be combined easily and efficiently with machinery records. But keeping labor use records can become a very time-consuming process. The most important decision (after deciding to maintain labor use records at all) is how much detail is desired and how often the labor use should be reported. For example, are daily records desired or will weekly estimates be sufficient? Monthly reporting will very likely stretch memories too much and produce fairly poor estimates. Further, should workers report time spent on each crop, on each livestock enterprise, on repairs or maintenance, and on other activities to the nearest quarter-hour, to the nearest hour, or to the nearest day?

The records system chosen and the record forms designed should be flexible enough to accommodate changes in detail desired and should be easy to summarize. One example of a labor record form is presented in Figure 10–1. However, it might fit workers better to develop an index-card-style labor record such as Figure 10–2, which could later be transferred to a permanent summary record. This system can be used both as a labor diary and as an incidental expense record.

The goal, once again, of the labor use records is to determine the crop/livestock enterprise efficiency of labor use. The daily, weekly, or monthly labor records kept by each worker need to be added together to summarize total labor used on the farm or ranch and labor used by each crop or livestock enterprise. The labor summaries will be used to develop enterprise budgets and to compare actual budgets to those published by universities, records associations, and farm/ranch magazines. The development of enterprise budgets will be discussed in Chapter 16.

Payroll

When nonoperator labor is hired, it becomes necessary to record wages paid and taxes withheld. Hired labor wages are tax-deductible expenses as are any associated employer taxes. Furthermore, the employee and employer taxes withheld must be deposited on a periodic basis with the IRS. In general, there are four types of employment taxes that an employer may be obliged to collect: social security (FICA), federal income tax withholding (FWT), state income tax withholding (SWT) and federal unemployment tax (FUTA). The rules and regulations surrounding each one of these taxes are different and likely to change from one year to the next. Consequently, to remain current, each agricultural employer should obtain and read the latest copy of two IRS publications (in addition to IRS publication 225, *Farmers Tax Guide*): Circular A, *Agricultural Employer's Tax Guide,* IRS Publication 51, and Circular E, *Employer's Tax Guide,* IRS Publication 15. The methods used to report these taxes will be analyzed in Chapter 20.

Before we describe the general rules for each of the four major withholding taxes, an assumption of this chapter must be stated. It will be assumed from here forward that only one person from each farm or ranch family will be designated as the operator. This person will generally be the boss or manager or failing that, the person who puts in the most work. Although many farms and ranches operate with a management team composed of husband and wife or as an informal partnership between spouses and sometimes in conjunction with children, there are major tax advantages to paying wages to family members. To have a basis for terminology, we will arbitrarily select the husband as the operator and assume the wife and children to be wage earners.

FARM LABOR RECORD [1/] — FIELD FORM 1 AE-4340

______________________'s Record of Man Labor Used From ______________ To ______________ 197__, Sheet No.________

WORKER	Time Period (day, wk., mo.)	Total Hours Worked	ON-FARM WORK						Off-Farm Work	
			Crop		Livestock			General Overhead	Farm	Non-Farm
			Grain	Other						
________										
	Total									
________										
	Total									
________										
	Total									
________										
	Total									

1. Designed for use with Illinois Farm Business Record prepared in Department of Agricultural Economics, University of Illinois in cooperation with Illinois FBFM Association.

FIGURE 10–1. Example labor record form.

EMPLOYEES RECORD
HOURS WORKED

Week ending ______________________

Monday	
Tuesday	
Wednesday	
Thursday	
Friday	
Saturday	
Sunday	

TOTAL HOURS ____________

I hereby certify that the above is an accurate statement of the hours worked by me for the period indicated.

Signed ______________________

EFFICIENCY® LINE 23-205 AN **AMPAD** PRODUCT

FIGURE 10–2. Example index card labor record.

Social Security (FICA)

A farmer or rancher who is also an employer will have to withhold and pay social security taxes if one or more persons (including parents or children age 21 or over) are employed and if an employee has been paid $150 or more in cash wages during the year or if an employee has been paid cash wages for agricultural labor performed on 20 or more days during the year. Family members, including the employer's spouse and children under the age of 21, are specifically excluded from coverage under the Social Security provisions, unless those family members work for and are paid by a corporation, partnership, or estate.[1]

Calculation of FICA taxes due is relatively simple. Cash wages are multiplied first by 7.15 percent (in 1986). That amount is the employee's contribution and must be withheld from his or her paycheck. Secondly, cash wages are multiplied by 7.15 percent (again for 1986). This second amount is the employer's contribution and is consequently a tax on the right to employ another person, over and above the wages paid. However, both the total cash wages paid and the employer's tax are deductible business expenses.

[1]The rules for partnerships and corporations are very specific. Consult the *Farmers Tax Guide,* an accountant, or an attorney.

Alternatively, an employer may determine FICA taxes by using the tables printed in IRS Publication 51, Circular A. A page of this table is shown in Figure 10–3.

To show a numerical example, suppose an employee was paid $250 cash wages in one week in April. To calculate FICA taxes, multiply $250 by 7.15 percent:

$$\$250 \times .0715 = \$17.88$$

The employee's contribution is $17.88 and the employer's contribution is also $17.88:

$$\$250 \times .0715 = \$17.88$$

The table page shown in Figure 10–3 does not reach $250 and the instructions require multiplication by .0715.

The taxes withheld plus the employer's contribution must be deposited with the IRS. The rules for these deposits will be discussed in Chapter 20.

Federal Income Tax Withholding

Agricultural employers and employees are allowed to withhold federal income taxes from an employee's paycheck in addition to FICA taxes. But unlike FICA taxes, federal withholding is not required. It is voluntary and *both* employee and employer must agree to withhold, neither is required to do so. Furthermore, federal income taxes may be withheld from a spouse's paycheck or from a minor's paycheck if she or he requests it.

Calculation of the amount of taxes to be withheld involves a series of equations. For most farmers and ranchers, it is much easier to use the tables provided in IRS Publication 15, Circular E. An example of these tables is shown in Figure 10–4. To carry the example forward, a weekly paycheck of $250 to a married employer with three withholding allowances would require that $18.00 be withheld.

There is no matching employer contribution for withholding taxes like that for FICA tax.

State Income Tax Withholding

Each state has different income tax laws and those laws should be checked regarding withholding.

Federal Unemployment Tax (FUTA)

Most farmers and ranchers do not fall into the category of employers required to pay unemployment taxes to the federal government, although they may be required to pay state unemployment taxes. To be required to pay FUTA taxes, the farmer or rancher must meet either one of two IRS tests: (1) paid cash wages of $20,000 or more in any calendar quarter during the current year or preceding year,

Social Security Employee Tax Table for 1986

7.15% employee tax deductions

Wages at least	But less than	Tax to be withheld	Wages at least	But less than	Tax to be withheld	Wages at least	But less than	Tax to be withheld	Wages at least	But less than	Tax to be withheld
50.84	50.98	3.64	63.57	63.71	4.55	76.30	76.44	5.46	89.03	89.17	6.37
50.98	51.12	3.65	63.71	63.85	4.56	76.44	76.58	5.47	89.17	89.31	6.38
51.12	51.26	3.66	63.85	63.99	4.57	76.58	76.72	5.48	89.31	89.45	6.39
51.26	51.40	3.67	63.99	64.13	4.58	76.72	76.86	5.49	89.45	89.59	6.40
51.40	51.54	3.68	64.13	64.27	4.59	76.86	77.00	5.50	89.59	89.73	6.41
51.54	51.68	3.69	64.27	64.41	4.60	77.00	77.14	5.51	89.73	89.87	6.42
51.68	51.82	3.70	64.41	64.55	4.61	77.14	77.28	5.52	89.87	90.00	6.43
51.82	51.96	3.71	64.55	64.69	4.62	77.28	77.42	5.53	90.00	90.14	6.44
51.96	52.10	3.72	64.69	64.83	4.63	77.42	77.56	5.54	90.14	90.28	6.45
52.10	52.24	3.73	64.83	64.97	4.64	77.56	77.70	5.55	90.28	90.42	6.46
52.24	52.38	3.74	64.97	65.11	4.65	77.70	77.84	5.56	90.42	90.56	6.47
52.38	52.52	3.75	65.11	65.25	4.66	77.84	77.98	5.57	90.56	90.70	6.48
52.52	52.66	3.76	65.25	65.39	4.67	77.98	78.12	5.58	90.70	90.84	6.49
52.66	52.80	3.77	65.39	65.53	4.68	78.12	78.26	5.59	90.84	90.98	6.50
52.80	52.94	3.78	65.53	65.67	4.69	78.26	78.40	5.60	90.98	91.12	6.51
52.94	53.08	3.79	65.67	65.81	4.70	78.40	78.54	5.61	91.12	91.26	6.52
53.08	53.22	3.80	65.81	65.95	4.71	78.54	78.68	5.62	91.26	91.40	6.53
53.22	53.36	3.81	65.95	66.09	4.72	78.68	78.82	5.63	91.40	91.54	6.54
53.36	53.50	3.82	66.09	66.23	4.73	78.82	78.96	5.64	91.54	91.68	6.55
53.50	53.64	3.83	66.23	66.37	4.74	78.96	79.10	5.65	91.68	91.82	6.56
53.64	53.78	3.84	66.37	66.51	4.75	79.10	79.24	5.66	91.82	91.96	6.57
53.78	53.92	3.85	66.51	66.65	4.76	79.24	79.38	5.67	91.96	92.10	6.58
53.92	54.06	3.86	66.65	66.79	4.77	79.38	79.52	5.68	92.10	92.24	6.59
54.06	54.20	3.87	66.79	66.93	4.78	79.52	79.66	5.69	92.24	92.38	6.60
54.20	54.34	3.88	66.93	67.07	4.79	79.66	79.80	5.70	92.38	92.52	6.61
54.34	54.48	3.89	67.07	67.21	4.80	79.80	79.94	5.71	92.52	92.66	6.62
54.48	54.62	3.90	67.21	67.35	4.81	79.94	80.07	5.72	92.66	92.80	6.63
54.62	54.76	3.91	67.35	67.49	4.82	80.07	80.21	5.73	92.80	92.94	6.64
54.76	54.90	3.92	67.49	67.63	4.83	80.21	80.35	5.74	92.94	93.08	6.65
54.90	55.04	3.93	67.63	67.77	4.84	80.35	80.49	5.75	93.08	93.22	6.66
55.04	55.18	3.94	67.77	67.91	4.85	80.49	80.63	5.76	93.22	93.36	6.67
55.18	55.32	3.95	67.91	68.05	4.86	80.63	80.77	5.77	93.36	93.50	6.68
55.32	55.46	3.96	68.05	68.19	4.87	80.77	80.91	5.78	93.50	93.64	6.69
55.46	55.60	3.97	68.19	68.33	4.88	80.91	81.05	5.79	93.64	93.78	6.70
55.60	55.74	3.98	68.33	68.47	4.89	81.05	81.19	5.80	93.78	93.92	6.71
55.74	55.88	3.99	68.47	68.61	4.90	81.19	81.33	5.81	93.92	94.06	6.72
55.88	56.02	4.00	68.61	68.75	4.91	81.33	81.47	5.82	94.06	94.20	6.73
56.02	56.16	4.01	68.75	68.89	4.92	81.47	81.61	5.83	94.20	94.34	6.74
56.16	56.30	4.02	68.89	69.03	4.93	81.61	81.75	5.84	94.34	94.48	6.75
56.30	56.44	4.03	69.03	69.17	4.94	81.75	81.89	5.85	94.48	94.62	6.76
56.44	56.58	4.04	69.17	69.31	4.95	81.89	82.03	5.86	94.62	94.76	6.77
56.58	56.72	4.05	69.31	69.45	4.96	82.03	82.17	5.87	94.76	94.90	6.78
56.72	56.86	4.06	69.45	69.59	4.97	82.17	82.31	5.88	94.90	95.04	6.79
56.86	57.00	4.07	69.59	69.73	4.98	82.31	82.45	5.89	95.04	95.18	6.80
57.00	57.14	4.08	69.73	69.87	4.99	82.45	82.59	5.90	95.18	95.32	6.81
57.14	57.28	4.09	69.87	70.00	5.00	82.59	82.73	5.91	95.32	95.46	6.82
57.28	57.42	4.10	70.00	70.14	5.01	82.73	82.87	5.92	95.46	95.60	6.83
57.42	57.56	4.11	70.14	70.28	5.02	82.87	83.01	5.93	95.60	95.74	6.84
57.56	57.70	4.12	70.28	70.42	5.03	83.01	83.15	5.94	95.74	95.88	6.85
57.70	57.84	4.13	70.42	70.56	5.04	83.15	83.29	5.95	95.88	96.02	6.86
57.84	57.98	4.14	70.56	70.70	5.05	83.29	83.43	5.96	96.02	96.16	6.87
57.98	58.12	4.15	70.70	70.84	5.06	83.43	83.57	5.97	96.16	96.30	6.88
58.12	58.26	4.16	70.84	70.98	5.07	83.57	83.71	5.98	96.30	96.44	6.89
58.26	58.40	4.17	70.98	71.12	5.08	83.71	83.85	5.99	96.44	96.58	6.90
58.40	58.54	4.18	71.12	71.26	5.09	83.85	83.99	6.00	96.58	96.72	6.91
58.54	58.68	4.19	71.26	71.40	5.10	83.99	84.13	6.01	96.72	96.86	6.92
58.68	58.82	4.20	71.40	71.54	5.11	84.13	84.27	6.02	96.86	97.00	6.93
58.82	58.96	4.21	71.54	71.68	5.12	84.27	84.41	6.03	97.00	97.14	6.94
58.96	59.10	4.22	71.68	71.82	5.13	84.41	84.55	6.04	97.14	97.28	6.95
59.10	59.24	4.23	71.82	71.96	5.14	84.55	84.69	6.05	97.28	97.42	6.96
59.24	59.38	4.24	71.96	72.10	5.15	84.69	84.83	6.06	97.42	97.56	6.97
59.38	59.52	4.25	72.10	72.24	5.16	84.83	84.97	6.07	97.56	97.70	6.98
59.52	59.66	4.26	72.24	72.38	5.17	84.97	85.11	6.08	97.70	97.84	6.99
59.66	59.80	4.27	72.38	72.52	5.18	85.11	85.25	6.09	97.84	97.98	7.00
59.80	59.94	4.28	72.52	72.66	5.19	85.25	85.39	6.10	97.98	98.12	7.01
59.94	60.07	4.29	72.66	72.80	5.20	85.39	85.53	6.11	98.12	98.26	7.02
60.07	60.21	4.30	72.80	72.94	5.21	85.53	85.67	6.12	98.26	98.40	7.03
60.21	60.35	4.31	72.94	73.08	5.22	85.67	85.81	6.13	98.40	98.54	7.04
60.35	60.49	4.32	73.08	73.22	5.23	85.81	85.95	6.14	98.54	98.68	7.05
60.49	60.63	4.33	73.22	73.36	5.24	85.95	86.09	6.15	98.68	98.82	7.06
60.63	60.77	4.34	73.36	73.50	5.25	86.09	86.23	6.16	98.82	98.96	7.07
60.77	60.91	4.35	73.50	73.64	5.26	86.23	86.37	6.17	98.96	99.10	7.08
60.91	61.05	4.36	73.64	73.78	5.27	86.37	86.51	6.18	99.10	99.24	7.09
61.05	61.19	4.37	73.78	73.92	5.28	86.51	86.65	6.19	99.24	99.38	7.10
61.19	61.33	4.38	73.92	74.06	5.29	86.65	86.79	6.20	99.38	99.52	7.11
61.33	61.47	4.39	74.06	74.20	5.30	86.79	86.93	6.21	99.52	99.66	7.12
61.47	61.61	4.40	74.20	74.34	5.31	86.93	87.07	6.22	99.66	99.80	7.13
61.61	61.75	4.41	74.34	74.48	5.32	87.07	87.21	6.23	99.80	99.94	7.14
61.75	61.89	4.42	74.48	74.62	5.33	87.21	87.35	6.24	99.94	100.00	7.15
61.89	62.03	4.43	74.62	74.76	5.34	87.35	87.49	6.25			
62.03	62.17	4.44	74.76	74.90	5.35	87.49	87.63	6.26			
62.17	62.31	4.45	74.90	75.04	5.36	87.63	87.77	6.27			
62.31	62.45	4.46	75.04	75.18	5.37	87.77	87.91	6.28			
62.45	62.59	4.47	75.18	75.32	5.38	87.91	88.05	6.29			
62.59	62.73	4.48	75.32	75.46	5.39	88.05	88.19	6.30			
62.73	62.87	4.49	75.46	75.60	5.40	88.19	88.33	6.31			
62.87	63.01	4.50	75.60	75.74	5.41	88.33	88.47	6.32			
63.01	63.15	4.51	75.74	75.88	5.42	88.47	88.61	6.33			
63.15	63.29	4.52	75.88	76.02	5.43	88.61	88.75	6.34			
63.29	63.43	4.53	76.02	76.16	5.44	88.75	88.89	6.35			
63.43	63.57	4.54	76.16	76.30	5.45	88.89	89.03	6.36			

Wages	Taxes
100	$7.15
200	14.30
300	21.45
400	28.60
500	35.75
600	42.90
700	50.05
800	57.20
900	64.35
1,000	71.50

FIGURE 10–3. Social security employer tax table.

Source: IRS Publication 51, *Agricultural Employer's Tax Guide,* Circular A.

or (2) employed ten or more farm workers for some part of at least one day during each of twenty different calendar weeks during the current or preceding year.

Several rules have been enacted that specify how the number of workers is calculated, especially for employers who use crew laborers. Those rules, specified in the *Farmers Tax Guide,* should be studied carefully if either of the two tests *might* be met, especially by employers of crew laborers.

If either of the two tests is met, all employees of that employer fall under FUTA. To calculate FUTA tax, multiply 6.2 percent times the cash wages paid (up to $7,000 of wages).

Payroll Records

Payroll records can take many forms, but they should meet two major requirements: (1) a place to record *each* pay check that you issue and (2) sufficient space to record all of the information required by the IRS, as listed in Box 10–1.

Example payroll records are shown in Figures 10–5 and 10–6. The previous example is recorded in Figure 10–7. Assuming no state income tax withholding, Joe Mayberry earned $250.00 in gross wages, but his paycheck was written for $214.12 after withholdings. An amount equal to $53.76 ($17.88 employee FICA contribution, $17.88 employer FICA contribution and $18.00 FWT withholding) will eventually be deposited with the IRS.

Box 10–1

Payroll Record Requirements

16. Records You Should Keep

Every employer subject to employment taxes must keep all related records available for inspection for at least 4 years after the due date of the return period to which the records relate, or the date the taxes are paid, whichever is later. You may keep the records in whatever form you choose.

For each employee to whom you pay, or expect to pay, taxable wages for farmwork, keep a record of:

- the employee's name and social security number.
- cash payments to the employee for farmwork.
- any amount deducted as employee social security tax.
- if the 20 days test applies, the number of days the employee did farmwork for cash wages figured on a time basis.
- the amount, if any, of income tax withheld.
- the amount of noncash wages paid (for income tax purposes only).

If you did not keep records because you did not expect an employee to meet the $150 test or the 20 days test and later find the wages will be taxable, you should begin keeping records for the employee. Estimate as best you can what you paid the employee previously during the year.

If a crew leader furnished you with farmworkers, you must keep a record of the name, permanent mailing address, and employer identification number of the crew leader. If the crew leader has no permanent mailing address, record his or her present address.

Source: IRS Publication 51, *Agricultural Employer's Tax Guide,* Circular A.

MARRIED Persons–WEEKLY Payroll Period

(For Wages Paid After December 1985)

And the wages are–		And the number of withholding allowances claimed is–										
At least	But less than	0	1	2	3	4	5	6	7	8	9	10
		The amount of income tax to be withheld shall be–										
$0	$54	$0	$0	$0	$0	$0	$0	$0	$0	$0	$0	$0
54	56	1	0	0	0	0	0	0	0	0	0	0
56	58	1	0	0	0	0	0	0	0	0	0	0
58	60	1	0	0	0	0	0	0	0	0	0	0
60	62	1	0	0	0	0	0	0	0	0	0	0
62	64	1	0	0	0	0	0	0	0	0	0	0
64	66	2	0	0	0	0	0	0	0	0	0	0
66	68	2	0	0	0	0	0	0	0	0	0	0
68	70	2	0	0	0	0	0	0	0	0	0	0
70	72	2	0	0	0	0	0	0	0	0	0	0
72	74	3	0	0	0	0	0	0	0	0	0	0
74	76	3	0	0	0	0	0	0	0	0	0	0
76	78	3	1	0	0	0	0	0	0	0	0	0
78	80	3	1	0	0	0	0	0	0	0	0	0
80	82	3	1	0	0	0	0	0	0	0	0	0
82	84	4	1	0	0	0	0	0	0	0	0	0
84	86	4	2	0	0	0	0	0	0	0	0	0
86	88	4	2	0	0	0	0	0	0	0	0	0
88	90	4	2	0	0	0	0	0	0	0	0	0
90	92	5	2	0	0	0	0	0	0	0	0	0
92	94	5	2	0	0	0	0	0	0	0	0	0
94	96	5	3	0	0	0	0	0	0	0	0	0
96	98	5	3	1	0	0	0	0	0	0	0	0
98	100	5	3	1	0	0	0	0	0	0	0	0
100	105	6	4	1	0	0	0	0	0	0	0	0
105	110	6	4	2	0	0	0	0	0	0	0	0
110	115	7	5	2	0	0	0	0	0	0	0	0
115	120	8	5	3	1	0	0	0	0	0	0	0
120	125	8	6	3	1	0	0	0	0	0	0	0
125	130	9	6	4	2	0	0	0	0	0	0	0
130	135	9	7	5	2	0	0	0	0	0	0	0
135	140	10	8	5	3	1	0	0	0	0	0	0
140	145	11	8	6	3	1	0	0	0	0	0	0
145	150	11	9	6	4	2	0	0	0	0	0	0
150	160	13	10	7	5	2	0	0	0	0	0	0
160	170	14	11	8	6	4	1	0	0	0	0	0
170	180	15	12	10	7	5	2	0	0	0	0	0
180	190	17	14	11	8	6	3	1	0	0	0	0
190	200	18	15	12	10	7	5	2	0	0	0	0
200	210	20	17	14	11	8	6	3	1	0	0	0
210	220	21	18	15	12	9	7	4	2	0	0	0
220	230	22	19	17	14	11	8	6	3	1	0	0
230	240	24	21	18	15	12	9	7	4	2	0	0
240	250	26	22	19	16	14	11	8	6	3	1	0
250	260	27	24	21	18	15	12	9	7	4	2	0
260	270	29	25	22	19	16	13	11	8	5	3	1
270	280	30	27	24	21	18	15	12	9	7	4	2
280	290	32	29	25	22	19	16	13	10	8	5	3
290	300	34	30	27	24	21	18	15	12	9	7	4
300	310	35	32	28	25	22	19	16	13	10	8	5
310	320	37	33	30	27	23	20	18	15	12	9	6
320	330	39	35	32	28	25	22	19	16	13	10	8
330	340	40	37	33	30	27	23	20	17	14	12	9
340	350	42	38	35	32	28	25	22	19	16	13	10
350	360	44	40	37	33	30	27	23	20	17	14	11
360	370	46	42	38	35	31	28	25	22	19	16	13
370	380	48	44	40	36	33	30	26	23	20	17	14
380	390	49	46	42	38	35	31	28	25	21	19	16
390	400	51	47	44	40	36	33	30	26	23	20	17
400	410	53	49	46	42	38	35	31	28	25	21	18
410	420	55	51	47	44	40	36	33	29	26	23	20
420	430	58	53	49	45	42	38	34	31	28	24	21
430	440	60	55	51	47	43	40	36	33	29	26	23
440	450	62	57	53	49	45	42	38	34	31	28	24
450	460	64	60	55	51	47	43	40	36	33	29	26
460	470	66	62	57	53	49	45	41	38	34	31	27
470	480	69	64	60	55	51	47	43	39	36	32	29
480	490	71	66	62	57	53	49	45	41	37	34	31
490	500	73	68	64	59	55	51	47	43	39	36	32

Page 24

(Continued on next page)

FIGURE 10–4. Income tax withholding table.

Source: IRS Publication 15, *Employer's Tax Guide,* Circular E.

MARRIED Persons–WEEKLY Payroll Period

(For Wages Paid After December 1985)

And the wages are–		And the number of withholding allowances claimed is–										
At least	But less than	0	1	2	3	4	5	6	7	8	9	10
		The amount of income tax to be withheld shall be–										
$500	$510	$76	$71	$66	$62	$57	$52	$49	$45	$41	$37	$34
510	520	78	73	68	64	59	55	50	47	43	39	35
520	530	81	76	71	66	61	57	52	48	45	41	37
530	540	83	78	73	68	64	59	54	50	46	43	39
540	550	86	81	75	70	66	61	57	52	48	45	41
550	560	88	83	78	73	68	63	59	54	50	46	43
560	570	91	86	80	75	70	66	61	56	52	48	44
570	580	93	88	83	78	72	68	63	59	54	50	46
580	590	96	91	85	80	75	70	65	61	56	52	48
590	600	98	93	88	83	77	72	68	63	59	54	50
600	610	101	96	90	85	80	75	70	65	61	56	52
610	620	104	98	93	88	82	77	72	67	63	58	54
620	630	106	101	95	90	85	80	75	70	65	61	56
630	640	109	103	98	93	87	82	77	72	67	63	58
640	650	112	106	100	95	90	85	80	74	70	65	60
650	660	115	109	103	98	92	87	82	77	72	67	63
660	670	118	112	106	100	95	90	85	79	74	69	65
670	680	120	115	109	103	97	92	87	82	77	72	67
680	690	123	117	112	106	100	95	90	84	79	74	69
690	700	126	120	114	109	103	97	92	87	82	76	71
700	710	129	123	117	111	106	100	95	89	84	79	74
710	720	132	126	120	114	108	103	97	92	87	81	76
720	730	135	129	123	117	111	105	100	94	89	84	79
730	740	139	132	126	120	114	108	102	97	92	86	81
740	750	142	135	128	123	117	111	105	99	94	89	84
750	760	145	138	131	125	120	114	108	102	97	91	86
760	770	148	142	135	128	122	117	111	105	99	94	89
770	780	152	145	138	131	125	119	114	108	102	96	91
780	790	155	148	141	134	128	122	116	111	105	99	94
790	800	158	151	145	138	131	125	119	113	108	102	96
800	810	162	155	148	141	134	128	122	116	110	105	99
810	820	165	158	151	144	138	131	125	119	113	107	102
820	830	168	161	155	148	141	134	128	122	116	110	104
830	840	172	165	158	151	144	137	130	125	119	113	107
840	850	175	168	161	154	147	141	134	127	122	116	110
850	860	178	171	164	158	151	144	137	130	124	119	113
860	870	181	175	168	161	154	147	140	133	127	121	116
870	880	185	178	171	164	157	150	144	137	130	124	118
880	890	188	181	174	167	161	154	147	140	133	127	121
890	900	191	184	178	171	164	157	150	143	137	130	124
900	910	195	188	181	174	167	160	154	147	140	133	127
910	920	198	191	184	177	171	164	157	150	143	136	130
920	930	201	194	188	181	174	167	160	153	146	140	133
930	940	205	198	191	184	177	170	163	157	150	143	136
940	950	208	201	194	187	180	174	167	160	153	146	139
950	960	212	204	197	191	184	177	170	163	156	149	143
960	970	216	208	201	194	187	180	173	166	160	153	146
970	980	220	212	204	197	190	183	177	170	163	156	149
980	990	223	216	208	200	194	187	180	173	166	159	153
990	1,000	227	219	212	204	197	190	183	176	170	163	156
1,000	1,010	231	223	215	208	200	193	187	180	173	166	159
1,010	1,020	234	227	219	211	204	197	190	183	176	169	162
1,020	1,030	238	230	223	215	207	200	193	186	179	173	166
1,030	1,040	242	234	226	219	211	203	196	190	183	176	169
1,040	1,050	245	238	230	222	215	207	200	193	186	179	172
1,050	1,060	249	241	234	226	218	211	203	196	189	182	176
1,060	1,070	253	245	237	230	222	214	207	199	193	186	179
1,070	1,080	257	249	241	234	226	218	210	203	196	189	182
1,080	1,090	260	253	245	237	230	222	214	206	199	192	186
1,090	1,100	264	256	249	241	233	226	218	210	203	196	189
1,100	1,110	268	260	252	245	237	229	222	214	206	199	192
1,110	1,120	271	264	256	248	241	233	225	218	210	202	195
1,120	1,130	275	267	260	252	244	237	229	221	214	206	199
1,130	1,140	279	271	263	256	248	240	233	225	217	210	202
1,140	1,150	282	275	267	259	252	244	236	229	221	213	206
		37 percent of the excess over $1,150 plus–										
$1,150 and over		284	277	269	261	254	246	238	231	223	215	207

NEW MEXICO FRBR

INDIVIDUAL PAYROLL RECORD

Year ________

Form 6D

Page ________

Name ______________________ Address ______________________

Ln No	TJ Ref Pg-Ln	Time Worked	Pay Rate	Ent. Code	Gross Wage	Deductions: FICA	Deductions	Deductions	Deductions	Total Deductions	Net Amount Paid
	Balances Forward										
1											
2											
3											
4											
5											
6											
7											
8											
9											
10											
11											
12											
13											
14											
15											
16											
17											
18											
19											
20											
Totals											
					6D1	6D2	6D3	6D4	6D5	6D6	6D7

Social Security Number ________

Date Employed ________

Date Discontinued ________

Number of Exemptions ________

Pay Rate ________

FIGURE 10–5. Payroll record.

Source: Michael L. Jones and James D. Libbin, *New Mexico Farm and Ranch Business Record Book,* New Mexico Cooperative Extension Service, New Mexico State University, 1986.

PAYROLL DEDUCTIONS

Period Ending ____________ Soc. Sec. No. ____________

Name ________________________

Earnings ________________________ $ ________

Fed. Ins. Cont. Act ____________ ________

Fed. Withholding Tax ____________ ________

State Withholding Tax ____________ ________ ________

____________ ________

____________ ________

____________ ________ ________

Total Deductions ________

Net Earnings $ ________

Employer ________________________

EFFICIENCY® LINE 23-206 AN **AMPAD** PRODUCT

FIGURE 10–6. Payroll record.

Ideal **PAYROLL RECORD** FORM Q812R

YEAR ENDING Dec. 31, 1986 EMPLOYEE Joe Mayberry ADDRESS Route 1, Box 100 TELEPHONE

DATE OF BIRTH MARITAL STATUS married SOCIAL SEC. NO. 123-45-6789 EXEMPTIONS

	WEEK ENDING	HOURS WORKED			RATE	EARNINGS		TOTAL SALARY & WAGES	DEDUCTIONS					NET PAY	CHECK NO.	
		TOTAL	REG.	O.T.		REGULAR	OVERTIME		FICA	FWT	SWT		TOTAL			
1	1-8	50	50	0	5.00	250 00		250 00	17 88	18 00			35 88	214 12	1108	1
2																2
3																3
4																4
5																5
6																6
7																7
8																8
9																9
10																10
11																11
12																12
13																13
14																14
15																15
16	1st QTR. TOTAL															16
17																17
18																18
19																19
20																20
21																21
22																22
23																23
24																24
25																25
26																26
27																27
28																28
29																29
30																30
31																31
32	2nd QTR. TOTAL															32
33	6 MO. TOTAL															33

THE IDEAL SYSTEM, REG. U.S. PAT. OFFICE. MADE IN U.S.A.

FIGURE 10–7. Ideal system payroll record.

Source: Ideal Bookkeeping and Tax Record: Farm and Ranch, Esselte Pendaflex Corporation, Garden City, NY, 1982.

Summary

Given all of the different tax possibilities, it should be obvious that payroll records are an absolute necessity. In addition to payroll records, business owners should consider the value to their operation of maintaining labor use records. Since labor is a major input in farming or ranching, good business analysis requires some idea of labor requirements so that adequate enterprise comparisons can be made. Whether to keep labor use records becomes an individual operation decision, especially with respect to detail and reporting periods.

Taxes once again rule the decisions about the precision and form of payroll records. Employers must be aware that they are potentially liable to withhold social security, federal income, state income, and federal unemployment taxes. Records of the calculations performed for each paycheck should be maintained and collected taxes must be deposited.

CHAPTER 11

Machinery Records

Like labor records, there are several different reasons for keeping machinery records and several different types of records to be kept. Probably the first and foremost records needed on machinery are the terms of purchase of each item, that is,

1. The date of purchase
2. The purchase price
3. Whether the machine was new when purchased

The primary purpose of this information is to justify the depreciation taken and to justify capital gains claimed upon resale of the machine. All of this information is best kept within the depreciation schedule as discussed in Chapter 7. In addition to depreciation records, the types of information needed about machinery include machinery use by purpose as well as repair and maintenance schedules. This chapter will concentrate on these two purposes.

Machinery Use

Also like the labor records, machinery use records are not required by any outside source or agency, but they can become an important management tool. It is useful for any farm or ranch manager to know how his available resources are being used, especially the limiting resources. The manager should attempt to gain maximum economic efficiency from those resources. Although machinery can be purchased when needed, on a typical farm the cost of owning and operating machinery represents a very large portion of total operating expenses. However, on ranches and other single-enterprise livestock operations machinery costs are usually relatively minor, and since few items of machinery are actually owned on these operations, individual machinery use records might not be necessary.

FARM LABOR RECORD[1]—FIELD FORM 1 AE-4340

____________'s Record of Man Labor Used From ______ To ______ 197__, Sheet No. ______

Worker	Time Per.	Mach #	Total Hours Worked	On-Farm Work						Off-Farm Work	
				Crop		Livestock			General Overhead	Farm	Non-Farm
				Grain	Other						
	Total										
	Total										
	Total										
	Total										

1. Designed for use with Illinois Farm Business Record prepared in Department of Agricultural Economics, University of Illinois, in cooperation with Illinois FBFM Association.

FIGURE 11–1. Example labor–machine record form.

Probably the most effective way to maintain machinery records is to combine machinery and labor use onto one form. If the example forms of Figures 10–1 and 10–2 were extended to include the machine and implement used, such as those of Figures 11–1 and 11–2, the enterprise, operator, and hours used could be recorded together fairly easily. Although laborers can work without machines, seldom can a machine work without an operator.

This type of machinery record can be easily summarized and provides virtually complete detail for assessing how much of each resource is used by each enterprise.

EMPLOYEES RECORD
HOURS WORKED

Week ending ____________

	HOURS	MACHINE & IMPLEMENT
Monday		
Tuesday		
Wednesday		
Thursday		
Friday		
Saturday		
Sunday		

TOTAL HOURS ____________

I hereby certify that the above is an accurate statement of the hours worked by me for the period indicated.

Signed ____________

EFFICIENCY® LINE 23-205 AN **AMPAD** PRODUCT

FIGURE 11–2. Example index-card labor–machine record form.

Machinery Record

Item ____________ Identification no. ____________

Year ____________ Date purchased ____________

Month or date	Hours of use	Fuel		Oil and lubricant		Repairs and maintenance	
		Gallons	Cost	Qty.	Cost	Cost (including labor)	Description
Totals							

Depreciation ____________ Property taxes ____________ Total fixed cost ____________

Interest ____________ Insurance ____________ Total variable cost ____________

Total cost ____________

Average fixed costs per hour ____________

Average variable cost per hour ____________

Average total cost per hour ____________

FIGURE 11–3. Office-based Machinery Record form.

Machinery Fuel, Maintenance, and Repairs

In addition to the machinery use records, it is very useful to record the consumption of fuel and lubricants as well as any maintenance operations and repairs performed on each machine. The type of record system developed for machinery must be both useful and easy to maintain. For some operations, this may mean a record book or an index card that stays with each machine. Or the record may be kept in the office such as the example in Figure 11–3.

Another system for maintaining machinery fuel, lubricants, and repair records is a two-part system. A chart can be placed at the fuel tank. There, date, machine number, machine hours, amount of fuel put into the tank, and minor maintenance information can be recorded fairly easily. The second part of this system might involve an index card or the stick-on record forms provided by oil distributors, like the example shown in Figure 11–4. Major repair notes could be added to the labor record.

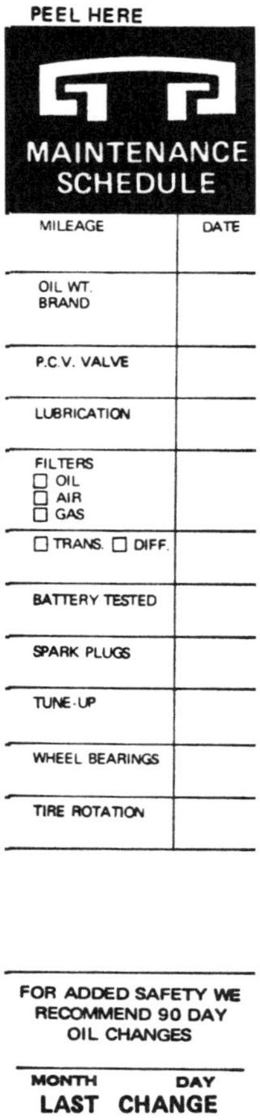
PEEL HERE

MAINTENANCE SCHEDULE

MILEAGE	DATE
OIL WT. BRAND	
P.C.V. VALVE	
LUBRICATION	
FILTERS ☐ OIL ☐ AIR ☐ GAS	
☐ TRANS. ☐ DIFF.	
BATTERY TESTED	
SPARK PLUGS	
TUNE-UP	
WHEEL BEARINGS	
TIRE ROTATION	

FOR ADDED SAFETY WE RECOMMEND 90 DAY OIL CHANGES

MONTH DAY
LAST CHANGE

FIGURE 11–4. Stick-on lubrication and maintenance record.

Summary

Machinery use records, like labor use records, can be a very important part of overall business analysis because machinery costs are a major component of total farming costs. It is also important to allocate machinery costs to the various crop and livestock commodities produced in order to obtain an accurate reading of which enterprises make money and which ones lose.

Machinery records come in three parts: depreciation records, use records, and repair records. Depreciation records are mathematically complex but they are required by the IRS. Use records are simple in concept, but can be cumbersome and time consuming. Repair records are vital to decisions about when to replace old machinery.

CHAPTER 12

Family Records

Most farms and ranches are operated as family-owned and family-run businesses. The predominance of the family involvement implies strong ties between personal financial concerns and business financial concerns, ties so strong that they are often inseparable. The bookkeeper for the business is often the family bookkeeper as well; the family savings are also the business savings, family living expenditures are often reflected on the business cash flow budget; family assets, such as vehicles, furniture, and the residence usually appear on a consolidated personal-business balance sheet; and family investments are often made directly in the business.

Despite the fact that many business affairs are inseparably intertwined with personal finances, some distinctions must be made. There are two primary reasons for separating purely family affairs from purely business affairs: income tax reporting differences and business analysis. The manner in which income taxes are reported requires that personal and business income and expenditures be handled differently. For example, farm and ranch business receipts are initially reported in a different tax form than are off-farm or off-ranch wage income of the owner and family members, deductible personal expenses are much fewer than deductible business expenses and are subject to different restrictions, and nonbusiness assets are not depreciable. Furthermore, to properly and accurately analyze farm or ranch business performance, personal financial concerns must be set aside.

For these reasons personal financial records need to be separated from business records. But for the same reasons advanced in earlier chapters, personal financial records are as necessary as business records. However, the degree to which some personal records should be maintained can vary significantly from family to family and between the family and the business.

Family Records Needed

Necessary family records fall into the following general categories:

1. Off-farm or off-ranch income
2. Off-farm or off-ranch expenditures
3. Investments and returns on investments
4. Farm or ranch products used by the family
5. Family liabilities

Detailed records of all income from off-farm or off-ranch jobs, other businesses (such as a welding business or seed dealerships), savings accounts, and investments (such as corporate stocks and bonds) are necessary because each one of these types of income is taxable to some extent. Also these receipts fit into the cash flow needs of the business-family unit.

Records of certain expenditures, those that can legally be used to reduce taxable income, are the minimum necessary. Often those records are called upon to substantiate claims during an IRS audit. For the cash flow budget to be complete, it is also necessary to maintain at least a total amount for family living expenditures on a periodic basis. It is also necessary for cash flow purposes to have summaries of savings accounts and investment transactions and liabilities.

Finally, it is necessary to maintain records of farm or ranch products consumed by the family, in terms of both amount and value. The cost of raising these products was very likely deducted from the farm or ranch income as a business expense. If the expenses were deducted, then the value of the products consumed and thus not sold must be reported as business taxable income.

Example Family Record Forms

The big decision in family records is not so much which records to keep as in how much detail should the records be maintained? For income records, it is generally best to record receipts by type of income in as much detail as possible. There are a myriad of personal finance record and budget books available both commercially and from state extension services. Examples of two sufficient income record forms are shown in Figures 12–1 and 12–2.

The amount of detail in family expenditures records is generally a matter of personal preference, desire to budget family expenditures, or possibly a requirement of a particular lender. Again, a myriad of record books are available, but an example of a general family expenditure record is shown in Figure 12–3. A more detailed and specific record form is shown in Figure 12–4. Regardless of the choice of general or detailed records, the important point is to maintain records to meet the four needs listed in the previous section and to provide a periodic (usually a monthly) summary of total cash income and total cash expenditures.

Products furnished by the farm or ranch do not require greatly detailed records. The example shown in Figure 12–5 should be sufficient for most needs.

NON FARM INCOME AND EXPENSES

DATE	DESCRIPTION	Taxable Non Farm Income				Non-Taxable Receipts	DATE	DESCRIPTION	Taxable Non Farm Income				Non-Taxable Receipts
				Rental Property							Rental Property		
		Income	Expenses	Income	Expenses				Income	Expenses	Income	Expenses	

FIGURE 12–1. New Mexico Farm Record Book income record.

Date	Source of income	Wages and salaries[a]		Transfers from business[b]	Dividends, interest	Commis-sions, tax refunds, misc.	Gifts, inherit-ance
		Gross wages	Net wages				
	Total						

[a] The difference between gross and net wages may represent income tax and Social Security deductions, insurance, savings, or other forms of expenditures. These expenditure items should be entered in the proper expense account either monthly or at the end of the year. Perquisites subject to income tax should be included with gross wages.

[b] Transfers from the farm business or other business enterprises.

FIGURE 12–2. Illinois Family Account Book income record.

Source: Marilyn Dunsing and A. G. Mueller, *Illinois Family Account Book,* Illinois Cooperative Extension Service, University of Illinois.

PERSONAL EXPENDITURES

(Items on this page are necessary for tax returns and cash flow statements.)

DATE	DESCRIPTION	Tax Deductible Contributions	Medical Expenses	Medical Miles	Drugs Medicines	Insurance Premiums	Family Living Expendables	Family Living Capital
	Total							

DATE	DESCRIPTION	Tax Deductible Contributions	Medical Expenses	Medical Miles	Drugs Medicines	Insurance Premiums	Family Living Expendables	Family Living Capital
	Total							

FIGURE 12–3. Example general family expenditures record.

FAMILY LIVING EXPENDITURES for month of ________________

Enter amount of purchases in Column C if paid by cash or in Column B if charged; also enter the purchase in the proper Column, 1-12.

Charge accounts[a]		Cash purchases and payments[a]	Date	Item or kind of expense	Food purchases	Household operating expenses	Household equipment, furnishings
Payments on accounts A	Purchases charged B	C			1	2	3

[a] Column C ($________) plus Column B ($________) minus Column A ($________) equals the sum of Columns 1 through 12

FIGURE 12–4. Illinois Family Account Book expenditures record.

Source: Marilyn Dunsing and A. G. Mueller, *Illinois Family Account Book,* Illinois Cooperative Extension Service, University of Illinois.

Savings, life insurance, debts, interest, and business expenses (nonfarm) are entered directly on pages 31-34.

House repairs, rent	Clothing	Personal items	Entertainment, recreation, rec. equip.	Education, reading	Medical care, drugs	Church, welfare contributions	Gifts outside home	Transportation, automobile
4	5	6	7	8	9	10	11	12

($__________). For accounts on a cash basis, the total of Column C (cash purchases) will equal the sum of Columns 1 through 12.

NOTES-SUMMARY AND INVENTORY SECTION

Products Furnished by Farm for

Product	Family Consumption: Amount (pounds, gallons, etc.)	Family Consumption: Value	Hired Labor: Amount (pounds, gallons, etc.)	Hired Labor: Value
Milk				
Eggs				
Beef				
Pork				
Lamb				
Other				
Total value				

Inventory of Fertilizer, Fuels, Spray Materials, and Supplies

Description	Beginning of year		End of year	
	Weight	Value	Weight	Value
Total				
Operator share				

FIGURE 12–5. Furnished products record.

CHARGE ACCOUNTS

Date	Amount	Account Name & Number	Date	Amount	Account Name & Number

SUMMARY OF CHARGE ACCOUNTS

	Unpaid balance, first of month		Purchases charged this month, Col. B		Paid on charge accounts this month, Col. A		Unpaid balance, end of month
January		+		−		=	
February		+		−		=	
March		+		−		=	
April		+		−		=	
May		+		−		=	
June		+		−		=	
July		+		−		=	
August		+		−		=	
September		+		−		=	
October		+		−		=	
November		+		−		=	
December		+		−		=	

FIGURE 12–6. Charge accounts record.

Savings and investment records for the business were discussed in Chapter 8. Again, because business and personal investments and savings are not separated, and do not need to be, the business record book can serve both functions. If separate records are desired, the example shown in Figure 8–4 should fit most needs.

Personal liabilities should be recorded, but they are usually split into two parts: charge accounts and amortized loans. Summaries of charge accounts should be maintained with a record form such as the examples seen in Figures 12–6 and 8–8. Longer-term loans should be recorded in the same fashion as business loans. Those records were also discussed in Chapter 8 and a loan record was presented in Figure 8–9.

Summary

Family records are a very important part of the farm or ranch records system if the business is family owned and operated. A farm or ranch manager who thoroughly plans for everything except his family's food, clothing, housing, and personal needs has failed. An alternative to family records for the farm or ranch business is to pay the manager a consistent salary. But the dedication and purpose of keeping detailed business records should be carried over to family financial planning as well.

Frequently family financial information is requested on loan applications, both asset and liability records. The next step, tracking expenditures and receipts, can also be very helpful to family planning and to business planning. The principles of business record keeping discussed in the first eleven chapters of this book can be just as easily applied to family records as to business records. The pain of record keeping can be greatly reduced if two checking accounts, one business and one personal, are maintained.

PART III

Credit Application and Financial Statements

Obtaining credit and using it wisely are two parts of one of the most important financial problems faced by modern farmers and ranchers, primarily because of the great explosion of debt financing and increased production costs that occurred during the 1970s. That trend has continued into the 1980s. Because of increased risks of bankruptcy and increased cash flow problems caused by high debt loads and debt servicing requirements, lenders now require more financial documentation than ever before they will extend credit.

The financial documentation most commonly required by lenders is the set of three basic financial statements: the *balance sheet,* the *income statement,* and the *cash flow budget.* These three financial statements comprise a concise and nearly complete set of information necessary to analyze the business as a whole. Furthermore, they are prepared in a form that is recognized by all financial professionals; simply stated, they are written in the language of business.

While each lender has his own preference in terms of format for these statements (most lenders do not use a standardized format), one set of forms is rapidly becoming the standard: *The Coordinated Financial Statements for Agriculture.*[1] This system, which will be referred to as CFS, provides a complete, coordinated package of financial statements, supporting schedules, and budget forms that is applicable to a broad range of agricultural operations.

Part III will use the CFS system exclusively. The first three chapters describe the balance sheet, the income statement, and the cash flow statement. These three statements have very different uses and are built upon very different concepts, but they are still very closely related. That relationship will be explored in Chapter 16. Additionally, Chapter 16 will discuss projected (or pro forma) statements.

Individuals who need to prepare their own financial statements will find the *You and Your Financial Statements* series very helpful, especially if the CFS forms are used.[2] Two recent texts, *Understanding Financial Statements in Agriculture* and *How*

[1]Thomas L. Frey and Danny A. Klinefelter, *Coordinated Financial Statements for Agriculture,* 2nd ed. (Skokie, IL: AgriFinance, 1980).

[2]Thomas L. Frey and A. Gene Nelson with Elaine Cull, *You and Your Balance Sheet* (Skokie, IL: AgriBusiness Communications Division, Century Communications, 1981).

to Analyze Financial Statements in Agriculture, will also be very helpful in preparation and analysis of farm or ranch financial statements.[3]

Because of the nature of the various financial statements, a priority order for their completion can significantly reduce work load and confusion. The following list presents the order that best accomplishes the task of preparing financial statements.

1. Complete record book inventories of all assets and liabilities (see Chapters 6, 7, and 8) as of the beginning of the year.
2. Prepare a balance sheet at the beginning of the year.
3. Maintain cash transactions, physical production, labor, machinery and other records through the year (Chapters 5 and 9 through 12).
4. Complete record book inventories of all assets and liabilities as of the end of the year.
5. Estimate income and social security tax liability for the year and calculate income tax rate.
6. Prepare a balance sheet at the end of the year.
7. Prepare an income statement for the year.

At the end of the first year, the manager will have a complete record book and a set of *historical* (or actual) financial statements for his operation. The manager can then turn his attention to planning for the future. To finish the financial statement/records process, the following steps should be taken in order.

8. Project costs and returns for each commodity enterprise.
9. Develop a whole-farm or whole-ranch plan for the second year.
10. Project cash inflows and outflows and prepare the cash flow budget.[4]
11. Project a balance sheet (modified cash column only) to the end of the second year.
12. Project an income statement for the second year.

Projected financial statements are often called *pro forma* statements. Once Step 12 has been performed, a complete plan for the business (for the next year) has been developed. Then, the task of the records process turns back to maintaining the record book throughout the second year.

13. Maintain cash transactions, physical production, labor, machinery and other records through the year.
14. Monitor cash flow comparing projected values from Step 10 to actual values from Step 13.

The process then repeats itself beginning with Step 4.

Thomas L. Frey and A. Gene Nelson with Elaine Cull, *You and Your Income Statement* (Skokie, IL: Century Communications, 1983).

A. Gene Nelson and Thomas L. Frey, *You and Your Cash Flow* (Skokie, IL: Century Communicaitons, 1983).

[3]John B. Penson, Jr., and Clair J. Nixon. *Understanding Financial Statements in Agriculture* (College Station, TX: Agri-Information Corporation, 1983).

John B. Penson, Jr., and Clair J. Nixon. *How to Analyze Financial Statements in Agriculture* (College Station, TX: Agri-Information Corporation, 1985).

[4]Often it is useful to understand the working of the statement and to gain insight into past business practices. However, this will likely be done only once.

CHAPTER 13

The Balance Sheet

A balance sheet is often estimated by farmers and ranchers as they apply for operating loans or longer-term credit. If the balance sheet is prepared hurriedly (often it is estimated in the lender's office), many important details can be overlooked. Although that estimate may serve the purpose of obtaining the loan, a detailed balance sheet prepared in advance will portray a more serious, businesslike attitude on the part of the potential borrower. Prepared statements should also increase the probability of a successful loan application, will definitely shorten the time required to approve the loan and actually obtain the funds, will ensure that the proper amount of funds are requested, and will provide an accurate picture of the business for later analysis.

Preparation of a complete, accurate balance sheet (a good example is shown in Figure 13–1), especially one as complex and detailed as the CFS balance sheet, can be an awesome task if started without a complete, accurate set of records as described in Chapters 5 through 12 of this text. But if such a records system has been maintained, balance sheet preparation is a nearly mechanical process. In fact, that preparation can be programmed into a personal computer and the balance sheet developed with very little trouble or additional input.

A Snapshot

The balance sheet is called a snapshot statement or a *stock statement,* because it presents a snapshot picture of the business as of one particular date. It is not a moving picture, nor does it present information for a period of time. Rather, it states: "This is the status of the business on this day."

Usually a balance sheet is prepared on the last day of the calendar year (December 31) or fiscal year. However, a balance sheet can be developed for any day during the year and some companies (especially large publicly held corporations) will prepare a statement monthly or at least quarterly.

B1 B1

ASSETS		
Current Assets	**Modified Cost**	**Market Value**
Cash (on hand) or in checking	$ 102,000	$ 102,000
Savings accounts & time certificates	43,000	43,000
Hedging account equity		
Marketable bonds & securities (Sch. 1)	10,000	15,000
Notes & accounts receivable (good)		
Livestock & poultry to be sold: Kind / No. / Avg. Wt. / $/Unit / Value		
CALVES / 40 / 500# / $63 / 12,600	12,600	12,600
Crops and feed: Item / Quant. / Unit / $/Unit / Value		
84 COTTON / 150 / BALES / 290 / 43,500		
HAY / 100 / T / 80 / 8,000	51,500	51,500
Cash investment in growing crops: Crop / Acres / $/Ac. / Value		
Supplies (Sch. 2)	4,000	4,000
Prepaid expenses	5,000	5,000
Other		
TOTAL CURRENT ASSETS	$ 228,100	$ 233,100
Intermediate Assets		
Notes & accounts receivable (good)	$	$
Machinery, equipment, trucks (Sch. 9):		203,000
Cost or Basis $		
Less accumulated dep $	150,423	
Breeding stock (Sch. 4)	25,000	25,000
Retirement accounts (IRA, HR-10)		
Cash value of life insurance (Sch. 7)	6,000	6,000
Securities not readily mktable (Sch. 1)		
Personal & recreational vehicles (Sch. 3)	7,000	7,000
Household goods & personal effects	4,000	4,000
Other		
TOTAL INTERMEDIATE ASSETS	$ 192,423	$ 245,000
Fixed Assets		
Contracts & notes receivable	$	$
Farm real estate (Sch. 5):		1,200,000
Cost or Basis $		
Less accumulated dep $	600,000	
Non-farm real estate (Sch. 6)		
Other		
TOTAL FIXED ASSETS	$ 600,000	$ 1,200,000
TOTAL ASSETS	$ 1,020,523	$ 1,678,100

LIABILITIES AND NET WORTH		
Current Liabilities	**Modified Cost**	**Market Value**
Accounts payable: Labor $ ___ Feed & seed $ ___ Repairs ___ Fert. & chem ___ Storage ___ Fuel & oil ___ Other ___ Mach. hire ___	$ 8,000	$ 8,000
Medical & other personal		
Notes payable within 12 months (to whom, maturity, purpose) FIRST NATIONAL	75,000	75,000
Amount delinq. on notes payable		
Principal portion of longer term debts due within 12 months (Sch. 8):		
Intermediate	10,000	10,000
Long term	75,000	75,000
Principal delinq. (interm. & long term)		
Estimated accrued interest on:		
Accounts $ ___ Notes $ ___		
Intermediate liabilities $		
Long term liabilities $	14,265	14,265
Estimated accrued tax liability:		
Property	3,400	3,400
Real estate		
Employer payroll withholdings	5,120	5,120
Income & Soc. Sec.	35,677	35,677
Interest delinq. (interm. & long term)		
Accrued rents & lease payments		
Other (including relatives)		
Contingent income tax liability:		
Current assets		23,718
Marketable securities		1,000
TOTAL CURRENT LIABILITIES	$ 226,462	$ 251,180
Intermediate Liabilities (Sch. 8)		
(Principal due beyond 12 months)		
Notes payable	$ 20,000	$ 20,000
Sales contracts		
Life insurance policy loans		
Other		
Contingent income tax liability:		
Machinery		26,289
Breeding stock		5,000
Securities not readily mktable		
Contingent income tax liability & interest penalty on retirement accts.		
TOTAL INTERMEDIATE LIABILITIES	$ 20,000	$ 51,289
Long Term Liabilities (Sch. 8)		
(Principal due beyond 12 months)		
Mortgage on farm real estate	$ 350,000	$ 350,000
Land contracts		
Mortgage on non-farm real estate		
Other		
Contingent capital gains tax liability:		
Real estate		120,000
TOTAL LONG TERM LIABILITIES	$ 350,000	$ 470,000
TOTAL LIABILITIES	$ 596,462	$ 772,469
NET WORTH	$ 424,061	$ 905,631
TOTAL LIABILITIES AND NET WORTH	$ 1,020,523	$ 1,678,100

Working capital (CA÷CL) ______________ Debt to worth ratio (TL ÷NW) ______________

Current ratio (CA ÷ CL) ______________

FIGURE 13–1. Example CFS balance sheet.

Source: Thomas L. Frey and Danny A. Klinefelter, *Coordinated Financial Statements for Agriculture,* 2nd ed. (Skokie, IL: AgriFinance, 1980).

Almost by definition, a balance sheet can be prepared nearly instantaneously if a double-entry accrual accounting system is maintained. Closing and adjusting journal and ledger entries must be made in order to wrap the income and expense accounts into the net worth account (see Figure 2–1 for the account structure) and inventory changes into the inventory asset and the expense accounts, but once those operations are performed, the balance sheet is ready to be prepared. The process of adjusting and closing accounts is called the *trial balance*.

If an accrual accounting system is not maintained, the process of preparing a balance sheet is somewhat more difficult. But that process can be shortened significantly if all of the inventories described in Chapters 6, 7, and 8 have been developed first.

Business versus Consolidated Balance Sheets

A balance sheet may be prepared to portray just the business assets and liabilities or it may be prepared to portray the total of business plus personal/family assets and liabilities. Often lenders prefer a *consolidated* balance sheet for family farms and ranches, that is, they wish to see both the business and non-business affairs put together on one balance sheet. This desire arises from the intertwined nature of business and family financial concerns as noted in Chapter 2. (Think of savings accounts, automobiles, farm or ranch residences and their mortgages as a few examples among many.) Family assets are also often called upon to meet business debt obligations. As discussed in earlier chapters, the accounting entity for which financial statements are prepared must be carefully considered before work begins.

A = L + NW

The balance sheet is the classical representation of the fundamental accounting equation:

$$Assets = Liabilities + Net\ Worth$$

or

$$A = L + NW$$

The general structure of most balance sheets is common: assets are listed in the left-hand column and liabilities and net worth are listed in the right-hand column (Figure 13–2). This format comes directly from the underlying fundamental equation (A is on the left and L and NW are on the right) and has a solid basis in accounting theory. *Assets* represent everything owned by the business, family, or business/family unit. *Liabilities* and net worth represent all of the claims against those assets. In other words, assets are all of the physical or financial items *owned* by the business, liabilities are the portion of those assets committed or *owed* to other individuals, and net worth is the portion of the assets owned outright by the

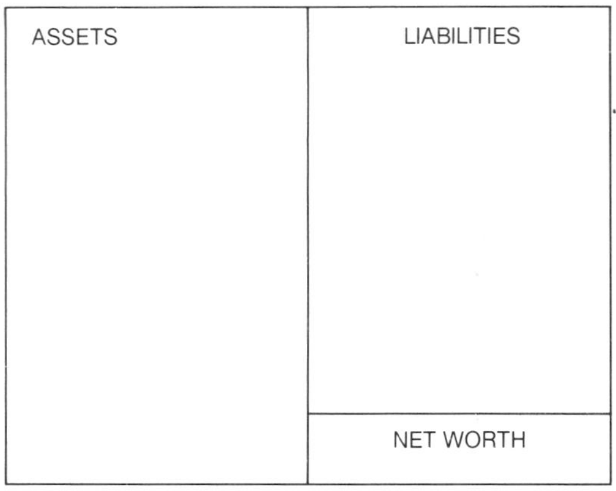

FIGURE 13–2. Generalized balance sheet format.

business owner. *Net worth* is often called the *residual claim,* that is, what the owner would have left over if all assets were converted to cash and all liabilities were paid.

Balance Sheet Classification

In addition to the left/right structure, most balance sheets are further structured, or *classified,* as depicted in Figure 13–3. Assets are normally classified into *current, intermediate* and *long-term* categories, as are the liabilities.

Asset Classification

Each asset is classified into one of the three categories based on two general rules: *convertibility* and *useful life*. Convertibility relates to *liquidity,* a financial term that describes how quickly the asset can be sold for cash. The most liquid assets are cash, checking accounts and savings accounts. Those funds are immediately available for use. Marketable securities too can be sold and their cash value received very quickly, as can grain inventories or market livestock. Consequently each of these assets would be classified as current assets. Other assets, such as inventories of feed, seed, chemicals, fertilizers, and other supplies; calves under weaning weight; expenses for insurance or grazing leases paid in advance; accounts receivable; and cash invested in growing crops are also classified as current assets even though in many cases they cannot be sold quickly. Each of these assets, however, will be used or ready for sale within the next year. In summary, assets are usually classified as current if they can be converted to cash very quickly or if they have a useful life of one year or less.

Subtle problems can arise in classifying certain assets as current even though they may have short lives and are easily converted into cash. Hay for feed, for example, is the same physically as hay for sale and both are almost always classified as current assets. Yet hay for feed is really a resource needed for

ASSETS	LIABILITIES AND NET WORTH
Current Assets	Current Liabilities
Intermediate Assets	Intermediate Liabilities
Long-Term Assets	Long-Term Liabilities
Total Assets	Total Liabilities
	Net Worth
	Total Liabilities and Net Worth

FIGURE 13–3. Classified balance sheet.

production of livestock. If it is sold to obtain cash in distress situations, a disruption of normal business and an overstatement of liquidity could occur. There is no easy answer to this type of classification problem. Convention is often the only answer.

Intermediate assets usually are somewhat more difficult to convert to cash or else they have a useful life between one year and about seven to ten years. Such items as machinery and equipment may take some time to sell. Furthermore, they are productive assets which are normally not held for resale but rather are owned in order to produce other commodities for sale. Similarly breeding, work, and dairy livestock are usually held as productive assets rather than primarily for sale and are thus classified as intermediate assets, even though these animals can often be sold at a local auction within a week or two. Other intermediate assets include notes receivable (money owed to you but not due within a year), personal vehicles, household furnishings, retirement accounts, cash value of whole-life insurance policies, and securities that are not readily marketable. Each of these assets requires a few months or so to sell for cash or has a relatively long useful life. Sometimes the classification process is somewhat arbitrary and is guided by tradition rather than a strict adherence to the definition.

Two of the intermediate assets listed above need a little more explanation: cash value of life insurance and secdurities not readily marketable. Whole-life insurance policies normally carry two investment components: the face value and the cash value. When a whole-life insurance contract is signed, the insurance company agrees to pay a certain amount of money (say $20,000 or $100,000) upon the death of the policyholder. This amount is called the *face value;* the policy remains in effect as long as premiums are paid. In addition to the face value, whole-life policies build a cash reserve and act just like savings accounts. A value, called the *cash value,* builds through time. That amount can be borrowed by the policyholder from the insurance company and it can be withdrawn entirely from the insurance company if the owner chooses to discontinue the policy. The other major type of life insurance policy is called *term life insurance*. Term insurance carries only a face value. If the policyholder discontinues payment of the premium for any reason, the entire insurance contract is canceled and all previous benefits and premium payments are forfeited. Consequently term life insurance builds no cash value that can be classified as an asset.

Securities not readily marketable are usually ownership stock in cooperatives (such as the local grain elevator, cotton gin, Farm Credit System (FCS), or herd improvement associations) or stock in closely held or family corporations and partnerships. Stock ownership is sometimes required for users of cooperative facilities and is always required for borrowers from the FCS. Cooperative stocks can usually be sold, but generally the owner must personally find someone else to buy them rather than sell them through registered stockbrokers. Some FCS stock must always be held when money is owned to them. Similarly, closely held corporation stock and partnership shares or other assets definitely have value, but often it is very difficult to find a buyer.

The final asset category is long-term (or fixed) assets. The assets that are classified as long-term normally are difficult to sell, require a significant amount of time to be sold, or have a useful life longer than 7 to 10 years. Like intermediate assets, the typical examples are assets used in the production of other commodities, but include such items as land and buildings. Notes receivable due beyond 7 to 10 years are also classified as long-term assets.

Liability Classification

Parallel to assets, liabilities are also classified into current, intermediate, and long-term categories. But the classification of liabilities is generally easier than that of assets. Current liabilities include accounts and notes payable and accrued liabilities that are or will be due within one year. Usually the amount of principal or longer-term notes due within one year is also classified as a current liability.

Any liability due beyond one year is classified as either an intermediate or a long-term liability. While the 7-to-10-year distinction between intermediate and long-term liabilities can be used, liabilities are normally classified according to their purpose. For example, a loan taken out to buy land is classified as a long-term liability because land is a long-term asset, regardless of the length of the loan.

Net Worth

Net worth is virtually never classified as current, intermediate, or long-term. It is calculated as one lump sum figure.

Valuation of Assets

Before turning to preparation of the balance sheet, one major issue needs to be discussed: the valuation of assets. Whether assets are valued on the basis of their original purchase price minus accumulated depreciation (*book value* or *adjusted basis*) or on the basis of their *current fair market value* has great consequences for credit application, for potential taxes, and for later analysis. The inventory valuation methods discussed in Chapter 6 (LIFO and FIFO) have little bearing on this problem. The real concern is the valuation of the big-ticket items such as land, other real estate, and major pieces of machinery. These items do not turn over (are not bought, sold, or used up in the production process) as rapidly as feed, grain, or supplies inventories. Consequently, LIFO-versus-FIFO valuation theory helps very little.

The valuation problem has grown primarily as a result of the inflation in the general economy and the agricultural economy experienced throughout the 1960s, 1970s, and 1980s. In a noninflationary economy, the fair market value of an asset is relatively close to its book value. But during inflationary times, many assets appreciate in dollar value. Consequently, the declining depreciated value carried on the business records has very little relevance to current conditions. The inflation experienced in the U.S. has pointed out clearly that depreciation is solely a tax concept and no longer offers a true theoretical basis for representing value. The declining values common to U.S. farmland in the early 1980s showed the counter side of this situation. Loans evaluated solely on a fair market value that was falling caused many financial difficulties for agricultural producers and for lending institutions as well.

The Valuation Difference

As the difference between fair market value and book value grows, the consequences of valuation on the balance sheet grow. Again, there are three primary consequences: credit, taxes, and analysis.

Lenders will generally prefer that the big-ticket assets be listed on the balance sheet with their fair market value. When extending credit, one of the primary concerns of the lender is to protect himself, his institution, and his depositors. As a result, the lender is concerned with the potential borrower's ability to repay the loan and also, in the worst possible case, if the borrower defaults on his loan, whether there is sufficient value in assets (called *collateral*) that can be sold to recover the lender's capital. In the event of this sale (called *foreclosure*), the value that can be realized is the fair market value, not the book value.

Quite differently from lenders, tax authorities are interested in both the market value and the book value. The main reason for this concern is that the difference between what an asset is sold for and its depreciated book value is a taxable gain. Consequently, for tax purposes, both values should be used, and the potential tax on the gain should be calculated. This issue will be discussed further in a later section of this chapter (Contingent Liabilities) and in Chapter 19.

When analyzing balance sheets for business analysis and management needs, both values are again needed. The market value represents a true picture of the current worth of the business owner, for purposes of analysis just as for those of the lender as discussed previously. But, when analyzing the pattern of growth or change in business assets, liabilities and net worth, the influences of real reinvestment growth must be segregated from apparent dollar "growth" caused solely by inflation. Real growth in a business comes primarily from reinvested profits. To prove this, we really only need to study the relationships among the major types of accounts described in Chapter 2 and shown in Figure 2–1. Only through earnings (the excess of revenues over expenses) can net worth be increased. This point will not be labored here, but should become clear in Chapter 16.

Double-Column Balance Sheets

The issue of whether market value or book value should be used in balance sheet preparation can actually be solved very easily: record them *both*. The CFS balance sheet and others have adopted one of the simplest solutions to the dual valuation problem by requiring that two values be recorded for each asset and each liability, and consequently for net worth. Two data columns are provided as shown in Figure 13–4, one for modified cost or book value, and one for market value.

Technically, book value is not always used in the modified cost column. Some items, such as raised livestock and raised crops inventories, have a zero book value since the costs of producing these commodities has already been deducted from income as an expense. Furthermore, the book value of personal assets, such as household furniture, cannot be depreciated for tax purposes; consequently its original purchase price remains its book value forever. But seldom are those items sold at a gain, and a loss on the sale of personal assets is not deductible either. A compromise that is commonly used records market value under both columns for raised commodities and for personal assets. Any changes in value for business inventories will be accounted for in accrual net income and any value changes for personal assets will be adjusted for in determining family withdrawals for the business. These concepts will also be explained in greater detail in Chapter 16.

Valuation Methods

For virtually all current assets, the market value and the modified cost value will be the same. This condition occurs because of the nature of current assets: they are in cash or have been purchased recently or can be converted into cash quickly. The major exception under current assets is marketable securities. If they were purchased more than a few months previously, the purchase price of these securities is very likely to be different from their current market value. They can also be sold relatively quickly. Furthermore, the gain on sales of marketable securities is taxable and any loss (market value less than purchase price) is deductible on the tax return.

To value current assets, published lists of current grain and livestock prices can be consulted. Cash and savings accounts already have distinct dollar values.

ASSETS	Modified Cost	Market Value	LIABILITIES AND NET WORTH	Modified Cost	Market Value
Current Assets			Current Liabilities		
Intermediate Assets			Intermediate Liabilities		
Long-Term Assets			Long-Term Liabilities		
Total Assets			Total Liabilities		
			Net Worth		
			Total Liabilities and Net Worth		

FIGURE 13–4. Double-column classified balance sheet.

Expenses paid in advance and classified as prepaid expense assets will be valued at the price paid. Most supplies inventories will be used within a production process in a short time and typically will represent only a minor portion of farm and ranch assets, so their value can be stated as the price actually paid or current prices can be obtained from input suppliers.

The valuation of intermediate and long-term assets is somewhat more difficult and usually more important. The modified-cost side is relatively easy if a complete depreciation schedule has been maintained. The value to be entered is the depreciated book value from the depreciation schedule. The market value must be estimated by appraisal; for machinery and equipment, the valuation guides mentioned in Chapter 7 may be used (see p. 94). Financial assets such as accounts and notes receivable are recorded at face value, since they are always dollar-denominated. The cash value of life insurance can be determined easily through the insurance agent or company. And finally, personal vehicles and household goods can usually be determined through valuation guides or personal experience. Each of these last four categories of assets (accounts and notes receivable, cash value of life insurance, personal vehicles, and household goods) normally carries the same value in both the modified-cost column and the market value column.

Similarly, the same value of securities not readily marketable is recorded under both columns, since such securities (especially FCS stock) are usually carried at par value (the original stated price). For ownership rights in closely held or family corporations and partnerships, however, the valuation problem can be severe and may only be solved by an independent appraisal or an adjustment of a past appraisal. Typically, a family corporation or a partnership that extends beyond the immediate family will prepare its own balance sheet. The owned portion of the corporation's net worth will be transferred to the individual's balance sheet.

The value of breeding, work, and dairy livestock is often taken to be current fair market value, again for both columns. The primary reason for this compromise is that much of this type of livestock was actually raised rather than purchased and consequently has a zero book value. Practice dictates that the market value be listed even in the modified-cost column.

Liability Valuation

Since virtually all liabilities are dollar-denominated, that is, the borrower is required to repay in cash, the valuation problem found for assets vanishes. Liabilities will consequently carry the same value in the modified cost column as they do in the market value column. However, some liabilities, the *contingent tax liabilities,* will appear only in the market value column because they arise solely as a result of the difference in values of the underlying assets. Contingent tax liabilities are simply potential taxes that will be due *if* a specific action occurs. For example, assume that land was puchased for $1,000 and is now worth $3,000. According to current income tax law, the $2,000 value gain is *not* taxable until the land is sold. So there is no liability due. But *if* the land is sold, *then* taxes on the $2,000 gain will be due. Consequently, the tax liability is contingent upon sale of the land. It occurs solely because of the inflated market value and will be entered in the market value column only. The calculation of contingent liabilities will be discussed further in a later section.

Net Worth

Net worth is calculated as the residual difference between assets and liabilities. As a result, two values of net worth must be calculated, the modified-cost net worth and the market value net worth. But, since they are calculated as residual differences, there is no difficulty in calculating the two values. There is a problem in interpreting what the two values mean, though. That interpretation will be discussed in Chapter 16.

Preparation of the CFS Balance Sheet

Many lenders, farmers, ranchers and other business people use preprinted balance sheet forms. The primary advantages of these forms are that they have been prepared by accounting experts to maintain theoretical consistency with the most common accounting practice, they are usually readily available, and they provide a common format and language for credit and management analysis.

As noted earlier, one of the best sets of preprinted financial statements for agriculture is the *Coordinated Financial Statements for Agriculture*.[1] This set of financial statements presents a format for recording and summarizing most of the information needed for overall business management and credit analysis. The format is particularly appealing since it has been designed specifically for American farms and ranches, is theoretically sound, and presents a dual valuation analysis. The CFS has been adopted by lenders, professional farm and ranch managers, farmers, ranchers and others in recent years.

The Balance Sheet and Its Schedules

The CFS balance sheet is prepared on two basic pages with ten supporting schedules. The balance sheet and those schedules are presented in Figures 13–5 through 13–12.[2] Each of the supporting schedules provides a place to record the information necessary to develop and to justify the information recorded on the front side of the balance sheet.

Much of the information called for by the schedules has already been developed within the overall farm or ranch records system, as discussed in Chapters 5–12. Consequently, the schedules can be prepared easily if the records systems have been maintained. In fact, much of the work involved in preparing these schedules may seem redundant since the data is already recorded elsewhere. However, transferring that information to the CFS schedules can be very useful as it can provide background data and justification in an organized fashion for the lender or management analyst.

Since information is presented in detail in the schedules and summarized for transfer to the balance sheet, preparation of the balance sheet can be simplified significantly if the schedules are prepared *first*. After the schedules have been prepared, the summary information can be transferred to the balance sheet as it is needed.

Schedule 1: Securities. A complete listing of all marketable securities and not-readily-marketable securities should be gathered in Schedule 1 (see Figure 13–7). The listing should include purchase date, whether the securities are owned outright or shared with someone else, the purchase price (cost or basis), and the current fair market value. As noted earlier, the value of marketable securities will generally differ between the cost and market value approaches and will be listed as a current asset. Securities which are not readily marketable may or may not have different values depending upon the type of security. These securities will later be recorded as intermediate assets.

Schedule 2: Supplies. Supplies inventories were discussed in detail in Chapter 6 and should be summarized in Schedule 2 (see Figure 13–7). While it may not be

[1]Thomas L. Frey and Danny A. Klinefelter, *Coordinated Financial Statements for Agriculture*, 2nd ed. (Skokie, IL: AgriFinance, 1980).
[2]Additional assistance in preparation of CFS balance sheets can be found in: Frey, Nelson, and Cull, *You and Your Balance Sheet*, (Skokie, IL: Century Communications, 1981).

B1 **B1**

☐ Personal
☐ Business
☐ Consolidated

BALANCE SHEET

As of ____________, 19____

Name ____

Address ________________________________ Phone ____________

ASSETS

Current Assets	Modified Cost	Market Value
Cash (on hand) or in checking	$	$
Savings accounts & time certificates . . .		
Hedging account equity		
Marketable bonds & securities (Sch. 1)		
Notes & accounts receivable (good)		
Livestock & poultry to be sold:		

Kind	No.	Avg. Wt.	$/Unit	Value

Crops and feed:

Item	Quant.	Unit	$/Unit	Value

Cash investment in growing crops:

Crop	Acres	$/Ac.	Value

	Modified Cost	Market Value
Supplies (Sch. 2)		
Prepaid expenses		
Other		
TOTAL CURRENT ASSETS	$	$
Intermediate Assets		
Notes & accounts receivable (good)	$	$
Machinery, equipment, trucks (Sch. 9):		
Cost or Basis . . . . $______		
Less accumulated dep . . . . $______		
Breeding stock (Sch. 4)		
Retirement accounts (IRA, HR-10)		
Cash value of life insurance (Sch. 7)		
Securities not readily mktable (Sch. 1) . .		
Personal & recreational vehicles (Sch. 3)		
Household goods & personal effects		
Other		
TOTAL INTERMEDIATE ASSETS	$	$
Fixed Assets		
Contracts & notes receivable	$	$
Farm real estate (Sch. 5):		
Cost or Basis . . . . $______		
Less accumulated dep . . . . $______		
Non-farm real estate (Sch. 6)		
Other		
TOTAL FIXED ASSETS	$	$
TOTAL ASSETS	$	$

LIABILITIES AND NET WORTH

Current Liabilities	Modified Cost	Market Value
Accounts payable:		
Labor $______ Feed & seed $______		
Repairs ______ Fert. & chem ______		
Storage ______ Fuel & oil ______		
Other ______ Mach. hire ______	$	$
Medical & other personal		
Notes payable within 12 months (to whom, maturity, purpose)		
Principal portion of longer term debts due within 12 months (Sch. 8):		
Intermediate		
Long term		
Estimated accrued interest on:		
Accou[illegible] $______ Notes $______		
In[illegible] liabilities . . . . . $______		
[illegible]ilities . . . . $______		
[illegible]ccrued tax liability:		
[illegible]ty		
[illegible]al estate		
Employer payroll withholdings		
Income & Soc. Sec.		
Accrued rents & lease payments		
Other (including relatives)		
Contingent income tax liability:		
Current assets		
Marketable securities		
TOTAL CURRENT LIABILITIES	$	$
Intermediate Liabilities (Sch. 8)		
(Principal due beyond 12 months)		
Notes payable	$	$
Sales contracts		
Life insurance policy loans		
Other		
Contingent income tax liability:		
Machinery		
Breeding stock		
Securities not readily mktable		
Contingent income tax liability & interest penalty on retirement accts. . . .		
TOTAL INTERMEDIATE LIABILITIES	$	$
Long Term Liabilities (Sch. 8)		
(Principal due beyond 12 months)		
Mortgage on farm real estate	$	$
Land contracts		
Mortgage on non-farm real estate		
Other		
Contingent capital gains tax liability:		
Real estate		
TOTAL LONG TERM LIABILITIES	$	$
TOTAL LIABILITIES	$	$
NET WORTH	$	$
TOTAL LIABILITIES AND NET WORTH	$	$

SAMPLE

Developed by Thomas L. Frey and Danny A. Klinefelter, Dept. of Agricultural Economics, Cooperative Ext. Service, University of Illinois. ©1980 AGRI FINANCE. Available from Agri Finance, 5520-G Touhy Ave., Skokie, IL 60077.

FIGURE 13–5. CFS balance sheet.

Source: Thomas L. Frey and Danny A. Klinefelter, *Coordinated Financial Statements for Agriculture,* 2nd ed. (Skokie, IL: AgriFinance, 1980).

1. Your age __________ Physical condition ________________

2. Ages of dependents ______________________________

3. Marital status: (Do not complete if this is an application for individual unsecured credit not primarily for agricultural purposes, unless you reside in a community property state or property upon which you are relying as a basis for repayment of the credit requested is located in a community property state.)
☐ married ☐ separated
☐ unmarried (single, divorced, or widowed)

4. Spouse or other person: Complete only if (check as applicable) a spouse or other person is a;
☐ co-applicant, or ☐ the applicant is relying upon the income or assets of a spouse or other person for repayment of the loan.

Name ______________________________

Address ______________________________

Employer ______________________________

Involvement in farm operation ________________

5. Form of business organization:
☐ proprietorship ☐ partnership ☐ corporation

6. If applicant is or includes a partnership or corporation:

Explain type and percent ownership ________________

____

Have you provided the lender a current borrow...
☐ Yes ☐ No

7. Have you ever or are you now involved in bankruptcy
☐ Yes ☐ No

8. Have you filed all necessary state & federal income tax returns?
☐ Yes ☐ No

9. Are any taxes delinquent or under dispute? ☐ Yes ☐ No

10. Are you a defendant in any suits or legal actions or are any judgments or mechanic's liens outstanding? ☐ Yes ☐ No

11. Have you prepared a will and/or estate plan? ☐ Yes ☐ No

How recently? ______________________________

12. List off-farm employment, if any: ________________

____

13. List any contingent liabilities as endorser, co-signor or guarantor (names and amounts):

____

____

____

14. Insurance on buildings $ __________

15. Insurance on machinery, equipment, and livest... $ __________

16 ...ops $ __________

...insurance coverage $ __________

...you carry:

Life insurance (Sch. 7)	☐ Yes	☐ No
Medical and hospital insurance	☐ Yes	☐ No
Disability insurance	☐ Yes	☐ No
Workmens' compensation	☐ Yes	☐ No

If yes
number of paid employees . . . . __________

STATEMENT OF OWNER EQUITY

Net worth, beginning of period (modified cost)	$	________
Net income for period	+	________
Gifts and inheritances received for period	+	________
Additions to paid-in capital (for partnership or corporation)	+	________
TOTAL AVAILABLE	$	________
Less: Gifts made for estate transfer	−	________
Net worth, end of period (modified cost)	−	________
EQUALS NET WITHDRAWALS (transfer to schedule 10, 12, Income Statement)	$	________

SUMMARY

Period	Market Value Net Worth	Cost or Basis Net Worth	Net Worth Change Caused by Value Change
Ending	$________	$________	
Beginning	________	________	
Change	(a) $________	(b) $________	(a-b) $________

The above information and the statement on the reverse are furnished for the purpose of securing and maintaining credit and are certified to be complete and correct. The undersigned authorize the below named lender to make all inquiries deemed necessary to verify the accuracy of the information contained herein to determine my (or our) creditworthiness and to answer questions about his credit experience with me (or us). I (or we) agree to notify said lender promptly of any material change herein.

Prepared for (lender): ______________________ Date Signed __________, 19____

Applicant ______________________ Co-applicant ______________________

FIGURE 13–6. CFS balance sheet (supporting and summary information).

Source: Thomas L. Frey and Danny A. Klinefelter, *Coordinated Financial Statements for Agriculture,* 2nd ed. (Skokie, IL: AgriFinance, 1980).

B2 **B2**

SCHEDULE 1 — SECURITIES

No. of Shares or Face Value	Description	Date Acquired	Percent Owned	Cost or Basis Per Unit	Cost or Basis Total	Current Market Value Per Unit	Current Market Value Total
			MARKETABLE				
				$	$	$	$
				TOTAL	$		$
			NOT READILY MARKETABLE				
				$	$	$	$
				TOTAL	$		$

SCHEDULE 2 — SUPPLIES

Item	Quantity	Value
	TOTAL	$

SCHEDULE 3 — PERSONAL & RECREATIONAL VEHICLES

Make & Model	Year	Value
		$
	TOTAL	$

SCHEDULE 4 — BREEDING STOCK

Number	Description	Age	Wt.	$/Unit	Value
				$	$
				TOTAL	$

FIGURE 13–7. Schedules 1 through 4, CFS balance sheet.

Source: Thomas L. Frey and Danny A. Klinefelter, *Coordinated Financial Statements for Agriculture,* 2nd ed. (Skokie, IL: AgriFinance, 1980).

SCHEDULE 5 — FARM REAL ESTATE

OWNED:	Legal Description	Title Held in Name of	Percent Owned	Date Acquired	Acres	Cost or Basis	Accumulated Depreciation	Book Value	Estimated Mkt. Value
Tract 1 Land						$	$	$	$
Residence									
Service bldgs.									
Improvements									
Tract 2:									
				TOTAL		$	$	$	$

RENTED: Legal Description	Landlord	Type of Lease	Expires	Acres	Annual Cash Rent
					$
			TOTAL		$

SAMPLE

SCHEDULE 6 — NON-FARM REAL ESTATE

Description	Title Held in Name of	Percent Owned	Date Acquired	Cost or Basis	Accumulated Depreciation	Book Value	Estimated Mkt. Value
				$	$	$	$
			TOTAL	$	$	$	$

FIGURE 13–8. Schedules 5 and 6, CFS balance sheet.

Source: Thomas L. Frey and Danny A. Klinefelter, *Coordinated Financial Statements for Agriculture,* 2nd ed. (Skokie, IL: AgriFinance, 1980).

B3 B3

SCHEDULE 7 — LIFE INSURANCE

Insurance Company	Beneficiary	Type of Policy	Face Value	Cash Value
			$	$
		TOTAL	$	$

SCHEDULE 8 — INTERMEDIATE AND LONG TERM LIABILITIES

	To Whom	Purpose and/or Security	Due Date	Interest Rate	Payment Dates	Accrued Interest	Current Principal Balance	Portion of Principal	
								Due within 12 months	Due beyond 12 months
INTERMEDIATE									
Notes						$	$	$	$
					TOTAL	$	$	$	$
Sales Contracts						$	$	$	$
					TOTAL	$	$	$	$
Life Insurance Policy Loans						$	$	$	$
					TOTAL	$	$	$	$
Other						$	$	$	$
					TOTAL	$	$	$	$
LONG TERM									
Farm Real Estate Mortgages						$	$	$	$
					TOTAL	$	$	$	$
Land Contracts						$	$	$	$
					TOTAL	$	$	$	$
Non-Farm Real Est. Mortgages						$	$	$	$
					TOTAL	$	$	$	$
Other						$	$	$	$
					TOTAL	$	$	$	$

SAMPLE

FIGURE 13–9. Schedules 7 and 8, CFS balance sheet.

Source: Thomas L. Frey and Danny A. Klinefelter, *Coordinated Financial Statements for Agriculture,* 2nd ed. (Skokie, IL: AgriFinance, 1980).

SCHEDULE 9 — MACHINERY, EQUIPMENT AND TRUCKS

Item	Year, Make & Model	Date Acquired	Percent Owned	Cost or Basis	Accumulated Depreciation	Book Value	Estimated Mkt. Value
				$	$	$	$
					$	$	$

FIGURE 13–10. Schedule 9, CFS balance sheet.

Source: Thomas L. Frey and Danny A. Klinefelter, *Coordinated Financial Statements for Agriculture,* 2nd ed. (Skokie, IL: AgriFinance, 1980).

B4 **B4**

SCHEDULE 10 — CONTINGENCY WORKSHEET

WORKSHEET TABLE FOR DETERMINING AMOUNT OF CURRENT ASSETS FOR CONTINGENT TAX LIABILITY CALCULATION

CURRENT ASSETS (from Balance Sheet)

Hedging account equity **minus** deposits		$ ________
Notes and accounts receivable (good)		________
Livestock and poultry to be sold: Balance sheet value $________		
Less purchase cost $________		
Net Value		________
Crops and feed		________
Cash investment in growing crops		________
Supplies		________
Prepaid expenses		________
TOTAL CURRENT ASSETS THAT WO'	**(a)**	**$** ________

SAMPLE

CURRENT LIABILITIES (from Balan\

Accounts payable		$ ________
Medical and other personal (only those deductible on Federal income tax schedule A)		________
Estimated accrued interest (total)		________
Estimated accrued tax liability:		
Property		________
Real estate		________
Employer payroll withholdings		________
Income (**only** state or local)		________
Accrued rents and lease payments		________
TOTAL CURRENT LIABILITIES THAT COULD BE DEDUCTED	(b)	$ ________
Amount to calculate contingent tax liability on: (a-b)	(c)	$ ________
ORDINARY TAX RATE (d)	X	________%
CONTINGENCY TAX (e)	$	________

FIGURE 13–11. Contingent tax liability worksheet for current assets.

Source: Thomas L. Frey and Danny A. Klinefelter, *Coordinated Financial Statements for Agriculture,* 2nd ed. (Skokie, IL: AgriFinance, 1980).

**WORKSHEET FOR DETERMINING AMOUNT OF CONTINGENT TAX LIABILITIES
FOR ALL ITEMS EXCEPT CURRENT ASSETS**

MARKETABLE SECURITIES

Market value	$______
Book value	$______
Difference	$______
Capital gain rate	X______ %
Capital gain	$______
Ordinary tax rate	X______ %
Contingency tax	$______

MACHINERY

Market value	$______
Book value	$______
Difference	$______
Ordinary tax rate	X______ %
Contingency tax	$______

BREEDING STOCK (RAISED)

Market value	$______
Capital gain rate	X__
Capital gain	[illegible]
Ordinary tax rate	[illegible]
Contingency tax	___

…CK (PURCHASED)

[illegible]	$______
…e	$______
…ifference	$______
Ordinary tax rate	X______ %
Contingency tax	$______

REAL ESTATE

Market value	$______
Book value	$______
Difference	$______
Capital gain rate	X______ %
Capital gain	$______
Ordinary tax rate	X______ %
Contingency tax	$______

RETIREMENT ACCOUNTS

Market value	$______
Less int. penalty	$______
Taxable amount	$______
Ordinary tax rate	X______ %
Contingency tax	$______
Plus int. penalty	$______
Cont. tax & int. penalty	$______

NOT READILY
MARKETABLE SECURITIES
(Applicable only when cost and market values differ.)

Market value	$______
Book value	$______
Difference	$______
Capital gain rate	X______ %
Capital gain	$______
Ordinary tax rate	X______ %
Contingency tax	$______

FIGURE 13–12. Contingent tax liability worksheet or capital assets.

Source: Thomas L. Frey and Danny A. Klinefelter, *Coordinated Financial Statements for Agriculture,* 2nd ed. (Skokie, IL: AgriFinance, 1980).

necessary to list every last screw, bolt, nail, and welding rod, the amounts of supplies on hand do indicate materials available for future use without future expenditure.

Schedule 3: Personal and Recreational Vehicles. Schedule 3 (see figure 13–7) should contain a listing of all personal and recreational vehicles and their current fair market value. Only the personal share of these vehicles should be listed here; the business portion will be depreciated and listed in Schedule 9. For example, many farmers and ranchers use an automobile both for personal purposes and for business purposes. Consequently, they will allocate a portion of that vehicle (say 50 percent) to the business and depreciate that percentage of the value of the car. The other half will be listed in Schedule 2.

The value of cars and recreational vehicles can be estimated fairly easily through personal experience or through such valuation guides as the *NADA Official Used Car Guide* and the *Recreational Vehicle Appraisal Guide*.[3] Usually the wholesale value of these vehicles will be listed since that is the value most likely to be received in a trade.

Schedule 4: Breeding Stock. The current market value of breeding, dairy, and work livestock should be listed in Schedule 4 (see Figure 13–7) along with a breakdown by type, age category, and weight. These values can be taken directly from the inventories described in Chapter 6. Caution should be exercised that these animals are not overvalued. Oftentimes a good producing cow seems to be "worth" more to her owner than she will actually bring at a sale. Sale barns, auctions, terminal markets, and production sales can provide general current market values for these animals.

Given the current structure of the CFS balance sheet and schedules, it is probably best to list only *raised* animals in Schedule 4. Purchased animals are depreciable and would be more properly listed along with machinery and equipment in Schedule 9.

Schedules 5 and 6: Real Estate. Valuation of real estate is a difficult task, as was discussed earlier in this chapter and in Chapter 7. Real estate inventories and values should be transferred from the record book to Schedules 5 and 6 (see Figure 13–8) with great care since the value of these items will typically far outweigh the accumulated values of other assets for most farms and ranches.

A benchmark for real estate values should be established through a professional appraisal. This benchmark is usually established when the ownership of the property is transferred through a sale, gift, inheritance, or transfer to a family partnership or corporation. Once a benchmark appraisal has been made, most lenders can assist in updating that appraisal.

When recording or transferring real estate descriptions and values to the CFS schedules, care should be taken to separate the real estate into one of five categories:

[3]The National Automobile Dealers Used Car Guide Company (McLean, VA) publishes several valuation guides, including *NADA Official Used Car Guide, Recreational Vehicle Appraisal Guide, Boat Appraisal Guide,* and *Mobile Home Appraisal Guide*.

1. Owned farm or ranch land
2. Owned personal real estate (such as the residence)
3. Owned farm or ranch depreciable property
4. Rented farm or ranch real estate
5. Nonfarm real estate

The first four property categories are entered in Schedule 5 and the fifth is entered in Schedule 6. The primary reasons for assigning real estate to one of these five categories are that:

1. Owned land and personal real estate are not depreciable.
2. Owned personal real estate on the farm or ranch is generally tied to the rest of the land and may have very little resale value without the rest of the property.
3. Rented real estate is not an asset to the renter, but it does indicate the size of the productive resource base controlled by the business.
4. Nonfarm real estate, such as apartment buildings or retirement/vacation homes, can be sold or otherwise separated from the farm or ranch without affecting the productive base of the business.

In addition to a description of the property (a legal description is generally preferable), the title holder and location of the title, the date acquired, and the cost or basis and market value of each item or parcel should be listed in Schedule 5 or 6. Accumulated depreciation and book value (cost minus accumulated depreciation) should also be listed for each piece of depreciable real estate (barns, corrals, fences, shops, storage facilities).

Schedule 7: Life Insurance. Every life insurance policy maintained by the owners of the farm or ranch business should be listed in Schedule 7 (see Figure 13–9). Both term and whole-life policies (see Chapter 8) should be included even though their asset values are widely different. A term life insurance policy has no asset value, except in the extreme catastrophe of loss of life, whereas a whole-life insurance policy carries a cash value that can be classified as an intermediate asset.

The face values of life insurance policies are noted in Schedule 7 primarily because they indicate cash available to transfer and maintain operation of the business upon death of the current owner. In the same vein, major disability insurance policies should also be listed in Schedule 7 since they too indicate liquidity in the event of a disabling catastrophe. Lenders, business analysts, and estate planners usually view life and disability insurance as a major positive factor for continued operation of the business.

Schedule 8: Intermediate and Long-Term Liabilities. One of the most difficult tasks in preparing a complete balance sheet is the calculation and allocation of intermediate and long-term liabilities. The principles described at length in Chapter 8 must be used to adequately complete Schedule 8 (see Figure 13–9). Intermediate and long-term notes are generally amortized (scheduled for repayment in a series of payments over a period of time). The amortization process most commonly used specifies that interest is calculated on the outstanding principal balance for each payment, all current interest is due with each payment so that immediately after the payment is made no interest is carried over or due, and there is a principal and an interest component in each payment.

The amortization process thus generates three types of liabilities:

1. Interest due as of the payment date or as of the date of balance sheet preparation (called accrued interest)
2. Principal due within 12 months
3. Principal due beyond 12 months

All interest accrued is classified as a current liability; principal due within 12 months is also classified as a current liability since it is due within a year; and principal due beyond 12 months is classified as either an intermediate or a long-term liability depending upon the life of the loan.

All information surrounding a loan should be recorded in Schedule 8 as follows:

1. The designation between intermediate and long-term should be made.
2. The type of loan should be designated (whether the loan is a note, sales contract, life-insurance policy loan, farm real estate mortgage, land contract, nonfarm mortgage, or other).
3. The contract terms (such as note mortgagor, due date, interest rate and payment dates) should be recorded.
4. The calculation of accrued interest and principal portions should be performed as discussed in Chapter 8 and as shown in Figure 8–10.

The separation of the types of loans and the classification (current, intermediate, or long-term) provides information for later credit and management analysis. Financial flexibility is reduced by large amounts of payments due within a short time and by large commitments against nonproductive assets.

Schedule 9: Machinery, Equipment, and Trucks. Schedule 9 (Figure 13–10) provides a format for recording original purchase price (basis), accumulated depreciation, remaining book value, and current fair market value for each item of machinery, equipment, and trucks. Schedule 9 is a direct representation of the depreciation schedule described in Chapter 7. It should also include purchased breeding, dairy, or work livestock and the business portion of automobiles as discussed in earlier sections of this chapter. But it should specifically *exclude* the depreciable real estate portion of the depreciation schedule since those items are listed in Schedule 5.

Because the depreciation schedule for a typical farm or ranch can be very lengthy, an acceptable alternative to rewriting all of the information called for is to attach a photocopy or carbon copy of the depreciation schedule to the CFS balance sheet package.

Schedule 10: Contingency Worksheet. The last schedules in the CFS package (Schedule 10, Figures 13–11 and 13–12) provide worksheets for computing contingent tax liabilities. These forms will be discussed in a later section in this chapter.

The Balance Sheet Form

Once the first nine schedules of the CFS Balance Sheet have been completed, attention can be directed toward preparation of the front side of the balance sheet. Much of the most difficult work has already been accomplished. The schedules plus inventories of market livestock, crops, feed, and accounts payable or accrued as described in Chapters 6, 9, and 10 will facilitate completion of the balance sheet.

Current Assets

Liquid assets such as cash, bank accounts, notes receivable, livestock and poultry to be sold, crops and feed, cash investment in growing crops, and prepaid expenses should be listed in the Current Assets section of the balance sheet (see Figure 13–5). Each of these items was recorded in the inventory section of the record book. The modified cost and the market value of each item will almost always be the same. In addition to these items, the total value of marketable securities should be transferred from Schedule 1 and the total value of supplies should be transferred from Schedule 2.

Intermediate and Long-term Assets

Assets that are less liquid or have longer useful lives than current assets are listed as intermediate and long-term assets. Intermediate assets such as notes and accounts receivable, retirement accounts, deferred annuities, and household goods need to be recorded in the intermediate assets sections. These three types of assets almost always carry the same value in the modified cost and market value columns. In addition to these three values, modified cost and market value totals should be drawn from the appropriate schedules as listed on the balance sheet and recorded in the proper column for the remaining intermediate assets: machinery, equipment and trucks (Schedule 4), breeding stock (Schedule 9), life insurance (Schedule 7), securities not readily marketable (Schedule 1), and personal and recreational vehicles (Schedule 3).

Similarly, long-term (or fixed) assets include any notes receivable, which again will carry the same value in both columns. Farm and nonfarm real estate summaries should be drawn from Schedules 5 and 6.

Liabilities

As noted earlier, the difference between modified cost and market value vanishes on the liabilities side of the balance sheet, except for the contingent liabilities. The CFS format (Figure 13–5) blocks out the contingent liabilities under the modified-cost column so that for each liability entry, the same value will be entered under each column when a blank line is presented.

Most of the information required under the liabilities sections comes from one of two sources: the records discussed in Chapter 8 or Schedule 8 of the CFS balance sheet (see Figure 13–9). The remaining information must be calculated: accrued liabilities and contingent liabilities.

Before proceeding to accrued and contingent liabilities, note the information necessary. *Accounts payable,* short-term credit usually from nonbank sources, should be listed. Accounts payable are often overlooked when balance sheets are prepared, but sometimes their amounts can be sizeable. All bills due to garages, elevators, supply companies, fuel distributors, custom operators, neighbors, and any others should be included. Similar to an account payable, a *note payable* is due within a short period of time and usually is not amortized. Usually, though, a note payable is due to a financial institution such as a commercial bank or the Farm Credit System and is more structured (usually an operating loan and/or line of credit) than a credit account with a supply business.

The principal and accrued interest on longer-term debts, regardless of when they are due, can be taken from Schedule 8.

Accrued Liabilities

Accrued liabilities generally must be calculated as of the day the balance sheet is prepared and usually cause difficulties because of the terminology. However, The concept of an accrued liability is not that difficult. Simply stated, an accrued liability is an amount of money actually owed, or an expense actually incurred, which does not yet technically need to be paid in cash.

To illustrate the concept, take an example of an hourly employee. Assume that this employee works eight hours per day five days per week, has agreed to wages of $15.00 per hour, and is paid by check once each week at the close of business on Friday. In other words, that employee earns $120.00 per day or $600.00 per week. To understand the concept of the accrued liability, consider the position of the employer at the end of the day on Wednesday. How much does he owe the employee? Technically the employer owes him nothing, because they have agreed to a weekly check on Friday. *But* the employee has actually earned $360.00, even though it is not payable until Friday. Consequently the employer has an accrued liability of $360.00 as of the close of business on Wednesday.

That example can be carried one step farther to illustrate other accrued liabilities. The employer must withhold both Social Security taxes and income taxes from this employee's wages. (Most farmers and ranchers do not have to withhold income taxes. More on this in Chapters 10 and 21.) The 1986 rates specify that the employer must withhold 7.15 percent of wages paid for social security taxes, the employer must contribute an additional 7.15 percent and the income tax withholding for a married employee with two children is $48.00 (see Figure 10–4). In summary, the employer has the following accrued liabilities as of the end of Wednesday:

Employee social security tax accrued liability:	.0715 × 360 =	$ 25.74
Employer social security tax accrued liability:	.0715 × 360 =	25.74
Income tax withholding tax accrued liability:	$80 × 3/5 =	48.00
Accrued employer payroll withholdings:		$ 99.48
Wages accrued liability ($360.00 − 25.74 − 48.00):		$286.26

In addition to wage and withholding tax accrued liabilities, typical types of accrued liabilities on farms and ranches include property and real estate, interest, rents and lease payments, and owner's income and social security taxes. The calculation of the amount of accrued liability for each of these types is similar to that of wages illustrated above. But a simple example for each may help solidify the concept.

Assume Fred Farmer pays $2,400 per year in property and real estate taxes. His bill is delivered by the County Assessor in October of each year. Fred pays his bill on November 1, 1986 for the county tax year, which runs from November 1, 1985 to October 31, 1986. Thus, he pays his real estate and property taxes in arrears. If Fred completes a balance sheet on December 31, 1986, and expects the same bill next year, he must account for the accrued liability since he has incurred taxes for two months but has not yet paid them. How much is his accrued liability on December 31? Fred has accrued taxes for 2 months at the rate of $200 per month ($2,400 ÷ 12 months). His accrued liability is thus $400.

Accrued rents and lease payments would work exactly the same, if the payment is made in arrears, or after the production year. Consider Ron Rancher, who pays his State Land Office $6,000 per year on October 1 for grazing rights on state lands.[4] As of December 31, Ron has an accrued liability of $1,500 ($6,000 ÷ 12 months = $500 per month; $500 per month × 3 months = $1,500).

Interest expenses were explained in detail in Chapter 8. Accrued interest on short-term notes is calculated in the same manner as interest on longer-term debts (interest rate × outstanding balance × portion of the year). Income and social security taxes accrued by the operator are somewhat more difficult to calculate because to adequately accomplish this task, the manager must either complete his tax return or estimate his taxes using a worksheet similar to the example shown in Figure 13–13.

Contingent Tax Liabilities

Contingent tax liabilities are the last major type of liabilities that must be calculated to complete a CFS balance sheet. The concept of the contingent tax liability arises solely because of the structure of the U.S. income tax system. Under that tax system, if an item is purchased for one price and later sold at a higher price, the difference between the two prices is called a *gain* and is taxable. If the item purchased was an investment asset (such as diamonds, gold, or IBM

[4] If Ron paid his grazing fees in advance, instead of in arrears, he would perform a similar analysis to calculate his prepaid expense rather than accrued liability. For example, assume Ron pays $12,000 each year to the Bureau of Land Management on October 1. He calculates a prepaid expense of $9,000 (12,000 ÷ 12 months = $1,000. $1,000 per month × 9 months remaining = $9,000).

RECEIPTS		Amount to Date	Estimated Rest of Year	Estimated Year's Total
Sales of products raised(*) and miscellanous receipts:				
Cattle, hogs, sheep and wool		$ ______	______	______
Poultry, eggs and dairy products		______	______	______
All crop sales		______	______	______
Custom work, prorations, refunds and agriculture program payments		______	______	______
Total sales and other income	(1)	______	______	______
Sales of purchased market livestock. . . . (**)$ ______				
Cost of purchased market livestock . . . .(***)$ ______				
Gross profits on sales of purchased livestock (sales - cost) . . .	(2)			______
Gross farm profits (Item 1 + 2).	(3)			______

EXPENSES

Labor hired	$ ______	Veterinary, medicine .	$ ______
Repairs, maintenance.	______	Gasoline, fuel, oil. .	______
Interest.	______	Storage, warehousing .	______
Rent.	______	Taxes.	______
Feed purchased.	______	Insurance.	______
Seed, plants purchased. . . .	______	Utilities.	______
Fertilizers	______	Freight, trucking. . .	______
Chemicals	______	Conservation expenses.	______
Machine hire.	______	Other.	______
Supplies purchased	______	Other.	______
Breeding fees.	______	Other.	______

Total cash expenses.	(4)	$ ______	______	______
Depreciation on machinery, improvements, dairy and breeding stock	(5)			______
Total deductions (Item 4 + 5).	(6)	______	______	______
Self employment income (Item 3 - 6)	(**)(7)	______	______	______
Net taxable gain from Schedule D (sales of dairy and breeding stock, machinery and other capital exchanges)	(8)			______
Taxable non-business income.	(9)	______	______	______
Adjusted gross income (Item 7 + 8 + 9)	(10)	______	______	______
Less: Excess itemized deductions, from Schedule A . . . . $ ______				
Personal exemptions. . . . . $ ______				
Total non-business deductions and exemptions	(11)	______	______	______
Taxable income (Item 10 - 11).	(12)	______	______	______
Estimated income tax (from applicable tax rate schedules).	(13)			______
Estimated self employment tax (Item 7 x current rate).	(14)			______
TOTAL TAX (Item 13 + 14)	(15)			______
Credits: allowable investment credit and carryover, gas tax, income tax withheld and estimated tax paid	(16)			______
Estimated tax due (Item 15 - 16)	(17)			______

----

* Omit for accrual method.
** For accrual method, adjust for change in inventory and new livestock purchases.
*** For accrual method, includes sales of all livestock.

FIGURE 13–13. Income tax estimate worksheet.

stock) or a business asset (such as land or buildings or machinery), the gain is called a *capital gain*. Capital gains are also taxable, but if they meet certain requirements (which will be discussed further in Chapter 19), they may qualify for special tax treatment (60 percent of the gain is excluded from taxation).[5]

Contingent tax liabilities should be calculated on each asset item that has a value in the modified cost column different from its value in the market value column. Figures 13–11 and 13–12 provide worksheets for calculating contingent tax liabilities on the various categories of assets. Throughout the calculation of these contingent tax liabilities, remember the following points:

1. Contingent tax liabilities arise because gains in value are taxable.
2. Contingent liabilities will not become due until another action is taken. In other words, a contingent tax liability must be paid *if* the business is sold.
3. Contingent tax liabilities apply only to the difference between market value and modified cost.
4. The capital gains tax rate is currently 40 percent of the ordinary income tax rate (60 percent exclusion).
5. The appropriate ordinary income tax rate is the rate found by estimating ordinary income (Figure 13–13).
6. Different sets of tax rules apply to different categories of gain.

To calculate the contingent tax liabilities, the worksheets presented in Figures 13–11 and 13–12 should be followed closely. Because different rules apply to different situations regarding basis, depreciation, depreciation recapture and qualification for capital gains treatment, the worksheets should be used. Additional explanation of the tax rules will be presented in Chapter 19. Before leaving contingent tax liabilities, let us walk through one example to drive home the concept and the calculations.

Assume that Phil Producer purchased his farm in 1971 for $100,000. He allocated $80,000 of the purchase price to land and $20,000 to existing buildings. Since 1971, he has deducted $14,000 in depreciation on those buildings. He had the farm appraised in 1986; the market value was estimated to be $200,000. He estimates his income tax bracket to be 35 percent. His contingent tax liability on this real estate is $15,960:

Market value	$200,000	
Book value	86,000	(80,000 + (20,000 − 14,000))
Difference	114,000	
Capital gain rate	× .40	
Capital gain	45,600	
Ordinary tax rate	× .35	
Contingency tax	$15,960	

[5]The Tax Reform Act of 1986 will change the method of calculating and taxing capital gains beginning in 1987 with the goal of eventually eliminating the capital gains exclusion.

Contingent tax liabilities may turn out to be very large amounts on commercial farms and ranches, especially if the major land resources were purchased or inherited several years ago. While they may seem like a fictitious liability or paper games, it must be remembered that the goal of calculating net worth on the balance sheet is to estimate a value that the owner would have left over if he were to sell all assets and pay all debts. Certainly the tax authorities will collect their share before the owner-seller gets to pocket his net worth.

Net Worth and Balance Sheet Summary

After all of the various asset and liability categories are evaluated, all that remains is to finish the front side of the balance sheet (Figure 13–5), calculate net worth, and complete the back side (Figure 13–6). Subtotals should be calculated for current, intermediate, and fixed assets and liabilities. These subtotals can be used for later analysis. Totals for both assets and liabilities should be calculated next, and net worth should be computed as the difference between assets and liabilities ($NW = A - L$). There will be different values for the cost and market value columns for assets and for liabilities and thus for net worth.

The top half of the back side of the CFS balance sheets is important since it contains quite a bit of information that lenders need to evaluate a loan application. A great part of the loan processing procedure depends upon the personal interview with the potential borrower. Consequently, truthfulness, a businesslike attitude, thoroughness, and accuracy are all very important to a successful loan application.

The bottom half of the balance sheet requires information from two successive years' balance sheets and an income statement. These relationships will be discussed in Chapter 16.

Summary

The balance sheet is one of the most important parts of a farm or ranch records system because it is one of the big three financial statements used by lenders and financial analysts. The balance sheet is a primary communication of financial information about a specific business; it is written in the language of business. There are several primary features to a balance sheet:

1. It is a snapshot, or stock, statement prepared for a point in time, not a period.
2. It can be prepared for the business or it can be consolidated to include business and family finances.
3. $A = L + NW$ sets the basic structure.
4. Most items are classified into current, intermediate, or long-term assets and liabilities.
5. Assets are classified based on useful life and convertibility (or liquidity).

6. Liabilities are classified according to a matching process—they are matched to an asset.
7. Net worth is the residual difference between assets and liabilities.

Valuation of assets is a large problem when preparing a balance sheet. This valuation problem is evident in two areas: choosing a market value for current assets such as grain or feeder livestock inventories and estimating a value for intermediate and long-term assets that appreciate. The latter area is more troublesome since it involves a lengthy and often difficult process of searching for appraisals and guide books. But since tax authorities will tax much of the gain (the difference between market value and book value) on the eventual sale of these assets, this valuation process is necessary.

Since liabilities are always financial obligations and thus dollar-denominated, a valuation problem does not arise. A market-value balance sheet, however, should include contingent liabilities, those liabilities that would come due *if* the operation was sold or some other specific action took place.

One of the best sets of preprinted financial statement forms for farm and ranch use is the *Coordinated Financial Statements for Agriculture* system. CFS consists of a dual valuation (double column) balance sheet, a form for calculation or change in net worth caused solely by value changes, a statement of owner equity, and a set of ten supporting schedules to list some of the data that support the summary figures on the front side of the balance sheet.

CHAPTER 14

The Income Statement

In addition to a balance sheet, many farmers and ranchers have also prepared an *income statement*. If an income statement is prepared on the cash (or modified cash) basis of accounting, then virtually every farmer and rancher has prepared this statement through Schedule F of his annual federal income tax return. To provide the best information for credit application and management analysis purposes, however, the income statement should be prepared on an accrual basis.

Although most farmers and ranchers maintain their records systems on a modified cash accounting basis, an accrual income statement is actually quite simple to develop. We will cover the step-by-step preparation later in this chapter. To develop an accrual income statement from a modified cash accounting system we will need:

1. A beginning balance sheet
2. A summary of receipts and expenditures (an income tax Schedule F or an estimate would be sufficient)
3. An ending balance sheet

The CFS income statement lays out this process in a simplified way.

As with the balance sheet, development of the accrual income statement from an accrual-based double-entry bookkeeping system is nearly instantaneous. The trial balance procedure, which adjusts and summarizes individual income and expense accounts, provides all of the information necessary to develop an income statement.

The income statement, if prepared with care, can be one of the most important building blocks in a farm or ranch records/business management analysis system. An example statement is shown in Figure 14-1. A series of income statements over a period of years provides a track record of the profitability of the business. Since it is developed on an accrual basis, it will not show cash difficulties, but it can point out the reasons for cash flow problems. Furthermore, the income statement is intricately tied to a set of balance sheets. In fact, it describes almost all of the

1 USDA-FmHA
Form FmHA 1930-3
(5-12-83)

INCOME STATEMENT

FORM APPROVED
OMB No. 0575-0061
Expires 4/86

☒ Actual
☐ Projected

For 12 Month Period
Ending December 31, 19 85

Name Frank Farmer

Address Star Route 2, Box 18 Anytown, AB 88061

REVENUE			
Crops and feed:			
cash sales . . . $	243,330 (1a)		
inventory change (Sch. 1) . . . (±)	+14,000 (1b)	$ 257,330 (1)	
Livestock & poultry sold:			
cash sales . . . $	112,074 (2a)		
inventory change (Sch. 1) . . . (±)	−10,000 (2b)	102,074 (2)	
Breeding stock:			
cash sales . . . $	750 (3a)		
inventory change (Sch. 1) . . . (±)	+2,000 (3b)	2,750 (3)	
Products: livestock and poultry — cash sales		(4)	
Custom work: cash		(5)	
Government payments and patronage dividends		4,000 (6)	
Income from hedging transactions (Sch. 2) . . . (±)		5,450 (7)	
Other		(8)	
Adjustment in notes and accounts receivable (Sch. 1) . . . (±)		(9)	
Gross revenue (Add lines 1 thru 9) . . . $		371,604 (10)	
Less livestock & poultry purchases (feeder & breeding) . . . (−)		55,000 (11)	
Less feed purchased . . . (−)		(12)	
VALUE OF FARM PRODUCTION			$ 316,604 (a)
EXPENSES			
Cash operating expenses (Sch. 3) . . . $		121,318 (13)	
Expense adjustment (unused assets) (Sch. 4) . . . (±)		+2,000 (14)	
Expense adjustment (unpaid items) (Sch. 5) . . . (±)		(15)	
Depreciation:			
Machinery and equipment . . . $	(16a)		
Fixed farm improvements . . . $	(16b)	60,167 (16)	
Total operating expenses (Add lines 13 thru 16)			183,485 (b)
Income from farm operations . . . (a−b)			$ 133,119 (c)
Less interest expense (Sch. 6) . . . (−)			63,431
Gain or loss on disposal of farm assets (Sch. 7) . . . (±)			+39,220
NET FARM INCOME			$ 108,908 (d)
NON-FARM INCOME			
Operator's wage off farm (net of expenses incurred) . . . $		5,000 (17)	
Spouse's wage off farm (net of expenses incurred)		28,000 (18)	
Interest and dividends (including int. on retirement accts.)		1,075 (19)	
Gain or loss on sale of non-farm assets (Sch. 8) . . . (±)		+230 (20)	
Net income — other farms		(21)	
Net income — non-farm real estate		(22)	
Other		(23)	
NET NON-FARM INCOME (Add lines 17 thru 23)			$ 34,305 (e)
Income before income taxes and extraordinary items . . . (d+e)			$ 143,213 (f)
Provision for income and Soc. Sec. taxes (all income) (Sch. 9)			38,677 (g)
Income before extraordinary items . . . (f−g)			$ 104,536 (h)
Extraordinary items (explain)			(i)
NET INCOME			$ 104,536

The information requested on this form is being collected in order (1) for FmHA to make loan and/or servicing decisions and (2) to provide management assistance to borrowers. The FmHA retains the form in the borrower's file for future reference. The information collected on this form is required in order for a borrower to obtain the benefits of a farmer program loan from FmHA.

FIGURE 14–1. Example of completed farm and income statement.

change in net worth from one balance sheet to the next. That relationship will be explored in greater detail in Chapter 16.

The Flow Statement

The income statement is often called a *flow statement* because it presents information over a period of time. In contrast to the balance sheet, the income statement is a movie rather than a still photograph. The balance sheet is prepared as of a given date, say December 31. It tells the position of the business on that day. On the other hand, the income statement is prepared for the year ending December 31 and describes the net results of the events that occurred throughout the year.

Usually, a farm or ranch income statement is developed once each year for the period of time between two successive balance sheets. It could be prepared more frequently, such as monthly or quarterly, but because the balance sheet is usually prepared only once each year and because of the seasonal patterns of agricultural production, an annual income statement is generally considered to be sufficient.

Purpose of the Income Statement

The primary purpose of the income statement is to calculate (and justify) the net income earned by the business over the course of the year. To accomplish this task, the general format of the income statement (represented in Figure 14–2) lays out the major sources of farm or ranch income. Income comes from two primary places: receipts and increases in marketable crop and livestock inventories.

From income, farm or ranch expenses are subtracted. Those expenses come from three primary places: expenditures, decreases in supplies inventories, and depreciation on machinery, breeding or dairy livestock, and buildings.

The difference between income and expenses from farm or ranch operations is called *net farm or ranch income*. This net income figure represents the net earnings from the farm or ranch for the year. In other words, it represents the net returns, which can be used for famly living expenses, for paying taxes, for further asset acquisition, or for repayment of debt. Net income is the return on the farm or ranch operator's labor, management, assets, and the risk taken in production.

Generally a farm or ranch income statement will continue to show any off-farm or off-ranch income and the expenses incurred to generate that income (but *not* family living expenses) and to calculate a net income figure for all activities of the farm or ranch family.

Accrual versus Cash Basis

As indicated previously, an income statement should be prepared on an accrual basis rather than on a cash basis. A cash-based income statement such as the

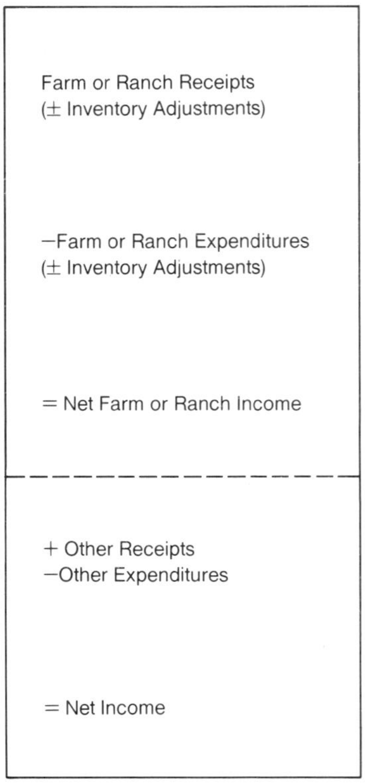

FIGURE 14–2. General income statement format.

Schedule F (Figure 6–5) or the example shown in Figure 14–3, is more properly called a cash flow summary rather than an income statement. Throughout the rest of this chapter and text, it will be assumed that every income statement is based on an accrual accounting foundation.

Adequate adjustments can be made to a cash accounting system to develop an accrual income statement, *if* inventory records are maintained. The CFS income statement lays out the inventory adjustment process very well and makes income statement calculations nearly painless. But before we describe the mechanics, a few brief examples of the reasoning should help convince us that the process is correct.

Inventory Change for Income

Consider a change in the inventory of corn raised by Fred Farmer. On January 1, Fred had 10,000 bushels of No. 2 yellow corn in storage. The prevailing market price was $3.00 per bushel. During the year, Fred sold those 10,000 bushels for $3.00 per bushel, raised 25,000 bushels on 250 acres and sold 5,000 of those new crop bushels for $3.25 per bushel. Consequently, at the end of the year, Fred held 20,000 bushels. The prevailing market price on December 31 was $3.25 per bushel.

The calculation of Fred's income from corn for the year would proceed as follows:

SUMMARY OF CASH FARM RECEIPTS

Line	Description	Source	A Raised op. share	B Purchased op. share	Operator share (p. 10, 11) C	Landlord share (p. 12, 13) D	Total farm E
101	Cattle	RA-RB					
102	Breeding cattle	RH-RI					
103	Total cattle			+	=		
104	Hogs	RC-RD					
105	Breeding hogs	RH-RI					
106	Total hogs			+	=		
107	________ livestock	RE-RF					
108	Breeding ________	RH-RI					
109	Total ________			+	=		
110	Eggs, milk, or wool	RG					
111	Total sales livestock and products (total lines 101 thru 110)						
112	Corn	RK					
113	Soybeans	RL					
114	Oats	RM					
115	Wheat	RM					
116	Hay and other crops	RN					
117	Other cash crops	RO					
118	Total sales crops (total lines 112 thru 117)						
119	Machine hire	RP					
120	Patronage dividends and refunds	RQ					
121	Other farm income	RS					
122		RT					
123	Timber, etc.	RU					
124	Rent received by landlord						
125	Total miscellaneous income (total lines 119 thru 124)						
126	Total cash sales	(Total lines 111, 118, 125)					
127	Capital sales: machinery and equipment	RJ					
128	Total farm receipts	(Total lines 126 and 127)					
129	Money borrowed	RW					
130	Total dollars received	(Total lines 128 and 129)					
131	Cash on hand January 1 (p. 78)						
132	Non-farm income, gifts, etc.						
133	Total dollars available (130 + 131 + 132)						
134	Total dollars accounted for (184)						
135	Cash on hand December 31 (133 − 134)						
			A	B	C	D	E

SUMMARY OF CASH FARM EXPENDITURES

Line	Description	Source	Operator share (p. 38, 39) F	Landlord share (p. 40, 41) G	Total farm H
151	Purchases of breeding stock: Cattle	EC			
152	Hogs	EC			
153		EC			
154	Purchases of feeder livestock: Cattle	ED			
155	Hogs	EE			
156		EF			
157	Feed purchased	EG			
158	Total purchases livestock and feed (lines 151 thru 157)				
159	Livestock expense	EH			
160	Hired labor	EJ			
161	Machine repairs	EK			
162	Gas, fuel, and oil	EL			
163	Machine hire	EM			
164	Auto expense	EN			
165	Electricity and telephone	EO			
166	Fertilizer and lime	EP			
167	Seeds and plants	EQ			
168	Crop supplies	ER			
169	Building repairs	ES			
170	Rent	ET			
171	Taxes	ET			
172	Insurance	EU			
173		EV			
174	Miscellaneous	EW			
175	Interest paid	EX			
176	Total cash operating expenses (lines 159 thru 175)				
177	Capital items: Building and improvements	EA			
178	Machinery and equipment	EB			
179	Total capital items purchased (lines 177 and 178)				
180	Total farm expenditures (lines 158 + 176 + 179)				
181	Principal payments	EY			
182	Total farm dollars spent (lines 180 and 181)				
183	Family and non-farm withdrawals				
184	Total dollars accounted for (182 and 183)				
			F	G	H

FIGURE 14–3. Summary of cash receipts and expenditures.

Source: Robert B. Schwart, *Illinois Farm Record Book*, Illinois Cooperative Extension Service, University of Illinois.

Corn Sales (Example 1)

10,000 bushels @ $3.00	$30,000	
5,000 bushels @ 3.25	+16,250	
Cash corn sales		$46,250
+ Inventory adjustment		
(Ending inventory) 20,000 bushels @ $3.25	$65,000	
(Beginning inventory) 10,000 bushels @ 3.00	−30,000	
Inventory increase		+$35,000
Corn income		$81,250

To calculate his income, Fred added together both sales to get $46,250. But the income from the 10,000-bushel sale actually belonged to the previous year, since the corn was produced in the previous year and placed in storage. Everything belongs to the previous year except the cash from the sale. Consequently, Fred should subtract the beginning inventory value of corn available for sale. By the same reasoning in the opposite direction, the 20,000-bushel ending inventory belongs to *this* year despite the fact that it has not yet been sold for cash, and it should be added to sales to calculate income. A slightly different process would calculate sales ($46,250) and adjust for inventory change (*ending inventory − beginning inventory*, $65,000 − 30,000 = $35,000, to calculate income [$81,250]). Inventory change for salable commodities should always add ending inventory and subtract beginning inventory, since ending inventory increases income and beginning inventory decreases income.

Fred's example assumed no price difference between sale and inventory, either for the beginning inventory or the ending inventory. If sale prices are different from inventory values, income would be calculated in a similar manner. For example, if the beginning inventory was valued at $3.00 per bushel, the 10,000 bushels sold @ $3.10, the 5,000 bushels sold @ $3.25, and ending inventory was valued at $3.15, income would be as follows:

Corn Sales (Example 2)

10,000 bushels @ $3.10	$31,000	
5,000 bushels @ 3.25	+16,250	
Cash corn sales		$47,250
+ Inventory adjustment		
(Ending inventory) 20,000 bushels @ $3.15	$63,000	
(Beginning inventory) 10,000 bushels @ 3.00	−30,000	
Inventory increase		$33,000
Corn income		$80,250

The analysis of Example 2 is similar to that for Example 1. However, in keeping with commonly accepted valuation methods for agricultural commodities, market price is always used. Consequently, Fred gained $.10 per bushel by storing his beginning inventory of corn after January 1.[1] But he lost $.10 per bushel for

[1] Florence R. Sneed, Clair J. Nixon, and Steven D. Grossman, "Auditing Agricultural Producers Cooperatives," *CPA Journal,* December 1983.

storing corn after harvest. In summary, Fred gained $1,000 ($31,000 − 30,000 or $.10 × 10,000 bushels) by storing his beginning inventory and lost $2,000 ($.10 × 20,000) by storing grain harvested during the year. Consequently, Fred realized a net loss of $1,000 from price movements throughout the year.

Both Examples 1 and 2 were based on an inventory increase. Using the prices from Example 1, assume that Fred sold 20,000 bushels of his 25,000-bushel new corn crop and ended the year with 5,000 bushels. His net income would again be $81,250:

Corn Sales (Example 3)		
10,000 bushels @ $3.00	$30,000	
20,000 bushels @ 3.25	+65,000	
Cash corn sales		$95,000
+ Inventory adjustment		
5,000 bushels @ $3.25	$16,250	
10,000 bushels @ 3.00	−30,000	
Inventory decrease		−$13,750
Corn income		$81,250

There will be *no* difference in income between Examples 1 and 3 because the only difference between the two examples is the timing of the sale of new crop corn. (However, a difference in income *could* result if prices changed between the inventory valuation date and the sale date.)

Inventory Change for Expenses

In addition to having inventories of corn, Fred also holds inventories of fertilizer because he likes to buy early to capture price advantages. Assume Fred had three tons of fertilizer on hand January 1, which he applied during the year. At the beginning of the year, fertilizer cost $248 per ton. During the year, Fred purchased seven tons of fertilizer for $245 per ton, spread three, and maintained the other four tons in ending inventory. Fred calculated his fertilizer cost for the year to be $1,479:

Cash expenditure		
7 tons @ $245		$1,715
+ Inventory adjustment		
(Beginning inventory) 3 tons @ $248	$744	
(Ending inventory) 4 tons @ 245	−980	
Inventory decrease		−236
Fertilizer expense		$1,479

Fred added his beginning inventory to cash purchases to represent the amount available for use in the year and subtracted the ending inventory since it is really a cost attributable to the following year. The inventory adjustment process adds cash expenditures to the difference between beginning inventory (+) and ending inventory (−) since beginning inventory increases costs for the year and ending inventory decreases costs for the year.

An opposite effect occurs if the expense item was carried on credit, on an account payable. Fred purchased feed on account from his supplier during the year; he had account balances with the dealer at both the beginning and at the end of the year. Assume Fred owed $3,000 at the beginning of the year, paid that amount during the year, and charged $4,000 that he had *not* paid by the end of the year. His feed expense for the year would be $4,000 (assuming no change in asset inventory of feed on hand):

Cash expenditure (Paid on Account)		$3,000
+ Accounts payable adjustment		
Ending balance	$4,000	
Beginning balance	−3,000	
Accounts payable increase		$1,000
Feed expense		$4,000

Ending accounts payable balances increase expenses for the year (since they actually belong to this year) and beginning accounts payable balances decrease expenses for the year (since they actually belong to the previous year).

The CFS income statement and its schedules automatically adjust for these types of inventory changes.

Preparation of the CFS Income Statement

Like the balance sheet, the CFS income statement[2] is theoretically sound and is particularly appropriate for American farms and ranches. Also like the balance sheet, the CFS Income Statement is developed on a primary form with several supporting schedules. The primary form is shown in Figure 14–4 and the schedules are presented in Figures 14–5 and 14–6.

All of the information needed to complete the income statement has already been developed in the records systems plus a balance sheet for the beginning of the year and a balance sheet for the end of the year. The primary information needed from the records systems included the cash receipts and cash expenditures (Chapter 5) and depreciation records (Chapter 7). As with the balance sheet, prepare the schedules first and then transfer the necessary information to the income statement.

Schedules 1, 4 and 5—Adjustments

All of the numbers needed to fill out Schedules 1, 4, and 5 (Figure 14-5) can be drawn directly from the beginning balance sheet and the ending balance sheet.

[2]Additional assistance for preparation of CFS income statements can be found in Frey, Nelson, and Cull, *You and Your Income Statement* (Skokie, IL: Century Communications, 1983).

INCOME STATEMENT

☐ Actual
☐ Projected

For 12 Month Period
Ending __________, 19____.

Name ____

Address ____

REVENUE			
Crops and feed:			
cash sales	$ ______ (1a)		
inventory change (Sch. 1)	(±) ______ (1b)	$ ______ (1)	
Livestock & poultry sold:			
cash sales	$ ______ (2a)		
inventory change (Sch. 1)	(±) ______ (2b)	______ (2)	
Breeding stock:			
cash sales	$ ______ (3a)		
inventory change (Sch. 1)	(±) ______ (3b)	______ (3)	
Products: livestock and poultry — cash sales		______ (4)	
Custom work: cash		______ (5)	
Government payments and patronage dividends		______ (6)	
Income from hedging transactions (Sch. 2)	(±)	______ (7)	
Other		______ (8)	
Adjustment in notes and accounts receivable (Sch. 1)	(±)	______ (9)	
Gross revenue (Add lines 1 thru 9)		$ ______ (10)	
Less livestock & poultry purchases (feeder & bre…	(−)	______ (11)	
Less feed purchased	(−)	______ (12)	
VALUE OF FARM PRODUCTION			$ ______ (a)
EXPENSES			
Cash operating expenses (Sch. 3)		$ ______ (13)	
Expense adjustment (unused assets) (Sch. 4)	(±)	______ (14)	
Expense adjustment (unpaid items) (Sch. 5)	(±)	______ (15)	
Depreciation:			
Machinery and equipment	$ ______ (16a)		
Fixed farm improvements	$ ______ (16b)	______ (16)	
Total operating expenses (Add lines 13 thru 16)			______ (b)
Income from farm operations		(a−b)	$ ______ (c)
Less interest expense (Sch. 6)		(−)	______
Gain or loss on disposal of farm assets (Sch. 7)		(±)	______
NET FARM INCOME			$ ______ (d)
NON-FARM INCOME			
Operator's wage off farm (net of expenses incurred)		$ ______ (17)	
Spouse's wage off farm (net of expenses incurred)		______ (18)	
Interest and dividends (including int. on retirement accts.)		______ (19)	
Gain or loss on sale of non-farm assets (Sch. 8)	(±)	______ (20)	
Net income — other farms		______ (21)	
Net income — non-farm real estate		______ (22)	
Other		______ (23)	
NET NON-FARM INCOME (Add lines 17 thru 23)			$ ______ (e)
Income before income taxes and extraordinary items		(d+e)	$ ______ (f)
Provision for income and Soc. Sec. taxes (all income) (Sch. 9)			______ (g)
Income before extraordinary items		(f−g)	$ ______ (h)
Extraordinary items (explain)			______ (i)
NET INCOME			$ ______

FIGURE 14–4. CFS income statement.

Source: Thomas L. Frey and Danny A. Klinefelter, *Coordinated Financial Statements for Agriculture,* 2nd ed. (Skokie, IL: AgriFinance, 1980).

SCHEDULE 1 — REVENUE ADJUSTMENT

	Transfer Totals from Balance Sheets: Beginning Inventory (−)	Transfer Totals from Balance Sheets: Ending Inventory (+)	Change (+ or −)
Notes and accounts receivable	−$______	+$______	$______
Livestock & poultry to be sold	−$______	+$______	$______
Crops and feed	−$______	+$______	$______
Breeding stock	−$______	+$______	$______

SCHEDULE 2 — HEDGING TRANSACTIONS ADJUSTMENT

Hedging account equity, from beginning balance sheet	−	$______
Deposits during period	−	______
Withdrawals during period	+	______
Hedging account equity, from ending balance sheet	+	______
TOTAL GAIN OR LOSS ON HEDGING TRANSACTIONS	±	$______

SCHEDULE 3 — CASH OPERATING EXPENSES (excluding interest, feed, livestock)

Item	Cash Expense
Labor hired (including all taxes and employee benefits)	$______
Repairs — machinery and equipment	______
Repairs — building and improvements	______
Rents and leases	______
Seed	______
Fertilizer and lime	______
Chemicals	______
Custom machine hire	______
Supplies	______
Livestock expenses	______
Gas, fuel, oil	______
Storage and custom drying	______
Taxes (real estate and personal property)	______
Insurance (property, liability, crop)	______
Utilities (farm share)	______
Marketing and transportation expense	______
Auto (farm share)	______
Miscellaneous	______
______	______
______	______
______	______
TOTAL CASH OPERATING EXPENSES	$______

SAMPLE

SCHEDULE 4 — EXPENSE ADJUSTMENT (unused assets)

	Transfer Totals from Balance Sheet: Beginning Balance (+)	Transfer Totals from Balance Sheet: Ending Balance (−)	Change (+ or −)
Cash investment in growing crops	+$______	−$______	$______
Supplies	+$______	−$______	______
Prepaid expenses	+$______	−$______	______
TOTAL	+$______	−$______	$______

SCHEDULE 5 — EXPENSE ADJUSTMENT (unpaid items)

	Transfer Totals from Balance Sheets: Beginning Balance (−)	Transfer Totals from Balance Sheets: Ending Balance (+)	Change (+ or −)
Farm accounts payable	−$______	+$______	$______
Accrued property taxes	−$______	+$______	______
Accrued real estate taxes	−$______	+$______	______
Accrued employer payroll withholdings	−$______	+$______	______
Accrued rent and lease payments	−$______	+$______	______
TOTAL	−$______	+$______	$______

FIGURE 14–5. Schedules 1 through 5, CFS income statement.

Source: Thomas L. Frey and Danny A. Klinefelter, *Coordinated Financial Statements for Agriculture,* 2nd ed. (Skokie, IL: AgriFinance, 1980).

SCHEDULE 6 — INTEREST EXPENSE

Amount paid in cash or by renewal			$
Accrued interest:			
Beginning of period	(−)	$	
End of period	(+)	$	
Adjustment for change in accrued interest			±
TOTAL INTEREST EXPENSE			$

SCHEDULE 7 — GAIN OR LOSS ON DISPOSAL OF FARM ASSETS

	Net Sale Price (+)	Original Cost Less Accum. Dep. (book value) (−)	Change (+ or −)
Machinery and equipment	+$	−$	$
Fixed assets	+$	−$	
TOTAL	+$	−$	$

SCHEDULE 8 — GAIN OR LOSS ON SALE OF NON-FARM ASSETS

	Net Sale Price (+)	Original Cost Less Accum. Dep. (book value) (−)	Change (+ or −)
Securities	+$	−$	$
Fixed assets	+$	-$	
Other	+$	$	
TOTAL	[illegible]	−$	$

[illegible]ION FOR INCOME TAX

Amount paid in cash			$
Accrued income and Soc. Sec. taxes:			
Beginning of period	(−)	$	
End of period	(+)	$	
Adjustment for change in accrued income and Soc. Sec. taxes			±
Income tax refund due:			
Beginning of period	(+)	$	
End of period	(−)	$	
Adjustment for change in income tax refund due			±
TOTAL			$

SCHEDULE 10 — WITHDRAWALS FOR FAMILY LIVING

	Transfer Totals from Balance Sheets		Change (+ or −)
	Beginning Balance	Ending Balance	
Net family living withdrawals (from Statement of Owner Equity on B1)			$
Marketable bonds & securities −	$	+$	
Cash value of life insurance −	$	+$	
Personal & recreational vehicles −	$	+$	
Household goods & personal effects −	$	+$	
Non-farm real estate −	$	+$	
Other non-farm investments −	$	+$	
Medical & other personal accounts payable +	$	−$	
Plus retirement account deposits			+
GROSS FAMILY LIVING WITHDRAWALS			$

FIGURE 14–6. Schedules 6 through 10, CFS income statement.

Source: Thomas L. Frey and Danny A. Klinefelter, *Coordinated Financial Statements for Agriculture,* 2nd ed. (Skokie, IL: AgriFinance, 1980).

Following the plus and minus signs within these three schedules will help to keep the adjustment direction correct, which was discussed in the previous section of this chapter.

Schedules 2 and 3—Hedging Adjustments and Cash Expenses

The hedging transactions adjustment schedule (see Figure 14–5) is also simple to complete. Account equity values are listed on the two balance sheets. Deposits and withdrawals are recorded in the cash transactions records and are totaled either in the farm income summary, income tax estimate worksheet (Figure 13–13) or the Schedule F tax return (Figure 6–5), as are the other cash operating expenses.

Schedules 6 and 9—Interest and Income Tax Expenses

Both Schedules 6 and 9 (see Figure 14–6) adjust the amount of expenditure paid in cash during the year by the beginning and ending account balances. Once again, the amounts paid in cash can be transferred from the records system or Schedule F, and the account balances can be found on the beginning and ending balance sheets.

Schedules 7 and 8—Gain or Loss on Assets

The depreciation schedule is a key element in the completion of Schedule 7 (see Figure 14–6). For each capital asset sold, sale price and current depreciated book value must be identified. Similarly, for nonfarm or ranch assets, sale price and book value must be identified regardless of whether the item was depreciated. These assets were totaled in CFS Balance Sheet Schedules 1 and 6. For both farm and nonfarm assets, if the sale price exceeds the book value, a gain is realized. That gain will increase income for the year. On the other hand, if the book value exceeds the sale price, a loss is realized that will reduce income.

Schedule 10—Withdrawals for Family Living

Discussion of Schedule 10 will be delayed to Chapter 16, primarily because Schedule 10 is extremely important to the correct correlation of two successive balance sheets. In fact, Schedule 10 cannot be finished until the back side of the CFS Balance Sheet has been completed.

The Income Statement Form

Once the first nine schedules of the CFS Income Statement have been completed, attention can be directed toward preparation of the income statement itself

(Figure 14–4). Similar to the balance sheet, some information is taken from the records system (or in this case the income tax return) and adjustments to those numbers are transferred from the schedules.

Revenue

The revenue portion of the income statement develops the value of farm or ranch production on an accrual basis by drawing *business* receipts (cash sales and other miscellaneous farm or ranch income) from revenue summaries (such as the records system or tax return) and adjusting those receipts for inventory changes. If left there, the receipts and adjustments would indicate gross income from the farming or ranching operation. Although that number might be interesting, a much more useful number is value of production. Value of farm or ranch production is obtained by subtracting purchased livestock and purchased feed from gross farm income. This indicates the amount of income actually generated by this business rather than including the value generated by another farm or ranch business.

Expenses

The expenses portion of the income statement lists all remaining farm or ranch expenditures and account balance adjustments. All items can be pulled from the income statement schedules plus the summary of the depreciation schedule.

Nonfarm Income and Net Income Summary

Revenue minus expenses yields net farm income, the accrual net earnings of the farm or ranch business for the year. But, since most farm or ranch businesses are family operated and often consolidate farm/ranch business affairs with family matters, the CFS Income Statement includes a separate section to add in nonfarm or nonranch income (income from other businesses or employment). Further, since taxes are a function of both business and nonbusiness income, a provision for taxes paid follows the calculation of net farm or ranch income and net nonfarm or nonranch income. The bottom line, including farm or ranch income, nonfarm or nonranch income, and taxes and extraordinary items (truly extraordinary events, such as the sale of the entire farm or ranch or major insurance receipts, are included here) thus calculates net income for the business/family unit for the year.

For farm and ranch business management analysis, the most important numbers are the total value of farm or ranch production, total operating expenses, and net farm or ranch income. But for matching the income statement to the balance sheet, the bottom line net income is also vital.

Summary

An income statement, in contrast to the balance sheet, is a flow statement. It portrays specific financial information for a period of time, usually one year for a farm or ranch. To be consistent and most useful, the income statement must be prepared on an accrual basis. The purpose of the income statement is to develop and justify a few net return numbers, the most important of which is net farm or ranch income. This figure represents the amount of income earned during the period regardless of the timing of sales and purchases.

An accrual income statement can be developed very easily from a cash-based accounting system if beginning and ending balance sheets are available. The cash flow summary taken from the record books must be adjusted for inventory changes taken from the two balance sheets. An increase in inventory of a saleable commodity increases net income whereas an increase in inventory of a supply commodity decreases expense (which in turn increases income also).

CHAPTER 15

The Cash Flow Statement

The third major financial statement, the *cash flow statement,* can become an extremely valuable financial management tool for the modern farmer or rancher. The cash flow statement clearly addresses one of the most serious financial problems faced today—cash flow control. The cash flow statement is also a primary element in the credit application process because it directly shows how and when borrowed capital (especially short-term operating loans) will be repaid. In both cases, the farmer or rancher develops a cash flow statement because he recognizes the cyclical flows of cash into and out of an agricultural business. Typically, cash needs are greater than the amount of cash available during productive stages of animal and plant growth. Moreover, sales tend to be lumpy; they come in bunches. As a result of both cash inflow and outflow patterns, periods of cash deficits exist that cause many farmers and ranchers either to draw down savings accounts or to borrow to cover the deficits.

But, perhaps the most important feature of the cash flow statement is that it can be used as the culmination of the annual farm or ranch marketing/financial planning process. If the cash flow statement is prepared with a great deal of thoughtful planning, it can wrap together the producer's marketing plans, optimal enterprise selection and input needs, family living needs, feed requirements, inventory sources and uses, capital expansion plans, and debt repayment capacity.

There is no doubt that the cash flow statement is a very important and useful financial management tool. However, it requires a lot of careful work (as can be seen in the example of Figure 15–1) and clearly defined plans. Conceptually, the cash flow statement is easy to understand. Practically, it is difficult to draw up, not because it involves many difficult or abstract concepts, but because it does require careful planning.

The cash flow statement addresses a primary fact of life—that negative bank balances are not very popular with bankers. In fact, the major objective of the cash flow statement is to show the sources of cash generated by the business and the cash needs or requirements of the business. In periods when cash needs exceed cash sources, money must be borrowed. When cash sources exceed

1985 CASH FLOW BUDGET

	TOTAL	Jan	Feb	Mar	Apr	May	Jun	Jul	Aug	Sep	Oct	Nov	Dec
CASH SOURCES													
1. Beginning cash balance (demand plus currency)		800	1,000	1,000	1,000	1,000	1,000	1,000	1,000	1,000	1,000	1,000	1,000
OPERATING RECEIPTS													
2. Crops and feeds	213,605	0	13,500	0	0	0	0	47,500	0	42,500	39,825	70,280	0
3. Livestock and poultry	34,930	15,834	19,096	0	0	0	0	0	0	0	0	0	0
4. Products: Livestock and poultry	0	0	0	0	0	0	0	0	0	0	0	0	0
5. Custom work	7,000	0	0	0	0	0	0	3,000	0	4,000	0	0	0
6. Government payments and paytronage dividends	0	0	0	0	0	0	0	0	0	0	0	0	0
7. Hedging account withdrawals	0	0	0	0	0	0	0	0	0	0	0	0	0
CAPITAL RECEIPTS													
8. Breeding stock	850	850	0	0	0	0	0	0	0	0	0	0	0
9. Machinery and equipment	0	0	0	0	0	0	0	0	0	0	0	0	0
NON FARM INCOME													
10. Off farm wages	9,000	0	1,500	0	1,500	0	1,500	0	1,500	0	1,500	0	1,500
11. Interest and dividends	0	0	0	0	0	0	0	0	0	0	0	0	0
12.		0	0	0	0	0	0	0	0	0	0	0	0
13.		0	0	0	0	0	0	0	0	0	0	0	0
14.		0	0	0	0	0	0	0	0	0	0	0	0
15.		0	0	0	0	0	0	0	0	0	0	0	0
16. TOTAL CASH AVAILABLE	277,185	17,484	35,096	1,000	2,500	1,000	2,500	51,500	2,500	47,500	42,325	71,280	2,500
OPERATING EXPENSES													
17. Labor hired (including employer taxes)	18,000	1,500	1,500	1,500	1,500	1,500	1,500	1,500	1,500	1,500	1,500	1,500	1,500
18. Repairs - machinery and equipment	8,000	0	0	1,000	2,500	0	0	0	1,500	1,000	2,000	0	0
19. Repairs - building and improvements	2,500	0	0	0	0	1,200	0	0	0	0	0	1,300	0
20. Rents and leases	62,400	0	0	31,200	0	0	0	0	0	31,200	0	0	0
21. Seed	11,860	0	0	9,540	1,920	0	0	0	0	400	0	0	0
22. Fertilizer and lime	40,300	0	0	32,260	3,200	0	0	0	0	0	0	4,840	0
23. Chemicals	11,730	0	0	0	10,557	1,173	0	0	0	0	0	0	0
24. Custom machine hire	0	0	0	0	0	0	0	0	0	0	0	0	0
25. Supplies	0	0	0	0	0	0	0	0	0	0	0	0	0
26. Livestock expenses (breeding, vet., etc.)	5,000	1,250	0	0	1,250	0	0	1,250	0	0	1,250	0	0
27. Gas, fuel and oil	12,000	500	0	1,000	2,000	2,000	3,000	0	0	500	1,000	2,000	0
28. Storage and Custom Drying	4,500	0	0	0	0	0	0	0	0	0	3,500	1,000	0
29. Taxes (real estate and personal property)	6,000	0	0	0	0	0	0	0	0	6,000	0	0	0
30. Insurance (property, liability, crop)	1,400	0	0	0	0	0	0	0	0	0	0	0	1,400
31. Utilities	6,000	500	500	500	500	500	500	500	500	500	500	500	500
32. Marketing and Transportation expense	0	0	0	0	0	0	0	0	0	0	0	0	0
33. Auto (farm share)	1,800	150	150	150	150	150	150	150	150	150	150	150	150
34.		0	0	0	0	0	0	0	0	0	0	0	0
35.		0	0	0	0	0	0	0	0	0	0	0	0
36. TOTAL CASH OPERATING EXPENSES	191,490	3,900	2,150	77,150	23,577	6,523	5,150	3,400	3,650	41,250	9,900	11,290	3,550
LIVESTOCK AND FEED PURCHASES													
37. Feeder livestock		0	0	0	0	0	0	0	0	0	0	0	0
38. Breeding Livestock		0	0	0	0	0	0	0	0	0	0	0	0
39. Feed Purchased	18,000	0	0	18,000	0	0	0	0	0	0	0	0	0
CAPITAL EXPENDITURES													
40. Machinery and Equipment	35,000	0	0	0	0	0	0	0	0	0	0	35,000	0
41. Building and Improvements		0	0	0	0	0	0	0	0	0	0	0	0
42.		0	0	0	0	0	0	0	0	0	0	0	0
OTHER EXPENDITURES													
43. Hedging account deposits		0	0	0	0	0	0	0	0	0	0	0	0
44. Gross family living withdrawals	21,600	1,800	1,800	1,800	1,800	1,800	1,800	1,800	1,800	1,800	1,800	1,800	1,800
45. Non-farm business and investments		0	0	0	0	0	0	0	0	0	0	0	0
46. Income tax and social security	10,659	0	0	10,659	0	0	0	0	0	0	0	0	0
47.		0	0	0	0	0	0	0	0	0	0	0	0
48. Intermediate/long term loan payments - principal	27,861	0	0	0	0	0	0	3,800	19,883	0	0	4,178	0
49. - interest	15,493	0	0	0	0	0	0	8,185	5,164	0	0	1,894	250
50. TOTAL CASH REQUIRED	320,103	5,700	3,950	107,609	25,377	8,323	6,950	17,185	30,497	43,050	11,700	54,162	5,600
51. CASH AVAILABLE LESS CASH REQUIRED	(42,918)	11,784	31,146	(106,609)	(22,877)	(7,323)	(4,450)	34,315	(27,997)	4,450	30,625	17,118	(3,100)
52. Inflows from savings	67,390	1,000	0	0	0	0	0	0	28,997	0	0	11,939	25,454
53. CASH POSITION BEFORE BORROWING	24,472	12,784	31,146	(106,609)	(22,877)	(7,323)	(4,450)	34,315	1,000	4,450	30,625	29,057	22,354
MONEY TO BE BORROWED													
54. ¦Intermediate and long term loans	25,000	0	0	0	0	0	0	0	0	0	0	25,000	0
¦OPERATING LOAN with minimum balance...	166,326	10,894	10,173	107,609	23,877	8,323	5,450	0	0	0	0	0	0
55. Operating loan payments - principal	118,069	18,000	39,978	0	0	0	0	0	0	0	0	40,000	20,091
56. - interest	19,339	4,678	341	0	0	0	0	0	0	0	0	13,057	1,263
57. Outflows to savings	66,390	0	0	0	0	0	0	33,315	0	3,450	29,625	0	0
58. ENDING CASH BALANCE	12,000	1,000	1,000	1,000	1,000	1,000	1,000	1,000	1,000	1,000	1,000	1,000	1,000
LOAN BALANCE: (at the end of period)													
59. Current year's operating loans	166,326	10,894	21,067	128,676	152,553	160,876	166,326	166,326	166,326	166,326	166,326	126,326	106,235
60. Previous year operating loan ¦ 57,978		39,978											
61. Total intermediate/long term loans ¦ 193030	190,169	193,030	193,030	193,030	193,030	193,030	193,030	189,230	169,347	169,347	169,347	190,169	190,169
62. TOTAL LOANS	356,495	243,902	214,097	321,706	345,583	353,906	359,356	355,556	335,673	335,673	335,673	316,495	296,404
63. Savings		0	0	0	0	0	0	33,315	4,318	7,768	37,393	25,454	0
	TOTAL	Jan	Feb	Mar	Apr	May	Jun	Jul	Aug	Sep	Oct	Nov	Dec

FIGURE 15–1. Example of a completed farm cash flow statement.

needs, the excess can be carried over to cover future needs or deficits, or it can be used to purchase additional supplies or capital assets, increase savings, or retire existing debts. Regardless of whether there is an excess of cash sources over requirements and regardless of which purposes the excess is committed to, the summation of cash sources must be equal to the summation of cash uses. In other words, every dollar flowing into the business must be allocated to some use, *and,* regardless of whether money is earned or borrowed, there must be enough of it available to cover all cash needs.

Projected versus Actual Cash Flow

To be truly useful, a cash flow statement must be prepared on a projected basis. In other words, it must represent a *plan* for the future. Usually, cash flow statements are prepared as plans or guidelines for each month of the next calendar year. Consequently this plan is often called a *budget.* Cash flow plans can be prepared further in advance than one year, but the farther into the future the plan is projected, the more tentative become the results because of the increasing likelihood of unanticipated events. Consequently, cash flow plans projected more than one year in advance typically are used for major capital expansion plans and will usually related only to the purchase of that capital asset, along with the additional cash sources and uses generated by the expansion. This chapter will be confined to a typical, short-term (12-month) cash flow statement.

An *actual* cash flow statement (a statement of the monthly cash inflows and outflows for the year just past) can be prepared with little work by transferring totals from cash receipts and cash expenditures from the transactions records discussed in Chapter 5. Or if cash flows are regularly *monitored* (see page 230), then the actual cash flow records have already been done. An actual cash flow statement can be a very useful device to "get the hang of" the mechanics of the cash flow statement and to get an idea of the purchasing and sales patterns of the business. In other words, the work of planning for the future can be reduced if there is a base to work from.

Cash Flow and Profitability

Despite the recent emphasis by agricultural lenders on cash flow planning and debt repayment capacity, the cash flow statement represents only one part of the financial management picture—a very important part of that picture, but nonetheless only one part. As argued throughout this text, much of the farm or ranch records system must be maintained on the accrual accounting system or converted to the accrual system through inventory adjustments. The primary purpose for this argument is that the cash accounting system cannot adequately reflect how profitable a business is because the timing of purchases and sales should not affect the underlying use of inputs or the production of income.

Within the cash flow statement, only cash inflows and outflows count. Consequently, the cash flow statement says *nothing* about the profitability of the business. Profitability information is the task of the income statement, the cash flow statement describes cash position, and the balance sheet shows overall financial position and strength. All three statements are necessary.

The cash flow statement holds a unique position within the set of financial statements because it forces recognition of the manipulations necessary to maintain a positive bank balance. The world moves on the basis of cash. Bankers and other suppliers no longer accept payment in commodities. A net loss in one year will not necessarily force a business to fail, but a cash deficit or the inability to raise enough cash to meet obligations will cause failure. And yet a positive cash flow is not sufficient by itself. Eventually, the business must be profitable for it to continue. More will be discussed about the management uses of financial statements in Part V of this text.

Structure of the Cash Flow Statement

A cash flow statement typically can be broken into five major parts:

1. *Beginning cash balance* (the amount of cash in currency or checking accounts ready for use)
2. *Receipts* (the money inflows generated by sales, off-farm employment, and other income)
3. *Expenditures* (the money outflows needed to operate the business)
4. *Adjustments* (additional cash inflows to cover deficits or cash control maneuvers)
5. *Ending cash balance*

A typical example of a general cash flow statement is shown in Figure 15–2. The beginning cash balance is added to receipts to calculate the total cash available for the period. From the cash available, expenditures are subtracted. The balance will be either positive or negative. If negative, a cash deficit exists and somehow must be covered. Typically, cash will either be withdrawn from savings or it will be borrowed to cover the deficit and provide some operating reserves for the next period. (If additional income is generated through early sales to cover the projected deficit, those sales would be reported in the receipts section.) If cash available less cash required results in a cash surplus, or positive balance, then the manager must decide whether he wants to place the excess into savings (an interest-bearing account), repay outstanding operating loans (or possibly prepay longer-term loans) or carry the positive balance into the next period. After these savings/borrowing/repayment adjustments, the amount left over is the ending cash balance. That ending cash balance will then become the beginning cash balance for the next period (shown as A in Figure 15–2).

Each farmer and rancher will want to develop his own cash balance strategy for determining how much cash is to be carried over from one period to the next. A certain minimum amount will usually be determined to provide a cushion at the beginning of the next period. Within a period, as well as between periods, cash

	Period 1	Period 2	...
Beginning Cash Balance		A	
+ Receipts			
= Total Cash Available			
− Expenditures			
= Cash Available Less Cash Required			
+ Inflows from Savings			
+ Cash Borrowed			
− Short-term Loan Repayments			
− Outflows to Savings Accounts			
= Ending Cash Balance	A		

FIGURE 15–2. General structure of a cash flow statement.

flow inequities can occur. For example, Period 2 sales may not be slated until the middle of the period, but some of the bills are due at the beginning of the period.

In addition to this type of within-period cash flow problem, the cash balance strategy should reflect any notice required before savings can be withdrawn, whether a cash excess or deficit is projected for the next few periods, and the type of operating loan structure under which short-term borrowings are made. If short-term money is borrowed via a revolving line of credit, repayments should be made as quickly as possible. A revolving line of credit specifies the maximum amount of borrowed money that can be outstanding at any one time. Money can be borrowed when needed, repaid, and at a later date borrowed again, as long as the maximum outstanding limit is not reached. Since interest charges are always assessed against the outstanding balance; it will be to the borrower's advantage to keep that outstanding balance as low as possible. If money is not borrowed under a revolving line of credit (either through a nonrevolving line or a fixed amount loan), the funds cannot be reborrowed. Consequently, such a borrower will want to be more conservative in his repayment plans to protect against future contingencies.

The CFS Cash Flow Statement

Beyond the general structure of the cash flow statement outlined in Figure 15–2, commercially prepared or personalized cash flow statement formats simply add whistles and bells—extra detail lines—to separate types of receipts and types of expenditures and to make the task of projecting individual items easier. The CFS Cash Flow Statement[1] (Figure 15–3) is an imposing statement because of its size, but the concept remains the same as the general example discussed in previous sections of this chapter.

[1] Additional assistance for preparation of CFS cash flow statements can be found in Nelson and Frey, *You and Your Cash Flow* (Skokie, IL: Century Communications, 1983).

C1 **CASH FLOW STATEMENT**

C1 (CUT ALONG DOTTED LINE TO PLACE IN LEDGER)

☐ Projected for 19____

☐ Actual for 19____ Name______________________ Address______________________

Date completed ______________, 19____

1	Beginning cash balance (demand deposits plus currency)															1
2	**Operating Receipts:** Crops and feed															2
3	Livestock and poultry															3
4	Products: livestock and poultry															4
5	Custom work															5
6	Government payments and patronage dividends															6
7	Hedging account withdrawals															7
8																8
9	**Capital Receipts:** Breeding stock															9
10	Machinery and equipment															10
11																11
12	**Non-Farm Income:** Off-farm wages															12
13	Interest and dividends															13
14																14
15																15
16	**TOTAL CASH AVAILABLE** (add lines 1 thru 15)															16
17	**Operating Expenses:** Labor hired (including employer taxes)															17
18	Repairs — machinery and equipment															18
19	Repairs — buildings and improvements															19
20	Rents and leases															20
21	Seed															21
22	Fertilizer and lime															22
23	Chemicals															23
	Custom machine hire															24
25	Supplies															25
26	Livestock expense (breeding, vet., etc.)															26
27	Gas, fuel, oil															27
28	Storage and custom drying															28
29	Taxes (real estate and personal property)															29
30	Insurance (property, liability, crop)															30
31	Utilities (electricity, gas, telephone)															31
32	Marketing and transportation expense															32
33	Auto (farm share)															33
34																34
35																35
36	**Total Cash Operating Expenses**															36
37	**Livestock and Feed Purchases:** Feeder livestock															37
38	Breeding livestock															38
39	Feed purchased															39
40	**Capital Expenditures:** Machinery and equipment															40
41	Building and improvements															41
42																42
43	**Other Expenditures:** Hedging account deposits															43
44	Gross family living withdrawals															44
45	Non-farm business and investments															45
46	Income tax and social security															46
47																47
48	Intermediate and long term loan payments — principal															48
49	— interest															49
50	**TOTAL CASH REQUIRED** (add lines 36 thru 49)															50
51	**CASH AVAILABLE LESS CASH REQUIRED** (16-50)															51
52	Inflows from savings															52
53	Cash position before borrowing															53
54	Money to be borrowed — operating loans															54
	— intermediate and long term loans															
55	Operating loan payments — principal															55
56	— interest															56
57	Outflows to savings															57
58	Ending cash balance															58
59	**Loan Balances:** (at end of period) Current year's operating loans															59
60	Previous year's operating loans															60
61	Intermediate and long term loans															61
62	Total Loans															62
63	**Consistency Check:** Total inflows including borrowed money (16+52+54)															63
64	Total outflows (50+55+56+57+58)															64
65	Budgeting error (63−64)															65

SAMPLE

FIGURE 15–3. CFS cash flow statement.

Source: Thomas L. Frey and Danny A. Klinefelter, *Coordinated Financial Statements for Agriculture,* 2nd ed. (Skokie, IL: AgriFinance, 1980).

The CFS statement separates operating receipts, capital receipts, and nonfarm income within the receipts section (lines 2–15) and separates the expenditures section (lines 17–49) into operating expenses (note that lines 17–36 match Schedule 3 of the income statement in Figure 14–5), livestock and feed purchases, capital expenditures, and other expenditures (which include family living withdrawals and scheduled loan payments).

Like the balance sheet and the income statement, the cash flow statement is supported by a series of schedules (Figures 15–4 through 15–10). These schedules provide a convenient place to plan crop and livestock input needs, capital expenditure plans, loan repayment requirements, and a family living budget.

However, unlike the other two statements, the cash flow schedules are not required for an actual statement and the mechanisms for transferring information from the schedules to the primary statement are not as precise.

Crop Production and Costs

The primary agronomic characteristics of each field should be listed in the Budget Sheet for Crop Production (Figure 15–4). To make this budget sheet most useful, each field should be numbered and correlated to the farm map (Figure 9–1). For each field, list the crop, acreage, and expected per-acre yield. Expected production from the field can then be calculated. Similarly, the variety, analysis, and kind of seed, fertilizers, and chemicals planned for each crop should be listed, as well as the per-acre rate and total needs of each of these three major inputs.

After each field has been listed and expected input needs and output have been calculated, the information should be summarized to project how much production of each major crop can be expected and to project the likely total needs of each major input (by variety, by analysis, and by type of chemical). These totals are then transferred to other schedules.

The entire amount of each fertilizer formulation should be transferred to the top half of the Summary of Crop Costs (Figure 15–5), along with total requirements of each seed variety and chemical. Each formulation should be priced and the total cost should be calculated. The final decision to be made is when to purchase each type. Once the purchase month has been recorded (in the Mo. column), the totals are ready for transfer to the cash flow statement.

The Summary of Crop Costs also allows for scheduling amounts, unit prices, and total costs by month for three other primary inputs: fuels and lubricants, custom machine hire, and hired labor.

Livestock, Crop Usage, and Feed

The two budget sheets for livestock and for crop usage and feed requirements (Figures 15–6 and 15–7) have a substantially similar format. Both budget sheets could be called sources and uses schedules since they both attempt to balance a primary flow equation:

$$Sources = Uses$$

C2

BUDGET SHEET FOR CROP PRODUCTION

C2

Crop To Be Grown	Field		Yield Per Acre	Total Production	Seed			Fertilizer			Chemicals		
	No.	Acres			Variety	Rate Per Acre	Total	Analysis	Rate Per Acre	Total	Kind	Rate Per Acre	Total

FIGURE 15–4. Budget sheet for crop production, CFS cash flow statement.

Source: Thomas L. Frey and Danny A. Klinefelter, *Coordinated Financial Statements for Agriculture,* 2nd ed. (Skokie, IL: AgriFinance, 1980).

C3 **SUMMARY OF CROP COSTS** C3

FERTILIZER					SEED					CHEMICALS				
Analysis	Quantity	$/Unit	Total Cost	Mo.	Variety	Quantity	$/Unit	Total Cost	Mo.	Kind	Quantity	$/Unit	Total Cost	Mo.
		$	$				$	$				$	$	

GAS, FUEL, OIL					CUSTOM MA					LABOR HIRE				
Kind	Quanity	$/Unit	Total Cost	Mo.	Job	Quantity	$/Unit		Mo.	Job	Quantity	$/Unit	Total Cost	Mo.
		$	$				$	$				$	$	

FIGURE 15–5. Summary of crop costs, CFS cash flow statement.

Source: Thomas L. Frey and Danny A. Klinefelter, *Coordinated Financial Statements for Agriculture,* 2nd ed. (Skokie, IL: AgriFinance, 1980).

C4 **BUDGET SHEET FOR LIVESTOCK** C4

Kind	Beginning Inventory			Purchases					No. Raised	No. Died	Sales					Ending Inventory		
	No.	Wt/Hd	Value	Mo.	No.	Wt/Hd	$/Unit	Cost			Mo.	No.	Wt/Hd	$/Unit	Value	No.	Wt/Hd	Value
FEEDER STOCK—LIVESTOCK AND POULTRY																		
			$				$	$						$	$			$
BREEDING AND LAYING STOCK																		
			$				$	$						$	$			$

SAMPLE

FIGURE 15–6. Budget Sheet for Livestock, CFS cash flow statement.

Source: Thomas L. Frey and Danny A. Klinefelter, *Coordinated Financial Statements for Agriculture,* 2nd ed. (Skokie, IL: AgriFinance, 1980).

C5 **BUDGET SHEET FOR CROP USAGE AND FEED REQUIREMENTS** C5

	Beginning Inventory			Purchases				Farm Use		Sales				Ending Inventory	
Crop	Quantity	Value	Produced	Mo.	Quantity	$/Unit	Cost	Feed	Seed	Mo.	Quantity	$/Unit	Value	Quantity	Value
		$				$	$					$	$		
															$

SAMPLE

LIVESTOCK FEED REQUIREMENTS

Livestock			Corn, bu.				…nt		Hay, tons		Silage, tons	
Kind	No.	Period On Feed	Old	New	Old	New	Old	New	Old	New	Old	New

FIGURE 15–7. Budget Sheet for Crop Usage and Feed Requirements, CFS cash flow statement.

Source: Thomas L. Frey and Danny A. Klinefelter, *Coordinated Financial Statements for Agriculture,* 2nd ed. (Skokie, IL: AgriFinance, 1980).

C6

CAPITAL EXPENDITURES AND REPAIRS

C6

Description	Number or Quantity	Month	Purchases & Capital Repairs		Operating Repairs	
			Machinery and Equipment	Buildings and Improvements	Machinery and Equipment	Buildings and Improvements
			$		$	$

SAMPLE

FIGURE 15–8. Capital expenditures and repairs, CFS cash flow statement.

Source: Thomas L. Frey and Danny A. Klinefelter, *Coordinated Financial Statements for Agriculture,* 2nd ed. (Skokie, IL: AgriFinance, 1980).

C7 **INTERMEDIATE AND LONG TERM LOAN PAYMENTS** C7

Beginning Balance &/or New Loans	Date of Note	Int. Rate	Payments	Jan	Feb	Mar	Apr	May	Jun	Jul	Aug	Sep	Oct	Nov	Dec	Ending Principal Balance
$			Principal													$
			Interest													

SAMPLE

Developed by Thomas L. Frey and Danny A. Klinefelter, Dept. of Agricultural Economics, Cooperative Ext. Service, University of Illinois. ©1980 AGRI FINANCE. Available from Agri Finance, 5520-G Touhy Ave., Skokie, IL 60077

FIGURE 15–9. Intermediate and long-term loan payments, CFS cash flow statement.

Source: Thomas L. Frey and Danny A. Klinefelter, *Coordinated Financial Statements for Agriculture,* 2nd ed. (Skokie, IL: AgriFinance, 1980).

C8

FAMILY LIVING BUDGET

C8

	Jan	Feb	Mar	Apr	May	Jun	Jul	Aug	Sep	Oct	Nov	Dec	TOTAL
Food													
Household operating expenses													
Household equipment and furnishings													
House repairs													
Rent													
Clothing													
Personal													
Entertainment and recreation													
Education and reading materials													
Medical care and drugs													
Church and charity													
Personal gifts													
Utilities (non-farm)													
Transportation and auto (non-farm)													
Personal & recreational vehicle purchases													
Medical and disability insurance													
Life insurance													
Additions to personal investments													
Additions to retirement accounts													
Gross Family Living Withdrawals													

SAMPLE

Developed by Thomas L. Frey and Danny A. Klinefelter, Dept. of Agricultural Economics, Cooperative Ext. Service, University of Illinois. ©1980 AGRI FINANCE. Available from Agri Finance, 5520-G Touhy Ave., Skokie, IL 60077.

FIGURE 15–10. Family living budget, CFS cash flow statement.

Source: Thomas L. Frey and Danny A. Klinefelter, *Coordinated Financial Statements for Agriculture,* 2nd ed. (Skokie, IL: AgriFinance, 1980).

For any particular commodity, the sources must equal the uses; nothing vanishes into thin air or is created by magic. For livestock, the *sources* = *uses* equation can be expanded into:

Beginning inventory
+ *Purchases*
+ *Number of animals weaned* (or *transferred from* another class)
– *Death loss*
– *Sales* (or *transferred to* another class)
= *Ending inventory*

For crops and feed, a similar expansion, obviously without the death loss, is possible.

Beginning inventory
+ *Production*
+ *Purchases*
– *Use for feed or seed*
– *Sales*
– *Losses* (due to drying, handling, or pests)
= *Ending inventory*

In addition to physical amounts, both budget sheets provide locations for inventory values as well as values for cash transactions.

The bottom half of the Budget Sheet for Crop Usage and Feed Requirements should be used by livestock producers. Planned numbers of livestock and their feed requirements should be listed so that the feed column of the top half of the page will reflect the best estimates of feed use.

Capital Expenditures, Debt Payments, and Family Living

The remaining three cash flow schedules (Figures 15–8 through 15–10) provide an opportunity to budget planned capital expenditures (purchases of new long-term assets) and repairs, principal and interest components of longer term loan payments, and family living expenditures. It is usually very difficult to plan the timing of capital expenditures without the benefit of a completed cash flow. An estimate of cash balances, usually late in the year or production cycle, may be necessary before realistic capital expenditure plans can be made. So there may be a need to look back and forth more than once between the capital expenditures budget and the cash flow statement.

Operating repairs are extremely difficult to predict, except that they will obviously occur when least expected or when the machine is needed the most. Planning estimates can be based to some extent on judgment and on past repair history, but enough flexibility should be built in to cover contingencies.

In contrast to capital expenditures and repairs, intermediate and long-term loan payments are easy to project. The information necessary to complete this schedule has actually already been located and prepared in Schedule 8 of the balance sheet (Figure 13–9).

The level of detail in the family living budget (Figure 15–10) is an individual matter. Chapter 12 discussed the need to maintain family records at length. Depending upon the individual and family, this budget might be prepared in great detail or a bottom line total withdrawal projection might be sufficient. For the cash flow statement itself, the only information transferred is found in the gross withdrawals totals at the bottom.

The CFS Statement

Development of a projected cash flow statement will normally begin by taking beginning cash balances and loan balances from the balance sheet. From there, the operating and capital receipts projected on the budget sheets and other cash flow schedules can be distributed across the page by month. Nonscheduled amounts, such as custom work and nonfarm income, should be budgeted in carefully. Last year's receipts adjusted for planned changes and price changes are the best guide.

The large operating expenses, such as fertilizer, feed, and livestock purchases, can be transferred from the cash flow schedules as well. The remaining items also need to be planned carefully. Again, adjusted estimates built from the previous year's experience will be the easiest to develop.

After each applicable line has been completed for the budgeting period, the mechanics of the cash flow statement, as discussed earlier in this chapter, can be addressed.

Cash Flow Monitoring

Cash flow control does not end with the preparation of the cash flow budget. The budget is really only the first step. To develop any managerial control over the business cash position, future events must be anticipated and adjustments to new information made before they occur. One of the most useful means to keep on top of cash position is to periodically monitor actual cash movements as compared to budgeted estimates. The Monitoring Worksheet displayed in Figure 15–11 provides a big step to *controlling* cash flow. The worksheet lays out three columns: amount budgeted (copied line for line from the current cash flow budget), actual results (the actual amounts received or paid, taken from the record book) and the difference between planned and actual amounts.

It should be clear that cash flow monitoring is a year-long task. The budget or plan should not be taken as the final word on how money is to be spent or how commodities will be sold. Rather, it should be used as a guide. If differences are great, adjustments in selling, purchasing and borrowing plans should be made to avoid problems before they occur.

The cash flow budget should be a living, breathing document that is changed as conditions change or new information is gathered. As conditions change, the cash flow budget for the remainder of the year can be modified and recalculated to see the effects of the projected cash balances later in the year. If the cash flow budget

C9 **MONITORING WORKSHEET** C9

Item from Cash Flow Statement	Month of: ______			Year-to-Date Totals		
	Amount Budgeted	Actual Results	Variance (+ or −)	Amount Budgeted	Actual Results	Variance (+ or −)
1 Beginning cash balance						
Operating Receipts: 2 Crops and feed						
3 Livestock and poultry						
4 Products: livestock and poultry						
5 Custom work						
6 Government payments and patronage dividends						
7 Hedging account withdrawals						
8						
Capital Receipts: 9 Breeding stock						
10 Machinery and equipment						
11						
Non-Farm Income: 12 Off-farm wages						
13 Interest and dividends						
14-15						
16 **TOTAL CASH AVAILABLE** (add lines 1 thru 15)						
Operating Expenses: 17 Labor hired (including employer taxes)						
18 Repairs — machinery and equipment						
19 Repairs — buildings and improvements						
20 Rents and leases						
21 Seed						
22 Fertilizer and lime						
23 Chemicals						
24 Custom machine hire						
25 Supplies						
26 Livestock expense (breeding, vet., etc.)						
27 Gas, fuel, oil						
28 Storage and custom drying						
29 Taxes (real estate and personal property)						
30 Insurance (property, liability, crop)						
31 Utilities (electricity, gas, telephone) [farm share]						
32 Marketing and transportation expense						
33 Auto (farm share)						
34-35						
36 Total Cash Operating Expenses						

SAMPLE

CONTINUED ON OTHER SIDE

FIGURE 15–11(a). CFS cash flow monitoring worksheet.

Source: Thomas L. Frey and Danny A. Klinefelter, *Coordinated Financial Statements for Agriculture,* 2nd ed. (Skokie, IL: AgriFinance, 1980).

MONITORING WORKSHEET CONTINUED

Item from Cash Flow Statement	Month of: ______ Amount Budgeted	Actual Results	Variance (+ or −)	Year-to-Date Totals Amount Budgeted	Actual Results	Variance (+ or −)
Livestock and Feed Purchases: 37 Feeder livestock						
38 Breeding livestock						
39 Feed purchased						
Capital Expenditures: 40 Machinery and equipment						
41 Building and improvements						
42						
Other Expenditures: 43 Hedging account deposits						
44 Gross family living withdrawals						
45 Non-farm business and investments						
46 Income tax and social security						
47						
48 Term loan payments — principal						
49 — interest						
50 **TOTAL CASH REQUIRED** (add lines 36 thru						
51 **CASH AVAILABLE LESS CASH REQUIRED**						
52 Inflows from savings						
53 Cash position before borrowing						
Money to be borrowed 54 — operating loans						
— intermediate and long term loans						
55 Operating loan payments — principal						
56 — interest						
57 Outflows to savings						
58 Ending cash balance						
Loan Balances: (at end of period) 59 Current year's operating loans						
60 Previous year's operating loans						
61 Intermediate and long term loans						
62 Total Loans						

FIGURE 15–11(b). CFS cash flow monitoring worksheet, continued.

Source: Thomas L. Frey and Danny A. Klinefelter, *Coordinated Financial Statements for Agriculture,* 2nd ed. (Skokie, IL: AgriFinance, 1980).

is maintained on a personal computer, this process is not difficult, primarily because all of the time-consuming mechanics can be programmed easily.

Summary

Cash flow planning and control are vital areas of farm and ranch business management. These areas really have always been important for farmers and ranchers but only recently has the importance of cash flow as the lifeblood of a business become so clear.

Cash flow statements can be prepared in two different manners: projected or actual. A projected statement, or a budget, should be thought of as a plan for the near future. This plan must be monitored throughout the time period, that is, it should be checked to see how closely actual conditions compare with projections. The purpose of this comparison is not to test the skills of the projector but rather to decide whether further adjustments and/or plans are needed to respond to changing conditions. An actual cash flow statement is easy to prepare, completes or closes the cash transactions portion of the record books, and provides an excellent starting point for preparing the next cash flow budget.

The underlying structure of the cash flow statement can be summarized in one simple equation: *sources of funds = uses of funds*. *Sources of funds* include beginning balances, cash receipts from business or nonbusiness sources, and loan receipts. *Uses of funds* include expenditures, family withdrawals, investments, and cash carryovers. Adjustments, such as savings account balance and operating loan changes, even out the flow funds.

Cash flow planning is a long, difficult, tedious process if it is done seriously and is really intended to reflect all of the near future plans of the business. Many estimates and outright guesses must be made which may seem to negate the value of the exercise. No prognonsticator, however, can see into the future with perfect accuracy. A good plan, or road map into the future, will help to define both goals and paths to meet those goals.

A careful understanding of the difference between the two major financial flow statements, the income statement and the cash flow statement, is absolutely necessary. The income statement is an accrual statement that portrays profitability. The cash flow statement is totally cash oriented and depicts liquidity or the ability to meet cash obligations as they come due.

CHAPTER 16

Relationships Among the Financial Statements

Early in this part of the book it was argued that each one of the three primary financial statements was built upon different assumptions and each serves a very different purpose. Although that argument is very true, it should also be evident that these three statements are not developed in a vacuum, each relates to the same business, and in fact the same numbers from the records system are often used in two separate statements and sometimes in all three.

The primary purpose of this chapter is to explore the relationships among the three statements. Once established, the use of these relationships will be extended to the development of projected or pro-forma statements. But first we should review the basic assumptions of each of the statements.

The balance sheet is the classic representation of the fundamental accounting equation at *one point in time*. It is a *stock* statement that describes the financial position of the business and/or family unit on a particular day. Because of the heavy reliance in agriculture on fixed, expensive, long-term assets such as land, buildings, and machinery, assets whose values have been greatly influenced by inflation, a double-column (market value and modified cost) balance sheet is necessary to maintain relationships with the income statement and to recognize the different needs of different users of the statement. Often the balance sheet is developed on a consolidated family-business basis, usually because the owner will prepare only one balance sheet, because the owner often operates the farm or ranch as an integral part of his family life, and because the assets and liabilities of a typical farm or ranch may often be difficult to split between business and family.

In contrast to the balance sheet, the income statement is a *flow* statement, that is, it is prepared to represent a *period of time* rather than a point in time. A farm or ranch income statement is usually prepared to cover a period of one year because the normal production cycle in agriculture makes more frequent statements difficult to estimate. However, for more continuous production patterns such as dairy or year-round-confinement hog or poultry production, more frequent income statements become easier if they are needed. The income statement must be developed on an *accrual* basis to be useful. A cash-based receipts-expenditures records system can be converted to an accrual income statement through the recognition of changes in inventories.

The cash flow statement is strictly *cash* based. It represents a monthly summary of cash receipts and cash expenditures as well as the steps taken, such as new borrowing and repayment of operating loans or flows of cash reserves between the savings and checking accounts, to adjust for differences between cash inflows and outflows. Like the income statement, the cash flow statement is a *flow* statement; it is prepared to cover a period of time. Once again, that period is usually one year, although it can be projected further into the future. A typical monthly cash flow statement is most useful if projected into the future and used as a budget or plan for cash flows for the next production year, and then monitored during the year.

The Balance Sheet and the Income Statement

To best understand the relationship between the balance sheet and the income statement, let us once again refer to the fundamental accounting equation:

$$A = L + NW$$

Also refer to the discussion of the fundamental accounting equation, the effect of financial transactions on the equation and the use of T-accounts in Chapter 2. The theoretical discussion of the fundamental accounting equation culminated in a pictorial representation of the relationship of types of accounts. That representation is shown again in Figure 16–1.

The formalized, formatted summary of the activity recorded in the individual income accounts and expense accounts is the income statement. Although that step may be clear, the next step of the relationship between the income statement and the balance sheet may not be so evident. By the position of the income and expense accounts in Figure 16–1, it is easy to see that changes in income and expenses directly translate into changes in net worth. In fact, *all of the changes in net worth* occur as a result of activity in the income and expense accounts, with the exception of money withdrawn from the business or money added from an outside source.

To confirm that this statement is true, ask yourself the question: How can net worth be increased? The most often heard reactions are to buy new assets or to pay off existing debts. But if we analyze these two reactions using the T-account strategy, both can be shown to have *no* effect on net worth.

If new assets are purchased, say 100 acres of land are added to an existing farm, the land account increased by the value of the 100 acres. Temporarily the asset side of the fundamental accounting equation increases. But what paid for that land? If cash was paid, then the cash account decreased by the same amount as the land account increased. Consequently assets net out the same as they were before the land was purchased and more importantly, net worth has not changed. If the land purchased was financed, then the mortgages payable account increases by the same amount as the land account increased. Consequently the fundamental equation remained in balance with no effect on net worth.

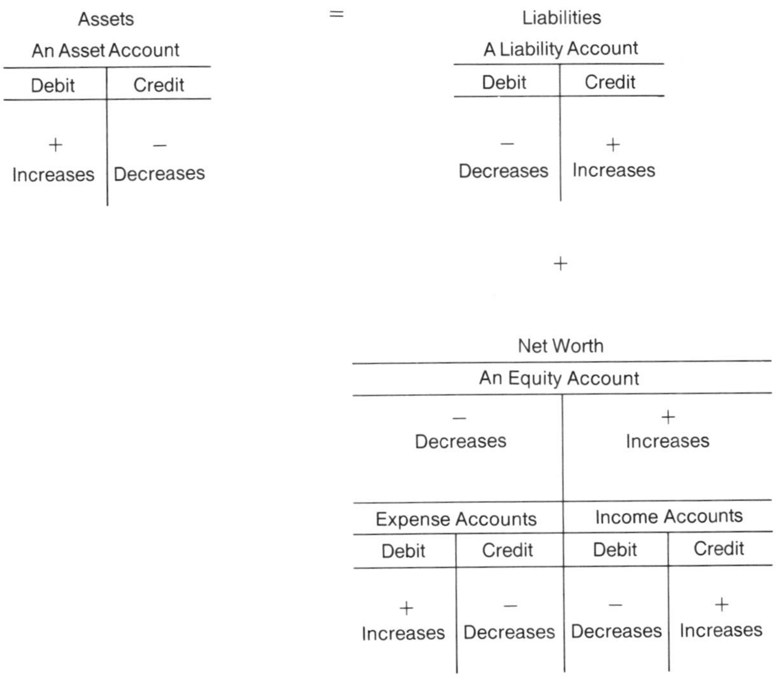

FIGURE 16–1. Relationship of types of accounts to the fundamental accounting equation.

Next consider what happens when a debt is paid off. The note payable account will decrease by the amount of the principal balance due. Although liabilities decrease, the cash account will decrease by the same amount and the equation will balance without effect on net worth. These two reactions, or cases, are set out in Figure 16–2.

In contrast to these two cases, which have no effect on net worth, consider a year in which crop income exceeded expenses by $20,000 (Net Farm Income on the income statement is $20,000). The crop income account increases and the cash account increases (if the crop is sold). Income is wrapped into net worth through the chart of Figure 16–1 and the net result is that assets increase by $20,000 and net worth increases by $20,000 (Case 3 in Figure 16–2). No other activity *within* the business can increase net worth!

There are other ways to affect net worth, but each of these ways implies transactions between the business and an outside agent. Net worth can be increased through inheritances or gifts or through outside investment dollars flowing into the business. Net worth will be decreased by gifts to others or through owner withdrawals for family purposes. A gift of $10,000 to the business is shown as Case 4 in Figure 16–2.

Statement of Owner Equity

The Statement of Owner Equity, the back side of the CFS Balance Sheet (Figure 16–3) presents a convenient location to summarize the relationship be-

Case 1: Buy assets (land)

a. Cash purchase of $200,000

Land		Cash		NW	
$200,000			$200,000	0	0

b. Financed purchase

Land		Mortgages Payable		NW	
$200,000			$200,000	0	0

Case 2: Repay Existing Debt ($100,000)

Cash		Mortgages Payable		NW	
	$100,000	$100,000		0	0

Case 3: Net Farm Income is $20,000

Crop Expenses		Crop Income		=	NW	
	$80,000	$100,000				$20,000

Cash	
$20,000	

Case 4: Gift of $10,000 to the Business

Cash		Capital-NW	
$10,000			$10,000

FIGURE 16–2. Effect of transactions on net worth.

tween the income statement and change in net worth. Mathematically, that relationship is:

Net worth at the beginning of the period
+ *Net investments, gifts, or inheritances*
+ *Net income*
− *Withdrawals for family and gifts*
= *Net worth at the end of the period*

Intuitively, the relationship is that net income describes *all* of the change in net worth between two successive balance sheets, with the two exceptions of outside money coming into or withdrawals from the business. Stated another way, the only way to increase net income is to *earn* it (or inherit it).

The mechanics of the CFS Statement of Owner Equity, as shown in Figure 16–3, generate a figure which represents net owner withdrawals. To finish the process, Schedule 10 of the CFS can now be completed (Figure 16–4). Schedule 10 adjusts net withdrawals for changes in personal asset inventory changes to calculate *gross* withdrawals. Gross withdrawals represent the *total* cash amount withdrawn from the business. Those gross withdrawals are allocated to general family expenditures such as food or utilities or to the purchase of items that are long-lived and that will be carried as personal assets on the balance sheet. Non-cash perquisites (such as a provided pickup or the gift of a slaughter steer) also represent withdrawals and thus reduce inventory and net worth. Although usually small relative to the value of cash withdrawals, they must also be recognized.

STATEMENT OF OWNER EQUITY

Net worth, beginning of period (modified cost)	$	______
Net income for period	+	______
Gifts and inheritances received for period	+	______
Additions to paid-in capital (for partnership or corporation)	+	______
TOTAL AVAILABLE	$	______
Less: Gifts made for estate transfer	–	______
Net worth, end of period (modified cost)	–	______
EQUALS NET WITHDRAWALS (transfer to sc[illegible] statement)	$	______

SAMPLE

SUMMARY

Period	Market Value Net Worth	Cost or Basis Net Worth	Net Worth Change Caused by Value Change
Ending	$______	$______	
Beginning	______	______	
Change	(a) $______	(b) $______	(a-b) $______

The above information and the statement on the reverse are furnished for the purpose of securing and maintaining credit and are certified to be complete and correct. The undersigned authorize the below named lender to make all inquiries deemed necessary to verify the accuracy of the information contained herein to determine my (or our) creditworthiness and to answer questions about his credit experience with me (or us). I (or we) agree to notify said lender promptly of any material change herein.

Prepared for (lender): ______ Date Signed ______, 19____

Applicant ______ Co-applicant ______

FIGURE 16–3. CFS statement of owner equity.

Source: Thomas L. Frey and Danny A. Klinefelter, *Coordinated Financial Statements for Agriculture,* 2nd ed. (Skokie, IL: AgriFinance, 1980).

The Balance Sheet Summary

The final block of the CFS Balance Sheet is the Summary (lower part of Figure 16–3) whose task is to segregate the effects of earned net worth change from the effects of value change. In addition to being relatively easy to interpret, the Summary is also easy to calculate. The change in market value is composed of two parts: (1) change due to earnings (change in cost net worth) and (2) change due to

SCHEDULE 10 — WITHDRAWALS FOR FAMILY LIVING

	Transfer Totals from Balance Sheets: Beginning Balance	Transfer Totals from Balance Sheets: Ending Balance	Change (+ or –)
Net family living withdrawals (from Statement of Owner Equity on B1)			$______
Marketable bonds & securities –	$______	______	______
Cash value of life insurance		______	______
Personal & recreational vehicles		[illegible]______	______
Household goods & personal effects	______	+$______	______
Non-farm real estate –	$______	+$______	______
Other non-farm investments –	$______	+$______	______
Medical & other personal accounts payable +	$______	–$______	______
Plus retirement account deposits			+______
GROSS FAMILY LIVING WITHDRAWALS			$______

SAMPLE

FIGURE 16–4. CFS income statement, schedule 10.

Source: Thomas L. Frey and Danny A. Klinefelter, *Coordinated Financial Statements for Agriculture,* 2nd ed. (Skokie, IL: AgriFinance, 1980).

value change. Subtracting the two change amounts as indicated will isolate the change in market value net worth resulting from inflation (or value change).

This segregation can be important to business management analysis. It is quite conceivable that market value net worth will increase from year to year while the cost net worth decreases. In other words, inflation has totally clouded the fact that more has been withdrawn from the business than has been earned. The farm or ranch has actually been disinvesting, but that fact is overshadowed by the increase or appreciation in asset values. Many U.S. farms and ranches found themselves in this position during the late 1970s and early 1980s. Each year they refinanced their operations at increasingly higher debt levels. Unfortunately, those new loans were collateralized by nothing more solid than appreciating asset values.

Projecting Financial Statements

Preparing financial statements that are projected into the future can be an extremely valuable management exercise. A complete set of projected (also called pro-forma) financial statements will give a good picture of where the farm or ranch is headed financially. Having this picture before the production year begins can help to revise previous decisions concerning marketing and production plans, to determine the best plan for purchases of new or replacement capital assets, and to determine existing debt repayment capabilities for the coming year. The bottom line is that the farmer or rancher will be able to begin the year with a detailed plan or outline of where he is going; in other words, he has a road map to follow in getting from one point to another. He has some idea of the obstacles that will have to be faced as well.

In taking a long trip, a traveler may change his mind about the direct route to be taken, about side excursions, and sometimes even about the final destination. Similarly, a farmer or rancher may change his mind after the production year has started. Changing his mind simply implies that the manager is updating or revising his plans based on new information; it does not imply that the planning process is useless. Going through the planning process will point out unrealistic goals and will help to avoid obstacles and potential problems. Knowing that a problem exists will alert the manager to find a way around it *before* it occurs.

The best, most useful plan or road map for a farmer or rancher is the cash flow budget because in that budget (if all of the CFS forms and worksheets have been completed all of the inflows and outflows of cash, feed requirements, crops produced, livestock, and other major inputs are summarized. But as has been argued many times in this text, cash flow provides an incomplete picture of the business. An accrual income statement is also needed.

The best planning process can be summarized as an iterative (or circular) pattern that begins with the market outlook for the coming year. The steps in this pattern are:

1. Determine the marketing and price outlook or projections for each of the major commodities.
2. Develop a preliminary marketing plan.

3. Determine optimal levels of inputs and output from each commodity, based upon production function principles.[1]
4. Combine production plans for each commodity and develop an optimal whole-farm or whole-ranch plan based upon projected input and output prices and resource constraints.[2]
5. Prepare a projected whole-farm or whole-ranch budget (or preliminary projected income statement) from the production plan prepared in Step 4.
6. Prepare a projected cash flow budget based on the plan and budget developed in Steps 4 and 5.
7. Check the levels of cash borrowing requirements projected by the cash flow budget.
   a. Are the requirements reasonable in relation to borrowing capacity?
   b. Can short-term cash flow problems be averted by modifying purchasing patterns?
   c. Can short-term cash flow problems be averted by modifying the output marketing plans?
8. Revise the marketing plan developed in Step 2. Repeat the planning process beginning with Step 3 based on the new marketing plan and new restrictions on borrowing.

Once a cash flow budget has been prepared after as many repetitions of this process as necessary, we can turn our attention to preparation of a financial statement package that can be used to obtain financing and that can be used as a road map throughout the year. The plan can also be used at a later date as a measuring stick to test how well the management of the business actually performed as compared to plans.

After a complete CFS Cash Flow Statement has been prepared, the planning process continues, with a pro-forma income statement. But first, a projected balance sheet should be prepared.

9. Prepare a projected balance sheet (cost basis only).
10. Prepare a projected income statement.

Balance Sheet Projection

A projected balance sheet should be prepared for the last day of the calendar or fiscal year so that it can be directly compared to past actual statements. It is necessary to project only the cost side of the CFS balance sheet, as the only two real goals of projecting a balance sheet are to provide all of the information needed for the income statement and to provide a check based on the Statement of Owner Equity.

[1] An excellent source of further background material concerning production economics principles can be found in R. D. Kay, *Farm Management: Planning, Control and Implementation,* 2nd ed. (New York: McGraw-Hill, 1986). Read especially Chapter 2 through 4.

[2] Kay (1981), Chapters 5 and 6, presents one means of obtaining the optimal whole-farm plan. Linear programming may also be used as can programmed budgeting.

The impression from Chapter 13 is undoubtedly that balance sheet preparation is a difficult, laborious process. While that may be true in some circumstances, especially for an actual statement, completing a projected balance sheet is relatively easy for two reasons: all of the calculations of market values and contingent tax liabilities are avoided by ignoring the market value column; and most of the year-end inventory information has already been prepared in the crop, feed, and livestock budget sheets (see Figures 15–5 and 15–6). Furthermore, the longer-term loan payments have been prepared (Figure 15–9) and capital purchases have been projected (Figure 15–8). Beyond these schedules or budget sheets, the cash flow statement itself will state the ending cash, savings, and loan balances. The most difficult task that remains is to complete the depreciation schedule for the coming year and to estimate accrued interest liabilities (see Figure 13–9). All other balance sheet entries will normally be estimated to remain the same as they were at the beginning of the year. Modified cost values for marketable and not readily marketable securities, machinery, breeding stock, retirement accounts, personal vehicles, household assets, and real estate will all remain the same except for adjustments of purchases, sales, and depreciation.

Income Statement Projection

Once the year-end projected balance sheet has been prepared, a projected income statement can be completed very quickly. Follow the same principles as laid out in Chapter 14 and use the actual beginning balance, projected revenues, and expenditures from the cash flow budget and the projected ending balance sheet. No additional information is needed to prepare the pro-forma income statement.

Statement of Owner Equity

The final step in the projection process is to prepare the satement of owner equity as discussed earlier in this chapter. This statement and Schedule I–10 from the income statement should be prepared to double-check the accuracy of the figures. The gross family living withdrawals calculated in Schedule I–10 should correspond to the annual total projected in the cash flow statement.

Summary

Although each of the big three financial statements has a unique function and assumptions very different from its counterparts, all three are closely tied together and represent three different views of the same business story.

All financial transactions throughout the year can be analyzed in an $A = L + NW$ format including income earned and expenses incurred since income and expense accounts are subsets of the net worth account. A summary of these accounts can thus provide an income statement. And yet, if filtered through a trial balance at any point in time, these same accounts would provide a balance sheet. If noncash transactions are ignored and only the transactions with an entry in the cash account are totaled, the cash flow statement would result.

The Balance Sheet Summary and Statement of Owner Equity show many of these relationships and also show the effect of value change relative to earned change. A very important relationship that falls out very quickly is $NW_{beg} + I + NI - W = NW_{\text{end}}$. This relationship drives home the fact that the only way to increase net worth (with the exception of gifts and inheritances) is to earn it through a positive net income.

An iterative planning sequence is necessary to fully develop a comprehensive, logical forward plan. In outline form, the steps of a planning process include:

1. Price projections
2. Marketing plan
3. Optimal combination of inputs and outputs
4. Optimal whole-farm plan
5. Projected whole-farm budget
6. Projected cash flow budget
7. Assess borrowing requirements
8. Revise marketing plan and other projections

These chapters have just scratched the surface of understanding financial statements. Many good reference materials are available for further study; two of the very best are the Penson and Nixon books referenced in the introduction to Part III on page 170.

PART IV

Income Tax Reporting and Management

Income tax reporting, is one of the most critical records/financial management problems in any business, critical in the sense that *everyone* must do it, it *must* be done within specified time constraints every year, the paperwork is very structured and intimidating, and there is a club hanging over our heads (in the form of the tax audit) if the job is not done correctly. The complexity of the current federal income tax legal structure scares most people, including most farmers and ranchers. This structural complexity is so severe that many taxpayers admit defeat before they even begin the reporting process. Admitting defeat before starting can grow into several financial problems and managerial weaknesses—to begin with, the fact that most farmers and ranchers dislike the work involved in maintaining records and have a fairly short tolerance for any paperwork. To carry that dislike onward, federal income tax returns are complex, someone else is going to fill out the tax return anyway, so why not just turn over the entire record-keeping process to an accountant?

Previously in this text, we have argued that although a farm or ranch manager may not need to do all of the pencil and paper work himself, he does need to maintain diligent control over all of his financial affairs. It has also been argued several times that monitoring the records system, and especially the cash flow position, is one of the very best ways to monitor the financial affairs of the business.

Few farmers and ranchers will, or probably should, complete their own final income tax return. But having a basic understanding of the federal income tax structure can be a great benefit to cash flow monitoring, to recognition of income tax opportunities before transactions are complete, and to general financial awareness. Furthermore, a basic knowledge of taxes can help the producer estimate his tax position before the tax year is over so that he can accomplish two short-term goals: accurate tax liability estimates for financial statement preparation, and tax planning to gain maximum benefits (or to minimize losses). Finally, a farmer or rancher might prepare a preliminary version of his return for the professional tax preparer, thereby reducing his bill and increasing accuracy.

Like the preceding three parts of this text, the following five chapters attempt to provide a basic understanding of the mechanics of tax reporting requirements.

These chapters are organized in a start-to-finish order rather than in an order of what is easiest or hardest to complete. The tax reporting process should obviously begin with two primary elements: (1) collection of all necessary financial information to complete the return and (2) collection of all necessary forms, schedules, instructions, and other reference materials. If the method laid out in the first sixteen chapters of this text has been followed, all of the necessary financial information has already been prepared and is readily available in the records summaries and depreciation schedule. All farmers and ranchers should obtain copies of several Internal Revenue Service publications, including the *Farmers Tax Guide* (IRS Publication 225) and *Your Federal Income Tax* (IRS Publication 17) as well as copies of a farm forms package before beginning the process.

CHAPTER 17

Asset Sales

The most burdensome federal income tax problems are encountered in preparation of the asset transactions forms. The primary difficulties in preparing these forms are determining which form to complete first, which form is the correct one for each transaction, and which section of each form is appropriate.

Internal Revenue Code Asset Sections

To begin the asset sales reporting process, an understanding of the *Internal Revenue Code* (IRC) sections dealing with asset sales is crucial. There are two basic types of assets, as recognized by the IRC: *capital assets* and *noncapital* (or *ordinary*) *assets*.[1] At this stage, we are most concerned with the assets that will qualify for *capital gains treatment,* that is, those assets that normally would be listed as intermediate or long-term on a consolidated balance sheet. But according to the IRC, we cannot simply call all of these items capital assets. Let's start with a definition of a capital asset and work from there.

Capital Assets: Section 1221

Capital assets are defined by Section 1221 of the IRC through an exclusion process. Section 1221 defines capital assets to be *all property* owned by the taxpayer *except:*

1. Stock in trade or other property included in inventory.
2. Property held by the taxpayer primarily for sale to customers in the ordinary course of his trade or business, such as grain or market livestock.

[1] A thorough review of Chapter 7 will greatly help in understanding the capital asset transactions discussed in this chapter.
Also, be aware of earlier warnings about the effect of the Tax Reform Act of 1986 on all tax concerns, especially investment tax credit and capital gains.

3. Trade or business property that is subject to depreciation.
4. Real property used in a trade or business.
5. Accounts and notes receivable acquired in normal business operations.
6. Copyrights and certain compositions held by the one who created them.
7. Certain short-term, non-interest-bearing government securities. (Bock and Harris)[2]

For most farmers and ranchers, most of their assets fall into one of the first four categories. The first two categories generally contain all of the normal short-term or current asset inventories, such as those items discussed in Chapter 6. Those items are normally labelled "ordinary income items. The sale of these assets will be covered in Chapter 19.

Categories 3 and 4 cover almost all business assets other than the inventories held primarily for sale. Consequently, such assets as tractors, combines, all machinery and equipment, breeding livestock, dairy and work stock, buildings, wells, fences, irrigation systems, and land are *not* classified as capital assets.

The remaining three categories are not typically encountered by farmers and ranchers. Accounts and notes receivable may be found, however, especially when payment is due from a cooperative or when a parcel of land is sold on an installment contract.

Despite the impression that all normal farm and ranch assets are excluded from Section 1221, there are several types of assets that must be recognized under this section. Examples of property owned by a farmer or rancher that would qualify as Section 1221 assets include:

1. Personal residence
2. Stocks, bonds, and other investments such as mutual funds
3. Commodity futures contracts held for speculative rather than hedging purposes
4. Household goods
5. Personal automobile
6. Sand, gravel, rock, and minerals on farm or ranch land
7. Standing timber[3]

Each of these items is, by a strict definition, nonbusiness and thus qualifies as a Section 1221 asset.

Business Assets: Section 1231

Although specifically excluded from the definition of a capital asset by Section 1221, many intermediate or long-term farm and ranch assets are treated *as if* they were capital assets. This treatment is offered by IRC Section 1231. The distinction is not a purely theoretical one, although on the surface it does seem rather foolish to separate an item from capital assets and then give it the same treatment as capital assets. The separation is made primarily because Section 1231 assets are treated as if they were Section 1221 assets *only* when there is a net gain on the sale

[2] C. Allen Bock and Phillip E. Harris, *1985 Farm Income Tax Schools Workbook* (University of Illinois, November 1985), p. 273.

[3] Bock and Harris, p. 273.

of these Section 1231 assets during the tax year. *Gain* means that the sale price is greater than the current adjusted basis on the date of the sale (see Chapter 7).

Before we go much further, we need to define specifically what is involved in Section 1231. This section applies only to *business tangible or real property* that is land or is normally depreciable. Consequently, it covers virtually all business intermediate or long-term assets. Section 1231 says that:

> Sales of assets used in the trade or business of farming, including drafts, breeding, dairy and sporting livestock, unharvested crops sold with the land, certain timber and other transactions receive a special tax benefit under Section 1231 of the Internal Revenue Code.
>
> The effect of Section 1231 is to tax the net gain on the aggregate of the qualifying transactions as a capital gain and to treat a net loss on the aggregate of the qualifying transactions as a capital loss.[4]

As noted by Bock and Harris,[5] this "rule" applies to property used in a trade or business or held for the production of income. This generally includes:

1. Depreciable property held for more than six months (one year for property acquired before June 23, 1984), or
2. Real property held for more than six months (one year for property acquired before June 23, 1984), that is *not* any of the following:
   a. Stock in trade or other property included in inventory.
   b. Property held by the taxpayer primarily for sale to customers in the ordinary course of business, such as gain and market livestock.

IRC Section 1231(b) definition of property used in a trade or business also includes the following:

> Livestock. (a) Cattle and horses, regardless of age, held by the taxpayer for draft, breeding, dairy or sporting purposes, and held for 24 months or more from the date of acquisition, and (b) livestock (but not including poultry), regardless of age, held by the taxpayer for draft, breeding, dairy or sporting purposes, and held for 12 months or more from the date of acquisition.
>
> Unharvested crops. In the case of an unharvested crop or land used in the trade or business and held for more than six months (one year if acquired before June 23, 1984), if the crop and the land are sold or exchanged (or compulsorily or involuntarily converted) at the same time and to the same person, the crop is considered as Section 1231 property.
>
> Timber.
> Coal or domestic iron ore.

Remember that assets held primarily for sale to customers in the ordinary course of business and assets that are properly included in inventory do *not* qualify for Section 1231 treatment. These items are considered to be ordinary sales and will be discussed in Chapter 19.

[4] John C. O'Bryne and Charles Davenport. *Farm Income Tax Manual,* 7th ed. (Indianapolis, IN: The Allen Smith Company, 1984.)

[5] Bock and Harris, pp. 283 and 284.

Long-Term versus Short-Term

As might have been noticed already, there is a major distinction as to how property is treated solely according to the length of time it has been owned prior to its sale. The distinction is a fairly simple one, although it is extremely crucial.

If property is sold within a time period of 6 months or less from the date it was purchased, the property is called *short-term*. If the property is held more than 6 months from the date it was purchased, it is called *long-term*.[6]

The more-than-6-months rule is interpreted very strictly. For example, an asset purchased January 12, 1986 and sold July 12, 1986 is ruled to have been held 6 months and is therefore a short-term asset. However, if that same asset had been sold on July 13, 1986, instead of July 12, it would have been classified as long-term.

The 6-month holding period requirement is thus based on calendar months rather than days (leap years do not provide an escape just because they have 366 days). The holding period is legally calculated by excluding the date of purchase and including the date of sale or exchange.

Special Holding Period Requirements

The 6-month holding period rule applies to all assets except for commodity futures contracts and certain livestock. Commodity futures contracts held for speculation (rather than hedging) purposes have no holding-period rule to distinguish between short-term and long-term. At the other extreme, horses and cattle held for draft, dairy, breeding, or sporting purposes must be held for at least 24 months to qualify for long-term status (and also to qualify as a Section 1231 asset) and hogs, goats, sheep, and other livestock must be held for 12 months to qualify for long-term status.

Importance of the Holding Period. The distinction between short-term and long-term status has very important consequences for the eventual taxation of the asset sale. The concept of the long-term versus short-term capital gain or loss fills many taxation textbooks and journal articles primarily because of the tax advantages given to the long-term capital gain or loss. There are two primary advantages of meeting or exceeding the holding period requirement: (1) Section 1231 treatment and (2) Capital gains exclusion.

The first advantage seems to be a circular argument: if the holding period requirement is met, the business asset meets Section 1231 requirements and receives Section 1221 treatment if there is a gain. This advantage will become clearer with a definition of the capital gains exclusion, but it must be noted at the outset that the Section 1231 treatment described before applies *only* if the holding period requirement is met.

[6]The holding period was one year rather than 6 months prior to the passage of the Tax Reform Act of 1984. The one-year period applies to all assets purchased prior to June 23, 1984. Note the special holding periods for all forms of livestock. The Tax Reform Act of 1986 phases out capital gains exclusions beginning in 1987.

The capital gains exclusion is a very prominent tax advantage offered to asset investors and, through Section 1231, to business owners. Simply stated, capital gains treatment *excludes* 60 percent of the gain from the sale of an asset from taxation. In other words, if a $1,000 gain resulted from the sale of an asset, only $400 of that gain will be taxed, if the holding period requirement had been met. Thus, $600 of income (60 percent of the $1,000 gain) is excluded from taxation. Conversely, if a $1,000 net loss results from the sale of a *Section 1231* asset, the entire $1,000 can be counted as an ordinary loss, thereby offsetting other income dollar for dollar.

Depreciation Recapture

Before we can turn to examples of the application of capital gains treatment, we need to add one more complication. That complication involves the IRC sections commonly referred to as the *recapture* sections: Sections 1245, 1250, 1252, 1254, and 1255. The basic recapture section is Section 1245; the others relate to specific depreciable assets beyond those covered by Section 1245. These sections relate *only* to depreciable assets, thus they do not apply to land or any other nondepreciable asset.

The primary purpose of the recapture provisions is to segregate the gain-realized from the sale of a depreciable asset into two components: the amount of gain realized because the asset was depreciated and the amount of gain realized because the sale price was greater than its purchase price. The amount of gain realized solely because the asset was depreciated must be repaid (or recaptured) as ordinary income rather than as capital gains.

Before we describe the tax effects, an example might help clear up this difficult language. Assume a tractor was purchased on January 1, 1978, and sold on January 2, 1986. The tractor originally cost $20,000 and a useful life of 8 years and a salvage value of $4000 were selected. At the end of 1984, $16,000 of depreciation had been taken; consequently the adjusted basis on the date of sale was $4000. Assume further that the tractor was sold for $25,000 in January 1985.

In determining how much capital gain can be claimed, we must first recognize that the tractor qualified for Section 1231 treatment since it was owned for more than 12 months. But further, Section 1245 applies. Section 1245 states that on the sale, exchange, or involuntary conversion of most Section 1245 property, the recognized gain is treated as ordinary income to the extent of the depreciation taken after December 31, 1967.[7]

What this means is that we must categorize all gain up to the amount of depreciation claimed as ordinary income. Reconsider our tractor example. The total gain realized after the 1986 sale was $21,000: the $25,000 sale price minus the $4000 adjusted basis. But of the $21,000 gain, $16,000 must be taxed as ordinary income since a total of $16,000 was taken in depreciation. The remaining $5000 is treated as long-term capital gain and qualifies for the 60 percent exclusion. The

[7]Recognized gain is treated as ordinary to the extent of depreciation taken after December 31, 1969, in the case of depreciable livestock and for single-purpose agricultural and horticultural structures built after 1980.

bottom line for this sale is that $18,000 is taxable ($16,000 recapture plus 40 percent of $5000 long-term capital gain or $2000).

If our tractor had been sold for $18,000 instead of $25,000, the total gain would be $14,000 ($14,000 sale price minus $4000 adjusted basis). The entire $14,000 would be taxed as ordinary income since that amount is less than the amount of depreciation taken.

A simple way to remember how Section 1245 works is to compare the sale price to the original purchase price. If the sale price is higher than the original purchase price, that difference will be treated as capital gain. If the sale price is lower than the original purchase price, there will be *no* capital gain treatment. The second step is to bring the adjusted basis into the analysis. If the sale price is higher than the original purchase price, the difference between the original purchase price and the basis is treated as ordinary gain; in other words, that amount is recaptured as ordinary income since the depreciation offset ordinary income. If the sale price is lower than the purchase price, the difference between the sale price and the adjusted basis is treated as ordinary income.

Reporting Asset Sales

Once the defintions and all of the applicable asset sale IRC sections have been identified, attention can be turned to applying these principles to real examples and to reporting those sales on Internal Revenue Service (IRS) forms. Figure 17–1 presents a schematic diagram of how asset sales are reported. There are two basic forms on which most asset sales are reported: Schedule D and Form 4797. Form 4797 (shown in Figure 17–2) handles business assets and Schedule D (Figure 17–3) primarily handles capital assets, but net long-term gains from Form 4797 are carried to Schedule D.

Form 4797

Starting in the upper left-hand corner of Figure 17–1, the first decision is whether the asset is to be classified as a capital asset or as a business asset. Taking the business asset side first, short-term gains and losses on both depreciable and nondepreciable business assets are reported on Form 4797, Part II (Figure 17–2[a]). But long-term assets must be separated into nondepreciable items (such as unimproved land and raised breeding or dairy livestock), which are reported on Form 4797, Part I (Figure 17–2[a]) and depreciable items subject to Section 1245 recapture which are reported on Form 4797, Part III (Figure 17–2[b]).

The recapturable depreciation taken on Section 1245 property is separated from long-term capital gain in a relatively easy fashion in Form 4797, Part III. Further, recapture amounts for the other sections (1250, 1252, 1254, and 1255) are also calculated. Although not a complete description, the following list categorizes assets according to the applicable recapture section:

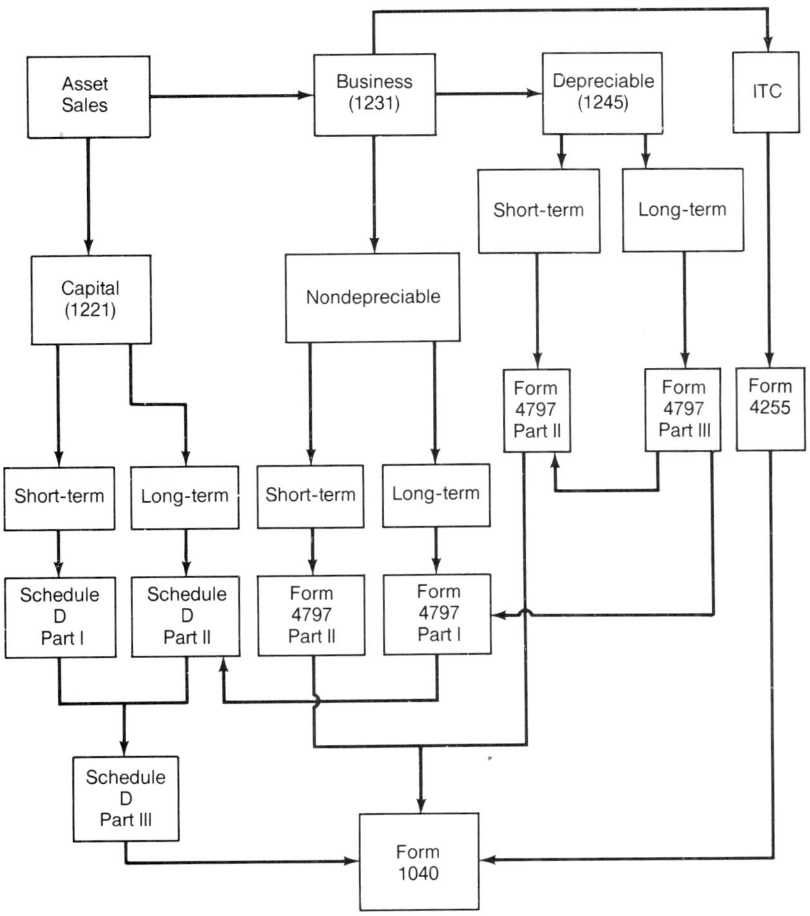

FIGURE 17–1. Schematic diagram of asset sales reporting.

Section 1245: All nonreal property including machinery, equipment, breeding livestock, and special purpose buildings.[8]
Section 1250: General-purpose buildings such as barns and machinery sheds.[9]
Section 1252: Soil and water conservation and land-clearing expenditures made within the last 10 years.
Section 1254: Intangible oil and gas drilling and development costs.
Section 1255: Property on which cost-sharing payments were received and were included from income such as Soil Conservation Service-approved and cost-shared structures.

The primary distinction between Section 1245 and the other recapture sections is that 1245 applies to nonreal property and the others apply to real property. However, there is another distinction that also carries a great impact. Section 1250 states that *only* the amount of depreciation claimed over the straight-line amount

[8] Section 1245 property now includes some real property as a result of the Economic Recovery Tax Act of 1981.
[9] Recapture on Section 1250 property is calculated *only* on the amount of depreciation claimed in *excess* of what is allowed under the straight-line method.

Form **4797**
Department of the Treasury
Internal Revenue Service (X)

Gains and Losses From Sales or Exchanges of Assets Used in a Trade or Business and Involuntary Conversions
(And Computation of Recapture Amounts Under Sections 179 and 280F)
▶ **Attach to your tax return. See Separate Instructions**

OMB No. 1545-0184
1985
27

Name(s) as shown on return | **Identifying number**

Part I **Sales or Exchanges of Property Used in a Trade or Business and Involuntary Conversions From Other Than Casualty and Theft—Property Held More Than 6 months (More Than 1 Year if Acquired Before 6/23/84)**

Notes:
- *Use Form 4684 to report involuntary conversions from casualty and theft.*
- *If you sold property that you claimed investment credit on, get Form 4255 to see if you are liable for recapture of the credit.*
- *File Form 6198 if you are reporting a loss and have amounts invested in the activity for which you are not at risk. (See instructions under "Special Rules.")*

(a) Description of property	(b) Date acquired (mo., day, yr.)	(c) Date sold (mo., day, yr.)	(d) Gross sales price	(e) Depreciation allowed (or allowable) since acquisition	(f) Cost or other basis, plus improvements and expense of sale	(g) LOSS ((f) minus the sum of (d) and (e))	(h) GAIN ((d) plus (e) minus (f))
1							

2 Gain, if any, from Form 4684, Section B, line 21 . . . . .
3 Section 1231 gain from installment sales from Form 6252, line 22 or 30 . . . . .
4 Gain, if any, on line 31 from other than casualty and theft . . . . .
5 Add lines 1 through 4 in column (g) and column (h) . . . . . ( )
6 Combine columns (g) and (h) of line 5. Enter gain or (loss) here, and on the appropriate line as follows (Partnerships see the instructions for your line references.): . . . . .
a If line 6 is zero or a loss, enter the amount on line 9 below and skip lines 7 and 8. (S corporations, enter the loss on Schedule K (Form 1120S), line 5.)
b If line 6 is a gain, see the instructions under Part I, **Nonrecaptured Net Section 1231 Losses**.
Note: *If you had no prior year section 1231 losses and line 6 is a gain, enter the gain from line 6 as a long-term capital gain on Schedule D.*
7 Nonrecaptured net section 1231 losses from prior years. (See instructions.) . . . . .
8 Subtract line 7 from line 6. If zero or less, enter zero . . . . .
a If line 8 is zero, enter the amount from line 6 on line 10 below.
b If line 8 is more than zero, enter the amount from line 7 on line 10 below, and enter the amount from line 8 as a long-term capital gain on Schedule D. See specific instructions for line 8b.

Part II **Ordinary Gains and Losses**

(a) Description of property	(b) Date acquired (mo., day, yr.)	(c) Date sold (mo., day, yr.)	(d) Gross sales price	(e) Depreciation allowed (or allowable) since acquisition	(f) Cost or other basis, plus improvements and expense of sale	(g) LOSS ((f) minus the sum of (d) and (e))	(h) GAIN ((d) plus (e) minus (f))

9 Loss, if any, from line 6 . . . . .
10 Gain, if any, from line 6 or amount from line 7 if applicable . . . . .
11 Gain, if any, from line 30, Part III . . . . .
12 Net gain or (loss) from Form 4684, Section B, lines 13 and 20a . . . . .
13 Ordinary gain from installment sales from Form 6252, line 21 or 29 (Applies only to sales before 6/7/84) . .
14 Recapture of section 179 deduction on property dispositions by partners and S corporation shareholders (see instructions) . . . . .
15 Other ordinary gains and losses (include property held 6 months or less (1 year or less if acquired before 6/23/84)):

16 Add lines 9 through 15 in column (g) and column (h) . . . . . ( )
17 Combine columns (g) and (h) of line 16. Enter gain or (loss) here, and on the appropriate line as follows: . . . . .
a For all except individual returns: Enter the gain or (loss) from line 17, on the return being filed. See instructions for Part II for specific line references.
b For individual returns:
(1) If the loss on line 9 includes a loss from Form 4684, Section B, Part II, column (b)(ii), enter that part of the loss here and on line 19 of Schedule A (Form 1040). Identify as from "Form 4797, line 17b(1)" . . . . .
(2) Redetermine the gain or (loss) on line 17, excluding the loss (if any) on line 17b(1). Enter here and on Form 1040, line 15.

For Paperwork Reduction Act Notice, see page 1 of separate instructions. Form **4797** (1985)

(309)

FIGURE 17–2(a). IRS Form 4797, parts I and II.

Form 4797 (1985) Page 2

Part III Gain From Disposition of Property Under Sections 1245, 1250, 1252, 1254, 1255

Skip section 1252 on line 26 and in the instructions, if you did not dispose of farmland, or if a partnership files this form.

18 Description of sections 1245, 1250, 1252, 1254, and 1255 property:	Date acquired (mo., day, yr.)	Date sold (mo., day, yr.)
A		
B		
C		
D		

Relate lines 18A through 18D to these columns ▶ ▶ ▶ ▶	Property A	Property B	Property C	Property D
19 Gross sales price				
20 Cost or other basis plus expense of sale				
21 Depreciation (or depletion) allowed (or allowable)				
22 Adjusted basis, subtract line 21 from line 20				
23 Total gain, subtract line 22 from line 19				
24 **If section 1245 property:**				
a Depreciation allowed (or allowable) (see instructions)				
b Enter smaller of line 23 or 24a				
25 **If section 1250 property:** (If straight line depreciation used, enter zero on line 25g unless you are a corporation subject to section 291.)				
a Additional depreciation after 12/31/75				
b Applicable percentage times the smaller of line 23 or line 25a (see instructions)				
c Subtract line 25a from line 23. If line 23 is not more than line 25a, skip lines 25d and 25e				
d Additional depreciation after 12/31/69 and before 1/1/76				
e Applicable percentage times the smaller of line 25c or 25d (see instructions)				
f Section 291 amount (For corporations only.)				
g Add lines 25b, 25e, and 25f				
26 **If section 1252 property:**				
a Soil, water, and land clearing expenses				
b Line 26a times applicable percentage (see instructions)				
c Enter smaller of line 23 or 26b				
27 **If section 1254 property:**				
a Intangible drilling and development costs deducted after 12/31/75 (see instructions)				
b Enter smaller of line 23 or 27a				
28 **If section 1255 property:**				
a Applicable percentage of payments excluded from income under section 126 (see instructions)				
b Enter the smaller of line 23 or 28a				

Summary of Part III Gains (Complete property columns A through D through line 28b before going to line 29)

29 Total gains for all properties (add columns A through D, line 23)	
30 Add columns A through D, lines 24b, 25g, 26c, 27b, and 28b. Enter here and in Part II, line 11	
31 Subtract line 30 from line 29. Enter the portion from casualty and theft on Form 4684, Section B, line 15; enter the portion from other than casualty and theft on Form 4797, Part I, line 4	

Part IV Complete This Part Only If You Elect Out of the Installment Method And Report a Note or Other Obligation at Less Than Full Face Value

☐ Check here if you elect out of the installment method.

Enter the face amount of the note or other obligation ▶ ____________

Enter the percentage of valuation of the note or other obligation ▶

Part V Computation of Recapture Amounts Under Sections 179 and 280F When Business Use Drops to 50% or Less (See instructions for Part V.)

	(a) Section 179	(b) Section 280F
1 Section 179 expense deduction or section 280F recovery deductions		
2 Depreciation (see instructions)		
3 Recapture amount. (Subtract line 2 from line 1.)		

(310)

FIGURE 17–2(b). IRS Form 4797, parts III and IV.

SCHEDULE D
(Form 1040)
Department of the Treasury
Internal Revenue Service (X)

Capital Gains and Losses and Reconciliation of Forms 1099–B

▶ Attach to Form 1040. ▶ See Instructions for Schedule D (Form 1040).

OMB No. 1545-0074
1985
12

Name(s) as shown on Form 1040 | Your social security number

Part I Summary of Forms 1099–B for Sales of Stocks, Bonds, Etc.

1a Report here and on line 37, Part VII, page 2, total sales of stocks, bonds, etc., reported for 1985 by your broker to you on Form(s) 1099–B or an equivalent substitute statement(s), such as a broker's confirmation statement	1a		
Note: *Also complete Part VII if you received one or more Form(s) 1099–B or equivalent statement(s) for 1985, for bartering income.*			

Part II Short-term Capital Gains and Losses-Assets Held Six Months or Less (one year or less if acquired before 6/23/84)

(a) Description of property (Example, 100 shares 7% preferred of "Z" Co.)	(b) Date acquired (Mo., day, yr.)	(c) Date sold (Mo., day, yr.)	(d) Gross sales price	(e) Cost or other basis (see instructions)	(f) LOSS If column (e) is more than (d) subtract (d) from (e)	(g) GAIN If column (d) is more than (e) subtract (e) from (d)
1b						

2 Short-term gain from sale or exchange of a principal residence from Form 2119, lines 6 or 12	2		
3 Short-term gain from installment sales from Form 6252, lines 22 or 30 . . .	3		
4 Net short-term gain or (loss) from partnerships, S corporations, and fiduciaries .	4		
5 Add lines 1b through 4 in columns (f) and (g)	5	( )	
6 Combine columns (f) and (g) of line 5 and enter the net gain or (loss)		6	
7 Short-term capital loss carryover from years beginning after 1969		7	( )
8 Net short-term gain or (loss), combine lines 6 and 7		8	

Part III Long-term Capital Gains and Losses-Assets Held More Than Six Months (more than one year if acquired before 6/23/84)

(a)	(b)	(c)	(d)	(e)	(f)	(g)
9						

10 Long-term gain from sale or exchange of a principal residence from Form 2119, lines 6, 8, or 12	10		
11 Long-term gain from installment sales from Form 6252, lines 22 or 30 . . .	11		
12 Net long-term gain or (loss) from partnerships, S corporations, and fiduciaries .	12		
13 Add lines 9 through 12 in columns (f) and (g)	13	( )	
14 Combine columns (f) and (g) of line 13 and enter the net gain or (loss)		14	
15 Capital gain distributions		15	
16 Enter gain from Form 4797, lines 6 or 8b		16	
17 Combine lines 14 through 16		17	
18 Long-term capital loss carryover from years beginning after 1969		18	( )
19 Net long-term gain or (loss), combine lines 17 and 18		19	

Note: *Complete the back of this form. However, if you have capital loss carryovers from years beginning before 1970, do not complete Parts IV or V. See Form 4798 instead.*

For Paperwork Reduction Act Notice, see Form 1040 Instructions. **Schedule D (Form 1040) 1985**

(77)

FIGURE 17–3(a). IRS Schedule D, parts I, II, and III.

Schedule D (Form 1040) 1985 12 Page 2

Name(s) as shown on Form 1040 (Do not enter name and social security number if shown on other side) | Your social security number

Part IV Summary of Parts II and III

20 Combine lines 8 and 19, and enter the net gain or (loss) here	20	
Note: *If line 20 is a loss, skip lines 21 through 23 and complete lines 24 and 25. If line 20 is a gain, complete lines 21 through 23 and skip lines 24 and 25.*		
21 If line 20 shows a gain, enter the smaller of line 19 or line 20. Enter zero if there is a loss or no entry on line 19 (21)		
22 Enter 60% of line 21	22	
If line 22 is more than zero, you may be liable for the alternative minimum tax. See Form 6251.		
23 Subtract line 22 from line 20. Enter here and on Form 1040, line 13	23	
24 If line 20 shows a loss, enter one of the following amounts: a If line 8 is zero or a net gain, enter 50% of line 20; b If line 19 is zero or a net gain, enter line 20; or c If line 8 and line 19 are net losses, enter amount on line 8 added to 50% of the amount on line 19	24	
25 Enter here and as a loss on Form 1040, line 13, the smallest of: a The amount on line 24; b $3,000 ($1,500 if married and filing a separate return); or c Taxable income, as adjusted	25	

Part V Computation of Post-1969 Capital Loss Carryovers from 1985 to 1986
(Complete this part if the loss on line 24 is more than the loss on line 25)

26 Enter loss shown on line 8; if none, enter zero and skip lines 27 through 30, then go to line 31	26	
27 Enter gain shown on line 19. If that line is blank or shows a loss, enter zero	27	
28 Subtract line 27 from line 26	28	
29 Enter smaller of line 25 or line 28	29	
30 Subtract line 29 from line 28. This is your short-term capital loss carryover from 1985 to 1986	30	
31 Subtract line 29 from line 25. (Note: If you skipped lines 27 through 30, enter amount from line 25.)	31	
32 Enter loss from line 19; if none, enter zero and skip lines 33 through 36	32	
33 Enter gain shown on line 8. If that line is blank or shows a loss, enter zero	33	
34 Subtract line 33 from line 32	34	
35 Multiply amount on line 31 by 2	35	
36 Subtract line 35 from line 34. This is your long-term capital loss carryover from 1985 to 1986	36	

Part VI Complete This Part Only If You Elect Out of the Installment Method and Report a Note or Other Obligation at Less Than Full Face Value

☐ Check here if you elect out of the installment method.
Enter the face amount of the note or other obligation. ▶ ....
Enter the percentage of valuation of the note or other obligation. ▶

Part VII Reconciliation of Forms 1099–B With Tax Return (Complete this part if you received one or more Form(s) 1099–B or an equivalent broker's substitute statement(s) reporting sales of stock, bonds, etc., or bartering income.)

SECTION A.—Reconciliation of Sales of Stocks, Bonds, etc.

37 Total sales of stock, bonds, etc., reported for 1985 by your broker to you, on Form(s) 1099–B or an equivalent substitute statement(s), such as a broker's confirmation statement. Also include on line 1a, page 1, Part I	37	
38 Proceeds from sale or exchange of capital assets reported on Schedule D, but not included in line 37	38	
39 Add lines 37 and 38	39	
40 Part of line 37 not reported on Schedule D this year (attach explanation)	40	
41 Subtract line 40 from line 39	41	

Note: *The amount on line 41 should be the same as the total of all amounts on page 1, lines 1b and 9 of column (d).*

SECTION B.—Reconciliation of Bartering Income Indicate below, the amount of bartering income reported on each form or schedule.		Amount of bartering from Form 1099–B or equivalent statement
42 Form 1040, line 22	42	
43 Schedule C (Form 1040)	43	
44 Schedule D (Form 1040)	44	
45 Schedule E (Form 1040)	45	
46 Schedule F (Form 1040)	46	
47 Other (identify) (if not taxable, indicate reason—attach additional sheets if necessary) ▶	47	
48 Total (add lines 42 through 47)	48	

Note: *The amount on line 48 should be the same as the total bartering on all Forms 1099–B or equivalent statements received.*

(78)

FIGURE 17–3(b). IRS Schedule D, parts IV, V, VI, and VII.

is recaptured. In other words, not all of the depreciation claimed in the past may be recaptured. In fact, if the straight-line depreciation method has been used, none of the depreciation claimed will be recaptured as ordinary income. The entire gain, defined as the difference between the sales price and the *adjusted* basis, would then be treated as capital gain.

Just when we think that we have all of the definitions under control, the law is changed. The Economic Recovery Tax Act of 1981 (ERTA) changed the status of some types of real property from Section 1250 status to Section 1245 status. At this writing, Section 1245 recapture rules apply to all 15-year, 18-year, or 19-year real property, that is, property acquired after December 31, 1980, to which ACRS rules apply. Two of the exceptions to this reclassification will apply to many farmers and ranchers: residential rental property remains Section 1250 property; but, more important, 15-year, 18-year, or 19-year property that has been depreciated by the straight-line alternative to ACRS is classified neither as Section 1245 nor as Section 1250—*all* of the gain is treated as capital gain!

Once the sales have been recorded in their proper place on Form 4797, Parts I, II, or III (see Figure 17–2[a] and [b]) and the appropriate calculations have been performed, the information is combined and transferred as indicated in Figure 17–1. Short-term gains (including depreciation recapture from Part III) are combined in Part II and then transferred to the summary tax report, IRS Form 1040. The long-term gain from Part III is combined with the gain on Part I and then is transferred to Schedule D (Figure 17–3) to be combined with nonbusiness capital gains.

Schedule D

We began the asset sales reporting process characterized in Figure 17–1 by taking the difficult decision first, the items destined for Form 4797. Now we can return to the upper left-hand corner and follow through a process for handling capital assets (Section 1221 assets). Short-term gains and losses are first reported on Schedule D, Part I, whereas long-term capital gains and losses are reported on Schedule D, Part II (Figure 17–3[a]). The total net long-term gains in Part II, Line 19 includes the gains transferred from Form 4797. Gains and losses offset each other on Schedule D to a certain extent. The basic rule is that short-term losses offset (or are subtracted from) short-term gains in Part I and long-term losses offset long-term gains in Part II, both on a dollar-for-dollar basis. Furthermore, the net short-term gain (loss) is added to the net long-term gain (loss) in Part III *before* the 60 percent capital gains exclusion is calculated (see Figure 17–3[b]).

From here, the calculation procedure is quite simple if both net short-term and net long-term *gains* are involved, but is significantly more complicated if one or the other involves a *loss*. To examine the netting procedure, we need to look at each of six combinations of short- and long-term gains and losses. Specific attention should be given to Lines 20–25 of Figure 17–3(b) as shown in Table 17–1.

Short-term, Long-term Gain. If both are gains, 60 percent of the long-term gain is subtracted from the total gain. The result is then transferred to Form 1040. That net result in this case is 100% of the short-term (ST) gain or 100 percent

TABLE 17–1 Example Calculations of Capital Gain Netting Procedures

Schedule D Line No.	Explanation	Case 1 ST Gain, LT Gain	Case 2 ST Gain, LT Loss (Net Gain)	Case 3 ST Loss, LT Gain (Net Gain)	Case 4 ST Loss, LT Gain (Net Loss)	Case 5 ST Gain, LT Loss (Net Loss)	Case 6 ST Loss, LT Loss
8	Net short-term gain	6,000	10,000	(6,000)	(10,000)	6,000	(10,000)
19	Net long-term gain	10,000	(6,000)	10,000	6,000	(10,000)	(10,000)
20	Net gain or (loss)	16,000	4,000	4,000	(4,000)	(4,000)	(20,000)
21	Smaller of long-term gain or net gain	10,000	0	4,000			
22	60% of Line 21	6,000	0	2,400			
23	Line 20 – line 22	10,000	4,000	1,600			
24(a) (b) (c)	Loss calculations				4,000	2,000	15,000
25(a) (b) (c)	Loss limitations				3,000	2,000	3,000
26	Short-term loss carryover				10,000		10,000
27					6,000		0
28					4,000		10,000
29					3,000		3,000
30					1,000		7,000
31	Long-term loss carryover						0
32							10,000
33							0
34							10,000
35							0
36							10,000

× $6,000 plus 40 percent of long-term (LT) gain or 40 percent × $10,000 for a taxable gain of $10,000. (Table 17–1, Case 1)

Short-term Gain, Long-term Loss (Net Gain). If short-term gains exceed long-term losses, no capital gains exclusion results. In fact, the long-term losses offset the short-term gains and the net gain is fully taxable. (Table 17–1, Case 2)

Short-term Loss, Long-term Gain (Net Gain). If long-term gains exceed short-term losses, the capital gain exclusion applies to the *overall* net gain. Thus, short-term losses offset or reduce long-term gains *before* the exclusion. (Table 17–1, Case 3)

Short-term Loss, Long-term Gain (Net Loss). When a net loss results, an entirely different set of regulations comes into force. If short-term losses are greater than long-term gains, the net loss is carried to Form 1040 to reduce other taxable income.

However, there is a maximum limit on the total amount of capital losses that can be used to reduce other income in any one tax year. That maximum limit is either \$3,000 or the taxable income without reference to capital losses, whichever is *less*. For example, assume a taxpayer earned \$2,500 in income but had a net capital loss (Line 24) of \$7,000. That maximum amount of loss he can use this year is \$2,500. In other words, taxable income cannot be reduced below zero.

Fortunately, the tax laws allow the unused \$4,500 loss (\$7,000 − 2,500) to be carried over to the next tax year (see Figure 17–3(b), Part V). Thus, although the loss cannot be used this year, it can eventually be used. Before we consider another loss-gain combination, let's assume our taxpayer earned \$8,000 in taxable income rather than \$2,500. Now the maximum amount of capital loss that can be claimed is \$3,000, the statutory limit. The remaining \$4,000 loss (\$7,000 − 3,000) is carried over to the next tax year. (Table 17–1, Case 4)

Short-term Gain, Long-term Loss (Net Loss). If long-term losses exceed short-term gains, a net loss again results, but now we see the flip side of the capital gains exclusion. That flip side is a long-term capital *loss exclusion*. Net losses caused by long-term capital losses are reduced by a 50% capital loss exclusion. Thus the taxable income-reducing value of a long-term capital loss is cut in half. The same maximum limits apply. (Table 17–1, Case 5)

Short-term Loss, Long-term Loss. When both short- and long-term losses are realized, the netting procedure remains substantially the same as in the previous case. Half of the long-term loss is excluded. The major difference between this case and previous cases appears in allocating the loss carryover to future tax years. (Table 17–1, Case 6)

Investment Credit Recapture

One further step must be considered in reporting the sale of an asset, but luckily the step is not nearly as complicated as the Schedule D–Form 4797 process. This extra step applies to all business assets that qualified for the *investment tax credit* when originally purchased. If the asset qualified for investment tax credit (ITC), there is a potential recapture of that credit. In other words, part of the credit may have to be repaid. Before we discuss the recapture amount, be careful to recognize that it does not matter whether the ITC was actually claimed. The Internal Revenue Code *assumes* that ITC was claimed; therefore recapture must be paid even if the original ITC amount was foregone.

Recapture of ITC is due whenever the property was held or owned for a period of time less than the useful life originally assigned to the item. In determining the amount to be recaptured, ITC allowable is recalculated based on the years that the asset was actually owned. The difference between that amount and the ITC originally claimed must be repaid or recaptured. The ITC recapture is calculated on IRS Form 4255 (Figure 17–4[a]) with the aid of the specific instructions on the back (Figure 17–4[b]).

Asset Sales Examples

Now that all of the ground rules have been set, it is time to consider a set of example asset sales to clear the waters and to view the mechanics of the asset sales reporting process. Consider the asset transactions and data presented in Table 17–2. Analyze each one and try to determine where and how each one should be reported.

Analysis of the Transactions

Eight of the ten transactions presented in Table 17–2 belong on the forms already discussed in this chapter: Schedule D, Form 4255, and Form 4797. However, Transactions 4 and 7 do *not* belong here. The completed forms for the transactions that *do* belong here are shown in Figures 17–5 through 17–7 in the order in which they should be completed.[10] Even though the forms are shown completed, it should be useful to analyze each individual transaction and the summaries.

Transaction 1, Land. The farmland was used in a farming business and was held for more than six months, therefore it qualifies for Section 1231 treatment. Since land is not depreciable and no soil or water conservation expenditures were made, none of the depreciation or ITC recapture provisions apply; consequently the point of original entry would be Form 4797, Part I, Line 1. The land was sold for a $5,000 long-term gain.

Transaction 2, Barn. The barn situated on the land in Transaction 1 was built shortly after the land was purchased. Since the barn was depreciated using a rapid depreciation method, Section 1250 recapture must be calculated. Consequently, the barn is entered on Form 4797, Part III, Line 18. (see Figure 17–5[b]). Lines 19–23 must be filled out with the data provided in Table 17–2 (taken from the owner's depreciation schedule). Furthermore, since Section 1250 applies, Lines 25(a) through 25(g) must also be completed. The only real entry in this block is on Line 25(a). Under the straight-line method, only $4,000 of depreciation is allowed. But the use of double-declining depreciation yielded $6,878 depreciation. The *excess* of double-declining over straight-line is $2,878, which is entered on Line 25(a). The net result of these calculations is that $2,878 of the $6,878 gain will be treated as ordinary income and the remaining $4,000 will be treated as capital gain. Finally, since general-purpose buildings do not qualify for ITC, there will be no recapture or entry on Form 4255.

[10]That is, in the order that reduces paper shuffling.

Form **4255** (Rev. September 1985) Department of the Treasury Internal Revenue Service

Recapture of Investment Credit
(Including Energy Investment Credit)
▶ **Attach to your income tax return.**

OMB No. 1545-0166 Expires 8-31-88

65

Name(s) as shown on return | **Identifying number**

Properties	Kind of property—State whether recovery or nonrecovery (see the Instructions for Form 3468 for definitions). If energy property, show type. Also indicate if rehabilitation expenditure property.
A	
B	
C	
D	
E	

Original Investment Credit

Computation Steps: (see Specific Instructions)	Properties A	B	C	D	E
1 Original rate of credit					
2 Date property was placed in service					
3 Cost or other basis					
4 Original estimated useful life or class of property					
5 Applicable percentage					
6 Original qualified investment (line 3 times line 5)					
7 Original credit (line 1 times line 6)					
8 Date property ceased to be qualified investment credit property					
9 Number of full years between the date on line 2 and the date on line 8					

Computation of Recapture Tax

	A	B	C	D	E
10 Recapture percentage (from instructions)					
11 Tentative recapture tax (line 7 times line 10)					

12 Add line 11, columns A through E

13 Enter tax from property ceasing to be at risk, or for which there was an increase in nonqualified nonrecourse financing (attach separate computation)

14 Total—Add lines 12 and 13

15 Portion of original credit (line 7) not used to offset tax in any year, **plus** any carryback and carryforward of credits you can now apply to the original credit year because you have freed up tax liability in the amount of the tax recaptured (Do not enter more than line 14—see instructions)

16 Total increase in tax—Subtract line 15 from line 14. Enter here and on the proper line of your tax return.

General Instructions

(Section references are to the Internal Revenue Code, unless otherwise noted.)

Paperwork Reduction Act Notice.—We ask for this information to carry out the Internal Revenue laws of the United States. We need it to ensure that taxpayers are complying with these laws and to allow us to figure and collect the right amount of tax. You are required to give us this information.

Purpose of Form.—Use Form 4255 to figure the increase in tax for the recapture of investment credit for regular and energy property. You must refigure the credit if you took it in an earlier year, but disposed of the property before the end of the recapture period or the useful life you used to figure the original credit. You must also refigure the credit if you changed the use of the property so that it no longer qualifies as regular or energy investment credit property. For example, you must refigure the credit if you change the use of property from business use to personal use, or if there is any decrease in the percentage of business use of investment credit property. See sections 47(a)(3) and 47(a)(5)(C) for information on recapture for progress expenditure property. Also, see the instructions for line 13 regarding recapture if property ceases to be at risk, or if there is an increase in nonqualified, nonrecourse financing related to certain at-risk property placed in service after July 18, 1984.

293

FIGURE 17–4(a). IRS Form 4255.

If certain "listed property" (i.e., transportation property, entertainment, recreation or amusement property, or certain home computer equipment) placed in service after June 18, 1984, ceases to be used predominantly for business, you may have to recapture investment credit taken on the property. See section 280F for details.

An election to be treated as an S corporation does not automatically trigger recapture of investment credit taken before the election was effective. However, the S corporation is liable for any recapture of investment credit taken before the election.

If property on which you took both regular and energy investment credit ceases to be energy credit property, but still qualifies as regular investment credit property, you need only refigure the energy investment credit. However, if you took both credits, and you dispose of the property, or the property ceases to be both energy and regular investment credit property, you must refigure both credits.

If you are an S corporation, a partnership, or an estate or trust that allocated any or all of the investment credit to the beneficiaries, you must give your shareholders, partners, or beneficiaries the information they need to refigure the credit. See regulations sections 1.47-4, 1.47-5 and 1.47-6.

Partners, shareholders and beneficiaries.—If your Schedule K-1 shows that you must recapture investment credit taken in an earlier year, you will need your retained copy of the original Form 3468 to complete lines 1 through 9 of this Form 4255.

Special Rules.—If you took a credit on the following kinds of property, see the sections listed below before you complete Form 4255:

Property	Section
Motion picture films and video tapes	47(a)(8)
Ships	46(g)(4)
Commuter highway vehicles	47(a)(4)

If you took any nonconventional source fuel credit, see section 29(b)(4).

If, before the end of the recapture period, you dispose of recovery property placed in service after 1982 (or the business use percentage decreases), increase the basis of this property by 50% of the recapture amount unless you originally made the section 48(q)(4) election to take a reduced credit instead of reducing the basis of the property.

For more information, see **Publication 572,** Investment Credit.

Specific Instructions

Note: *Do not figure the recapture tax on lines 1 through 12 for property ceasing to be at risk, or if there is an increase in nonqualified nonrecourse financing related to certain at-risk property placed in service after July 18, 1984. Figure the recapture tax for these properties on separate schedules and enter the recapture tax on line 13. Include any unused credit for these properties on line 15.*

Lines A through E.—Describe the property for which you must refigure the credit.

Fill in lines 1 through 11 for each property on which you are refiguring the credit. Use a separate column for each item. If you must recapture both the energy investment credit and the regular investment credit for the same item, use a separate column for each credit. If you need more columns, use additional Forms 4255, or other schedules with all the information shown on Form 4255. Enter the total from the separate sheets on line 12.

Line 1.—Enter the rate you used to figure the original credit from the tables below:

Regular Investment Property

Qualifying property acquired or constructed and placed in service	10%

Be sure to include any ESOP credit, or other credits listed under the special rules above.

See section 46(b)(4) for the rates for qualified rehabilitation expenditures made after 1981.

Energy Investment Property

Alternative energy property, specially defined energy property, recycling equipment, shale oil equipment, equipment for producing natural gas from geopressured brine, cogeneration equipment, and intercity buses	10%
Qualified hydroelectric generating equipment	11%
Solar and wind equipment acquired or constructed before 1980	10%
Solar and wind equipment, ocean thermal equipment, and geothermal equipment acquired or constructed after 1979	15%

Line 2.—For both recovery and nonrecovery property, enter the first day of the first month, and the year, that the property was available for service.

Line 3.—Enter the cost or other basis that you used to figure the original credit.

Line 4.—Enter the estimated useful life that you used to figure the original credit for nonrecovery property. Enter the class of property for recovery property.

Line 5.—Enter the applicable percentage that you used to figure the original qualified investment from the tables below:

Nonrecovery Property

Original estimated useful life	Applicable percentage
3 or more but less than 5 years	33⅓%
5 or more but less than 7 years	66⅔%
7 or more years	100%

Recovery Property

Class of property	Applicable percentage
3-year	60%
Other	100%

Section 48(q) Election Recovery Property

(Placed in service after 12/31/82)

Class of property	Applicable percentage
3-year	40%
Other	80%

Line 8.—Generally, this will be the date you disposed of the property. See regulations section 1.47-1(c) for more information.

Line 9.—Do not enter partial years. If the property was held less than 12 months, enter zero.

Line 10.—Enter the recapture percentage from the following tables:

Nonrecovery Property

If number of full years on line 9 of Form 4255 is:	The recapture percentage for property with an original useful life of: 3 or more but less than 5 years is	5 or more but less than 7 years is	7 or more years is
0	100	100	100
1	100	100	100
2	100	100	100
3	0	50	66.6
4	0	50	66.6
5	0	0	33.3
6	0	0	33.3

Recovery Property

If number of full years on line 9 of Form 4255 is:	The recapture percentage for: 3-year property is	Other than 3-year property is
0	100	100
1	66	80
2	33	60
3	0	40
4	0	20

Decrease in Business Use.—If you take investment credit for property and the percentage of business use in a later year falls below the percentage for the year placed in service, you are treated as having disposed of part of the property and may have to recapture part of the investment credit.

For example, if you are a calendar-year taxpayer who bought a $12,000 automobile in August 1984 and used it 90% for business, your investment credit for 1984 would have been $900 (90% of the $1,000 limit for automobiles placed in service from June 19, 1984, through April 2, 1985). If business use in 1985 fell to 60%, you are treated as having disposed of one-third ((90 − 60) ÷ 90) of the automobile on January 1, 1985 (since business use is computed on a tax-year basis and a decrease in business use is deemed to take place on the first day of the tax year). Multiply the recapture percentage from the table above (100%) by one-third and enter the result (33.3%) on line 10 of the form.

Line 12.—If you have used more than one Form 4255, or separate sheets to list additional items on which you figured an increase in tax, write on the dotted line "Tax from attached, $ ...." Include the amount in the total for line 12.

Line 13.—For certain taxpayers, the basis or cost of property placed in service after February 18, 1981, is limited to the amount the taxpayer is at risk for the property at year end. For property placed in service after July 18, 1984, the basis or cost must be reduced by the amount of any "nonqualified, nonrecourse financing" related to the property at year end. If property ceases to be at risk in a later year, or if there is an increase in nonqualified, nonrecourse financing, recapture may be required. See section 47(d) for details. Attach a separate computation schedule to figure the recapture tax and enter the total tax on line 13.

Line 15.—If you did not use all the credit you originally figured, either in the year you figured it or in a carryback or carryforward year, you do not have to recapture the amount of the credit you did not use. In refiguring the credit for the original credit year, be sure to take into account any carryforwards from previous years, plus any carrybacks arising within the 3 taxable years after the original credit year that are now allowed because the recapture and recomputation of the original credit made available some additional tax liability in that year. See regulations section 1.47-1(d), Revenue Ruling 72-221, and Publication 572 for more information.

Figure the unused portion on a separate sheet and enter it on this line. Do not enter more than the recapture tax on line 14.

REMINDER: Be sure to adjust your *current* unused credit to reflect any unused portion of the original credit that was recaptured on this form.

FIGURE 17–4(b). IRS Form 4255 instructions.

TABLE 17–2 Example Asset Sale Transactions

Transaction	Asset Description	Acquisition Date	Disposition Date	Original Basis	Depreciation Claimed	ITC Claimed	Adjusted Basis	Sales Price
1.	Farmland (no soil & water conservation improvements)	12/1/81	1/1/85	$10,000	$ 0	$ 0	$10,000	$15,000
2.	Barn[a]	12/31/80	1/1/85	20,000	6,878	0	13,122	20,000
3.	Raised breeding cow	12/15/83	6/1/85	0	0	0	0	400
4.	Feeder cattle	1/15/85	10/1/85	5,000	0	0	5,000	7,000
5.	Raised breeding cow	12/15/81	6/1/85	0	0	0	0	350
6.	Tractor	6/1/81	9/1/85	20,000	15,800	2,000	4,200	25,000
7.	Three-piece suit	12/1/84	6/1/85	80	0	0	80	10
8.	Dining room set	4/20/79	7/15/85	1,000	0	0	1,000	2,000
9.	International Trucking corporate stock (100 shares)	11/25/84	5/30/85	3,200	0	0	3,200	2,500
10.	Combine	5/1/75	6/1/85	15,000	14,000	1,500	1,000	9,000

[a] Depreciated with double declining balance method. Useful life originally specified as 20 years, no salvage value.

Transaction 3, Raised Breeding Cow. Raised breeding livestock has no basis, therefore it does not qualify for depreciation or ITC and thus there will be no recapture. Although this cow was held for 17 months, she is still classified as a short-term asset and is entered in Form 4797, Part II, Line 15 (see Figure 17–5[a]), because of the special 24-month holding-period requirement for cattle.

Transaction 4, Feeder Cattle. Feeder cattle are held primarily for sale and thus do not qualify for capital asset treatment. Their sales will be discussed in Chapter 19.

Transaction 5, Raised Breeding Cow. This cow is treated in exactly the same manner as the cow in Transaction 3 except for the fact that she was held for well over 24 months. Consequently, her sale is entered on Form 4797, Part I, Line 1 (see Figure 17–5[a]).

Transaction 6, Tractor. All machinery held for use in a business for over twelve months qualifies for Section 1231 treatment and Section 1245 recapture. The tractor sale is entered on Form 4797, Part III, Line 18 (see Figure 17–5[b]). Because the sale price of $25,000 was greater than both the purchase price and the adjusted basis, a gain was realized. The total gain was $20,800, but $15,800 of that amount is recapture of depreciation and consequently ordinary income, whereas the remaining $5,000 (the difference between the sale price and the original basis) is long-term gain. One further element arises with the sale of the tractor. Under ACRS rules, a tractor is considered to be 5-year property for depreciation and ITC. But, since the tractor was not owned for five full years, the ITC recapture provisions will come into play. The data from table 17–2 must therefore be entered also on Form 4255 as shown in Figure 17–6. Using the applicable percentages and instructions from Figure 17–4(b), a $400 recapture tax is imposed.

Form **4797** | **Gains and Losses From Sales or Exchanges of Assets Used in a Trade or Business and Involuntary Conversions** | OMB No. 1545-0184

Department of the Treasury
Internal Revenue Service (X)

(And Computation of Recapture Amounts Under Sections 179 and 280F)

▶ **Attach to your tax return. See Separate Instructions**

1985
27

Name(s) as shown on return: FRANK FARMER — Identifying number: 123-45-6789

Part I **Sales or Exchanges of Property Used in a Trade or Business and Involuntary Conversions From Other Than Casualty and Theft—Property Held More Than 6 months (More Than 1 Year if Acquired Before 6/23/84)**

Notes:
- *Use Form 4684 to report involuntary conversions from casualty and theft.*
- *If you sold property that you claimed investment credit on, get Form 4255 to see if you are liable for recapture of the credit.*
- *File Form 6198 if you are reporting a loss and have amounts invested in the activity for which you are not at risk. (See instructions under "Special Rules.")*

(a) Description of property	(b) Date acquired (mo., day, yr.)	(c) Date sold (mo., day, yr.)	(d) Gross sales price	(e) Depreciation allowed (or allowable) since acquisition	(f) Cost or other basis, plus improvements and expense of sale	(g) LOSS ((f) minus the sum of (d) and (e))	(h) GAIN ((d) plus (e) minus (f))
1 Land	12-1-81	1-1-85	15,000	0	10,000		5,000
Breeding Cow	12-15-81	6-1-85	350	0	0		350

Line		(g)	(h)
2	Gain, if any, from Form 4684, Section B, line 21		
3	Section 1231 gain from installment sales from Form 6252, line 22 or 30		
4	Gain, if any, on line 31 from other than casualty and theft		9,000
5	Add lines 1 through 4 in column (g) and column (h)	( )	14,350
6	Combine columns (g) and (h) of line 5. Enter gain or (loss) here, and on the appropriate line as follows (Partnerships see the instructions for your line references.):		14,350

a If line 6 is zero or a loss, enter the amount on line 9 below and skip lines 7 and 8. (S corporations, enter the loss on Schedule K (Form 1120S), line 5.)

b If line 6 is a gain, see the instructions under Part I, **Nonrecaptured Net Section 1231 Losses.**

Note: *If you had no prior year section 1231 losses and line 6 is a gain, enter the gain from line 6 as a long-term capital gain on Schedule D.*

7 Nonrecaptured net section 1231 losses from prior years. (See instructions.)

8 Subtract line 7 from line 6. If zero or less, enter zero . . . 14,350

a If line 8 is zero, enter the amount from line 6 on line 10 below.

b If line 8 is more than zero, enter the amount from line 7 on line 10 below, and enter the amount from line 8 as a long-term capital gain on Schedule D. See specific instructions for line 8b.

Part II **Ordinary Gains and Losses**

(a) Description of property	(b) Date acquired (mo., day, yr.)	(c) Date sold (mo., day, yr.)	(d) Gross sales price	(e) Depreciation allowed (or allowable) since acquisition	(f) Cost or other basis, plus improvements and expense of sale	(g) LOSS ((f) minus the sum of (d) and (e))	(h) GAIN ((d) plus (e) minus (f))
9 Loss, if any, from line 6							
10 Gain, if any, from line 6 or amount from line 7 if applicable							
11 Gain, if any, from line 30, Part III							26,678
12 Net gain or (loss) from Form 4684, Section B, lines 13 and 20a							
13 Ordinary gain from installment sales from Form 6252, line 21 or 29 (Applies only to sales before 6/7/84)							
14 Recapture of section 179 deduction on property dispositions by partners and S corporation shareholders (see instructions)							
15 Other ordinary gains and losses (include property held 6 months or less (1 year or less if acquired before 6/23/84)):							
Breeding cow	12-15-83	6-1-85	400	0	0		400
16 Add lines 9 through 15 in column (g) and column (h)						( )	27,078
17 Combine columns (g) and (h) of line 16. Enter gain or (loss) here, and on the appropriate line as follows:							27,078

a For all except individual returns: Enter the gain or (loss) from line 17, on the return being filed. See instructions for Part II for specific line references.

b For individual returns:

(1) If the loss on line 9 includes a loss from Form 4684, Section B, Part II, column (b)(ii), enter that part of the loss here and on line 19 of Schedule A (Form 1040). Identify as from "Form 4797, line 17b(1)"

(2) Redetermine the gain or (loss) on line 17, excluding the loss (if any) on line 17b(1). Enter here and on Form 1040, line 15. . . . 27,078

For Paperwork Reduction Act Notice, see page 1 of separate instructions. Form **4797** (1985)

(309)

FIGURE 17–5(a). Example filled-in Form 4797, parts I and II.

Form 4797 (1985) Page 2

Part III **Gain From Disposition of Property Under Sections 1245, 1250, 1252, 1254, 1255**

Skip section 1252 on line 26 and in the instructions, if you did not dispose of farmland, or if a partnership files this form.

18 Description of sections 1245, 1250, 1252, 1254, and 1255 property:	Date acquired (mo., day, yr.)	Date sold (mo., day, yr.)
A Barn	12-31-80	1-1-85
B Tractor	6-1-81	9-1-85
C Combine	5-1-75	6-1-85
D		

Relate lines 18A through 18D to these columns ► ► ► ►	Property A	Property B	Property C	Property D
19 Gross sales price	20,000	25,000	9,000	
20 Cost or other basis plus expense of sale	20,000	20,000	15,000	
21 Depreciation (or depletion) allowed (or allowable)	6,878	15,800	14,000	
22 Adjusted basis, subtract line 21 from line 20	13,122	4,200	1,000	
23 Total gain, subtract line 22 from line 19	6,878	20,800	8,000	
24 If section 1245 property:				
a Depreciation allowed (or allowable) (see instructions)		15,800	14,000	
b Enter smaller of line 23 or 24a		15,800	8,000	
25 If section 1250 property: (If straight line depreciation used, enter zero on line 25g unless you are a corporation subject to section 291.)				
a Additional depreciation after 12/31/75	2,878			
b Applicable percentage times the smaller of line 23 or line 25a (see instructions)	2,878			
c Subtract line 25a from line 23. If line 23 is not more than line 25a, skip lines 25d and 25e	4,000			
d Additional depreciation after 12/31/69 and before 1/1/76	0			
e Applicable percentage times the smaller of line 25c or 25d (see instructions)	0			
f Section 291 amount (For corporations only.)	0			
g Add lines 25b, 25e, and 25f	2,878			
26 If section 1252 property:				
a Soil, water, and land clearing expenses				
b Line 26a times applicable percentage (see instructions)				
c Enter smaller of line 23 or 26b				
27 If section 1254 property:				
a Intangible drilling and development costs deducted after 12/31/75 (see instructions)				
b Enter smaller of line 23 or 27a				
28 If section 1255 property:				
a Applicable percentage of payments excluded from income under section 126 (see instructions)				
b Enter the smaller of line 23 or 28a				

Summary of Part III Gains (Complete property columns A through D through line 28b before going to line 29)

29 Total gains for all properties (add columns A through D, line 23)	35,678
30 Add columns A through D, lines 24b, 25g, 26c, 27b, and 28b. Enter here and in Part II, line 11	26,678
31 Subtract line 30 from line 29. Enter the portion from casualty and theft on Form 4684, Section B, line 15; enter the portion from other than casualty and theft on Form 4797, Part I, line 4	9,000

Part IV **Complete This Part Only If You Elect Out of the Installment Method And Report a Note or Other Obligation at Less Than Full Face Value**

☐ Check here if you elect out of the installment method.

Enter the face amount of the note or other obligation ►

Enter the percentage of valuation of the note or other obligation ►

Part V **Computation of Recapture Amounts Under Sections 179 and 280F When Business Use Drops to 50% or Less (See instructions for Part V.)**

	(a) Section 179	(b) Section 280F
1 Section 179 expense deduction or section 280F recovery deductions		
2 Depreciation (see instructions)		
3 Recapture amount. (Subtract line 2 from line 1.)		

(310)

FIGURE 17–5(b). Example of filled-in Form 4797, part III.

Form **4255** (Rev. September 1985) Department of the Treasury Internal Revenue Service

Recapture of Investment Credit
(Including Energy Investment Credit)
► **Attach to your income tax return.**

OMB No. 1545-0166 Expires 8-31-88
65

Name(s) as shown on return: Frank Farmer

Identifying number: 123-45-6789

Properties	Kind of property—State whether recovery or nonrecovery (see the Instructions for Form 3468 for definitions). If energy property, show type. Also indicate if rehabilitation expenditure property.
A	Tractor
B	
C	
D	
E	

	Computation Steps: (see Specific Instructions)	**Properties** A	B	C	D	E
Original Investment Credit	1 Original rate of credit	10%				
	2 Date property was placed in service	6-1-81				
	3 Cost or other basis	20,000				
	4 Original estimated useful life or class of property	5				
	5 Applicable percentage	100				
	6 Original qualified investment (line 3 times line 5)	20,000				
	7 Original credit (line 1 times line 6)	2,000				
	8 Date property ceased to be qualified investment credit property	9-1-85				
	9 Number of full years between the date on line 2 and the date on line 8	4				
Computation of Recapture Tax	10 Recapture percentage (from instructions)	20				
	11 Tentative recapture tax (line 7 times line 10)	400				

Line		Amount
12	Add line 11, columns A through E	400
13	Enter tax from property ceasing to be at risk, or for which there was an increase in nonqualified nonrecourse financing (attach separate computation)	
14	Total—Add lines 12 and 13	400
15	Portion of original credit (line 7) not used to offset tax in any year, **plus** any carryback and carryforward of credits you can now apply to the original credit year because you have freed up tax liability in the amount of the tax recaptured (Do not enter more than line 14—see instructions)	
16	Total increase in tax—Subtract line 15 from line 14. Enter here and on the proper line of your tax return.	400

General Instructions

(Section references are to the Internal Revenue Code, unless otherwise noted.)

Paperwork Reduction Act Notice.—We ask for this information to carry out the Internal Revenue laws of the United States. We need it to ensure that taxpayers are complying with these laws and to allow us to figure and collect the right amount of tax. You are required to give us this information.

Purpose of Form.—Use Form 4255 to figure the increase in tax for the recapture of investment credit for regular and energy property. You must refigure the credit if you took it in an earlier year, but disposed of the property before the end of the recapture period or the useful life you used to figure the original credit. You must also refigure the credit if you changed the use of the property so that it no longer qualifies as regular or energy investment credit property. For example, you must refigure the credit if you change the use of property from business use to personal use, or if there is any decrease in the percentage of business use of investment credit property. See sections 47(a)(3) and 47(a)(5)(C) for information on recapture for progress expenditure property. Also, see the instructions for line 13 regarding recapture if property ceases to be at risk, or if there is an increase in nonqualified, nonrecourse financing related to certain at-risk property placed in service after July 18, 1984.

(293)

FIGURE 17–6. Example of filled–in Form 4255.

SCHEDULE D (Form 1040) — Department of the Treasury, Internal Revenue Service (X)

Capital Gains and Losses and Reconciliation of Forms 1099–B

▶ Attach to Form 1040. ▶ See Instructions for Schedule D (Form 1040).

OMB No. 1545-0074 — **1985** — 12

Name(s) as shown on Form 1040: Frank Farmer — Your social security number: 123 45 6789

Part I Summary of Forms 1099–B for Sales of Stocks, Bonds, Etc.

1a Report here and on line 37, Part VII, page 2, total sales of stocks, bonds, etc., reported for 1985 by your broker to you on Form(s) 1099–B or an equivalent substitute statement(s), such as a broker's confirmation statement	1a	2,500	00

Note: *Also complete Part VII if you received one or more Form(s) 1099–B or equivalent statement(s) for 1985, for bartering income.*

Part II Short-term Capital Gains and Losses-Assets Held Six Months or Less (one year or less if acquired before 6/23/84)

(a) Description of property (Example, 100 shares 7% preferred of "Z" Co.)	(b) Date acquired (Mo., day, yr.)	(c) Date sold (Mo., day, yr.)	(d) Gross sales price	(e) Cost or other basis (see instructions)	(f) LOSS If column (e) is more than (d) subtract (d) from (e)		(g) GAIN If column (d) is more than (e) subtract (e) from (d)	
1b International Trucking Stock 100 shares	11-25-84	3-30-85	2,500	3,200	700			

2 Short-term gain from sale or exchange of a principal residence from Form 2119, lines 6 or 12	2				
3 Short-term gain from installment sales from Form 6252, lines 22 or 30	3				
4 Net short-term gain or (loss) from partnerships, S corporations, and fiduciaries	4				
5 Add lines 1b through 4 in columns (f) and (g)	5	(700)			
6 Combine columns (f) and (g) of line 5 and enter the net gain or (loss)			6	(700	00)
7 Short-term capital loss carryover from years beginning after 1969			7	(	)
8 Net short-term gain or (loss), combine lines 6 and 7			8	(700	00)

Part III Long-term Capital Gains and Losses-Assets Held More Than Six Months (more than one year if acquired before 6/23/84)

(a)	(b)	(c)	(d)	(e)	(f)		(g)	
9 Dining Room Set	4-20-79	7-15-85	2,000	1,000			1,000	00

10 Long-term gain from sale or exchange of a principal residence from Form 2119, lines 6, 8, or 12	10				
11 Long-term gain from installment sales from Form 6252, lines 22 or 30	11				
12 Net long-term gain or (loss) from partnerships, S corporations, and fiduciaries	12				
13 Add lines 9 through 12 in columns (f) and (g)	13	( )		1,000	00
14 Combine columns (f) and (g) of line 13 and enter the net gain or (loss)			14	1,000	00
15 Capital gain distributions			15		
16 Enter gain from Form 4797, lines 6 or 8b			16	14,350	00
17 Combine lines 14 through 16			17	15,350	00
18 Long-term capital loss carryover from years beginning after 1969			18	(	)
19 Net long-term gain or (loss), combine lines 17 and 18			19	15,350	00

Note: *Complete the back of this form. However, if you have capital loss carryovers from years beginning before 1970, do not complete Parts IV or V. See Form 4798 instead.*

For Paperwork Reduction Act Notice, see Form 1040 Instructions. **Schedule D (Form 1040) 1985**

(77)

FIGURE 17–7(a). Example of filled-in Schedule D, parts I, II, and III.

Schedule D (Form 1040) 1985 — 12 — Page 2

Name(s) as shown on Form 1040 (Do not enter name and social security number if shown on other side) | Your social security number

Part IV Summary of Parts II and III

20	Combine lines 8 and 19, and enter the net gain or (loss) here	20	14,650	00
	Note: *If line 20 is a loss, skip lines 21 through 23 and complete lines 24 and 25. If line 20 is a gain, complete lines 21 through 23 and skip lines 24 and 25.*			
21	If line 20 shows a gain, enter the smaller of line 19 or line 20. Enter zero if there is a loss or no entry on line 19 — 21: 14,650 00			
22	Enter 60% of line 21	22	8,190	00
	If line 22 is more than zero, you may be liable for the alternative minimum tax. See Form 6251.			
23	Subtract line 22 from line 20. Enter here and on Form 1040, line 13	23	5,860	00
24	If line 20 shows a loss, enter one of the following amounts:			
	a If line 8 is zero or a net gain, enter 50% of line 20;			
	b If line 19 is zero or a net gain, enter line 20; or			
	c If line 8 and line 19 are net losses, enter amount on line 8 added to 50% of the amount on line 19	24		
25	Enter here and as a loss on Form 1040, line 13, the smallest of:			
	a The amount on line 24;			
	b $3,000 ($1,500 if married and filing a separate return); or			
	c Taxable income, as adjusted	25		

Part V Computation of Post-1969 Capital Loss Carryovers from 1985 to 1986

(Complete this part if the loss on line 24 is more than the loss on line 25)

26	Enter loss shown on line 8; if none, enter zero and skip lines 27 through 30, then go to line 31	26		
27	Enter gain shown on line 19. If that line is blank or shows a loss, enter zero	27		
28	Subtract line 27 from line 26	28		
29	Enter smaller of line 25 or line 28	29		
30	Subtract line 29 from line 28. This is your short-term capital loss carryover from 1985 to 1986	30		
31	Subtract line 29 from line 25. (Note: If you skipped lines 27 through 30, enter amount from line 25.)	31		
32	Enter loss from line 19; if none, enter zero and skip lines 33 through 36	32		
33	Enter gain shown on line 8. If that line is blank or shows a loss, enter zero	33		
34	Subtract line 33 from line 32	34		
35	Multiply amount on line 31 by 2	35		
36	Subtract line 35 from line 34. This is your long-term capital loss carryover from 1985 to 1986	36		

Part VI Complete This Part Only If You Elect Out of the Installment Method and Report a Note or Other Obligation at Less Than Full Face Value

☐ Check here if you elect out of the installment method.

Enter the face amount of the note or other obligation. ▶ ....

Enter the percentage of valuation of the note or other obligation. ▶

Part VII Reconciliation of Forms 1099–B With Tax Return

(Complete this part if you received one or more Form(s) 1099–B or an equivalent broker's substitute statement(s) reporting sales of stock, bonds, etc., or bartering income.)

SECTION A.—Reconciliation of Sales of Stocks, Bonds, etc.

37	Total sales of stock, bonds, etc., reported for 1985 by your broker to you, on Form(s) 1099–B or an equivalent substitute statement(s), such as a broker's confirmation statement. Also include on line 1a, page 1, Part I	37		
38	Proceeds from sale or exchange of capital assets reported on Schedule D, but not included in line 37	38		
39	Add lines 37 and 38	39		
40	Part of line 37 not reported on Schedule D this year (attach explanation)	40		
41	Subtract line 40 from line 39	41		

Note: *The amount on line 41 should be the same as the total of all amounts on page 1, lines 1b and 9 of column (d).*

SECTION B.—Reconciliation of Bartering Income

Indicate below, the amount of bartering income reported on each form or schedule.

			Amount of bartering from Form 1099–B or equivalent statement	
42	Form 1040, line 22	42		
43	Schedule C (Form 1040)	43		
44	Schedule D (Form 1040)	44		
45	Schedule E (Form 1040)	45		
46	Schedule F (Form 1040)	46		
47	Other (identify) (if not taxable, indicate reason—attach additional sheets if necessary) ▶	47		
48	Total (add lines 42 through 47)	48		

Note: *The amount on line 48 should be the same as the total bartering on all Forms 1099–B or equivalent statements received.*

(78)

FIGURE 17–7(b). Example of filled-in Schedule D, part IV.

Transaction 7, Three-piece Suit. The loss suffered on the sale of this suit, even if used in connection with a business, does not qualify for consideration anywhere on the tax return. General-purpose clothing is properly classified as a personal asset rather than an investment or a business asset.

Transaction 8, Dining-room Set. Like the suit, the dining-room set is a personal, noninvestment, nonbusiness asset. However, unlike the suit, it was sold at a gain. That gain is taxable even if a loss is not deductible. The sale is reported on Schedule D, Part III, Line 9 (see Figure 17–7[a]) since it was held for more than six months.

Transaction 9, Corporate Stock. All corporate stock is treated as a Section 1221 capital asset, whether sold at a gain or a loss. Consequently, all corporate stock sales or dispositions are reported on Schedule D. The International Harvester stock was held less than six-months and consequently belongs on Schedule D, Part I, Line 1b (see Figure 17–7[a]).

Transaction 10, Combine. The combine sale takes us back to Form 4797. It is treated in much the same way as the tractor and is reported on Form 4797, Part III, Line 18 (see Figure 17–5[b]). But there are two major differences: the relationship between the sale price and basis and the number of years owned. Because the sale price is less than the original basis but more than the adjusted basis, all of the gain is short-term recapture. And further, because the combine was held for more than seven years, the maximum requirement for full ITC benefits, there will be no ITC recapture and so no entry on Form 4255.

Analysis of the Form Summaries

After all of the transactions have been entered, the netting procedures and section totals must be completed. While the totaling and arithmetic manipulation instructions presented on the IRS forms are generally fairly easy to follow, it is wise to have a thorough understanding of what is being done so that the numbers can be analyzed and criticized at each step. The temptation to become a computer and just perform the functions mindlessly is strong, and if *everything* has been entered perfectly prior to the particular calculation in question, computer-like functions will be sufficient. But constant reflection on the meaning of each particular number will help discover errors before they are carried forward and will help to increase understanding of the tax implications of each action taken by the taxpayer. That understanding will become extremely valuable in forward planning for the next tax year.

Form 4797, Part III. Probably the best place to start the totaling and analyzing process is with the back of Form 4797 (Figure 17–5[b]). Once all of the Line 18 entries have been carried to Lines 19–23 and one of the recapture sections (Lines 24–28), the Summary can be completed. Line 29 sums the total gains for each property from Line 23. Total gain, once again, is the difference between the sale price and the adjusted basis. The total recapture amount is summed for all properties on Line 30. The difference between the total gain (Line 29) and the short-term or ordinary gain (Line 30) must then be the long-term gain (Line 31). Our example indicates a long-term gain of $9,000, which, if we think back to the example transactions, comes from the $4,000 straight-line depreciation on the barn (Transaction 2) and the $5,000 excess of the sale price over the original basis on the tractor (Transaction 6). The combine (Transaction 10) contributed no long-term gain. The $26,678 short-term gain is transferred to Part I to be netted with any other Section 1231 (other than recapture) transactions.

Form 4797—Part I. Turning back to the front side of Form 4797 (Figure 17–5[a]), we can now determine the net long-term gains from all Section 1231 properties. Ignoring casualty losses from Form 4684 (Line 2) and installment sale gains from Form 6252 (Line 3), the total long-term gain from Form 4797, Part III is entered on Line 4 and then added to any Line 1 entries to get total gain and total loss on Line 5. The net gain or loss is then computed on Line 6. If Line 6 shows a gain, that amount is transferred to Schedule D, Part III (see Figure 17–7[a]) to be added to other long-term Section 1221 gains. If Line 6 shows a loss, that amount is transferred to Form 4797, Part II, Line 9 to offset short-term gains.

Form 4797, Part II. Part II (Figure 17–5[a]) follows a pattern similar to that of Part I, but the long-term loss from Part I, Line 6 and the ordinary recapture gain from Part III, Line 11) are written in first, as well as ordinary casualty losses, installment sales gains, and recapture of capital expensing option begun in 1982.[11] Total gains and losses are summed on Line 16 and netted on Line 17. The amount on Line 17 must be recalculated to exclude losses due to casualty losses, but in our case, the entire amount is rewritten on Line 17(b)(2). Once that figure has been transferred to the umbrella Form 1040, we are finally finished with Form 4797. It is useful to note that this amount is transferred directly to Form 1040 rather than to Schedule D. This transfer implies that ordinary gains from business asset sales are not included in the long-term/short-term gain/loss netting procedure described earlier in this chapter. However, the long-term gain from Part I *is* included in that process.

Form 4255. The summary procedure on Lines 12–16 of Form 4255 (see Figure 17–6) is really quite simple if no out-of-the ordinary problems were encountered such as the items indicated on Lines 13 and 15. While there are cases in which there will be entries on these two lines, those cases are not typical for a family farm or ranch.

Schedule D, Parts II and III. Other than Lines 2, 3, and 4 on Part II and Lines 10, 11, 12, and 15 on Part III, the summary process for the front side of Schedule D (Figure 17–7[a]) is relatively straightforward. Items on those lines are either rare or beyond the scope of this text. Lines 7 and 18 are concerned with loss carryovers caused by the loss limitations described earlier in this chapter on page 260. In our case, the taxpaper can refer to his 1984 tax return (Lines 30 and 36, Schedule D, Part V) to pick up any loss carryover from prior tax years.

[11] See Chapter 18, page 275.

Schedule D, Part IV. Part IV (Figure 17–7[b]) processes the gain/loss netting procedure described at length previously. In our case, the calculations can be handled quite easily and there will be no entries in the carryover portion, Part V.

Summary

To avoid as much paper-shuffling as possible, a good place to begin the process of completing a federal income tax return is with the reporting of asset sales. Categorizing assets is difficult at first but soon becomes easy as experience is gained in federal income tax reasoning. Typically, assets, as defined for tax purposes, can be categorized in three ways: ordinary assets, capital assets, and business long-term (or productive) assets. *Ordinary assets* are those items such as grain and market livestock that are held primarily for sale. *Capital assets,* or nonbusiness investment assets, are held primarily for earnings in the short run (through interest or dividends) and/or to recognize a gain when they are sold (capital gains). Finally, *business productive assets* are held primarily to generate ordinary income through the sale of products or services. Sales of business assets are reported first on IRS Form 4797, investment asset sales on Schedule D, and ordinary income on Schedules C, F, or E or Form 4835.

The time period during which assets are owned is an important factor in determining taxability of capital gains. An exclusion of up to 60% of the gain recognized is available *if* the asset was held for more than six months (one year if purchased before June 23, 1984) for most assets, 12 months for hogs and certain other livestock, and 24 months for cattle and horses held for breeding, dairy, work, or sporting purposes.

The sale of a Section 1231 asset (a business asset) that is depreciable is potentially liable for a recapture tax. The depreciation recapture sections (IRC Sections 1245, 1250, 1252, 1254, and 1255) segregate the gain realized from the sale of a depreciable business asset into two components: the amount of gain realized because the asset was depreciated and the amount of gain realized because its sale price was greater than its purchase price. If the sale price is lower than the purchase price, but greater than the adjusted basis, the difference between the sale price and the adjusted basis is recaptured as ordinary income. If sale price is greater than purchase price, that difference is treated as a Section 1221 capital gain.

Although gains are more common than losses for depreciable business assets, losses are possible; for investments, losses are common. Capital gains and capital losses must follow a complex netting procedure before the capital gains exclusion is calculated. Losses offset gains before the exclusion in a gain situation. In a loss situation, the amount of losses that can be used to reduce other income is limited; any excess must be carried forward to other tax years. Further, the capital loss exclusion is 50 percent; that is, 50 percent of capital losses is excluded from use to reduce other income.

Similar to depreciation recapture, investment tax credit may also have to be repaid. If the business asset was held less than the length of its original estimated useful life, some of the ITC originally claimed may have to be repaid (using Form 4255).

CHAPTER 18

Asset Purchases and Depreciation

The analysis and reporting of asset purchases and depreciation logically follow asset sales for two, reasons: (1) We can handle all of the asset transactions (on Section 1221 and 1231 assets) and then move on to other topics, and (2) the information from the depreciation form is needed to complete the basic farm and ranch return. Asset purchase and depreciation tax analysis is *much* easier than sales analysis, again for two reasons: (1) we need only deal with depreciable assets and (2) almost all of the work is already done if the asset inventory records described in Chapter 7 have been prepared. Consequently another thorough review of Chapter 7 is needed before progressing to the reporting process.

Investment Tax Credit

Investment tax credit (ITC) criteria were discussed in detail in Chapter 7, so those criteria will not be repeated here, but a summary of the highlights might be helpful.

1. All depreciable property (with the exception of general purpose buildings and other 19-year recovery property) will qualify for ITC.
2. ITC rates vary based on class:

Property Class	*Full Rate*	*Reduced Rate*
3-year	6%	4%
5-year	10%	8%
10- and 19-year	0%	0%

3. The applicable ITC rate is multiplied by the depreciable basis in new property or boot paid in cash for used property.[1]

[1] Refer to Table 7–8 for a complete listing of ITC qualifying amounts.

4. ITC is a *credit* not a *deduction,* thus it *reduces taxes due* rather than taxable income.
5. ITC is not optional. Even if ITC is not taken, the Internal Revenue Service (IRS) assumes it was taken.
6. The amount of ITC that can be used in any one year is limited to taxes due. Refunds will not be made by IRS, but unused amounts can be carried to other years.
7. Beginning with the 1983 tax year, there are two ITC rates: the full rate and the reduced rate. If the full rate is elected, the basis in the property for depreciation purposes must be reduced by 50 percent of the ITC claimed. No such basis reduction is necessary if the reduced rate is selected. Depreciation recapture, as discussed in Chapter 17, works from the original basis, but is partially adjusted for the ITC recapture.

ITC is claimed on IRS Form 3468 (Figure 18–1). The entry process is very simple. All that is required is to allocate each qualifying purchase into one of the eight categories according to the diagram of Figure 18–2. Then the qualifying amounts within each one of the categories are added together and placed in Form 3468, Part II, Line 1 (see Figure 18–1). Total qualifying amounts are calculated in the remaining lines of Part II, tentative ITC is calculated in Part III, and limitations on the credit are calculated in Part IV.

Asset Depreciation

As was the case with ITC, Chapter 7 discussed most of the relevant criteria for the depreciation *deduction* (see pages 80–86). Once again, a summary of those criteria might be useful.

1. The purchase price of all business assets may be deducted as a cost of production over a period of years through depreciation (now referred to as *cost recovery*). There are two major exceptions to this rule. Land is assumed to have an infinite useful life and is therefore not depreciable (however, improvements on the land are depreciable). Zero-basis assets, such as raised breeding or dairy livestock, are not depreciable since there is no purchase price or basis to depreciate, *if* the taxpayer reports taxable income on the cash basis of accounting.
2. Assets purchased prior to 1981 were depreciated under one of three methods: straight-line, sum-of-years-digits, or declining-balance. It was necessary to specify useful life and salvage value for each asset and, for the declining-balance method, the rate as well. Each asset could be handled in a different manner.
3. Assets purchased after 1980 fall under the Accelerated Cost Recovery System (ACRS). ACRS separates all assets into one of four property classes: 3-year, 5-year, 10-year or 19-year. ACRS also offers two primary depreciation alternatives, the percentage system and the straight-line method. All

property that falls into the same property class must be handled using the same methods. ACRS also ignores salvage value and assumes the half-year convention for the year of purchase. No depreciation is allowed in the year of sale.

4. Depreciable basis is equal to the amount of money and/or value of any other goods and services paid or given in exchange.

Section 179 Capital Expense

The Economic Recovery Tax Act of 1981 offered a new alternative to business owners—expensing a portion of capital expenditures. Beginning in 1982, farmers, ranchers, and other business owners have been allowed through Internal Revenue Code Section 179 to deduct outright a portion of their capital expenditures in the year of purchase, which in essence gives them an additional first year write-off over and above normal first-year depreciation. The maximum amount that can be deducted is $5,000. The major reporting problem generated by this exclusion is that it must be calculated *before* ITC is determined and *before* regular depreciation is figured.

Consider a short example to clarify this process. Assume that a tractor was purchased in 1985 for $80,000. The maximum capital expensing option is elected, it is 5-year property, and reduced-rate ITC is selected. The capital expensing option is determined first: the maximum allowable is $5,000. Then ITC is determined at 8 percent of $75,000 ($80,000 basis minus $5,000 Section 179 capital expense) or $6,000. Then regular ACRS percentage depreciation is calculated as 15 percent of $75,000, or $11,250. In essence, the tractor buyer is allowed a $16,250 depreciation deduction (although it is not reported as one lump sum on Form 4562) and a $6,000 credit in 1985.[2]

Form 4562

Depreciation on both new assets and previously purchased items is claimed on IRS Form 4562 (Figure 18–3). Section 179 expense election is reported in Part I, Section A, ACRS property regular depreciation is reported in Part I, Section B, and non-ACRS property depreciation is reported in Part I, Section C, usually on Line 8.

The depreciable bases for each piece of property are added together for each year by property class. In other words, all items purchased in 1985 that are 3-year property are added together and entered onto Form 4562, Line 4(a); all items purchased in 1985 that are 5-year property are added together and entered onto Form 4562, Line 4(b); all items purchased in 1985 that are 10-year property are added together and entered onto Form 4562, Line 4(c), and so on.

[2]If the full ITC rate had been chosen, the ITC would be $7,500 and the regular depreciation would be $10,688 for a total depreciation deduction of $15,688 in 1985.

Form **3468**

Department of the Treasury
Internal Revenue Service (X)

Computation of Investment Credit

▶ Attach to your tax return.
▶ Schedule B (Business Energy Investment Credit) on back.

OMB No. 1545-0155
1985
24

Name(s) as shown on return | Identifying number

Part I Elections (Check the box(es) below that apply to you (See Instruction D).)

A I elect to increase my qualified investment to 100% for certain commuter highway vehicles placed in service before January 1, 1986 (section 46(c)(6)) ☐

B I elect to increase my qualified investment by all qualified progress expenditures made this and all later tax years ☐
Enter total qualified progress expenditures included in column (4), Part II ▶ ____

C I claim full credit on certain ships under section 46(g)(3) (See **Instruction B** for details.) ☐

Part II Qualified Investment (See instructions for rules on automobiles and other property with any personal use)

1 Recovery Property		Line	(1) Class of Property	(2) Cost or Other Basis	(3) Applicable Percentage	(4) Qualified Investment (Column 2 x column 3)
Regular Percentage	New Property	(a)	3-year		60	
		(b)	Other		100	
	Used Property	(c)	3-year		60	
		(d)	Other		100	
Section 48(q) Election to Reduce Credit (instead of adjusting basis)	New Property	(e)	3-year		40	
		(f)	Other		80	
	Used Property	(g)	3-year		40	
		(h)	Other		80	

2 Nonrecovery property—Enter total qualified investment (See instructions for line 2) . . . **2**

3 New commuter highway vehicle—Enter total qualified investment (See Instruction D(1)) . . . **3**

4 Used commuter highway vehicle—Enter total qualified investment (See Instruction D(1)) . . . **4**

5 **Total qualified investment in 10% property**—Add lines 1(a) through 1(h), 2, 3, and 4 (See instructions for special limits) . . . **5**

6 Qualified rehabilitation expenditures—Enter total qualified investment for:
 a 30-year-old buildings . . . **6a**
 b 40-year-old buildings . . . **6b**
 c Certified historic structures (You must attach NPS certification—see instructions) . . . **6c**

Part III Tentative Regular Investment Credit

7 10% of line 5 . . . **7**

8 15% of line 6a . . . **8**

9 20% of line 6b . . . **9**

10 25% of line 6c . . . **10**

11 Credit from cooperatives—Enter regular investment credit from cooperatives . . . **11**

12 Regular investment credit—Add lines 7 through 11 . . . **12**

13 Business energy investment credit—From line 11 of Schedule B (see back of this form) . . . **13**

14 Current year investment credit—Add lines 12 and 13 . . . **14**

Note: *If you have a 1985 jobs credit (Form 5884), credit for alcohol used as fuel (Form 6478), or employee stock ownership plan (ESOP) credit (Form 8007) in addition to your 1985 investment credit, or if you have a carryback or carryforward of any general business credit, stop here and go to* ***Form 3800****, General Business Credit, to claim your 1985 investment credit. If you have only a 1985 investment credit (which may include business energy investment credit), you may continue with lines 15 through 20 to claim your credit.*

Part IV Tax Liability Limitations

15 a Individuals—From Form 1040, enter amount from line 46
 b Corporations—From Form 1120, Schedule J, enter tax from line 3 (or Form 1120-A, Part I, line 1).
 c Other filers —Enter income tax before credits from return . . . **15**

16 a Individuals—From Form 1040, enter credit from line 47, plus any orphan drug, nonconventional source fuel, and research credits included on line 49
 b Corporations—From Form 1120, Schedule J, enter credits from lines 4(a) through 4(e) (Form 1120-A filers, enter zero)
 c Other filers—See instructions for line 16c . . . **16**

17 Income tax liability as adjusted (subtract line 16 from line 15) . . . **17**

18 a Enter smaller of line 17 or $25,000. (See instructions for line 18) . . . **18a**
 b If line 17 is more than $25,000—Enter 85% of the excess . . . **18b**

19 Investment credit limitation—Add lines 18a and 18b . . . **19**

20 Total allowed credit—Enter the smaller of line 14 or line 19. This is your **General Business Credit** for 1985. Enter here and on Form 1040, line 48; Form 1120, Schedule J, line 4(f); Form 1120-A, Part I, line 2 ; or the proper line of other returns . . . **20**

For Paperwork Reduction Act Notice, see separate instructions. Form **3468** (1985)

281

FIGURE 18–1(a) and (b). IRS Form 3468.

Form 3468 (1985) Page 2

Schedule B.—Business Energy Investment Credit

1 Enter on lines 1(a) through 1(e) your qualified investment in business energy property that is the kind listed in the instructions for line 1, column (2)

Type of Property	Line	(1) Class of Property or Life Years	(2) Code	(3) Basis	(4) Applicable Percentage	(5) Qualified Investment (Column 3 x column 4)
Recovery	(a)	3-year			60	
	(b)	Other			100	
Nonrecovery	(c)	3 or more but less than 5			33 ⅓	
	(d)	5 or more but less than 7			66 ⅔	
	(e)	7 or more			100	

2 Total 10% energy investment property—Add lines 1(a) through 1(e), column (5) 2

3 Enter on lines 3(a) through 3(e) the basis in qualified hydroelectric generating property. Enter nameplate capacity of the property (See instructions for line 3.) ▶

Type of Property	Line	(1) Class of Property or Life Years	(2) Code	(3) Basis	(4) Applicable Percentage	(5) Qualified Investment (Column 3 x column 4)
Recovery	(a)	3-year			60	
	(b)	Other			100	
Nonrecovery	(c)	3 or more but less than 5			33 ⅓	
	(d)	5 or more but less than 7			66 ⅔	
	(e)	7 or more			100	

4 Total 11% energy investment property—Add lines 3(a) through 3(e), column (5) 4

5 Enter on lines 5(a) through 5(e) the basis in energy property that is solar equipment, wind equipment, ocean thermal equipment, or geothermal equipment. (See instructions for line 5, column (2).)

Type of Property	Line	(1) Class of Property or Life Years	(2) Code	(3) Basis	(4) Applicable Percentage	(5) Qualified Investment (Column 3 x column 4)
Recovery	(a)	3-year			60	
	(b)	Other			100	
Nonrecovery	(c)	3 or more but less than 5			33 ⅓	
	(d)	5 or more but less than 7			66 ⅔	
	(e)	7 or more			100	

6 Total 15% energy investment property—Add lines 5(a) through 5(e), column (5) 6
7 Enter 10% of line 2 7
8 Enter 11% of line 4 8
9 Enter 15% of line 6 9
10 Cooperative credit—Enter business energy investment credit from cooperatives 10
11 Tentative business energy investment credit—Add lines 7 through 10. Enter here and on line 13 of page 1 11

Fiscal Year Filers

The business energy investment credits shown on this page may be claimed only for energy property placed in service by December 31, 1985 (except for certain hydroelectric generating property, which may be placed in service by December 31, 1988).

At the time we printed this form it was uncertain whether Congress would extend this deadline. We will use news media and other means available to provide you with information and instructions on any change.

Instructions for Schedule B (Form 3468)

Energy property must meet the same requirements as regular investment credit property, except that the provisions of sections 48(a)(1) and 48(a)(3) do not apply. See the separate Instructions for Form 3468 for definitions and rules regarding regular investment credit property.

Energy property must be acquired new. See sections 46(b)(2) and 48(l)(1) through (17) for details.

See section 48(l)(17) for special rules on public utility property, and section 48(l)(11) (as amended by the Crude Oil Windfall Profit Tax Act of 1980) for special rules on property financed by Industrial Development Bonds.

Specific Instructions

One Credit Only.—If property qualifies as more than one kind of energy property, you may take only one credit for the property.

Lines 1, 3, and 5—Type of Property.—For definition of recovery and nonrecovery property, see the separate Instructions for Form 3468.

Line 1—Column (2).—Use the code letters from the following list to indicate the kind of property for which you are claiming a credit. If you enter more than one kind of property on a line, enter the code letter for each kind of property in column (2) and the code letter and dollar amount of each kind of property in the right hand margin.

The code letters are:

a. Biomass property

b. Qualified intercity buses (see section 48(l)(16)(C) for the limitation on qualified investment for intercity buses based on the increase in operating seating capacity)

Line 3.—Figure your qualified investment in hydroelectric generating property. If the installed capacity is more than 25 megawatts, the 11% energy credit is allowed for only part of the qualified investment. See section 48(l)(13)(C).

In the space provided in line 3, enter the megawatts capacity of the generator as shown on the nameplate of the generator.

Line 5—Column (2).—Use the code letters from the following list to indicate the kind of property for which you are claiming a credit. Be sure to put the code or codes on the line for the correct recovery period or life years as explained in the instructions for line 1, column (2).

c. Solar equipment (but not passive solar equipment)

d. Wind equipment

e. Ocean thermal equipment

f. Geothermal equipment

See sections 48(l)(4) and 48(l)(3)(A)(viii) and (ix) for definitions and special rules that apply to these kinds of property.

282

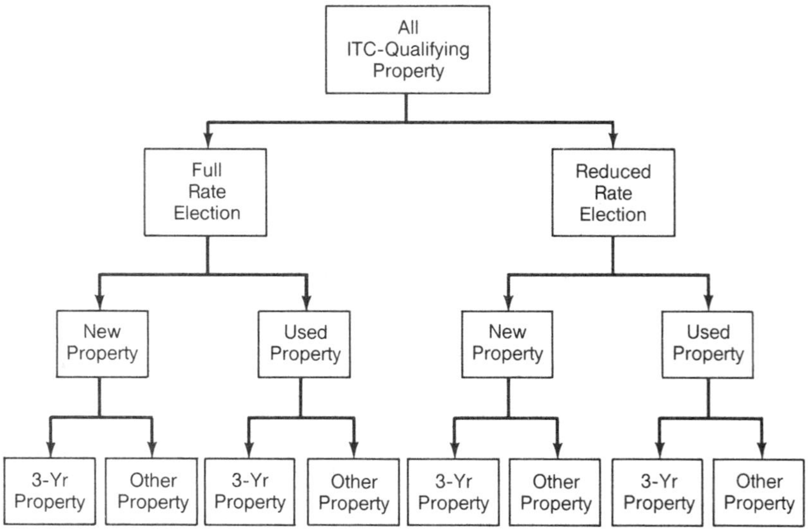

FIGURE 18–2. Schematic diagram of ITC categories.

For assets purchased prior to 1981, the depreciation reporting process is much simpler, *if* the calculations have already been performed in the owner's depreciation schedule. Simply write the total 1985 depreciation expense for all assets purchased prior to 1981 on Line 8.

Asset Purchases and Depreciation Example

As done in Chapter 17, a set of example purchases and other depreciating items is given in Table 18–1. Try to analyze each one before examining the filled-in forms or the analyses of each transaction that follow.

Analysis of the Depreciation Schedule

Eleven of the fourteen transactions in Table 18–1 are depreciable and belong on the depreciation schedule; the last three transactions do not belong. Consequently, these eleven also belong on Form 4562. But only the first five belong on Form 3468 since they were purchased in the year for which the tax return is being filled. All other items were purchased previously. The completed depreciation schedule for these transactions is shown in Figure 18–4, and completed tax forms are shown in Figures 18–5 and 18–6. Even though the forms have been completed, it should be useful to analyze each transaction and the summaries. The transactions will first be described for the depreciation schedule (Figure 18–4). Tax form analysis will follow later.

Transaction 1, Bull. The bull was purchased 8/1/85 for $2,000. All purchased breeding cattle are assigned to the 5-year ACRS property class. The election was made to claim the full 10 percent ITC rate, which yields $200 ITC. But the

Form **4562**

Department of the Treasury
Internal Revenue Service (L)

Depreciation and Amortization

▶ See separate instructions.
▶ Attach this form to your return.

OMB No. 1545-0172
1985
67

Name(s) as shown on return | Identifying number

Business or activity to which this form relates

Part I **Depreciation**

Use Part III, Specific Information Concerning Automobiles and other Listed Property, for certain transportation equipment (e.g. autos), amusement/recreation property, and computer/peripheral equipment.

Section A.—Election to Expense Recovery Property (Section 179)

	(a) Class of property	(b) Cost	(c) Expense deduction
1			
2	Listed property—Enter total from Part III, Section A, column (h)		
3	Total (see instructions for limitations). (Partnerships or S corporations—see the Schedule K and Schedule K-1 Instructions of Form 1065 or 1120S)		

Section B.—Depreciation of Recovery Property

(a) Class of property	(b) Date placed in service	(c) Cost or other basis	(d) Recovery period	(e) Method of figuring depreciation	(f) Deduction
4 Accelerated Cost Recovery System (ACRS) (see instructions): *For assets other than automobiles and other listed property placed in service* **ONLY** *during tax year beginning in 1985*					
a 3-year property					
b 5-year property					
c 10-year property					
d 15-year public utility property					
e Low-income housing					
f 15-year real property					
g 18-year real property					
(See "Items You Should Note")					

5	Listed property—Enter total from Part III, Section A, column (g)	
6	ACRS deduction for assets other than automobiles and other listed property placed in service prior to 1985 (see instructions)	

Section C.—Depreciation of Nonrecovery Property

7	Property subject to section 168(e)(2) election (see instructions)	
8	Other depreciation (see instructions)	

Section D.—Summary

9	Total (Add deductions on lines 3 through 8). Enter here and on the Depreciation line of your return (Partnerships and S corporations—Do NOT include any amounts entered on line 3.)	

Part II **Amortization**

(a) Description of property	(b) Date acquired	(c) Cost or other basis	(d) Code section	(e) Amortization period or percentage	(f) Amortization for this year
Total. Enter here and on Other Deductions or Other Expenses line of your return					

See Paperwork Reduction Act Notice on page 1 of the separate instructions.

Form **4562** (1985)

(225)

FIGURE 18–3. IRS Form 4562.

TABLE 18–1 Example Asset Purchase and Depreciation Transactions

Transaction	Asset Description	Acquisition Date	Original Basis	Depreciation Method	Useful Life	Salvage Value	New/ Used	ITC[a]
1.	Bull	8/1/85	2,000	ACRS	5	0	N	F
2.	Tractor	4/1/85	77,500	ACRS	5	0	N	F
3.	Bobtail truck	10/1/85	8,000	ACRS	5	0	U	F
4.	Boar	2/1/85	500	ACRS	3	0	N	F
5.	Light pickup	6/1/85	3,000	ACRS	3	0	U	F
6.	Bull	8/1/84	2,000	ACRS	5	0	N	R
7.	Pecan orchard	6/1/84	88,000	ACRS	5	0	U	R
8.	Tractor	8/1/82	50,000	ACRS	5	0	N	F
9.	Truck	1/1/80	20,000	SL	7	6,000	N	F
10.	Combine	1/1/79	75,000	SL	7	5,000	N	F
11.	Disc	1/1/79	5,000	DDB	10	1,000	U	F
12.	Land	6/1/85	50,000	—	—	—	—	—
13.	Feeder cattle	1/15/85	5,000	—	—	—	—	—
14.	Raised breedin cow	12/15/84	0	—	—	—	—	—

[a]F means full ITC rate selected, R means reduced ITC rate selected.

depreciable basis must then be reduced by one-half of the ITC claimed. Thus $1,900 is the depreciable basis. The applicable ACRS rate is 15 percent for the first year, so $285 (1,900 × 0.15) depreciation may be claimed in 1985.

Transaction 2, Tractor. The tractor is handled in much the same way as the bull, but Section 179 capital expensing was elected on the tractor. Consequently $5,000 is deducted from the $77,500 purchase price prior to ITC calculations, leaving an ITC basis of $72,500. The ITC at the full rate is $7,250. Reducing the depreciable basis by one-half the ITC ($3,625) leaves $68,875 to be depreciated over 5 years. The first-year normal ACRS depreciation would be $10,331 ($68,875 × 0.15). The net effect of all of this is really quite large: a $7,250 credit and a $15,331 deduction all in one year! It is really quite easy to see how the tax laws can be used to influence individuals and to stimulate the economy.

Transaction 3, Bobtail Truck. The bobtail truck is analyzed in exactly the same fashion as the bull, that is, 10 percent or $800 ITC, basis reduced by one-half the ITC or $400 to $7,600, and first-year depreciation of $1,140 ($7,600 × .15). However, because the truck was used, its purchase price will be entered under "Used Property" on Form 3468, Part II.

Transactions 4 and 5, Boar and Pickup. Although the analysis process is similar for the boar and the pickup as the first three transactions, breeding swine and light-duty trucks are assigned to the 3-year ACRS property class and consequently a different set of parameters applies. For the boar, the full 6 percent ITC rate was claimed, which yielded $30 ITC. The depreciable basis is reduced by one-half of that amount ($15) to $485. The applicable ACRS percentage of 25 percent yields $121 ($485 × .25) in the first year of ownership. The pickup is handled in the same manner, 6 percent or $180 ITC, basis reduced by $90 ($180 × .50) to $2,910, and first-year depreciation of $728 ($2,910 × .25).

Transactions 6 and 7, Bull and Orchard. Both purchased breeding cattle and existing pecan orchards are 5-year property. Since both of these properties were purchased in a previous year, and since no depreciable basis reduction was necessary because the reduced ITC rate was selected in the purchase year, all that is needed for the current tax year is the applicable ACRS pèrcentage. That rate, 22 percent in the second year of ownership, yields a $440 deduction on the bull ($2,000 × 0.22) and $19,360 on the orchard ($88,000 × .22).

Transaction 8, 1982 Tractor. Much like the 1983 bull and pecan orchard, the 1981 tractor was purchased in a previous year and although the full 10 percent ITC rate was applied, basis reduction was not required in 1982. For the current tax year, the applicable ACRS percentage rate is 21 percent, yielding a depreciation deduction of $10,500 ($50,000 × .21).

Transactions 9, 10, and 11: Truck, Combine, and Disc. Each of these three items was purchased prior to 1981, before ACRS was enacted. Therefore the depreciation method, useful life, and salvage value were determined by the owner. To calculate the current year's depreciation deductions of $2,000, $10,000, and $262, the formulas of Chapter 7 were applied.

Transactions 12, 13, and 14: Land, Feeder Cattle, and Raised Cow. None of the final three items is depreciable and consequently none is written either on the depreciation schedule or on Form 4562. Land has an infinitely useful life, feeder cattle are held for sale not for breeding purposes, and raised livestock have no basis for a cash-based taxpayer.

Analysis of Form 3468

Once the depreciation schedule has been completed, filling out the two tax forms, Forms 3468 and 4562, is really quite simple. For Form 3468 (Figure 18–5), all qualifying new purchases will be entered on Part II, Line 1. As noted earlier and as shown in Figure 18–2, there are eight categories of qualifying property based on new/used, full/reduced ITC rate, and 3-year/other ACRS property class.

In the example transactions of Table 18–1, we need to combine the ITC bases for the first five transactions. Line 1(a) covers full-rate new 3-year property, such as the boar. Consequently, the $500 purchase price is entered in Column 2. Line 1(b) covers full-rate new other property, like the bull and the tractor. Their bases ($2,000 and $77,500) are added, the $5,000 capital expense (Section 179) is deducted, and the resulting $74,500 is entered. Lines 1(c) and 1(d) are similar but apply to used property, which takes the light pickup to Line 1(c) and the bobtail truck to Line 1(d).

Calculation of the qualified investment involves only a multiplication of the unadjusted basis in Column 2 times the applicable percentages in Column 3. The results are summed on Line 5.

Part III calculates the actual credit by multiplying the investment by 10 percent. Note that the basis in 3-year property was reduced on Line 1 so that 10 percent does apply. For example, earlier discussions indicated the ITC on the

OPERATOR'S DEPRECIATION SCHEDULE FOR

Kind of machine or equipment	Date acquired	Cost or other basis	Investment credit	Salvage value	Depreciable balance	Depreciation, life & method		19 82 Remaining costs at beginning of year
Machinery & equipment								
Tractor #3 (N)	12-1-85	61,927	(10%) 5,693	—	5000+ 54,081	ACRS	5	
Tractor #2	11-1-75	12,000	1,200	1,000	11,000	DDB	7	1,518
Truck	6-1-77	5,000	500	1,000	4,000	SL	10	3,167
Swather	6-13-83	9,000	720	—	9,000	ACRS	5	
Baler	1-12-77	4,000	400	500	3,500	SYD	6	667
Disk #2 (O)	11-2-85	7,500	(8%) 560	—	7,500	ACRS	5	
Disk #1	7-1-75	2,500	250	500	2,000	SL	10	1,200
Planter	2-1-81	2,000	200	—	2,000	ACRS	5	1,700
Cotton Stripper	1-8-79	9,000	900	500	8,500	DDB	10	4,600
Trailer	10-1-69	2,000		200				200
Equipment	1-1-69	8,000		200				200
Pickup	6-1-75	3,000		300				300
Total Mach & Eq.								13,560
Auto	6-1-82	4,000	320	500	3,500	DDB	5	
Improvements								
Barns, Sheds	1-1-68	25,000		5,000	20,000	SL	20	12,000
Pipelines, Tanks	1-1-69	75,000		5,000	70,000	SL	20	29,500
Pecans	1-1-69	110,000		10,000	100,000	SL	25	58,000
Total Improvements								99,500
Breeding Stock								
1 Bull	6-1-80	1,150	115	350	800	SL	4	833
6 Bulls	6-1-80	4,500	450	2,400	2,100	SL	4	3,669
3 Bulls	6-1-81	2,700	270	—	2,700	ACRS	5	2,295
2 Bulls	7-1-85	3,000	240	—	3,000	ACRS	5	
Total Breeding Stock								6,797
TOTAL Total								119,857

DO NOT DEPRECIATE BELOW SALVAGE VALUE

FIGURE 18–4. Example of a filled-in depreciation schedule.

1982	1983		1984		1985		1986	
Depreciation for this year	Remaining costs at beginning of year	Depreciation for this year	Remaining costs at beginning of year	Depreciation for this year	Remaining costs at beginning of year	Depreciation for this year	Remaining costs at beginning of year	Depreciation for this year
						13,112	45,969	
434	1,084	84	1,000	0	1,000	0	1,000	
400	2,767	400	2,367	400	1,967	400	1,567	
		1,350	7,650	1,980	5,670	1,890	3,780	
167	500	0	500	0	500	0	500	$500
						1,125	6,375	
200	1,000	200	800	200	600	100	Traded 11-2-85	Final: ↓
440	1,260	420	840	420	420	420	0	$1927
922	3,686	737	2,949	590	2,359	432	Traded 12-1-85	Final: ↓
0	200	0	200	0	200	0	200	
0	200	0	200	0	200	0	200	
0	300	0	300	0	300	0	Sold	
2,563	10,997	3,191	16,806	3,590	13,216	17,479	59,591	
933	3,067	1,227	1,840	736	1,104	442	662	
1,000	11,000	1,000	10,000	1,000	9,000	1,000	8,000	
3,500	26,000	3,500	22,500	3,500	19,000	3,500	15,500	
4,000	54,000	4,000	50,000	4,000	46,000	4,000	42,000	
8,500	91,000	8,500	82,500	8,500	74,000	8,500	65,500	
								$350
200	633	200	433	83	350	0	Sold 6-25-85	Final: ↓
525	3,144	525	2,619	219	2,400	0	2,400	
594	1,701	567	1,134	567	567	567	0	
						450	2,550	
1,319	5,478	1,292	4,186	869	3,317	1,017	4,950	
13,315	110,542	14,210	105,332	13,695	91,637	27,438	130,703	

DO NOT DEPRECIATE BELOW SALVAGE VALUE

Form **3468**

Department of the Treasury
Internal Revenue Service (X)

Computation of Investment Credit

▶ Attach to your tax return.
▶ Schedule B (Business Energy Investment Credit) on back.

OMB No. 1545-0155
1985
24

Name(s) as shown on return: FRANK FARMER

Identifying number: 123-45-6789

Part I Elections (Check the box(es) below that apply to you (See Instruction D).)

A I elect to increase my qualified investment to 100% for certain commuter highway vehicles placed in service before January 1, 1986 (section 46(c)(6)) ☐

B I elect to increase my qualified investment by all qualified progress expenditures made this and all later tax years ☐
Enter total qualified progress expenditures included in column (4), Part II ▶

C I claim full credit on certain ships under section 46(g)(3) (See **Instruction B** for details.) ☐

Part II Qualified Investment (See instructions for rules on automobiles and other property with any personal use)

1 Recovery Property		Line	(1) Class of Property	(2) Cost or Other Basis	(3) Applicable Percentage	(4) Qualified Investment (Column 2 x column 3)
Regular Percentage	New Property	(a)	3-year	500	60	300
		(b)	Other	74,500	100	74,500
	Used Property	(c)	3-year	3,000	60	1,800
		(d)	Other	8,000	100	8,000
Section 48(q) Election to Reduce Credit (instead of adjusting basis)	New Property	(e)	3-year		40	
		(f)	Other		80	
	Used Property	(g)	3-year		40	
		(h)	Other		80	

2	Nonrecovery property—Enter total qualified investment (See instructions for line 2)	2	
3	New commuter highway vehicle—Enter total qualified investment (See Instruction D(1))	3	
4	Used commuter highway vehicle—Enter total qualified investment (See Instruction D(1))	4	
5	**Total qualified investment in 10% property**—Add lines 1(a) through 1(h), 2, 3, and 4 (See instructions for special limits)	5	84,600
6	Qualified rehabilitation expenditures—Enter total qualified investment for:		
	a 30-year-old buildings	6a	
	b 40-year-old buildings	6b	
	c Certified historic structures (You must attach NPS certification—see instructions)	6c	

Part III Tentative Regular Investment Credit

		Line	Amount
7	10% of line 5	7	8,460
8	15% of line 6a	8	
9	20% of line 6b	9	
10	25% of line 6c	10	
11	Credit from cooperatives—Enter regular investment credit from cooperatives	11	
12	Regular investment credit—Add lines 7 through 11	12	8,460
13	Business energy investment credit—From line 11 of Schedule B (see back of this form)	13	
14	Current year investment credit—Add lines 12 and 13	14	8,460

Note: *If you have a 1985 jobs credit (Form 5884), credit for alcohol used as fuel (Form 6478), or employee stock ownership plan (ESOP) credit (Form 8007) in addition to your 1985 investment credit, or if you have a carryback or carryforward of any general business credit, stop here and go to* **Form 3800**, *General Business Credit, to claim your 1985 investment credit. If you have only a 1985 investment credit (which may include business energy investment credit), you may continue with lines 15 through 20 to claim your credit.*

Part IV Tax Liability Limitations

15	a Individuals—From Form 1040, enter amount from line 46 b Corporations—From Form 1120, Schedule J, enter tax from line 3 (or Form 1120-A, Part I, line 1). c Other filers—Enter income tax before credits from return	15	44,246
16	a Individuals—From Form 1040, enter credit from line 47, plus any orphan drug, nonconventional source fuel, and research credits included on line 49 b Corporations—From Form 1120, Schedule J, enter credits from lines 4(a) through 4(e) (Form 1120-A filers, enter zero) c Other filers—See instructions for line 16c	16	
17	Income tax liability as adjusted (subtract line 16 from line 15)	17	44,246
18	a Enter smaller of line 17 or $25,000. (See instructions for line 18)	18a	25,000
	b If line 17 is more than $25,000—Enter 85% of the excess	18b	16,359
19	Investment credit limitation—Add lines 18a and 18b	19	41,359
20	Total allowed credit—Enter the smaller of line 14 or line 19. This is your **General Business Credit** for 1985. Enter here and on Form 1040, line 48; Form 1120, Schedule J, line 4(f); Form 1120-A, Part I, line 2 ; or the proper line of other returns	20	8 8,460

For Paperwork Reduction Act Notice, see separate instructions.

Form 3468 (1985)

(281)

FIGURE 18–5. Example of filled-in Form 3468.

Form **4562** | **Depreciation and Amortization** | OMB No. 1545-0172

Department of the Treasury
Internal Revenue Service (X)

▶ See separate instructions.
▶ Attach this form to your return.

1985
67

Name(s) as shown on return: FRANK FARMER
Identifying number: 123-45-6789

Business or activity to which this form relates: FARM

Part I Depreciation

Use Part III, Specific Information Concerning Automobiles and other Listed Property, for certain transportation equipment (e.g. autos), amusement/recreation property, and computer/peripheral equipment.

Section A.—Election to Expense Recovery Property (Section 179)

	(a) Class of property	(b) Cost	(c) Expense deduction
1	5-YEAR	77,500	5,000
2	Listed property—Enter total from Part III, Section A, column (h)		
3	Total (see instructions for limitations). (Partnerships or S corporations—see the Schedule K and Schedule K-1 Instructions of Form 1065 or 1120S)		5,000

Section B.—Depreciation of Recovery Property

(a) Class of property	(b) Date placed in service	(c) Cost or other basis	(d) Recovery period	(e) Method of figuring depreciation	(f) Deduction
4 Accelerated Cost Recovery System (ACRS) (see instructions): *For assets other than automobiles and other listed property placed in service* **ONLY** *during tax year beginning in 1985*					
a 3-year property		3,395	3	PRE	849
b 5-year property		78,375	5	PRE	11,756
c 10-year property					
d 15-year public utility property					
e Low-income housing					
f 15-year real property					
g 18-year real property (See "Items You Should Note")					
5 Listed property—Enter total from Part III, Section A, column (g)					
6 ACRS deduction for assets other than automobiles and other listed property placed in service prior to 1985 (see instructions)					30,300

Section C.—Depreciation of Nonrecovery Property

7	Property subject to section 168(e)(2) election (see instructions)	
8	Other depreciation (see instructions)	12,262

Section D.—Summary

9	Total (Add deductions on lines 3 through 8). Enter here and on the Depreciation line of your return (Partnerships and S corporations—Do NOT include any amounts entered on line 3.)	60,167

Part II Amortization

(a) Description of property	(b) Date acquired	(c) Cost or other basis	(d) Code section	(e) Amortization period or percentage	(f) Amortization for this year
Total. Enter here and on Other Deductions or Other Expenses line of your return					

See Paperwork Reduction Act Notice on page 1 of the separate instructions.

Form **4562** (1985)

(297)

FIGURE 18–6. Example of filled-in Form 4562.

$500 boar should be 6% or $30 ($500 × .06). Form 3468 calculates the ITC by multiplying the basis by 60 percent and then 10 percent ($500 × .60 × $300, $300 × .10 = $30) to reach the same answer.

Recall that ITC is limited, generally by the amount of tax due. Those limitations for the current year are calculated in Part IV. The example entry on Lines 17(a) and 18(a) will be explained in Chapter 20. In our case, the taxpayer is allowed to claim all of the tentative ITC shown on Line 16.

Analysis of Form 4562

Form 4562 (Figure 18–6) is also relatively easy to complete if the depreciation schedule has been computed. Part I, Section A asks for the property that will take the capital expensing option. Line 1 shows the 1985 maximum of $5,000 that was elected on Transaction 2, the tractor.[3] Section B lists all property, grouped by class/year of purchase, that was acquired in 1984. Referring again to Table 18–1, all 3-year property is grouped by year (only the 1985 boar and pickup). Their depreciable amounts are added ($485 + $2,910 = $3,395) and their depreciation amounts are added ($121 + $728 = $849) for Columns c and f. Similarly, the bull, tractor, and bobtail truck are grouped on Line 4(b) for 1985.

All data for assets purchased between 1981 and 1985 and prior to 1981 are taken directly from the depreciation schedule. The total 1985 depreciation claimed for pre-1985 ACRS property is written on Line 6 and the 1985 depreciation claimed for pre-ACRS property is written on Line 8 and is added to the totals from Sections A and B. The total of Line 9 should equal the amount shown for 1985 depreciation on the depreciation schedule.

Summary Transfers

Once all of the purchases and depreciation amounts have been entered and summed on Forms 3468 and 4562, we can set these two forms aside. The bottom line (Line 20) of Form 3468 is transferred to Form 1040 and the bottom line (Line 9) of Form 4562 will be transferred to Schedule F.

Summary

The tax reporting of asset purchases and depreciation is much simpler than the process necessary for asset sales. Analyzing asset purchases, however, requires a careful reading and understanding of the ITC rules, which in recent years have affected the level of depreciation that can be claimed.

A good depreciation schedule can significantly reduce the workload to complete the business asset depreciation form. In fact, it can now be seen that a depreciation schedule is a necessity. The relevant tax forms are IRS Form 3468 to claim ITC and IRS Form 4562 to claim the depreciation deduction.

[3]The $5,000 capital expensing option does not need to be taken entirely on one asset; it can be split among several. For example, $4,000 could be allocated to the tractor and $1,000 to either truck.

CHAPTER 19

The Farm and Ranch Return

It seems at times that asset sales, purchases, and depreciation tax issues cloud the central concern of reporting receipts and expenditures, the net returns of normal farm and ranch production. But analyzing the asset transactions points out three very important problems in federal income tax reporting and management: (1) The reporting process is long and difficult; there are many factors and concepts which complicate the process. (2) Many tax dollars are at stake with the "big ticket" purchases and sales, generally more dollars than are at stake with the normal, routine types of receipts and expenditures. (3) a logical process from start to finish that minimizes paper shuffling will minimize headaches for the tax preparer even if that does not seem possible at this point. Nevertheless, it is time to leave assets behind and proceed to other concerns.

Once the five forms discussed in Chapters 17 and 18 have been completed, the typical cash-flow types of items should be reported. This chapter will concentrate on three IRS forms: Schedule F, Schedule SE and Form 4136. Their places in the reporting process are shown in Figure 19–1. However, since Schedule SE and Form 4136 are relatively quick and easy to prepare, the real emphasis will be on Schedule F. Schedule F (Figures 19–2[a] and [b]) is the basic farm and ranch return for sole proprietors and the individual partners in a partnership. Schedule SE (Figure 19–3) computes the self-employment tax, which is a self-employed person's contributions to the Social Security system. Form 4136 (Figure 19–4(a) and (b) is an often overlooked yet easy-to-complete form which may or may not apply to every farmer and rancher. This form should be completed when gasoline and diesel fuel distributors charge the federal road-use fuel tax to farmers and ranchers for fuels that are not consumed on publicly owned or maintained roads.[1] The fuel tax applies *only* to road use, therefore the farmer or rancher who uses fuel (*and* pays the tax) for agricultural, off-highway purposes may use Form 4136 to claim a credit for the tax paid.

[1]Many distributors do *not* charge the tax for bulk deliveries to farms and ranchers. If the tax is not charged, this form should *not* be completed. In some parts of the country, this special handling of bulk deliveries is reflected in different coloring, commonly called "purple gas."

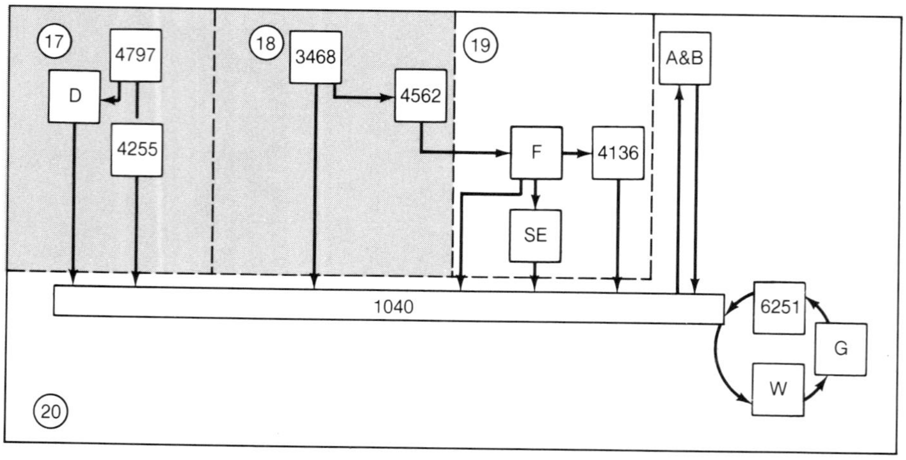

FIGURE 19–1. Farm and ranch tax reporting process.

Farm Income and Expenses

Normal (non-asset) farm and ranch taxable receipts and deductible expenditures are listed on Schedule F. The IRS (as well as the USDA) recognizes no difference between a farm and a ranch, consequently both report on the Schedule F. Similarly, other forms of agricultural businesses (such as fruit and nut orchards, dairies, fish farms, and other agricultural activities, tree and ornamental nurseries, and some timber producers) that normally are not referred to as *farms* are reported on Schedule F. Typical exceptions to Schedule F reporters are farm or ranch-related businesses owned and operated by farmers and ranchers, such as seed corn dealerships, welding or horseshoeing businesses, and custom farming and harvesting businesses. The income from those businesses should be reported on Schedule C. Other exceptions include landlords. A participating landlord who helps to manage and finance a farm or ranch will report on Schedule F, but a nonparticipating landlord will not. If cash rent is received, the landlord will report that rental income on Schedule E, but if the lease is written as a livestock-share or crop-share agreement, the landlord will report on Form 4835. Each of these three forms (Schedules C and E and Form 4835) will be discussed later in this chapter.

Accounting Systems

As discussed extensively in Chapters 2 and 5, farmers and ranchers may choose to report income tax liability using a cash method or accrual method of accounting. The decision between cash or accrual for income tax purposes is actually independent of the decision for business records and analysis purposes. It is easier to report tax liability on the same basis as the basis on which the business records are maintained, but there is no legal requirement concerning uniformity for sole proprietorships, partnerships, and even small corporations that are not publicly owned. Despite the ease of reporting taxes on the same basis of accounting as the business records, there may be distinct long-term advantages to maintaining an accrual-based records system for accuracy and period reporting (see

SCHEDULE F (Form 1040)

Department of the Treasury Internal Revenue Service (L)

Farm Income and Expenses

▶ Attach to Form 1040, Form 1041, or Form 1065.

▶ See Instructions for Schedule F (Form 1040).

OMB No. 1545-0074

1985

14

Name of proprietor(s) | Social security number (SSN)

A **Agricultural Activity Code.** (Write in the code that best describes your principal income-producing activity. The codes are listed on page 2 of this schedule.) ▶

B **Principal Product.** (Describe in one or two words your principal crop or output for the current tax year.) ▶

If you disposed of commodities received under the payments-in-kind (PIK) program, check the box(es) that apply:

☐ Feed for livestock ☐ Sold and reported in income

Employer ID number (NOT SSN)

Part I Farm Income—Cash Method—Complete Parts I and II

(Accrual method taxpayers complete Parts II and III, and line 12 of Part I.)

Do not include sales of livestock held for draft, breeding, sport, or dairy purposes; report these sales on Form 4797.

1 Sales of livestock and other items you bought for resale	1	
2 Cost or other basis of livestock and other items you bought for resale	2	
3 Subtract line 2 from line 1	3	
4 Sales of livestock, produce, grains, and other products you raised	4	
5 a Total distributions received from cooperatives (from Form 1099-PATR) 5a		
b **Less:** Nonincome items 5b		
6 Net distributions. Subtract line 5b from line 5a	6	
7 Agricultural program payments:		
a Cash	7a	
b Materials and services	7b	
8 Commodity credit loans under election (or forfeited)	8	
9 Crop insurance proceeds	9	
10 Machine work	10	
11 Other income, including Federal and state gasoline tax credit or refund (see instructions)	11	
12 **Gross income**. Add amounts on lines 3, 4, 6, and 7a through 11. If accrual method taxpayer, enter the amount from Part III, line 52 ▶	12	

Part II Farm Deductions—Cash and Accrual Method

Do not include personal or living expenses (such as taxes, insurance, repairs, etc. on your home), which do not produce farm income. Reduce the amount of your farm deductions by any reimbursements before entering the deduction below.

13 Breeding fees		26 Mortgage interest paid to financial institutions (see instructions)	
14 Chemicals		27 Other interest	
15 Conservation expenses		28 Pension and profit-sharing plans	
16 Depreciation, and section 179 expense deduction (from Form 4562)		29 Rent of farm, pasture	
17 Employee benefit programs other than on line 28		30 Repairs, maintenance	
		31 Seeds, plants purchased	
18 Feed purchased		32 Storage, warehousing	
19 Fertilizers and lime		33 Supplies purchased	
20 Freight, trucking		34 Taxes	
21 Gasoline, fuel, oil		35 Utilities	
22 Insurance		36 Veterinary fees, medicine	
23 a Labor hired		37 Other expenses (specify):	
b Jobs credit		a	
c Balance (subtract line 23b from line 23a)		b	
		c	
24 Land clearing (see instructions)		d	
25 Machine hire		e	

38 **Total deductions from Part II**. Add amounts in columns for lines 13 through 37e ▶	38	
39 **Net farm profit or (loss)** (subtract line 38 from line 12). If a profit, enter on Form 1040, line 19, and on Schedule SE, Part I, line 1. If a loss, you **MUST** go on to line 40. (Fiduciaries and partnerships, see the instructions.)	39	

40 If you have a loss, you **MUST** answer this question:
"Do you have amounts for which you are not at risk in this farm (see instructions)?" ☐ Yes ☐ No
If "Yes," you **MUST** attach **Form 6198**. If "No," enter the loss on Form 1040, line 19, and on Schedule SE, Part I, line 1.

For Paperwork Reduction Act Notice, see Form 1040 Instructions.

Schedule F (Form 1040) 1985

23

FIGURE 19–2(a). IRS Schedule F, parts I and II.

Schedule F (Form 1040) 1985 Page 2

Part III Farm Income—Accrual Method

Do not include sales of livestock held for draft, breeding, sport, or dairy purposes; report these sales on Form 4797 and do not include them on line 47 below.

41	Sales of livestock, produce, grains, and other products during year		41	
42	Agricultural program payments:			
	a Cash		42a	
	b Materials and services		42b	
43	Commodity credit loans under election (or forfeited)		43	
44	Machine work		44	
45	Other income, including Federal and state gasoline tax credit or refund (see instructions)		45	
46	**Total.** Add amounts on lines 41 through 45		46	
47	Inventory of livestock, produce, grains, and other products at beginning of year	47		
48	Cost of livestock, produce, grains, and other products purchased during year	48		
49	Add lines 47 and 48	49		
50	Inventory of livestock, produce, grains, and other products at end of year	50		
51	Cost of livestock, produce, grains, and other products sold. Subtract line 50 from line 49 *		51	
52	**Gross income.** Subtract line 51 from line 46. Enter here and on Part I, line 12 ▶		52	

* If you use the unit-livestock-price method or the farm-price method of valuing inventory and the amount on line 50 is larger than the amount on line 49, subtract line 49 from line 50. Enter the result on line 51 and add lines 46 and 51. Enter the total on line 52.

Principal Agricultural Activity Codes

Select one of the following codes and write the 4 digit number on line A on page 1 of this schedule. (**Note:** *If your principal source of income is from providing agricultural services such as soil preparation, veterinary, farm labor, horticultural, or management for a fee or on a contract basis, you should file Schedule C (Form 1040), Profit or (Loss) From Business or Profession.*)

0120 **Field crop,** including grains and nongrains such as cotton, peanuts, feed corn, wheat, tobacco, Irish potatoes, etc.

0160 **Vegetables and melons,** garden-type vegetables and melons, such as sweet corn, tomatoes, squash, etc.

0170 **Fruit and tree nuts,** including grapes, berries, olives, etc.

0180 **Horticultural specialty,** including the growing of ornamental floriculture and nursery products such as plants, shrubbery, bulbs, flowers, sod, etc.

0211 **Beefcattle feedlots**

0212 **Beefcattle,** except feedlots

0215 **Hogs, sheep, and goats**

0240 **Dairy**

0250 **Poultry and eggs,** including chickens, ducks, pigeons, quail, etc.

0260 **General livestock,** not specializing in any one livestock category

0270 **Animal specialty,** including fur bearing animals, pets, fish, horses, etc.

0300 **Agricultural production,** not specified

FIGURE 19–2(b). IRS Schedule F, part III.

Chapters 2 and 5).[2] For most operating commercial farms and ranches, however, the cash method of reporting tax liability offers more flexibility for managing or maneuvering receipts and expenditures to the advantage of the taxpayer.[3] The cash-based taxpayer will use Parts I and II of Schedule F (Figure 19–2[a]) whereas the accrual-based taxpayer will report expenditures in Part II also, but will report income in Part III (Figure 19–2[b]).

Income

Virtually every source of income to the farm or ranch business (other than nonfarm income or capital sales) is taxable and should be reported on Schedule F.

[2] As argued in earlier chapters, a well-maintained modified single-entry cash accounting system with inventory control can accomplish virtually everything that a double-entry accrual system does.

[3] Operations having a sizable proportion of their receipts attributable to capital gain income will benefit more from the cash method since here capital gains are based on the sale amount minus purchase price, whereas capital gains are based on the difference between sales amount and inventory value if the accrual method was used.

SCHEDULE SE (Form 1040)
Department of the Treasury
Internal Revenue Service (L)

Computation of Social Security Self-Employment Tax

▶ See Instructions for Schedule SE (Form 1040).
▶ Attach to Form 1040.

OMB No. 1545-0074
1985
18

Name of **self-employed** person (as shown on social security card) | Social security number of **self-employed** person ▶

Part I Regular Computation of Net Earnings From Self-Employment

Note: *If you performed services for certain churches or church-controlled organizations and you are not a minister or a member of a religious order, see the instructions.*

1 Net farm profit or (loss) from Schedule F (Form 1040), line 39, and farm partnerships, Schedule K-1 (Form 1065), line 13a . . .	1	
2 Net profit or (loss) from Schedule C (Form 1040), line 33, Schedule K-1 (Form 1065), line 13a (other than farming), and Form W-2 wages of $100 or more from an electing church or church-controlled organization. (See instructions for other income to report.) . . .	2	

Note: ☐ *Check here if you are* ***exempt*** *from self-employment tax on your earnings as a minister, member of a religious order, or Christian Science practitioner because you filed* ***Form 4361***.
See instructions for kinds of income to report. If you have other earnings of $400 or more that are subject to self-employment tax, include those earnings on line 2.

Part II Optional Computation of Net Earnings From Self-Employment (See "Who Can Use Schedule SE")

Generally, this part may be used **only** if you meet any of the following tests:

A Your gross farm income (Schedule F (Form 1040), line 12) was not more than $2,400; or

B Your gross farm income (Schedule F (Form 1040), line 12) was more than $2,400 and your net farm profits (Schedule F (Form 1040), line 39) were less than $1,600; or

C Your net nonfarm profits (Schedule C (Form 1040), line 33) were less than $1,600 and also less than two-thirds (2/3) of your gross nonfarm income (Schedule C (Form 1040), line 5).

See instructions for other limitations.

3 Maximum income for optional methods . . .	3	$1,600	00
4 Farm Optional Method—If you meet test A or B above, enter: the smaller of two-thirds (2/3) of gross farm income from Schedule F (Form 1040), line 12, and farm partnerships, Schedule K-1 (Form 1065), line 13b; or $1,600 . . .	4		
5 Subtract line 4 from line 3 . . .	5		
6 Nonfarm Optional Method—If you meet test C above, enter: the smallest of two-thirds (2/3) of gross nonfarm income from Schedule C (Form 1040), line 5, and Schedule K-1 (Form 1065), line 13c (other than farming); or $1,600; or, if you elected the farm optional method, the amount on line 5 . . .	6		

Part III Computation of Social Security Self-Employment Tax

7 Enter the amount from Part I, line 1, or, if you elected the farm optional method, Part II, line 4 . . .	7		
8 Enter the amount from Part I, line 2, or, if you elected the nonfarm optional method, Part II, line 6 . . .	8		
9 Add lines 7 and 8. If less than $400, do not fill in the rest of the schedule because you are not subject to self-employment tax. (**Exception:** If this line is less than $400 and you are an employee of an electing church or church-controlled organization, complete the schedule unless this line is a loss. See instructions.) . . .	9		
10 The largest amount of combined wages and self-employment earnings subject to social security or railroad retirement tax (Tier 1) for 1985 is . . .	10	$39,600	00
11 a Total social security wages and tips from Forms W-2 and railroad retirement compensation (Tier 1). **Note:** *U.S. Government employees whose wages are only subject to the 1.35% hospital insurance benefits tax (Medicare) and employees of certain church or church-controlled organizations should not include those wages on this line (see instructions)* . . . 11a			
b Unreported tips subject to social security tax from Form 4137, line 9, or to railroad retirement tax (Tier 1) . . . 11b			
c Add lines 11a and 11b . . .	11c		
12 a Subtract line 11c from line 10. . . .	12a		
b Enter your "qualified" U.S. Government wages if you are required to use the worksheet in Part III of the instructions. 12b			
c Enter your Form W-2 wages from an electing church or church-controlled organization. 12c			
13 Enter the smaller of line 9 or line 12a . . .	13		
If line 13 is $39,600, fill in $4,672.80 on line 14. Otherwise, multiply line 13 by .118 and enter the result on line 14 . . .		.118	
14 Self-employment tax. Enter this amount on Form 1040, line 51 . . .	14		

For Paperwork Reduction Act Notice, see Form 1040 Instructions. Schedule SE (Form 1040) 1985

(31)

FIGURE 19–3. IRS Schedule SE.

Form **4136**
Department of the Treasury
Internal Revenue Service (L)

Computation of Credit for Federal Tax on Gasoline and Special Fuels

(And Credit for Purchase of Qualified Diesel-Powered Highway Vehicles)

▶ Attach this form to your income tax return.

OMB No. 1545-0162

1985

26

Name (as shown on your income tax return)	Social security or employer identification number

Type of Use	Gasoline and Gasohol, Diesel Fuel, and Other Special Motor Fuels		
	A. Number of Gallons Used	B. Rate of Tax	C. Multiply Column A by Column B
1 Qualified diesel-powered highway vehicles (See instruction E.)			$
2 Nonhighway:			
a Farm (for farming purposes) (Ground applicators of agricultural substances, see instruction C.1.):			
(i) Gasoline		.09	
(ii) Diesel fuel		.15	$
(iii) Special fuels		.09	
b Commercial fishing vessel (See instructions C.2. and D.2.(b).):			
(i) Gasoline		.09	
(ii) Diesel fuel		.15	$
(iii) Special fuels		.09	
c Off-highway business (specify)			
(i) Gasoline		.09	
(ii) Diesel fuel		.15	$
(iii) Special fuels		.09	
3 a Intercity, local, and school buses (See instruction C.4.)		.09/.12	
b Qualified local buses (See instruction D.2.(c).)		.15	$
4 Gasoline/Alcohol Mixture		.0566	$
5 a Diesel/Alcohol Mixture		.06	
b Special Fuels/Alcohol Mixture		.06	$
6 Qualified taxicab		.04	$
Aviation Fuel			
7 Farm (for farming purposes) (Aerial applicators of agricultural substances, see instruction C.1.)		.09/.12/.14 *	$
8 Aviation (Applies only to commercial or certain helicopter uses as defined in instruction C.7.)		.09/.12/.14 *	$
9 Total income tax credit claimed (add lines 1 through 8). Enter here and on Form 1040, line 62; Form 1120, line 32f; Form 1120-A, line 28f; Form 1120S, line 26b; Form 1041, line 33b; or the proper line of other returns.			$

* *Use rate paid or allowable rate if less*

Instructions

(Section references are to the Internal Revenue Code)

Paperwork Reduction Act Notice.—We ask for this information to carry out the Internal Revenue laws of the United States. We need it to ensure that taxpayers are complying with these laws and to allow us to figure and collect the right amount of tax. You are required to give us this information.

Purpose of Form.—File this form if you claim credit for Federal excise tax on gasoline and gasohol, diesel fuel, and other special fuels used for certain types of business uses or the credit for diesel-powered highway vehicles.

Figure the credit by multiplying the number of gallons used by the rate of Federal excise tax paid or the allowable rate if lesser.

An individual, estate, trust, or corporation may file this form, but a partnership may not. However, when a partnership files **Form 1065**, it must include a statement showing how many gallons of the fuels are allocated to each partner, the types of use, and the applicable tax rate as shown on this form.

A. Refund Instead of Credit.—If you paid excise tax of at least $1,000 ($200 for gasoline/alcohol mixture and $50 for gasoline, diesel fuel, or special fuels used in a qualified taxicab) on any fuel category used in any one of the first three quarters of the tax year, you may file for a refund of the tax instead of claiming a credit for it on Form 4136. To claim the refund, file **Form 843**, Claim, before the end of the quarter following the one for which you are claiming the refund. Show separate computations for a claim on gasoline and special fuels. You may use Form 4136 as a worksheet to show your computations and attach it to the Form 843 you file.

You must claim a credit, rather than a refund, for any tax paid during the fourth quarter, and for any tax under $1,000 ($200 for gasoline/alcohol mixture and $50 for gasoline, diesel fuel, or special fuels used in a qualified taxicab) for any fuel used in the first three quarters. You must also claim the credit rather than a refund, for tax paid on gasoline or special fuels used on a farm for farming purposes.

Do not claim a credit on Form 4136 for any amount that you previously claimed as a refund.

B. Year To Include the Credit Or Refund In Income.—When you claim a credit or refund, you must include it in your gross income to the extent that it reduced your income tax. The year you include it depends on your method of accounting. This rule does not apply to the credit claimed on line 1.

1. Cash Method.—If you use the cash method, include the amount of the credit in gross income in the tax year that you file the tax return claiming the credit. If you file an amended return, or if you file a claim for credit or refund (and claim the credit for the tax), include the refund that results from the credit in your gross income for the tax year you receive it.

2. Accrual Method.—If you use the accrual method, figure the amount of the credit or refund due at the close of your tax year. Include it in income, even if you have not filed a claim for refund.

C. What Gasoline To Include.—You may claim credit for Federal excise tax on gasoline or gasohol for farm uses, commercial fishing vessel uses, off-highway business uses, intercity, local, or school bus uses, qualified taxicab use, and commercial or certain helicopter aviation uses.

1. Farm Use.—You may claim a credit for gasoline used for farming in a trade or business farm in the United States in the amount of 9¢ a gallon. (See instruction C. 7. for aviation fuel used for farming.) Such farms include the following: stock, dairy, poultry, fur-bearing animals, truck farms, plantations, ranches, nurseries, ranges, greenhouses, and orchards.

Gasoline is considered used for farming only if the owner, tenant, or operator of the farm uses the gasoline. If gasoline is used for the farming

215

FIGURE 19–4(a) and (b). IRS Form 4136.

tasks set forth below by anyone else, the owner, tenant, or operator is considered the user of the gasoline. There is an exception to this rule. An aerial or ground applicator of agricultural substances is considered to have used gasoline for farming when performing the farming tasks under the following conditions:

- The applicator is the ultimate purchaser of the gasoline; and
- The farm owner, tenant, or operator agrees not to be treated as the user and ultimate purchaser of the gasoline.

The following tasks are considered farming:

(a) Cultivating the soil, and raising or harvesting any horticultural or agricultural commodity. This includes raising livestock, bees, poultry, fur-bearing animals, and wildlife.

(b) Handling, drying, packing, grading, or storing any horticultural or agricultural commodity before manufacture, but only if the owner, tenant, or operator produced more than half of the commodity treated.

(c) Planting, cultivating, or cutting trees, or preparing them for market. This does not include milling the lumber.

(d) Operating, managing, conserving, improving, or maintaining the farm and its tools and equipment.

Gasoline is not considered used by you for farming if it is used:

(a) Off the farm to transport livestock, feed, crops, or equipment.

(b) In processing, packaging, freezing, or canning operations.

(c) For personal purposes.

(d) On the farm of another person, even though used for farming.

Often a vehicle is used on the farm both for farming and for nonfarming purposes. In these cases, show the number of gallons of gasoline used for each purpose when claiming a credit.

2. Commercial Fishing Vessel Use.—You may claim a credit for tax paid on gasoline used in a commercial fishing vessel in the amount of 9¢ a gallon. A commercial fishing vessel is one used in the fisheries or whaling business and used:

- Exclusively for catching shrimp and other types of aquatic life for sale commercially as bait; or
- On specific trips exclusively for catching fish to be sold commercially.

3. Off-highway Business Uses.—You may claim a credit for tax paid on gasoline used for off-highway business purposes in the amount of 9¢ a gallon. Off-highway business use does not include gasoline used in a motorboat that is not a commercial fishing vessel, or in a highway vehicle that either:

- Is registered or required to be registered for highway use under the laws of any state, the District of Columbia, or a foreign country; or
- Is owned by the United States and is used on the highway.

4. Intercity, Local, or School Bus Uses.— If you are the ultimate purchaser of gasoline for a bus that provides transportation (1) to the general public for a fee, or (2) to students and school employees, you may claim a credit for tax paid on gasoline used for this service in the amount of 9¢ a gallon.

If you provide transportation on intercity and local buses on nonscheduled irregular routes, you may take the credit only if the seating capacity of the bus is at least 20 adults, not including the driver.

5. Gasoline/Alcohol Mixture.—You may claim a credit for tax paid on gasoline that you used to produce a gasoline/alcohol mixture that was sold or used in your trade or business. The credit is 5⅔¢(.0566) a gallon if you paid 9¢ a gallon.

6. Qualified Taxicab Use.—The ultimate purchaser may claim a credit for tax paid on gasoline used before October 1, 1985, in a qualified taxicab providing qualified taxicab service in the amount of 4¢ a gallon.

A qualified taxicab is a land vehicle that holds fewer than 10 adults including the driver. For 1978 and later model years, the vehicle model must have a fuel economy rating better than the average fuel economy standard for the model year (18 mpg for 1978 models, 19 mpg for 1979, 20 mpg for 1980, 22 mpg for 1981, 24 mpg for 1982, 26 mpg for 1983, 27 mpg for 1984 and 27½ mpg for 1985). A taxicab is qualified if the vehicle received or could have received an exemption from the fuel economy standard for the model year. Fuel economy standards do not apply for vehicles acquired before 1979.

Qualified taxicab service means furnishing nonscheduled land transportation for a fixed fee in a qualified taxicab. The person operating the taxicab must be properly licensed as a taxicab driver and must not be prohibited from furnishing ride-sharing.

7. Farming, Commercial, or Helicopter Aviation Uses.—You may claim a credit for tax paid on aviation gasoline used in farming, commercial, or certain helicopter aviation uses in the amount of 9¢ or 12¢ a gallon, whichever is paid.

See C.1. for the definition of farm use.

Commercial aviation is the use of an aircraft in a business of carrying people or property by air for pay. The use of an aircraft may be considered noncommercial aviation if the aircraft:

(a) Has a maximum certificated takeoff weight of 6,000 pounds or less and is not operated on an established line; or

(b) Is owned or leased by a member of an affiliated group and is not available for hire to nonmembers.

Helicopter uses are:

(a) Transporting individuals, equipment, or supplies in the exploration for, or the development or removal of, hard minerals, or the exploration for oil or gas; or

(b) The planting, cultivation, cutting, or transportation of, or caring for, trees (including logging operations),

but only if the helicopter does not take off from, or land at, a facility eligible for assistance under the Airport and Airway Development Act of 1970, or otherwise use services provided pursuant to the Airport and Airway Improvement Act of 1982 during such use.

D. What Special Fuels To Include.—

1. Definitions of Special Fuels.—

(a) Diesel Fuel.—This fuel is a liquid that is:

(1) Sold for use in a diesel-powered highway vehicle; or

(2) Used for fuel in a diesel-powered highway vehicle unless there was a taxable sale of the fuel under (1) above.

(b) Special Motor Fuels.—These fuels include benzol, benzene, naphtha, liquefied petroleum gas, casing head and natural gasoline and other liquid, except kerosene, gas oil, fuel oil, diesel fuel, or any product taxable under section 4081. Special motor fuels must be:

(1) Sold for use as a fuel in a motor vehicle or motorboat; or

(2) Used for fuel in a motor vehicle or motorboat. (This use is not taxable if there was a taxable sale of the fuel under (1) above.)

(c) Diesel/Alcohol Mixture.—You may claim a credit for tax paid on diesel fuel that is later used in a mixture of diesel fuel and alcohol if it is at least 10% alcohol. The allowable credit is 6¢ a gallon.

(d) Special Fuel/Alcohol Mixture.—You may claim a credit for tax paid on special fuels that are later used in a mixture of a special fuel and alcohol if it is at least 10% alcohol. The allowable credit is 6¢ a gallon.

2. Uses of Special Fuels.—

(a) Farm Use.—You may claim a credit for tax paid on diesel fuel in the amount of 15¢ a gallon that was for farm use. You may claim a credit for tax paid on special motor fuels in the amount of 9¢ a gallon. See instruction C.1. for definition of farm use.

(b) Commercial Fishing Vessel Use.—You may claim a credit for tax paid on diesel fuel in the amount of 15¢ a gallon. You may claim a credit for tax paid on special motor fuels in the amount of 9¢ a gallon. See instruction C.2. for definition of commercial fishing vessel use.

(c) Intercity, Local, or School Bus Uses.—You may claim a credit for tax paid on diesel fuel in the amount of 12¢ a gallon.

You may claim a credit of 15¢ a gallon for tax paid on diesel fuel if the fuel is used in a qualified local bus. A "qualified local bus" is any local bus which has a seating capacity of at least 20 adults (not including the driver), and which is under contract (or is receiving more than a nominal subsidy) from any state or local government to furnish such transportation. You may claim a credit for tax paid on special motor fuels in the amount of 9¢ a gallon. See instruction C.4. for definition of intercity, local, and school bus uses.

(d) Qualified Taxicab Use.—You may claim a credit for tax paid on fuels used before 10/1/85 in the amount of 4¢ a gallon. See C.6. for definition of qualified taxicab use.

(e) Farming, Commercial, or Helicopter Aviation Uses.—You may claim a credit for tax paid on fuels in the amount of 14¢ a gallon that were for farming, commercial, or helicopter aviation uses. See instruction C.1. for definition of farming uses and instruction C.7. for definition of commercial or helicopter aviation uses.

(f) Nontaxable Uses.—Diesel and Special Motor Fuel.—These fuels (defined in D.1.) may qualify for a credit in any of the following cases:

- You resold the fuel.
- You used the fuel for a nontaxable purpose.
- You used the fuel for a purpose that was not the one you bought it for. This other purpose must make a lower tax rate apply.

If any of these three conditions exist, claim as a credit the figure you get by subtracting (a) the tax that applies to the actual use of the fuel, from (b) the tax imposed by section 4041 on the fuel sold to you.

E. Purchase of Qualified Diesel-Powered Highway Vehicles.—If you purchased a diesel-powered highway vehicle after 1/1/85, for a use other than resale and you are the first purchaser, you may claim a credit of $102 for an automobile and $198 for a light truck or van. A qualified diesel-powered highway vehicle is any diesel-powered highway vehicle which has at least 4 wheels, has a gross vehicle weight rating of 10,000 pounds or less, and is registered for highway use in the United States under the laws of any state. Enter the credit on line 1 of Form 4136. If you purchased more than one vehicle that qualifies for the credit, add the allowable credits and enter them on line 1.

F. What Fuels Not To Include.—Do not include gasoline lost or destroyed through spillage, accident, or shrinkage. Such gasoline is not considered to be used and therefore does not qualify for credit or refund.

G. Additional Information.—Publication 225, Farmer's Tax Guide, and **Publication 378,** Fuel Tax Credits, are available from the Internal Revenue Service.

216

The income from farming reported on Schedule F includes amounts received from:

cultivating the soil, or raising or harvesting any agricultural commodities. This includes the income from the operation of a stock, dairy, poultry, fish, bee, fruit, or truck farm, or plantation, ranch nursery, range, orchard, or oyster bed and income you receive in the form of crop shares you receive for the use of your land. (IRS Publication 225, *Farmers Tax Guide,* October 1985, p. 3)

On the cash method, gross income for the tax year includes:

1. The amount of cash and the value of merchandise or other property you receive during the tax year from the sale of raised livestock, poultry, vegetables, fruits, etc.
2. Your profits from the sale of all other livestock or other items. To find your profit from the sale of livestock or other items purchased, deduct the cost or other basis of this property, plus sales expenses, from the sales price. The cost of items you purchased for resale is not usually deducted in the year paid unless the payment and the sale happen in the same year.
3. All amounts you receive from breeding fees, fees from rent of teams, machinery, or land, and other incidental farm income.
4. All subsidy and conservation payments you receive that must be considered as income. (IRS Publication 225, *Farmers Tax Guide,* October 1985, p. 6)

Gross income as defined by the accrual method is similar except that adjustments for beginning and ending inventories must be made. To figure gross income on the accrual method, add the following items:

1. The selling price of all livestock and products you sell during the year;
2. The inventory value of livestock and products;
3. All miscellaneous items of income you earn during the year, such as breeding fees, fees from the rent of teams, machinery, or land, or other incidental farm income;
4. Any subsidy or conservation payments you receive that must be considered as income . . .

Then subtract the total of the following:

1. Inventory value of the livestock and products you had on hand at the beginning of the year; and
2. Cost of any livestock or products you purchased during the year. (IRS Publication 225, *Farmers Tax Guide,* October 1985, p. 6)

Although the passages from the *Farmers Tax Guide* presented above were written to be entirely comprehensive, there are a few items that should be highlighted as "other" income. Taxable farm income includes some timber sales; income from the sale of soil or other natural deposits; sod sales; gasoline, lubricating oil, and special fuels tax credits claimed in the previous tax year; refunds and reimbursements; easement and right-of-the-way income; machine work income; barter income; prizes and gains from commodity futures contracts entered for *business* (hedging) purposes. Note especially that barter income and hedging income are included.

Barter income is the value of any commodities, property, or services that are received in return for services performed for someone else. For example, if your neighbor pays you with a cow for helping him harvest his crops, the value of the cow must be reported as income. Similarly if your neighbor helps you harvest your crops in exchange for helping him, then the value of his services should be recognized as income. But in a practical sense, the trading of time, labor and machinery between neighbors is *not* reported as income if the efforts are reciprocal and roughly equal. Technically, the value of the services you receive are income but they are also expenses, so the two cancel.

Commodity futures contracts always pose problems because the contract must be used for hedging purposes. If the contract is a hedge, it is considered to be business insurance and is consequently reported on Schedule F rather than Schedule D. But the definitions are still tricky. A corn producer cannot consider any action in the corn futures market to be a hedge. If he buys corn on the futures market or holds a sell contract well beyond the date when he sold his cash corn, the contract is considered a speculative one rather than a hedge. Consequently, to be considered a hedge, a futures contract must be traded *within* the range of production.

Income reported on Schedule F, either under the cash method or the accrual method, specifically excludes gifts, inheritances, loans, the value of a dwelling furnished a tenant farmer by the landowner, and nonfarm income, as well as any gains on long-lived assets. Income on the cash method also specifically excludes any increases in the value of livestock or a standing crop.

Expenses

All normal business expenditures, with just a very few exceptions, are reported in Part II of Schedule F. Total expenses are subtracted from total income to calculate net farm profit (actually net taxable farm income). The primary expenses which may be included on Schedule F are listed in Part II (see Figure 19–2[b]). These expenses specifically *exclude* state and federal income and self-employment taxes, loan principal payments, losses of growing crops or raised livestock, fines and penalties, estate and gift taxes, capital expenditures (except through depreciation), and losses from hobby or pleasure farms. All interest on loan repayments is deductible as are such items as account books, depreciation from Form 4562, accounting fees, farm business travel, advertising, farm magazines, stamps and stationery, farm organization dues, and farm educational expenses. A complete list of deductible farm business expenses is included in the *Farmers Tax Guide,* Chapter 5. This guide should be studied carefully for expenses *and* income as well as other considerations before the tax return is attempted.

Schedule F Mechanics

The mechanics of completing Schedule F are relatively simple if taken from a cash receipts and expenditures summary such as the one shown in Figure 14–3.

Part I of Schedule F (Figure 19–2[a]) simply records all of the taxable ordinary farm receipts. But there is one division which must be treated in a special way. Recall that earlier in this text and this chapter, it was stated that livestock purchased for resale (such as feeder calves or pigs to be fed to slaughter weight) is deducted in the year of sale rather than the year of purchase. For example, the cost of a pen of calves purchased December 1, 1985, and sold on May 1, 1986, is deducted on the 1986 tax return rather than the 1985 return. Lines 1–4 separate the sales of livestock purchased for resale from other livestock sales and calculate the profit or gain over the cost of the feeder animals. This data should be maintained in a form like that of Figure 6–6.

Lines 1–11 itemize sales of livestock and crops as well as "other" income. Be careful to *exclude* sales of breeding livestock reported on Form 4797 from the totals entered on Schedule F.

Expense reporting in Part II is quite similar. Record the total dollar figures of deductible expenditures somewhere on Lines 13 through 37. If the categories used in a specific records system do not match those on Schedule F, record them where they seem to fit best. Notice that the bottom line depreciation figure from Form 4562 (Line 9 in Figure 18–3) is transferred to Line 16, Schedule F. This transfer marks the end of use for Form 4562.

Accrual income reporting on Part III (Figure 19–2[b]) is also relatively similar to that of Lines 1 through 11 of Part I and relatively simple if inventory records, such as those shown in Figures 6–2, 6–4, and 6–7 were maintained.

Self-Employment Tax

Since 1955 farmers and ranchers have been included under the umbrella of the Federal Insurance Contributions Act (FICA), more commonly known as Social Security. The Social Security tax on employed workers is based on a percentage of the worker's income. The employer contributes one portion (7.15 percent in 1986) and the employee contributes an additional portion (also 7.15 percent in 1986). For self-employed people, the tax is based on a percentage of net self-employed earnings (12.3 percent in 1986). The net effect of these tax percentages is that 14.3 percent of an employee's income (7.15 percent tax on the employee and 7.15 percent tax on the employer) is taxed for Social Security but only 12.3 percent is taxed for self-employed people. However those percentages have been changed through time, as indicated in Table 19–1.

The percentages noted above are multiplied by the income, but only up to a specified maximum level. Consequently there is a maximum Social Security contribution for employed and self-employed people alike. And like the percentages, that maximum earnings level (and thus the maximum contribution) has also been changed through time. Those figures are also shown in Table 19–1.

There is a great tendency on the part of farmers and ranchers to overlook the impact of the Social Security tax on their overall tax liability, at least relative to the income tax. But given the way in which the Social Security tax is calculated, this oversight can be damaging. A 12.3 percent slice of net farm income is a major bite—and there have been actual farm and ranch situations in which the Social Security tax liability actually exceeds the income tax liability.

Computation of the Social Security Self-Employment Tax on Schedule SE (Figure 19–3) is very simple. The information which will be needed to compute the tax includes:

1. Net farm profit
2. Net profit from any other businesses
3. Total wages earned in off-farm employment

Because the self-employment tax relates only to one person, *not* one firm, only the one individual who heads the business is considered. In other words, a spouse's income is not included in Item 3 above. The IRS takes a rather old-fashioned (or some may say a chauvinistic) view of business ownership. The ownership of a business operated by a husband-wife team is usually credited to the husband. Although this certainly will irritate many farm and ranch wives, a taxpaying couple should recognize that this is the way it is and use this distinction to their advantage.

The reason for adding all three sources of income together is that the maximum earnings limitation mentioned earlier applies to the sum of the three. Total wages earned are considered first because the gross percentage tax rate is higher for employed persons than it is for self-employed persons.

To figure the self-employment tax on Schedule SE (see Figure 19–3), simply transfer net farm profit from Schedule F (Line 39) to Schedule SE, Part I, Line 1. Any net profit from other (nonfarm) businesses should be transferred onto Line 2.

Part II of Schedule SE provides an optional computation method for low-income (or even loss) years. Since the Social Security benefits system is based upon years of contribution, it may be to the advantage of some taxpayers to contribute something. In particular, the optional method may be important to a young farmer or rancher in trying to establish Social Security coverage to protect his family against the possibility of his death or disability or it may be important to farmers and ranchers nearing retirement age who may wish to avoid having any zero income years in the calculation of their Social Security benefits. This optional computation method relates only to *gross* income (Schedule F, Line 12) under $2,400 and sets a figure only for the self-employment tax; it does not affect income tax liability.

The taxable earnings figures from Schedule SE, Parts I and II are written on Lines 7 and 8 and added on Line 9. Line 11 reports wages earned, which reduce the maximum taxable earnings figure on Line 10. The tax rate is multiplied by the taxable earnings figure on Line 13 to finish Schedule SE. This tax is then transferred to Form 1040, where it will be added to the income tax.

Fuels Tax Credit

Although it is somewhat unrelated to the other topics of this chapter, a few words need to be said about the tax credit that can be claimed by many farmers

TABLE 19–1 Social Security Minimum and Maximum Earnings and Contributions, 1980–86

		Contribution Rate (%)		Maximum Contribution ($)		
Year	Maximum Earnings ($)	Self-Employed	Employer and Employee	Self-employed	Employer or Employee	Employer and Employee Total
1980	25,900	8.10	6.13	2,097.90	1,587.67	3,175.34
1981	29,700	9.30	6.65	2,762.10	1,975.05	3,950.10
1982	32,400	9.35	6.70	3,029.40	2,170.80	4,341.60
1983	35,700	9.35	6.70	3,337.95	2,391.90	4,783.80
1984	37,800	11.30	7.00[a]	4,271.40	2,532.60[b]	5,178.60
1985	39,600	11.80	7.05	4,672.80	2,791.80	5,583.60
1986	42,000	12.30	7.15	5,166.00	3,003.00	6,006.00

[a]6.70% for employee, 7.00% for employer.
[b]at 6.70%; at 7.00%: $2,646.00.

and ranchers: the credit for federal tax on gasoline and special fuels. Often overlooked in the rush of the tax return filing process, this credit is easy to claim and can add up to fairly significant amounts. *If* the federal fuels tax was charged on gasoline and diesel fuel delivered to the farm or ranch *and if* the fuel was used for farming or ranching purposes and not on public roads, the farmer or rancher can reclaim the tax paid.

To compute the credit, the only information needed is:

1. Gallons purchased
2. Tax rate paid (on a per-gallon basis)
3. An estimate of the gallons not used on highways

Enter the number of nonhighway gallons on Form 4136 (Figure 19–4) and multiply by the tax rate paid. Since only Line 2 applies to farming and ranching, this figure should be carried down to Line 9.

Since the amount of fuel expenses listed on Schedule F, Part II includes the fuel taxes paid, the amount of tax refunded through Farm 4136 is actually a reduction in expenses and thus an increase in net profit. Consequently that amount is taxable income and must be reported on next year's Schedule F, Part I, Line 11.

Farm and Ranch Return Example

Table 19–2 lists a series of common farm and ranch transactions that should be analyzed and entered on Schedule F and eventually on Schedule SE and Form 4136. The filled-in forms for these transactions are presented in Figures 19–5, 19–6, and 19–7. In completing Schedule SE, it is assumed that our farmer holds a part-time off-farm job, but does not operate another business. It is further assumed that the farmer reports his tax liability on the cash method.

TABLE 19–2 Example Farm and Ranch Transactions

Transaction	Transaction Date	Description	Amount
1	1985	Hired labor	$18,255
2	12/1/84	Calves purchased	50,000
3	11/1/85	Hedging account closed (gain realized)	5,450
4	6/1/85	1984 Cotton crop sold	76,440
5	8/15/85	Vegetable crop sold	21,750
6	12/1/85	Hay sold	145,140
7	1985	Electricity and telephone (business)	2,502
8	1985	Fuel purchased[a]	40,788
9	2/15/85	Patronage dividends from 1984—Coop participation	4,000
10	12/16/85	Feeder cattle sold	72,000
11	1/15/85	Feeder cattle purchased	5,000
12	11/1/85	Raised calves sold	40,074
13	4/1/85	Fertilizer and chemical account paid	33,876
14	1985	Bills paid: Interest	57,266
		Insurance	2,589
		Property taxes	1,956
		Supplies	13,909
		BLM lease fees	5,755
		Machinery repairs	1,718
15	1985	Gross off-farm wages earned	5,000

[a]Fuel Purchases:

Purpose	Gallons	Dollars	Tax Paid
Auto—Business	664	797	$.09 per gallon
Truck—Highway	501	601	.09
Machinery—Off-Highway	4,844	5,768	.09
Machinery—Diesel	11,256	12,382	.15
Irrigation—Diesel	19,309	21,240	0

Analysis of Schedule F Transactions

Each one of Transactions 1–14 listed in Table 19–2 should first be entered on Schedule F, since each one is a normal operating receipt or expenditure. Because of their simple nature, at least relative to the asset transactions, each of the fourteen items can be directly entered onto Schedule F, as shown in Figure 19–5.

Transaction 1, Hired Labor. The gross wage bill paid for hired labor should be entered on Line 23a of Schedule F. This hired labor bill should include net wages paid in cash, deposits of employee income tax and social security taxes (more will be discussed on this process in Chapter 21), and deposits of the employer's portion of social security tax on employees.

Transaction 2, Calves Purchased. Although these calves were purchased in 1984, they were sold in 1985. Consequently, the $50,000 purchase price must be deducted from sales in 1985, on Line 2.

SCHEDULE F (Form 1040) Department of the Treasury Internal Revenue Service (X)	**Farm Income and Expenses** ► Attach to Form 1040, Form 1041, or Form 1065. ► See Instructions for Schedule F (Form 1040).	OMB No. 1545-0074 **1985** 14
Name of proprietor(s): FRANK FARMER		Social security number (SSN): 123 45 6789
A Agricultural Activity Code. (Write in the code that best describes your principal income-producing activity. The codes are listed on page 2 of this schedule.) ► 0120	**B Principal Product.** (Describe in one or two words your principal crop or output for the current tax year.) ► Hay	
If you disposed of commodities received under the payments-in-kind (PIK) program, check the box(es) that apply: ☐ Feed for livestock ☐ Sold and reported in income		Employer ID number (NOT SSN)

Part I Farm Income—Cash Method—Complete Parts I and II

(Accrual method taxpayers complete Parts II and III, and line 12 of Part I.)

Do not include sales of livestock held for draft, breeding, sport, or dairy purposes; report these sales on Form 4797.

1	Sales of livestock and other items you bought for resale			1	72,000
2	Cost or other basis of livestock and other items you bought for resale			2	55,000
3	Subtract line 2 from line 1			3	17,000
4	Sales of livestock, produce, grains, and other products you raised			4	283,404
5	a Total distributions received from cooperatives (from Form 1099-PATR)	5a	4,000		
	b **Less:** Nonincome items	5b			
6	Net distributions. Subtract line 5b from line 5a			6	4,000
7	Agricultural program payments:				
	a Cash			7a	
	b Materials and services			7b	
8	Commodity credit loans under election (or forfeited)			8	
9	Crop insurance proceeds			9	
10	Machine work			10	
11	Other income, including Federal and state gasoline tax credit or refund (see instructions)			11	5,450
12	**Gross income.** Add amounts on lines 3, 4, 6, and 7a through 11. If accrual method taxpayer, enter the amount from Part III, line 52 ►			12	309,854

Part II Farm Deductions—Cash and Accrual Method

Do not include personal or living expenses (such as taxes, insurance, repairs, etc. on your home), which do not produce farm income. Reduce the amount of your farm deductions by any reimbursements before entering the deduction below.

13	Breeding fees		26	Mortgage interest paid to financial institutions (see instructions)	57,266
14	Chemicals		27	Other interest	
15	Conservation expenses		28	Pension and profit-sharing plans	
16	Depreciation, and section 179 expense deduction (from Form 4562)	60,167	29	Rent of farm, pasture	5,755
17	Employee benefit programs other than on line 28		30	Repairs, maintenance	1,718
			31	Seeds, plants purchased	
18	Feed purchased		32	Storage, warehousing	
19	Fertilizers and lime	33,876	33	Supplies purchased	13,909
20	Freight, trucking		34	Taxes	1,956
21	Gasoline, fuel, oil	40,788	35	Utilities	2,502
22	Insurance	2,589	36	Veterinary fees, medicine	
23	a Labor hired 18,255		37	Other expenses (specify):	
	b Jobs credit			a	
	c Balance (subtract line 23b from line 23a)	18,255		b	
				c	
24	Land clearing (see instructions)			d	
25	Machine hire			e	

38	**Total deductions from Part II.** Add amounts in columns for lines 13 through 37e ►	38	238,781
39	**Net farm profit or (loss)** (subtract line 38 from line 12). If a profit, enter on Form 1040, line 19, and on Schedule SE, Part I, line 1. If a loss, you **MUST** go on to line 40. (Fiduciaries and partnerships, see the instructions.)	39	71,073

40 If you have a loss, you **MUST** answer this question:
"Do you have amounts for which you are not at risk in this farm (see instructions)?" ☐ Yes ☐ No
If "Yes," you **MUST** attach **Form 6198**. If "No," enter the loss on Form 1040, line 19, and on Schedule SE, Part I, line 1.

For Paperwork Reduction Act Notice, see Form 1040 Instructions. **Schedule F (Form 1040) 1985**

(81)

FIGURE 19–5. Example of filled-in Schedule F.

SCHEDULE SE (Form 1040)
Department of the Treasury Internal Revenue Service (X)

Computation of Social Security Self-Employment Tax

▶ See Instructions for Schedule SE (Form 1040).
▶ Attach to Form 1040.

OMB No. 1545-0074
1985
18

Name of **self-employed** person (as shown on social security card): FRANK FARMER
Social security number of **self-employed** person ▶ 123 45 6789

Part I **Regular Computation of Net Earnings From Self-Employment**

Note: *If you performed services for certain churches or church-controlled organizations and you are not a minister or a member of a religious order, see the instructions.*

1 Net farm profit or (loss) from Schedule F (Form 1040), line 39, and farm partnerships, Schedule K–1 (Form 1065), line 13a	1	71,073
2 Net profit or (loss) from Schedule C (Form 1040), line 33, Schedule K–1 (Form 1065), line 13a (other than farming), and Form W–2 wages of $100 or more from an electing church or church-controlled organization. (See instructions for other income to report.)	2	

Note: ☐ *Check here if you are* ***exempt*** *from self-employment tax on your earnings as a minister, member of a religious order, or Christian Science practitioner because you filed* ***Form 4361****.*
See instructions for kinds of income to report. If you have other earnings of $400 or more that are subject to self-employment tax, include those earnings on line 2.

Part II **Optional Computation of Net Earnings From Self-Employment (See "Who Can Use Schedule SE")**

Generally, this part may be used **only** if you meet any of the following tests:
A Your gross farm income (Schedule F (Form 1040), line 12) was not more than $2,400; or
B Your gross farm income (Schedule F (Form 1040), line 12) was more than $2,400 and your net farm profits (Schedule F (Form 1040), line 39) were less than $1,600; or
C Your net nonfarm profits (Schedule C (Form 1040), line 33) were less than $1,600 and also less than two-thirds (⅔) of your gross nonfarm income (Schedule C (Form 1040), line 5).
See instructions for other limitations.

3 Maximum income for optional methods	3	$1,600	00
4 Farm Optional Method—If you meet test A or B above, enter: the smaller of two-thirds (⅔) of gross farm income from Schedule F (Form 1040), line 12, and farm partnerships, Schedule K–1 (Form 1065), line 13b; or $1,600	4		
5 Subtract line 4 from line 3	5		
6 Nonfarm Optional Method—If you meet test C above, enter: the smallest of two-thirds (⅔) of gross nonfarm income from Schedule C (Form 1040), line 5, and Schedule K–1 (Form 1065), line 13c (other than farming); or $1,600; or, if you elected the farm optional method, the amount on line 5	6		

Part III **Computation of Social Security Self-Employment Tax**

7 Enter the amount from Part I, line 1, or, if you elected the farm optional method, Part II, line 4	7	71,073	
8 Enter the amount from Part I, line 2, or, if you elected the nonfarm optional method, Part II, line 6	8		
9 Add lines 7 and 8. If less than $400, do not fill in the rest of the schedule because you are not subject to self-employment tax. (**Exception:** If this line is less than $400 and you are an employee of an electing church or church-controlled organization, complete the schedule unless this line is a loss. See instructions.)	9	71,073	
10 The largest amount of combined wages and self-employment earnings subject to social security or railroad retirement tax (Tier 1) for 1985 is	10	$39,600	00
11 a Total social security wages and tips from Forms W–2 and railroad retirement compensation (Tier 1). **Note:** *U.S. Government employees whose wages are only subject to the 1.35% hospital insurance benefits tax (Medicare) and employees of certain church or church-controlled organizations should not include those wages on this line (see instructions)* — 11a: 5,000			
b Unreported tips subject to social security tax from Form 4137, line 9, or to railroad retirement tax (Tier 1) — 11b:			
c Add lines 11a and 11b	11c	5,000	
12 a Subtract line 11c from line 10	12a	34,600	
b Enter your "qualified" U.S. Government wages if you are required to use the worksheet in Part III of the instructions. 12b			
c Enter your Form W-2 wages from an electing church or church-controlled organization. 12c			
13 Enter the smaller of line 9 or line 12a	13	34,600	
If line 13 is $39,600, fill in $4,672.80 on line 14. Otherwise, multiply line 13 by .118 and enter the result on line 14		.118	
14 Self-employment tax. Enter this amount on Form 1040, line 51	14	4,083	

For Paperwork Reduction Act Notice, see Form 1040 Instructions.
Schedule SE (Form 1040) 1985
(89)

FIGURE 19–6. Example of filled-in Schedule SE.

Form **4136**

Department of the Treasury
Internal Revenue Service (X)

Computation of Credit for Federal Tax on Gasoline and Special Fuels

(And Credit for Purchase of Qualified Diesel-Powered Highway Vehicles)
▶ Attach this form to your income tax return.

OMB No. 1545-0162

1985
26

Name (as shown on your income tax return): FRANK FARMER
Social security or employer identification number: 123-45-6789

Type of Use	A. Number of Gallons Used	B. Rate of Tax	C. Multiply Column A by Column B
	Gasoline and Gasohol, Diesel Fuel, and Other Special Motor Fuels		
1 Qualified diesel-powered highway vehicles (See instruction E.)			$
2 Nonhighway:			
a Farm (for farming purposes) (Ground applicators of agricultural substances, see instruction C.1.):			
(i) Gasoline	4,844	.09	
(ii) Diesel fuel	11,256	.15	$ 2,124
(iii) Special fuels		.09	
b Commercial fishing vessel (See instructions C.2. and D.2.(b).):			
(i) Gasoline		.09	
(ii) Diesel fuel		.15	$
(iii) Special fuels		.09	
c Off-highway business (specify)			
(i) Gasoline		.09	
(ii) Diesel fuel		.15	$
(iii) Special fuels		.09	
3 a Intercity, local, and school buses (See instruction C.4.)		.09/.12	
b Qualified local buses (See instruction D.2.(c).)		.15	$
4 Gasoline/Alcohol Mixture		.0566	$
5 a Diesel/Alcohol Mixture		.06	
b Special Fuels/Alcohol Mixture		.06	$
6 Qualified taxicab		.04	$
Aviation Fuel			
7 Farm (for farming purposes) (Aerial applicators of agricultural substances, see instruction C.1.)		.09/.12/.14 *	$
8 Aviation (Applies only to commercial or certain helicopter uses as defined in instruction C.7.)		.09/.12/.14 *	$
9 Total income tax credit claimed (add lines 1 through 8). Enter here and on Form 1040, line 62; Form 1120, line 32f; Form 1120–A, line 28f; Form 1120S, line 26b; Form 1041, line 33b; or the proper line of other returns.			$ 2,124

* *Use rate paid or allowable rate if less*

Instructions

(Section references are to the Internal Revenue Code)

Paperwork Reduction Act Notice.—We ask for this information to carry out the Internal Revenue laws of the United States. We need it to ensure that taxpayers are complying with these laws and to allow us to figure and collect the right amount of tax. You are required to give us this information.

Purpose of Form.—File this form if you claim credit for Federal excise tax on gasoline and gasohol, diesel fuel, and other special fuels used for certain types of business uses or the credit for diesel-powered highway vehicles.

Figure the credit by multiplying the number of gallons used by the rate of Federal excise tax paid or the allowable rate if lesser.

An individual, estate, trust, or corporation may file this form, but a partnership may not. However, when a partnership files **Form 1065,** it must include a statement showing how many gallons of the fuels are allocated to each partner, the types of use, and the applicable tax rate as shown on this form.

A. Refund Instead of Credit.—If you paid excise tax of at least $1,000 ($200 for gasoline/alcohol mixture and $50 for gasoline, diesel fuel, or special fuels used in a qualified taxicab) on any fuel category used in any one of the first three quarters of the tax year, you may file for a refund of the tax instead of claiming a credit for it on Form 4136. To claim the refund, file **Form 843**, Claim, before the end of the quarter following the one for which you are claiming the refund. Show separate computations for a claim on gasoline and special fuels. You may use Form 4136 as a worksheet to show your computations and attach it to the Form 843 you file.

You must claim a credit, rather than a refund, for any tax paid during the fourth quarter, and for any tax under $1,000 ($200 for gasoline/alcohol mixture and $50 for gasoline, diesel fuel, or special fuels used in a qualified taxicab) for any fuel used in the first three quarters. You must also claim the credit rather than a refund, for tax paid on gasoline or special fuels used on a farm for farming purposes.

Do not claim a credit on Form 4136 for any amount that you previously claimed as a refund.

B. Year To Include the Credit Or Refund In Income.—When you claim a credit or refund, you must include it in your gross income to the extent that it reduced your income tax. The year you include it depends on your method of accounting. This rule does not apply to the credit claimed on line 1.

1. Cash Method.—If you use the cash method, include the amount of the credit in gross income in the tax year that you file the tax return claiming the credit. If you file an amended return, or if you file a claim for credit or refund (and claim the credit for the tax), include the refund that results from the credit in your gross income for the tax year you receive it.

2. Accrual Method.—If you use the accrual method, figure the amount of the credit or refund due at the close of your tax year. Include it in income, even if you have not filed a claim for refund.

C. What Gasoline To Include.—You may claim credit for Federal excise tax on gasoline or gasohol for farm uses, commercial fishing vessel uses, off-highway business uses, intercity, local, or school bus uses, qualified taxicab use, and commercial or certain helicopter aviation uses.

1. Farm Use.—You may claim a credit for gasoline used for farming in a trade or business farm in the United States in the amount of 9¢ a gallon. (See instruction C. 7. for aviation fuel used for farming.) Such farms include the following: stock, dairy, poultry, fur-bearing animals, truck farms, plantations, ranches, nurseries, ranges, greenhouses, and orchards.

Gasoline is considered used for farming only if the owner, tenant, or operator of the farm uses the gasoline. If gasoline is used for the farming

289

FIGURE 19–7. Example of filled-in Form 4136.

Transaction 3, Hedging Account. The gain realized by hedging agricultural production through commodity futures contracts is considered a business income item and is entered on Line 11.[4]

Transactions 4, 5, and 6: **Crop Sales.** Crop sales (minus expense of the sale) are reported in Part I, Line 4.

Transactions 7 and 8, Fuel and Utilities. All normal operating expenditures paid in cash during the tax year are reported in Part II. Only the portion of any expenditure that can be directly attributed to the business, that is, a cost incurred to generate income, can be listed on Schedule F. Frequently electricity and gas meters, telephone bills and fuel bills have both business and personal components. These mixed bills must be allocated before entering. Utilities are reported on Line 35 and fuel purchases are reported on Line 21. The footnote on fuel purchases and federal tax paid is not used on Schedule F.

Transactions 10 and 11, Feeder Sales and Purchases. As indicated by Transaction 1, feeder animal sales and the purchase of feeders are entered on Lines 1 and 2. But such purchases are entered only in the year of sale, which happens to also be the year of purchase for Transaction 11.

Transaction 12, Raised Animal Sales. The calves sold in Transaction 12 were born and raised on this operation, consequently there is no purchase cost or basis to report. These sales are then entered on Line 4.

Transaction 13, Fertilizer and Chemical Account. Because our farmer reports his tax liability on the cash method, he reports his bills when paid, not necessarily when incurred. Transaction 14 does not specify when the fertilizer and chemicals were contracted for or applied. But since the bill was paid in 1985, the amount paid is deducted in 1985.

Transaction 14, Other Expenditures. As indicated above, expenses paid in cash are reported on the appropriate lines in Part II. Note that *only* the amount paid in cash can be entered and that income and social security taxes are not deductible.

Transaction 15, Off-farm Wages. Schedule F considers only farm receipts and expenditures. Consequently off-farm wages are not listed on Schedule F; however, the wages will appear on Schedule SE.

Completion of Schedule F

After each receipt and expenditure transaction is reported on Schedule F, the form preparer should turn to completing the summaries to obtain net farm profit on Line 39. Specific attention should be paid to Line 3 to derive net profit on

[4] A hedging loss could be reported either as negative income on Line 11 or as an expense on Line 37.

livestock purchased for resale, Line 4 to derive total returns on raised and miscellaneous items, and Line 12 for gross profits. Specific attention should also be given to Line 23 to derive net hired labor costs, Line 16 to add in depreciation (see Figure 18–6), and Line 38 to calculate total deductions. Line 39 can then be computed by subtracting Line 38 from Line 12. Schedule F is then complete for a cash-basis taxpayer. However, this bottom line total must be transferred to Schedule SE and it must also be transferred to Form 1040.

Analysis of Schedule SE

For our simple example, in which we assume a part-time off-farm job and no other business, Schedule SE (Figure 19–6) is very easy to complete. Net farm profit is entered on Lines 1, 7, and 9. Any nonfarm business income would be entered on Line 2 and off-farm wages are entered on Line 11a. Line 13 reports taxable income (for social security purposes), which is always the smaller of net business profits or the maximum social security income base minus off-farm wages. The tax is calculated on Line 14 by multiplying taxable income by the current tax rate. The tax is then later transferred to Form 1040.

Analysis of Form 4136

If fuel purchase records have been maintained and categorized as shown in the footnote to Table 19–2, Form 4136 can be completed very quickly (see Figure 19–7). The *gallons* of fuel used should be entered on Line 1, categorized by rate of tax paid. Dollars spent on fuel are not considered by Form 4136 and any fuels used on highways or on which the tax was not paid are similarly ignored. Multiply the gallons of fuel eligible for the tax credit by the rate paid and enter the amount in Column C. That number is then carried to Line 9, and later to Form 1040.

Alternative Farm/Business Forms

Earlier in this chapter it was mentioned that not all farm or ranch income is reported on Schedule F and that farmers and ranchers do not report all of their self-employment income on Schedule F. Three other forms are common to farm and ranch returns although not nearly as common as the Schedule F: Schedule C, Schedule E, and Form 4835.

Schedule C

If an individual who normally considers himself a farmer or rancher operates another business as well, he must also complete a Schedule C (Figure 19–8). This schedule is filed even if that business is farm- or ranch-related, such as a seed corn dealership or a horseshoeing business. The accounting procedures for nonfarm or

SCHEDULE C (Form 1040)

Department of the Treasury Internal Revenue Service (L)

Profit or (Loss) From Business or Profession
(Sole Proprietorship)

Partnerships, Joint Ventures, etc., Must File Form 1065.

▶ Attach to Form 1040 or Form 1041. ▶ See Instructions for Schedule C (Form 1040).

OMB No. 1545-0074

1985
09

Name of proprietor | Social security number

A Principal business or profession, including product or service (see Instructions) | B Principal business code from page 2

C Business name and address ▶ | D Employer ID number

E Method(s) used to value closing inventory:
(1) ☐ Cost (2) ☐ Lower of cost or market (3) ☐ Other (attach explanation)

F Accounting method: (1) ☐ Cash (2) ☐ Accrual (3) ☐ Other (specify) ▶

	Yes	No
G Was there any change in determining quantities, costs, or valuations between opening and closing inventory? If "Yes," attach explanation.		
H Did you deduct expenses for an office in your home?		

Part I Income

1 a Gross receipts or sales	1a	
b Less: Returns and allowances	1b	
c Subtract line 1b from line 1a and enter the balance here	1c	
2 Cost of goods sold and/or operations (from Part III, line 8)	2	
3 Subtract line 2 from line 1c and enter the **gross profit** here	3	
4 a Windfall Profit Tax Credit or Refund received in 1985 (see Instructions)	4a	
b Other income	4b	
5 Add lines 3, 4a, and 4b. This is the **gross income** ▶	5	

Part II Deductions

6 Advertising		22 Pension and profit-sharing plans	
7 Bad debts from sales or services (Cash method taxpayers, see Instructions)		23 Rent on business property	
		24 Repairs	
8 Bank service charges		25 Supplies (not included in Part III below)	
9 Car and truck expenses		26 Taxes (Do not include Windfall Profit Tax here. See line 30.)	
10 Commissions			
11 Depletion		27 Travel and entertainment	
12 Depreciation and section 179 deduction from Form 4562 (not included in Part III below)		28 Utilities and telephone	
		29 a Wages	
		b Jobs credit	
13 Dues and publications		c Subtract line 29b from 29a	
14 Employee benefit programs		30 Windfall Profit Tax withheld in 1985	
15 Freight (not included in Part III below)		31 Other expenses (specify):	
16 Insurance		a	
17 Laundry and cleaning		b	
18 Legal and professional services		c	
19 Mortgage interest paid to financial institutions (see Instructions)		d	
		e	
20 Office expense		f	
21 Other interest		g	

32 Add amounts in columns for lines 6 through 31g. These are the **total deductions** ▶	32	
33 **Net profit or (loss).** Subtract line 32 from line 5 and enter the result. If a profit, enter on Form 1040, line 12, and on Schedule SE, Part I, line 2 (or Form 1041, line 5). If a loss, you **MUST** go on to line 34	33	

34 If you have a loss, you **MUST** answer this question: "Do you have amounts for which you are not at risk in this business (see Instructions)?" ☐ Yes ☐ No
If "Yes," you **MUST** attach **Form 6198.** If "No," enter the loss on Form 1040, line 12, and on Schedule SE, Part I, line 2 (or Form 1041, line 5).

Part III Cost of Goods Sold and/or Operations (See Schedule C Instructions for Part III)

1 Inventory at beginning of year (if different from last year's closing inventory, attach explanation)	1	
2 Purchases less cost of items withdrawn for personal use	2	
3 Cost of labor (do not include salary paid to yourself)	3	
4 Materials and supplies	4	
5 Other costs	5	
6 Add lines 1 through 5	6	
7 Less: Inventory at end of year	7	
8 **Cost of goods sold and/or operations.** Subtract line 7 from line 6. Enter here and in Part I, line 2, above.	8	

For Paperwork Reduction Act Notice, see Form 1040 Instructions. Schedule C (Form 1040) 1985

(17)

FIGURE 19–8. IRS Schedule C.

nonranch businesses can be quite similar to those used for farms and ranches throughout this text. A cash or an accrual accounting system may be used and Schedule C may be completed either on a cash or an accrual basis. Even though Schedule C looks quite different from Schedule F, it is completed in much the same way. Receipts are reported in Part I and deductions are reported in Part II.

Every reasonable attempt should be made to split receipts and deductions between the farm or ranch and the Schedule C business. But many of those allocations should be made primarily for business analysis purposes; the tax consequences may not be significant. For example, assume that Farmer Brown uses the same machinery on his farm and for a custom farming business. The depreciation claimed on that tractor will be the same whether it is used and reported as Schedule F or Schedule C, so the allocation between the two is not all that critical *if* both businesses show a positive profit. If the allocation process causes one or the other business to show a loss, the owner must be aware of the hobby business provisions of the IRC.

Schedule E and Form 4835

Many landlords get very confused when trying to determine how they should report their rental incomes from farms and ranches. Even if the individual considers himself a landlord, the IRS may not. Consequently, the choice among Schedule F, Schedule E, and Form 4835 must be made carefully. The choice may have a large impact on social security tax liability although the impact on income tax is usually not that great.[5]

The choice among Schedule F, Schedule E, or Form 4835 hinges on two elements:

1. Landlord participation in management
2. Form of payment received by the landlord

Schedule F is used by farm or ranch operators, both owner-operators and tenants, who are actively involved in the day-to-day operation and management of the business. A landlord who materially participates in the management and/or receives rent based on crop or livestock production or shares in production expenses, so that his interest in the operation extends beyond that of a landowner, is also treated as a farm operator for tax purposes and would use Schedule F to report his share of the income and expenses. The rents received in crop or livestock shares are reportable by the landlord when they are sold for cash or used for feed. Those rents are *also* subject to self-employment tax.

Form 4835 (Figure 19–9) and Schedule E (Figure 19–10) are used by the landlord who received a fixed rental (a fixed cash rent or a fixed quantity of the product) and/or does *not* materially participate in the management of the business. Cash rents are reported directly on Schedule E (Figure 19–10) in the year received, along with other nonfarm royalty or rental income such as mineral rights,

[5]The choice may also have a rather large impact on income tax through the alternative minimum tax (see Chapter 20).

Form **4835**

Department of the Treasury
Internal Revenue Service

Farm Rental Income and Expenses

(Crop and Livestock Shares Received by Nonparticipating Landowner (or Sub-lessor))
(Income not subject to self-employment tax)
▶ Attach to Form 1040.

OMB No. 1545-0187
1985
70

Name(s) as shown on Form 1040

Social security number

If you filed Form 943, enter employer identification number here

Purpose of Form.—Use this form to report farm rental income based on crops or livestock produced by the tenant if you were the landowner (or sub-lessor), and you did not materially participate in the operation or management of the farm.

Under both the cash and the accrual methods of reporting, you must report livestock or crop share rentals received in the year in which you convert them into money or its equivalent.

Landowners (or sub-lessors) must not use this form to report cash rent received for pasture, or farmland, if the amount is based on a flat charge. Report this income directly on Schedule E (Form 1040).

Part I Gross Farm Rental Income—Based on Production

(Include amount converted to cash or the equivalent)

1	Income from production of livestock, produce, grains, and other crops		1	
2a	Total distributions received from cooperatives (from Form 1099-PATR)	2a		
b	Less: Nonincome items	2b		
3	Net distributions. Subtract line 2b from line 2a		3	
4	Agricultural program payments:			
a	Cash		4a	
b	Materials and services		4b	
5	Commodity credit loans under election (or forfeited)		5	
6	Crop insurance proceeds		6	
7	Other income, including Federal and state gasoline tax credit or refund (see instructions)		7	
8	Gross farm rents. Add lines 1, 3, and 4a through 7		8	

Part II Deductions—Farm Rental Property

(Exclude personal and living expenses)

9	Breeding fees		22	Mortgage interest paid to financial institutions (see Schedule F instructions)	
10	Chemicals		23	Other interest	
11	Conservation expenses		24	Pension and profit-sharing plans (see Schedule F instructions)	
12	Depreciation and section 179 expense deduction (from Form 4562)		25	Rent of farm, pasture	
13	Employee benefit programs other than line 24 (see Schedule F instructions)		26	Repairs, maintenance	
14	Feed purchased		27	Seeds, plants purchased	
15	Fertilizer and lime		28	Storage, warehousing	
16	Freight, trucking		29	Supplies purchased	
17	Gasoline, fuel, oil		30	Taxes	
18	Insurance		31	Utilities	
19a	Labor hired (see Schedule F instructions)		32	Veterinary fees, medicine	
b	Jobs credit		33	Other expenses (specify):	
c	Balance (subtract line 19b from line 19a)		a		
20	Land clearing (see Schedule F instructions)		b		
21	Machine hire		c		
			d		
			e		

34	**Total deductions.** Add lines 9 through 33e	34	
35	Net farm rental profit or (loss) (subtract line 34 from line 8). If a profit, enter here and on Schedule E (Form 1040), Part I, line 25. If a loss, you **MUST** go on to line 36 before you enter the loss here and on Schedule E (Form 1040)	35	

36 If you have a loss, you **MUST** answer this question: "Do you have amounts for which you are not at risk in this farm?" (See Schedule F instructions.) ☐ **Yes** ☐ **No**
If you checked "Yes," you **MUST** attach **Form 6198.** If you checked "No," enter the loss on Schedule E, Part I, line 25.

For Paperwork Reduction Act Notice, see back of Form 4835. Form **4835** (1985)

(253)

FIGURE 19–9. IRS Form 4835.

SCHEDULE E (Form 1040)
Department of the Treasury Internal Revenue Service (L)

Supplemental Income Schedule

(From rents and royalties, partnerships, estates, and trusts, etc.)

▶ Attach to Form 1040. ▶ See Instructions for Schedule E (Form 1040).

OMB No. 1545-0074
1985
13

Name(s) as shown on Form 1040 | Your social security number

Part I Rental and Royalty Income or Loss

1 In the space provided below, show the kind and location of each rental property.

2 For each property listed, did you or a member of your family use for personal purposes any of the properties for more than the greater of 14 days or 10% of the total days rented at fair rental value during the tax year?

	Yes	No
Property A ▶		
Property B ▶		
Property C ▶		

Rental and Royalty Income		Properties A	Properties B	Properties C	Totals (Add columns A, B, and C)
3 a Rents received					3
b Royalties received					
Rental and Royalty Expenses					
4 Advertising	4				
5 Auto and travel	5				
6 Cleaning and maintenance	6				
7 Commissions	7				
8 Insurance	8				
9 Legal and other professional fees	9				
10 Mortgage interest paid to financial institutions (see Instructions)	10				10
11 Other interest	11				
12 Repairs	12				
13 Supplies	13				
14 Taxes (Do **not** include Windfall Profit Tax here. See Part III, line 34.)	14				
15 Utilities	15				
16 Wages and salaries	16				
17 Other (list) ▶					
18 Total expenses other than depreciation and depletion. Add lines 4 through 17	18				18
19 Depreciation expense (see Part V Instructions), or depletion	19				19
20 Total. Add lines 18 and 19	20				
21 Income or (loss) from rental or royalty properties. Subtract line 20 from line 3a (rents) or 3b (royalties)	21				

22 Add properties with profits on line 21, and write the total profits here	22	
23 Add properties with losses on line 21, and write the total (losses) here	23	( )
24 Combine amounts on lines 22 and 23, and write the net profit or (loss) here	24	
25 Net farm rental profit or (loss) from Form 4835, line 35	25	
26 Total rental or royalty income or (loss). Combine amounts on lines 24 and 25, and write the total here. If Parts II and III on page 2 do not apply to you, write the amount from line 26 on Form 1040, line 18. Otherwise, include the amount in line 36 on page 2 of Schedule E	26	

For Paperwork Reduction Act Notice, see Form 1040 Instructions.

Schedule E (Form 1040) 1985

(21)

FIGURE 19–10(a) and (b). IRS Schedule E.

Schedule E (Form 1040) 1985 13 Page 2

Name(s) as shown on Form 1040 (Do not enter name and social security number if shown on other side) | Your social security number

Part II Income or Losses from Partnerships, S Corporations, Estates, or Trusts

If you report a loss below, and have amounts invested in that activity for which you are not at risk, you may have to file Form 6198. See Instructions.

Partnerships and S Corporations

(a) Name	(b) Enter appropriate code [1]	(c) Check if foreign partnership	(d) Employer identification number	(e) Net loss (see instructions for at-risk limitations)	(f) Net income

[1] Codes: **P** for partnership; **S** for S corporation

27 Add amounts in columns (e) and (f) and write the total(s) here . . . . 27 ( )

28 Combine amounts in columns (e) and (f), line 27, and write the net income or (loss) here 28

29 Deduction for section 179 property (from Form 1065, Schedule K-1, and Form 1120S, Schedule K-1). (See Instructions for limitations.) . . . . 29 ( )

30 Total partnership and S corporation income or (loss). Combine amounts on lines 28 and 29. Write the total here and include in line 36 below 30

Estates and Trusts

(a) Name	(b)	(c)	(d) Employer identification number	(e) Net loss (see instructions for at-risk limitations)	(f) Net income

31 Add amounts in columns (e) and (f) and write the total(s) here . . . . 31 ( )

32 Total estate and trust income or (loss). Combine amounts in columns (e) and (f), line 31. Write the total here and include in line 36 below 32

Part III Windfall Profit Tax Summary

33 Windfall profit tax credit or refund received in 1985 (see Instructions) 33

34 Windfall profit tax withheld in 1985 (see Instructions) . . . . 34 ( )

35 Combine amounts on lines 33 and 34. Write the total here and include in line 36 below 35

Part IV Summary

36 TOTAL income or (loss). Combine lines 26, 30, 32, and 35. Write total here and on Form 1040, line 18 ▶ 36

37 Farmers and fishermen: Write your share of GROSS FARMING AND FISHING INCOME applicable to Parts I and II (see Instructions) 37

Part V Optional Depreciation Worksheet.—You may use this worksheet to figure depreciation only on property placed in service before 1981. If you placed any property in service after 1980, you must use **Form 4562** for all property.

	(a) Description of property	(b) Date acquired	(c) Cost or other basis	(d) Depreciation allowed or allowable in prior years	(e) Depreciation method	(f) Life or rate	(g) Depreciation for this year
A							
	Total (Property A)						
B							
	Total (Property B)						
C							
	Total (Property C)						

(22)

royalties, and apartment building rentals. Rents received in the form of crops and/or livestock are reported on Form 4835 (Figure 19–9) and carried over to Schedule E in the year when they are sold for cash. Both Schedule E and Form 4835 rental income are *not* subject to self-employment tax and are *not* considered gross farm income for estimated tax purposes (see Chapter 21).

Summary

Schedule F provides an income tax reporting form for operating receipts and expenditures regardless of whether the business owner keeps his record book or wants to file his tax return on a cash or an accrual basis of accounting. One companion form, Schedule SE, begins with the end or bottom-line net farm profit figure of Schedule F to calculate social security tax due. A side form, Form 4136, can be completed very quickly to reclaim any federal fuels tax paid but not actually owed.

Two additional forms must be used, rather than Schedule F, if the taxpayer is a landlord rather than a farm or ranch operator. These two forms are Schedule E and Form 4835. While there is no hard rule, share landlords usually look to Form 4835, whereas cash rent landlords will usually look at Schedule E first.

CHAPTER 20

Form 1040 and Tax Computations

Virtually all of the business-related work has now been completed through the forms and schedules of Chapters 17 through 19, and it is now time to turn to personal income tax concerns and to a tax summary. Form 1040 is the final collector of all information prepared and reported on all of the other Internal Revenue Service forms and schedules. Each one of the 100-odd supplementary income tax forms and schedules maintained by the IRS has a fairly narrow or specific purpose, just as the schedules did for the Coordinated Financial Statements balance sheet, income statement and cash flow budget. In contrast to the schedules, Form 1040 is a general-purpose form that acts much like the first page of one of the three CFS financial statements.

Some new information is entered directly onto Form 1040 rather than onto one of the supplementary forms, but for the most part, Form 1040 simply gathers, categorizes, and summarizes information from the other forms and schedules. In a way, Form 1040 can be thought of as the body of a written report whereas the other forms and schedules are the appendixes, the attachments that provide the documentation and justification for the numbers that appear on Form 1040.

Individual versus Family Unit Taxation

Before proceeding to the individual forms, let's take a brief digression to note a basic change in philosophy about who is involved with the tax forms and the eventual tax liability. In each of the forms prepared and discussed in Chapters 17 through 19, one and only one person was assumed to be the taxpayer. That assumption does not necessarily hold for the forms and computaitons in this chapter.

Note that top information block of Form 1040, shown in Figure 20–1(a), including *Filing Status* and *Exemptions,* and the signature block on the bottom of the back side shown in Figure 20–1(b). Contrast this to the information blocks of

Form **1040** Department of the Treasury—Internal Revenue Service
U.S. Individual Income Tax Return **1985** (L)

For the year January 1-December 31, 1985, or other tax year beginning , 1985, ending , 19 | OMB No. 1545-0074

Use IRS label. Otherwise, please print or type.

Your first name and initial (if joint return, also give spouse's name and initial) | Last name | **Your social security number**

Present home address (number and street, including apartment number, or rural route) | **Spouse's social security number**

City, town or post office, state, and ZIP code | Your occupation | Spouse's occupation

Presidential Election Campaign ▶ Do you want $1 to go to this fund? . . . . . Yes No
If joint return, does your spouse want $1 to go to this fund? . . Yes No
Note: *Checking "Yes" will not change your tax or reduce your refund.*

For Privacy Act and Paperwork Reduction Act Notice, see Instructions.

Filing Status

Check only one box.

1 ☐ Single
2 ☐ Married filing joint return (even if only one had income)
3 ☐ Married filing separate return. Enter spouse's social security no. above and full name here. ________
4 ☐ Head of household (with qualifying person). (See page 5 of Instructions.) If the qualifying person is your unmarried child but not your dependent, write child's name here. ________
5 ☐ Qualifying widow(er) with dependent child (year spouse died ▶ 19). (See page 6 of Instructions.)

Exemptions

Always check the box labeled Yourself. Check other boxes if they apply.

6a ☐ Yourself ☐ 65 or over ☐ Blind
b ☐ Spouse ☐ 65 or over ☐ Blind
Enter number of boxes checked on 6a and b ▶

c First names of your dependent children who lived with you ________
Enter number of children listed on 6c ▶

d First names of your dependent children who did not live with you (see page 6). ________
(If pre-1985 agreement, check here ▶ ☐ .)
Enter number of children listed on 6d ▶

e Other dependents:

(1) Name	(2) Relationship	(3) Number of months lived in your home	(4) Did dependent have income of $1,040 or more?	(5) Did you provide more than one-half of dependent's support?

Enter number of other dependents ▶

f Total number of exemptions claimed (also complete line 36). . . . . . Add numbers entered in boxes above ▶

Income

Please attach Copy B of your Forms W-2, W-2G, and W-2P here.

If you do not have a W-2, see page 4 of Instructions.

Please attach check or money order here.

7 Wages, salaries, tips, etc. (Attach Form(s) W-2.) **7**
8 Interest income *(also attach Schedule B if over $400)* **8**
9a Dividends *(also attach Schedule B if over $400)* ________ , **9b** Exclusion ________
c Subtract line 9b from line 9a and enter the result **9c**
10 Taxable refunds of state and local income taxes, if any, from the worksheet on page 9 of Instructions. **10**
11 Alimony received **11**
12 Business income or (loss) *(attach Schedule C)* **12**
13 Capital gain or (loss) *(attach Schedule D)* **13**
14 40% of capital gain distributions not reported on line 13 (see page 9 of Instructions) **14**
15 Other gains or (losses) *(attach Form 4797)* **15**
16 Fully taxable pensions, IRA distributions, and annuities not reported on line 17 (see page 9). . **16**
17a Other pensions and annuities, including rollovers. Total received **17a**
b Taxable amount, if any, from the worksheet on page 10 of Instructions **17b**
18 Rents, royalties, partnerships, estates, trusts, etc. *(attach Schedule E)* **18**
19 Farm income or (loss) *(attach Schedule F)* **19**
20a Unemployment compensation (insurance). Total received . . **20a**
b Taxable amount, if any, from the worksheet on page 10 of Instructions **20b**
21a Social security benefits (see page 10). Total received **21a**
b Taxable amount, if any, from worksheet on page 11. {Tax-exempt interest ________} **21b**
22 Other income (list type and amount—see page 11 of Instructions) ________ **22**
23 Add lines 7 through 22. This is your **total income** . . . . . ▶ **23**

Adjustments to Income

(See Instructions on page 11.)

24 Moving expense *(attach Form 3903 or 3903F)* **24**
25 Employee business expenses *(attach Form 2106)* **25**
26 IRA deduction, from the worksheet on page 12 . . . . . **26**
27 Keogh retirement plan deduction **27**
28 Penalty on early withdrawal of savings **28**
29 Alimony paid (recipient's last name ________ and social security no. ________) **29**
30 Deduction for a married couple when both work *(attach Schedule W)* **30**
31 Add lines 24 through 30. These are your **total adjustments** . . . . . ▶ **31**

Adjusted Gross Income

32 Subtract line 31 from line 23. This is your **adjusted gross income.** *If this line is less than $11,000 and a child lived with you, see "Earned Income Credit" (line 59) on page 16 of Instructions. If you want IRS to figure your tax, see page 13 of Instructions* . . . . . ▶ **32**

(13)

FIGURE 20–1(a). IRS Form 1040, page 1.

Form 1040 (1985) Page 2

Tax Computation (See Instructions on page 13.)	33	Amount from line 32 (adjusted gross income)			33
	34a	If you itemize, attach Schedule A (Form 1040) and enter the amount from Schedule A, line 26. **Caution:** If you have unearned income and can be claimed as a dependent on your parents' return, check here ▶ ☐ and see page 13 of Instructions. Also see page 13 if you are married filing a separate return and your spouse itemizes deductions, or you are a dual-status alien.			34a
	b	If you do not itemize but you made charitable contributions, enter your cash contributions here. (If you gave $3,000 or more to any one organization, see page 14.)	34b		
	c	Enter your noncash contributions *(you must attach Form 8283 if over $500)*	34c		
	d	Add lines 34b and 34c. Enter the total	34d		
	e	Divide the amount on line 34d by 2. Enter the result here			34e
	35	Subtract line 34a or line 34e, whichever applies, from line 33			35
	36	Multiply $1,040 by the total number of exemptions claimed on line 6f (see page 14)			36
	37	**Taxable income.** Subtract line 36 from line 35. Enter the result (but not less than zero)			37
	38	Enter tax here. Check if from ☐ Tax Table, ☐ Tax Rate Schedule X, Y, or Z, or ☐ Schedule G			38
	39	Additional taxes. (See page 14 of Instructions.) Enter here and check if from ☐ Form 4970, ☐ Form 4972, or ☐ Form 5544			39
	40	Add lines 38 and 39. Enter the total ▶			40
Credits (See Instructions on page 14.)	41	Credit for child and dependent care expenses *(attach Form 2441)*	41		
	42	Credit for the elderly and the permanently and totally disabled *(attach Schedule R)*	42		
	43	Residential energy credit *(attach Form 5695)*	43		
	44	Partial credit for political contributions for which you have receipts	44		
	45	Add lines 41 through 44. These are your total personal credits			45
	46	Subtract line 45 from line 40. Enter the result (but not less than zero)			46
	47	Foreign tax credit *(attach Form 1116)*	47		
	48	General business credit. Check if from ☐ Form 3800, ☐ Form 3468, ☐ Form 5884, ☐ Form 6478	48		
	49	Add lines 47 and 48. These are your total business and other credits			49
	50	Subtract line 49 from line 46. Enter the result (but not less than zero) ▶			50
Other Taxes (Including Advance EIC Payments)	51	Self-employment tax *(attach Schedule SE)*			51
	52	Alternative minimum tax *(attach Form 6251)*			52
	53	Tax from recapture of investment credit *(attach Form 4255)*			53
	54	Social security tax on tip income not reported to employer *(attach Form 4137)*			54
	55	Tax on an IRA *(attach Form 5329)*			55
	56	Add lines 50 through 55. This is your **total tax** ▶			56
Payments Attach Forms W-2, W-2G, and W-2P to front.	57	Federal income tax withheld	57		
	58	1985 estimated tax payments and amount applied from 1984 return	58		
	59	Earned income credit (see page 16)	59		
	60	Amount paid with Form 4868	60		
	61	Excess social security tax and RRTA tax withheld (two or more employers)	61		
	62	Credit for Federal tax on gasoline and special fuels *(attach Form 4136)*	62		
	63	Regulated Investment Company credit *(attach Form 2439)*	63		
	64	Add lines 57 through 63. These are your **total payments** ▶			64
Refund or Amount You Owe	65	If line 64 is larger than line 56, enter amount **OVERPAID** ▶			65
	66	Amount of line 65 to be **REFUNDED TO YOU** ▶			66
	67	Amount of line 65 to be applied to your 1986 estimated tax ▶	67		
	68	If line 56 is larger than line 64, enter **AMOUNT YOU OWE.** Attach check or money order for full amount payable to "Internal Revenue Service." Write your social security number and "1985 Form 1040" on it ▶ Check ▶ ☐ if Form 2210 (2210F) is attached. See page 17. **Penalty:** $			68

Please Sign Here

Under penalties of perjury, I declare that I have examined this return and accompanying schedules and statements, and to the best of my knowledge and belief, they are true, correct, and complete. Declaration of preparer (other than taxpayer) is based on all information of which preparer has any knowledge.

▶ Your signature | Date | ▶ Spouse's signature (if filing jointly, BOTH must sign)

Paid Preparer's Use Only

Preparer's signature ▶ | Date | Check if self-employed ☐ | Preparer's social security no.

Firm's name (or yours, if self-employed) and address ▶ | E.I. No. | ZIP code

(14)

FIGURE 20–1(b). IRS Form 1040, page 2.

the forms considered in the three previous chapters. Form 1040 applies to a family unit (if that unit exists) but the other forms apply to one individual—the business owner. The Filing Status block (Lines 1–5) lists the various options that each person/family must fit into. The most commonly used blocks are single and married filing jointly.

Each taxpayer should determine his or her own filing status, which is governed by status on December 31 of the tax year. For example, if a couple was married during the evening of December 31, 1985, they would file as married for the *entire* 1985 tax year. A married couple may choose to file one return for both spouses (married filing jointly) or they may choose to file two returns separately. A single, separated, divorced, or widowed individual may qualify as a head of household or as a qualifying widow or widower. Generally, these two categories involve a dependent child and other specific requirements or tests must be met.[1]

Introduction to Form 1040

The biggest problem with completing the 1040 is that it cannot be completed from start to finish without a great deal of paper shuffling. There are quite a few detours that must be taken to fill out or finish other forms. Although the temptation is now great to get after the Form 1040, we are going to take one short side step and consider one last schedule before starting the 1040.

To complete the tax return for our relatively standard farm and ranch return we must consider several forms in addition to Form 1040: Schedule A: Itemized Personal Deductions; Schedule B: Interest and Dividend Income; Schedule G: Income Averaging; and Schedule W: Deduction for a married couple when both work. After this process is done we will take a quick look at one very complicated form (Form 6251: Alternative Minimum Tax) and one quite simple form (Form 1040X: Amended Return). Let's look just at Schedules A and B and then turn our attention to Form 1040 and the other schedules.

Schedules A and B

Probably more than any other tax form or concept, Schedule A has been the subject of intense debate concerning the fairness of the American income tax structure and concerning tax loopholes and fraud. Schedule A details most of the legally allowable deductions from taxable income that are personal (that is, nonbusiness). Schedule A has also been the target of many studies by the IRS to determine acceptable guidelines or norms and of many books about how to beat the tax game. On the other hand, Schedule B is a very simple and rather noncontroversial form that lists sources of income from interest-bearing accounts, notes, and bonds, as well as income from dividends declared on corporate stock. Sched-

[1]For further detail, consult Chapter 2 of IRS Publication 17, *Your Federal Income Tax.*

ule B is used only by persons who earned more than $400 during the tax year in interest or in dividends.

Schedule A Deductions[2]

Six broad categories of personal expenditures are *deductible* under current Internal Revenue Code regulations: medical and dental expenses, taxes, interest paid, charitable contributions, casualty and theft losses, and miscellaneous deductions. Each of these six categories is a basic element of Schedule A (Figure 20–2). Single persons and married persons filing jointly will list their entire expenses on the appropriate lines, but married persons filing separately must allocate their expenses between the two spouses.

All medical and dental expenses, including hospital and doctor bills, nurses, prescription drugs, eyeglasses, and health insurance premiums, among others, are deductible. But, they are deductible *only* to the extent to which they exceed 5 percent of *adjusted gross income*. Although we have not defined or computed adjusted gross income (it is basically total wage income, investment income including net capital gains, and business income with a few adjustments), the implication of this 5 percent limitation can be seen generally now. Take an example of a wage earner who earns $20,000 per year and has no adjustments. That wage earner must have paid medical and dental expenses in excess of $1,000 to deduct any part. If this person spent $2,500 on medical and dental care, $1,500 could be deducted. But if $900 was spent, none would be deductible.

All taxes paid to any governmental unit *other than* the federal government are deductible on Lines 6 through 10. State and local income taxes, property taxes, and sales taxes should be listed here. These tax claims must, like all other deductions, be backed up by receipts of payment. For most taxes this is not a problem, but for general sales tax, which is collected in most areas on virtually every purchase of food, clothing, and other consumer goods, it could quickly become a nightmare to collect, maintain, store, and add the sales tax from each purchase. An alternative to this process exists, however, which greatly reduces the record-keeping requirements. A set of general sales tax tables has been computed by IRS for residents of each state, as shown in the examples in Figure 20–3. These tables are based on the statutory rates established in each state. For example, an Alabama married couple filing a joint return with two children (thus a family of four) could claim $240 in general sales tax if they had an adjusted gross income of $20,000 in 1985. Their alternative to this claim is to collect *every* cash register receipt.

Any interest paid on home mortgages, installment loans (such as on a car or washing machine), credit cards, and charge accounts may be deducted on Lines 11 through 14. Recall that the interest expense on business loans was deducted on Schedule F or Schedule C; consequently, that interest cannot be deducted here. Schedule A applies only to personal, nonbusiness expense items.

Cash and property contributions to charitable organizations may be listed on Lines 15 through 18. IRS regulations require that these charities or other organiza-

[2]The Tax Reform Act of 1986 greatly changed some of the deductions on Schedule A including state and local taxes and interest on consumer loans. Most of the changes take effect in 1987.

SCHEDULES A&B (Form 1040) Department of the Treasury Internal Revenue Service (L)	**Schedule A—Itemized Deductions** (Schedule B is on back) ▶ Attach to Form 1040. ▶ See Instructions for Schedules A and B (Form 1040).	OMB No. 1545-0074 **1985** 07
Name(s) as shown on Form 1040		Your social security number

Medical and Dental Expenses **(Do not include expenses reimbursed or paid by others.)** (See Instructions on page 19.)	1	Prescription medicines and drugs; and insulin	1	
	2 a	Doctors, dentists, nurses, hospitals, insurance premiums you paid for medical and dental care, etc.	2a	
	b	Transportation and lodging	2b	
	c	Other (list—include hearing aids, dentures, eyeglasses, etc.) ▶	2c	
	3	Add lines 1 through 2c, and write the total here	3	
	4	Multiply the amount on Form 1040, line 33, by 5% (.05)	4	
	5	Subtract line 4 from line 3. If zero or less, write -0-. **Total** medical and dental ▶		5
Taxes You Paid (See Instructions on page 20.)	6	State and local income taxes	6	
	7	Real estate taxes	7	
	8 a	General sales tax (see sales tax tables in instruction booklet)	8a	
	b	General sales tax on motor vehicles	8b	
	9	Other taxes (list—include personal property taxes) ▶	9	
	10	Add the amounts on lines 6 through 9. Write the total here. **Total** taxes ▶		10
Interest You Paid (See Instructions on page 20.)	11 a	Home mortgage interest you paid to financial institutions	11a	
	b	Home mortgage interest you paid to individuals (show that person's name and address) ▶	11b	
	12	Total credit card and charge account interest you paid	12	
	13	Other interest you paid (list) ▶	13	
	14	Add the amounts on lines 11a through 13. Write the total here. **Total** interest ▶		14
Contributions You Made (See Instructions on page 21.)	15 a	Cash contributions. (If you gave $3,000 or more to any one organization, report those contributions on line 15b.)	15a	
	b	Cash contributions totaling $3,000 or more to any one organization. (Show to whom you gave and how much you gave.) ▶	15b	
	16	Other than cash. (You must attach Form 8283 if over $500.)	16	
	17	Carryover from prior year	17	
	18	Add the amounts on lines 15a through 17. Write the total here. **Total** contributions ▶		18
Casualty and Theft Losses	19	Total casualty or theft loss(es). (You must attach Form 4684 or similar statement.) (See page 21 of Instructions.) ▶		19
Miscellaneous Deductions (See Instructions on page 21.)	20	Union and professional dues	20	
	21	Tax return preparation fee	21	
	22	Other (list type and amount) ▶	22	
	23	Add the amounts on lines 20 through 22. Write the total here. **Total** miscellaneous ▶		23
Summary of Itemized Deductions (See Instructions on page 22.)	24	Add the amounts on lines 5, 10, 14, 18, 19, and 23. Write your answer here		24
	25	If you checked Form 1040 { Filing Status box 2 or 5, write $3,540; Filing Status box 1 or 4, write $2,390; Filing Status box 3 write $1,770 }		25
	26	Subtract line 25 from line 24. Write your answer here and on Form 1040, line 34a. (If line 25 is more than line 24, see the Instructions for line 26 on page 22.) ▶		26

For Paperwork Reduction Act Notice, see Form 1040 Instructions. Schedule A (Form 1040) 1985

(15)

FIGURE 20–2. IRS Schedule A.

1985 Optional State Sales Tax Tables

(If you kept records that show you paid more sales tax than the table for your state indicates, you may claim the higher amount on Schedule A, line 8a.)

Your itemized deduction for general sales tax paid can be estimated from these tables plus any qualifying sales taxes paid on the items listed on page 20.

To use the tables:

Step 1—Figure your total available income. Use the total of the amount on Form 1040, line 33, and nontaxable income such as veterans' benefits, workers' compensation, nontaxable part of unemployment compensation or long-term capital gains (however, do not include gains that are nontaxable because they were reinvested in similar property, such as a principal residence), nontaxable part of social security and railroad retirement benefits, dividend's exclusion, deduction for a married couple when both work, and public assistance payments.

Step 2—Count the number of exemptions for you and your family. Do not count exemptions claimed for being 65 or over or blind as part of your family size.

Step 3 A—If your total available income is not over $40,000, find the income line for your state on the tables and read across to find the amount of sales tax for your family size.

Step 3 B—If your income is over $40,000 but not over $100,000, find the deduction listed on the income line "$38,001-$40,000" for your family size and state. For each $5,000 (or part of $5,000) of income over $40,000, increase the deduction by the amount listed for the line "$40,001-$100,000."

Step 3 C—If your income is over $100,000, your sales tax deduction is limited to the deduction for income of $100,000. To figure your sales tax deduction, use Step 3 B, but don't go over $100,000.

Alabama [1]

Income	Family size 1	2	3	4	5	Over 5
$1-$8,000	91	113	120	130	141	160
$8,001-$10,000	107	129	140	151	164	185
$10,001-$12,000	121	144	159	171	185	207
$12,001-$14,000	135	158	176	190	204	227
$14,001-$16,000	148	170	192	207	223	246
$16,001-$18,000	160	182	208	224	240	265
$18,001-$20,000	172	193	222	240	257	282
$20,001-$22,000	183	204	236	255	273	298
$22,001-$24,000	194	214	250	269	288	314
$24,001-$26,000	204	224	263	283	303	329
$26,001-$28,000	214	234	276	297	317	344
$28,001-$30,000	224	243	289	310	331	358
$30,001-$32,000	234	252	301	323	344	372
$32,001-$34,000	244	261	313	336	357	385
$34,001-$36,000	253	269	324	348	370	398
$36,001-$38,000	262	277	335	360	383	411
$38,001-$40,000	271	285	346	372	395	423
$40,001-$100,000 (See Step 3B)	14	14	17	19	20	21

Arizona [2]

Income	Family size 1&2	3	4	5	Over 5
$1-$8,000	105	120	120	127	133
$8,001-$10,000	124	142	142	150	157
$10,001-$12,000	141	162	163	171	179
$12,001-$14,000	157	181	183	192	200
$14,001-$16,000	173	199	203	211	219
$16,001-$18,000	187	217	222	229	238
$18,001-$20,000	201	233	240	247	256
$20,001-$22,000	215	249	258	264	274
$22,001-$24,000	228	265	275	281	291
$24,001-$26,000	241	280	292	297	307
$26,001-$28,000	253	295	308	313	323
$28,001-$30,000	265	309	324	328	338
$30,001-$32,000	277	323	340	343	353
$32,001-$34,000	289	337	355	358	368
$34,001-$36,000	300	350	370	372	382
$36,001-$38,000	311	363	385	386	396
$38,001-$40,000	322	376	399	400	410
$40,001-$100,000 (See Step 3B)	16	19	20	20	21

Arkansas [1]

Income	Family size 1	2	3	4	5	Over 5
$1-$8,000	102	127	134	143	153	173
$8,001-$10,000	119	145	156	167	178	199
$10,001-$12,000	135	161	176	188	201	223
$12,001-$14,000	150	176	195	208	222	245
$14,001-$16,000	164	190	212	227	242	266
$16,001-$18,000	177	203	229	245	261	285
$18,001-$20,000	189	215	245	262	279	304
$20,001-$22,000	201	226	260	279	296	321
$22,001-$24,000	213	237	275	295	313	338
$24,001-$26,000	224	248	289	310	329	355
$26,001-$28,000	235	258	303	325	344	371
$28,001-$30,000	246	268	316	339	359	386
$30,001-$32,000	256	278	329	353	374	401
$32,001-$34,000	266	287	342	367	388	415
$34,001-$36,000	276	296	354	380	402	429
$36,001-$38,000	286	305	366	393	416	443
$38,001-$40,000	295	313	377	405	429	456
$40,001-$100,000 (See Step 3B)	15	16	19	20	21	23

California [3]

Income	Family size 1&2	3&4	5	Over 5
$1-$8,000	125	147	155	164
$8,001-$10,000	147	173	183	193
$10,001-$12,000	167	198	208	219
$12,001-$14,000	186	220	232	243
$14,001-$16,000	204	242	255	266
$16,001-$18,000	222	263	276	288
$18,001-$20,000	238	282	297	309
$20,001-$22,000	254	301	317	330
$22,001-$24,000	270	320	336	349
$24,001-$26,000	285	338	355	368
$26,001-$28,000	299	355	373	386
$28,001-$30,000	313	372	391	404
$30,001-$32,000	327	389	408	422
$32,001-$34,000	341	405	425	439
$34,001-$36,000	354	421	441	455
$36,001-$38,000	367	436	457	471
$38,001-$40,000	380	451	473	487
$40,001-$100,000 (See Step 3B)	19	23	24	24

Colorado [2]

Income	Family size 1&2	3&4	5	Over 5
$1-$8,000	43	51	54	57
$8,001-$10,000	50	61	64	68
$10,001-$12,000	57	70	74	78
$12,001-$14,000	64	78	83	87
$14,001-$16,000	70	86	91	95
$16,001-$18,000	76	94	99	103
$18,001-$20,000	82	101	107	111
$20,001-$22,000	87	108	115	119
$22,001-$24,000	92	115	122	127
$24,001-$26,000	97	122	129	134
$26,001-$28,000	102	129	136	141
$28,001-$30,000	107	135	143	148
$30,001-$32,000	112	141	150	155
$32,001-$34,000	117	147	156	162
$34,001-$36,000	122	153	162	168
$36,001-$38,000	126	159	168	174
$38,001-$40,000	130	165	174	180
$40,001-$100,000 (See Step 3B)	7	8	9	9

Connecticut [4]

Income	Family size 1&2	3&4	5	Over 5
$1-$8,000	107	118	124	125
$8,001-$10,000	127	141	148	150
$10,001-$12,000	146	163	171	173
$12,001-$14,000	163	184	192	195
$14,001-$16,000	180	204	213	216
$16,001-$18,000	196	224	232	236
$18,001-$20,000	212	243	251	256
$20,001-$22,000	227	261	270	275
$22,001-$24,000	242	279	288	294
$24,001-$26,000	256	296	305	312
$26,001-$28,000	270	313	322	330
$28,001-$30,000	284	330	339	347
$30,001-$32,000	297	346	355	364
$32,001-$34,000	310	362	371	381
$34,001-$36,000	323	378	387	397
$36,001-$38,000	335	393	402	413
$38,001-$40,000	347	408	417	429
$40,001-$100,000 (See Step 3B)	17	20	21	21

Dist. of Columbia

Income	Family size 1	2	3	4	5	Over 5
$1-$8,000	94	112	125	125	132	140
$8,001-$10,000	110	129	145	146	155	164
$10,001-$12,000	125	145	164	166	177	186
$12,001-$14,000	139	159	182	186	197	206
$14,001-$16,000	152	173	198	205	216	225
$16,001-$18,000	165	186	213	223	234	244
$18,001-$20,000	177	198	228	240	252	261
$20,001-$22,000	189	210	242	256	268	278
$22,001-$24,000	200	221	256	272	284	294
$24,001-$26,000	211	232	269	287	300	310
$26,001-$28,000	222	242	282	302	315	325
$28,001-$30,000	232	252	295	317	330	340
$30,001-$32,000	242	262	307	332	345	354
$32,001-$34,000	252	272	319	346	359	368
$34,001-$36,000	262	281	330	360	373	382
$36,001-$38,000	271	290	341	373	387	396
$38,001-$40,000	280	299	352	386	400	409
$40,001-$100,000 (See Step 3B)	14	15	18	19	20	20

Florida [1]

Income	Family size 1&2	3	4	5	Over 5
$1-$8,000	104	123	123	130	139
$8,001-$10,000	124	145	146	155	164
$10,001-$12,000	142	166	168	178	187
$12,001-$14,000	159	186	189	200	209
$14,001-$16,000	176	204	210	221	230
$16,001-$18,000	192	222	229	241	250
$18,001-$20,000	207	239	248	260	269
$20,001-$22,000	222	255	267	278	288
$22,001-$24,000	236	271	285	296	306
$24,001-$26,000	250	287	302	314	324
$26,001-$28,000	264	302	319	331	341
$28,001-$30,000	277	317	336	348	357
$30,001-$32,000	290	331	353	364	373
$32,001-$34,000	303	345	369	380	389
$34,001-$36,000	315	359	385	396	405
$36,001-$38,000	327	373	400	412	420
$38,001-$40,000	339	386	415	427	435
$40,001-$100,000 (See Step 3B)	17	19	21	21	22

Georgia [1]

Income	Family size 1	2	3	4	5	Over 5
$1-$8,000	82	103	110	116	125	141
$8,001-$10,000	95	117	127	135	145	161
$10,001-$12,000	107	130	143	152	163	180
$12,001-$14,000	118	141	157	167	179	198
$14,001-$16,000	129	152	171	182	195	214
$16,001-$18,000	139	162	184	196	210	229
$18,001-$20,000	149	171	196	209	224	244
$20,001-$22,000	158	180	208	222	237	258
$22,001-$24,000	167	189	219	234	250	271
$24,001-$26,000	176	198	230	246	262	283
$26,001-$28,000	184	206	240	258	274	295
$28,001-$30,000	192	213	250	269	286	307
$30,001-$32,000	200	220	260	280	297	319
$32,001-$34,000	208	227	270	290	308	330
$34,001-$36,000	215	234	279	300	319	341
$36,001-$38,000	222	241	288	310	330	352
$38,001-$40,000	229	247	297	320	340	362
$40,001-$100,000 (See Step 3B)	11	12	15	16	17	18

Hawaii

Income	Family size 1&2	3	4	5	Over 5
$1-$8,000	158	180	183	190	204
$8,001-$10,000	181	206	209	219	235
$10,001-$12,000	201	228	234	245	263
$12,001-$14,000	220	249	256	269	288
$14,001-$16,000	238	268	277	292	312
$16,001-$18,000	254	286	297	313	335
$18,001-$20,000	270	303	315	333	356
$20,001-$22,000	285	319	333	352	377
$22,001-$24,000	299	335	350	371	397
$24,001-$26,000	313	350	366	389	416
$26,001-$28,000	326	364	382	406	434
$28,001-$30,000	339	378	397	423	451
$30,001-$32,000	351	391	412	439	468
$32,001-$34,000	363	404	426	454	485
$34,001-$36,000	375	416	440	469	501
$36,001-$38,000	386	428	454	484	517
$38,001-$40,000	397	440	467	499	533
$40,001-$100,000 (See Step 3B)	20	22	23	25	27

Idaho

Income	Family size 1	2	3	4	5	Over 5
$1-$8,000	91	111	119	129	141	159
$8,001-$10,000	107	127	139	151	164	184
$10,001-$12,000	122	142	158	170	185	206
$12,001-$14,000	136	156	175	189	204	227
$14,001-$16,000	149	168	191	206	222	246
$16,001-$18,000	161	180	206	223	239	264
$18,001-$20,000	173	191	220	239	256	281
$20,001-$22,000	185	202	234	254	272	298
$22,001-$24,000	196	212	248	268	287	314
$24,001-$26,000	206	222	261	282	301	329
$26,001-$28,000	217	231	274	296	315	344
$28,001-$30,000	227	240	286	309	329	358
$30,001-$32,000	237	249	298	322	343	372
$32,001-$34,000	247	257	310	335	356	385
$34,001-$36,000	256	266	321	347	369	398
$36,001-$38,000	265	274	332	359	381	411
$38,001-$40,000	274	282	343	371	393	423
$40,001-$100,000 (See Step 3B)	14	14	17	19	20	21

Illinois [5]

Income	Family size 1&2	3&4	5	Over 5
$1-$8,000	136	162	173	184
$8,001-$10,000	159	190	203	215
$10,001-$12,000	180	216	230	243
$12,001-$14,000	200	240	256	270
$14,001-$16,000	219	264	280	295
$16,001-$18,000	237	286	303	318
$18,001-$20,000	254	307	326	341
$20,001-$22,000	270	327	347	363
$22,001-$24,000	286	347	367	384
$24,001-$26,000	302	366	387	404
$26,001-$28,000	317	385	407	424
$28,001-$30,000	331	403	426	443
$30,001-$32,000	345	421	444	461
$32,001-$34,000	359	438	462	479
$34,001-$36,000	373	455	479	497
$36,001-$38,000	386	471	496	514
$38,001-$40,000	399	487	513	531
$40,001-$100,000 (See Step 3B)	20	24	26	27

Indiana

Income	Family size 1&2	3&4	5	Over 5
$1-$8,000	117	139	148	157
$8,001-$10,000	137	163	173	183
$10,001-$12,000	156	186	197	207
$12,001-$14,000	173	207	219	230
$14,001-$16,000	189	227	240	252
$16,001-$18,000	205	246	260	272
$18,001-$20,000	220	264	279	291
$20,001-$22,000	234	282	298	310
$22,001-$24,000	248	299	316	328
$24,001-$26,000	261	316	333	346
$26,001-$28,000	274	332	350	363
$28,001-$30,000	287	347	366	379
$30,001-$32,000	299	362	382	395
$32,001-$34,000	311	377	397	411
$34,001-$36,000	323	392	412	426
$36,001-$38,000	334	406	427	441
$38,001-$40,000	345	420	442	456
$40,001-$100,000 (See Step 3B)	17	21	22	23

Iowa

Income	Family size 1&2	3,4&5	Over 5
$1-$8,000	96	107	115
$8,001-$10,000	113	127	136
$10,001-$12,000	129	146	155
$12,001-$14,000	144	164	173
$14,001-$16,000	158	181	191
$16,001-$18,000	172	197	207
$18,001-$20,000	185	212	223
$20,001-$22,000	198	227	239
$22,001-$24,000	210	242	254
$24,001-$26,000	222	256	268
$26,001-$28,000	233	270	282
$28,001-$30,000	244	284	296
$30,001-$32,000	255	297	309
$32,001-$34,000	266	310	322
$34,001-$36,000	277	323	335
$36,001-$38,000	287	336	348
$38,001-$40,000	297	348	360
$40,001-$100,000 (See Step 3B)	15	17	18

Kansas [1]

Income	Family size 1	2	3	4	5	Over 5
$1-$8,000	72	90	95	102	108	122
$8,001-$10,000	85	104	112	120	127	142
$10,001-$12,000	96	116	127	136	144	160
$12,001-$14,000	107	128	141	151	160	177
$14,001-$16,000	117	138	155	165	176	193
$16,001-$18,000	126	148	168	179	191	208
$18,001-$20,000	135	158	180	192	205	222
$20,001-$22,000	144	167	192	205	219	236
$22,001-$24,000	153	176	203	217	232	249
$24,001-$26,000	161	184	214	229	245	262
$26,001-$28,000	169	192	225	240	257	274
$28,001-$30,000	177	200	235	251	269	286
$30,001-$32,000	185	208	245	262	281	298
$32,001-$34,000	192	215	255	273	293	309
$34,001-$36,000	199	222	265	284	304	320
$36,001-$38,000	206	229	275	294	315	330
$38,001-$40,000	212	236	284	304	325	340
$40,001-$100,000 (See Step 3B)	11	12	14	15	16	17

Kentucky

Income	Family size 1&2	3&4	5	Over 5
$1-$8,000	91	104	109	117
$8,001-$10,000	108	124	130	138
$10,001-$12,000	124	143	150	157
$12,001-$14,000	138	161	168	176
$14,001-$16,000	152	178	186	193
$16,001-$18,000	165	195	203	210
$18,001-$20,000	178	211	219	226
$20,001-$22,000	190	226	235	241
$22,001-$24,000	202	241	250	256
$24,001-$26,000	214	256	265	271
$26,001-$28,000	225	270	280	285
$28,001-$30,000	236	284	294	299
$30,001-$32,000	247	298	308	312
$32,001-$34,000	258	312	322	325
$34,001-$36,000	268	325	336	338
$36,001-$38,000	278	338	349	351
$38,001-$40,000	288	351	362	363
$40,001-$100,000 (See Step 3B)	14	18	18	18

Louisiana [6]

Income	Family size 1&2	3&4	5	Over 5
$1-$8,000	76	86	89	94
$8,001-$10,000	90	103	107	112
$10,001-$12,000	104	119	124	129
$12,001-$14,000	117	134	140	145
$14,001-$16,000	129	149	155	161
$16,001-$18,000	141	163	170	176
$18,001-$20,000	153	177	184	190
$20,001-$22,000	164	190	198	204
$22,001-$24,000	175	203	211	217
$24,001-$26,000	186	215	224	230
$26,001-$28,000	196	227	237	243
$28,001-$30,000	206	239	250	256
$30,001-$32,000	216	251	263	268
$32,001-$34,000	226	263	275	280
$34,001-$36,000	236	275	287	292
$36,001-$38,000	245	286	299	304
$38,001-$40,000	254	297	310	315
$40,001-$100,000 (See Step 3B)	13	15	16	16

Maine [7]

Income	Family size 1&2	3&4	5	Over 5
$1-$8,000	93	105	110	115
$8,001-$10,000	110	125	131	136
$10,001-$12,000	126	144	150	157
$12,001-$14,000	141	162	169	176
$14,001-$16,000	155	179	187	194
$16,001-$18,000	169	196	204	211
$18,001-$20,000	182	212	220	228
$20,001-$22,000	195	227	236	244
$22,001-$24,000	208	242	252	260
$24,001-$26,000	220	257	267	275
$26,001-$28,000	232	272	282	290
$28,001-$30,000	243	286	297	305
$30,001-$32,000	254	300	312	319
$32,001-$34,000	265	314	326	333
$34,001-$36,000	276	328	340	347
$36,001-$38,000	287	341	354	361
$38,001-$40,000	297	354	367	374
$40,001-$100,000 (See Step 3B)	15	18	18	19

Maryland

Income	Family size 1&2	3	4	5	Over 5
$1-$8,000	80	90	90	94	98
$8,001-$10,000	95	108	108	113	118
$10,001-$12,000	110	125	125	131	136
$12,001-$14,000	124	141	143	149	154
$14,001-$16,000	137	157	160	166	171
$16,001-$18,000	150	172	177	182	188
$18,001-$20,000	163	186	193	198	203
$20,001-$22,000	175	200	209	213	219
$22,001-$24,000	187	214	225	228	234
$24,001-$26,000	198	227	240	243	249
$26,001-$28,000	209	240	255	257	263
$28,001-$30,000	220	253	270	271	278
$30,001-$32,000	231	265	285	285	292
$32,001-$34,000	242	277	299	299	305
$34,001-$36,000	253	289	313	313	319
$36,001-$38,000	263	301	327	327	332
$38,001-$40,000	273	313	341	341	345
$40,001-$100,000 (See Step 3B)	14	16	17	17	17

Massachusetts [8]

Income	Family size 1 & 2	Over 2
$1-$8,000	58	63
$8,001-$10,000	69	76
$10,001-$12,000	79	89
$12,001-$14,000	89	101
$14,001-$16,000	98	113
$16,001-$18,000	107	124
$18,001-$20,000	116	135
$20,001-$22,000	124	146
$22,001-$24,000	132	157
$24,001-$26,000	140	168
$26,001-$28,000	148	178
$28,001-$30,000	155	188
$30,001-$32,000	162	198
$32,001-$34,000	169	208
$34,001-$36,000	176	217
$36,001-$38,000	183	227
$38,001-$40,000	190	236
$40,001-$100,000 (See Step 3B)	10	12

[1] Local sales taxes are not included. Add an amount based on the ratio between the local and state sales tax rates, considering the number of months each rate has been in effect.

[2] Local sales taxes are not included. Add the amount paid.

[3] The 1¼ percent local sales tax is included. If a ½ of 1 percent local sales tax for transportation is paid all year (Alameda, Contra Costa, Los Angeles, San Francisco, San Mateo, and Santa Cruz counties), add 8 percent of the table amounts. For Santa Clara county, add 15 percent of the table amounts.

[4] Sales tax paid on clothing purchased after March 31, 1985, may be added to the table amounts.

[5] Includes the sales tax on electricity, natural gas and telephone. The Illinois table is based on a combined 6 percent rate (5 state, 1 local). Residents of: Du Page, Kane, Lake, McHenry and Will counties may add 5 percent; Cook county may add 19 percent; Chicago may add an additional 19 percent of the table amount. For other local sales tax, see footnote 1.

[6] If your local sales tax applies to food for home consumption, check your local newspaper during mid-January for the correct deduction. Otherwise, see footnote 1.

[7] Sales tax paid on purchase of electricity of 750 KWH or more per month may be added to the table amounts.

[8] Sales tax paid on the purchase of any single item of clothing for $175 or more may be added to the table amounts.

[9] Sales tax paid on purchases of natural gas or electricity may be added to the table amounts. For local sales tax, see footnote 1.

[10] Local sales taxes are not included. If paid all year, add 26 percent of the table amount for each 1 percent of local sales tax rate. Otherwise, use a proportionate amount. For N.Y. City, add 107 percent of the table amount.

(Footnotes continued on next page)

Page 44

64

FIGURE 20–3. 1985 optional state sales tax tables.

tions (such as the Boy Scouts, colleges, and other foundations) be listed with the IRS. Consequently, a taxpayer cannot deduct money given to every organization, especially those set up for or by himself. The maximum amount of contributions that can be deducted is 50 percent of adjusted gross income.

Casualty losses result from property damaged, destroyed or lost due to fires, vandalism, theft, natural disasters such as earthquakes or floods, and accidents. These losses must first be detailed on IRS Form 4684 and the amount deductible is limited by three factors: $100 deductible, insurance reimbursements, and the 10 percent rule. The only part of a loss that is deductible is the amount which exceeds the sum of $100 plus insurance reimbursements plus 10 percent of adjusted gross income. Assume that a taxpayer earned $20,000 adjusted gross income in 1985 and suffered a $6,000 loss to his car due to a fire. Let's further assume that his insurance company paid him $3,500 for the loss. He can then deduct $400 on his tax return. The loss of $6,000 is reduced by the $100 deductible, the $3,500 insurance payment received, and $2,000 from the 10% rule ($6,000 − $100 − $3,500 − $2,000 = $400).

Miscellaneous deductions include tax return preparation fees paid and union or professional dues not claimed elsewhere on another form. Also included here are professional journals and books necessary to maintain an employees's professional status but which are not reimbursed by the employer.

Itemized Versus Standard Deduction

The sum of the deductions in the six deductible categories is written on Line 24. The next line then helps the taxpayer decide whether he just wasted his time completing Schedule A. The IRS sets up an amount called a *standard deduction* (commonly called the *zero-bracket amount*), which every taxpayer may claim regardless of whether he filled out Schedule A, that is, regardless of whether he *itemized* deductions. In 1985, the standard deduction was $3,540 for married persons filing jointly and qualifying widows, $2,390 for single persons and heads of household, and $1,770 for married persons filing separately (half of the married-filing-jointly deduction).[3]

If Line 24 is larger than Line 25, the difference is written on Line 26 and is called *excess itemized deductions*. If Line 24 is smaller than Line 25, Schedule A is *not* filed, since the taxpayer would be better off by claiming the standard deduction. This choice is quite legal and each taxpayer should select the option that helps him the most. In other words, claim Line 25 or Line 26, whichever is greater. If Line 25 is smaller than Line 26, do not rewrite any Schedule A figures to the final copy. Form 1040 will direct each taxpayer how to claim the standard deduction.

Schedule B

Relative to Schedule A, the process of completing Schedule B (Figure 20–4) is very easy. If a taxpayer receives more than $400 in *interest* income from bank

[3] In 1985, an indexing system authorized by 1981 ERTA was begun. These bracket amounts and others increase with the rate of inflation each year. Values in 1986 are $3,670, $2,480, and $1,835.

Schedules A&B (Form 1040) 1985 **Schedule B—Interest and Dividend Income** 08 OMB No. 1545-0074 Page 2

Name(s) as shown on Form 1040 (Do not enter name and social security number if shown on other side.) | **Your social security number**

Part I Interest Income

(See Instructions on pages 8 and 22.)

Also complete Part III.

If you received more than $400 in interest income, you must complete Part I and list ALL interest received. If you received interest as a nominee for another, or you received or paid accrued interest on securities transferred between interest payment dates, see page 22.

Interest income		Amount
1 Interest income from seller-financed mortgages. (See Instructions and show name of payer.) ▶	1	
2 Other interest income (list name of payer) ▶	2	
3 Add the amounts on lines 1 and 2. Write the total here and on Form 1040, line 8 . ▶	3	

Part II Dividend Income

(See Instructions on pages 8 and 22.)

Also complete Part III.

If you received more than $400 in gross dividends and other distributions on stock, or you are electing to exclude qualified reinvested dividends from a public utility, complete Part II. If you received dividends as a nominee for another, see page 23.

Dividend income				Amount
4 Dividend income (list name of payer—include on line 4 capital gain distributions, nontaxable distributions, etc.) ▶			4	
5 Add the amounts on line 4. Write the total here			5	
6 Capital gain distributions. Enter here and on line 15, Schedule D.*	6			
7 Nontaxable distributions. (See Schedule D Instructions for adjustment to basis.)	7			
8 Exclusion of qualified reinvested dividends from a public utility. (See page 23 of Instructions.)	8			
9 Add the amounts on lines 6, 7, and 8. Write the total here			9	
10 Subtract line 9 from line 5. Write the result here and on Form 1040, line 9a . . . ▶			10	

**If you received capital gain distributions for the year and you do not need Schedule D to report any other gains or losses, do not file that schedule. Instead, enter 40% of your capital gain distributions on Form 1040, line 14.*

Part III Foreign Accounts and Foreign Trusts

(See Instructions on page 23.)

If you received more than $400 of interest or dividends, OR if you had a foreign account or were a grantor of, or a transferor to, a foreign trust, you must answer both questions in Part III.	**Yes**	**No**
11 At any time during the tax year, did you have an interest in or a signature or other authority over a financial account in a foreign country (such as a bank account, securities account, or other financial account)? (See page 23 of the Instructions for exceptions and filing requirements for Form TD F 90-22.1.)		
If "Yes," write the name of the foreign country ▶		
12 Were you the grantor of, or transferor to, a foreign trust which existed during the current tax year, whether or not you have any beneficial interest in it? If "Yes," you may have to file Forms 3520, 3520-A, or 926.		

For Paperwork Reduction Act Notice, see Form 1040 Instructions. **Schedule B (Form 1040) 1985**

(16)

FIGURE 20–4. IRS Schedule B (found on back on Schedule A).

accounts, accounts or notes receivable, other notes, or other bonds, each source of income and the amount must be listed in Part I. If the total interest earned is $400 or less, the sum is written directly on Form 1040 without a detailed listing.

Similarly, if the taxpayer received more than $400 in *dividends* from a corporation, the sources and amounts must be listed individually in Part II. Dividends from all sources that total $400 or less are written directly on Form 1040 without a detailed listing.

Structure of Form 1040

Now that the most common business forms, investment schedules, and the personal schedules have been analyzed (Figure 20–5), we can finally attack the last step: Form 1040. The body of Form 1040 (after the information and status blocks) is divided into eight parts as shown in Figure 20–1. Because of the importance of each part, we must discuss the parts separately although only three or four really require much work.

Income

Lines 7–23 provide a listing of all of the various types of income available to taxpayers. But as has been noted several times already, much of the information that should be entered is actually a transfer from another form or schedule. To minimize confusion and avoid forgetting a critical element, it is best to approach the completion of Form 1040 on a line-by-line basis. Gather all official statements, such as W-2s and 1099s, and collect all of the forms and schedules already completed. It is generally also very useful to have a copy of IRS Publication 17, *Your Federal Income Tax,* at hand and opened to Part II—Income. Publication 17 contains a short chapter on each of the income lines not covered by another form or schedule.

Start filling in the income block on Line 7 with wages and salaries. On a worksheet it may be best to write a 0 on each line for which there is no entry. But the IRS prefers that a zero entry be left blank rather than filled in with a zero—it lessens the distractions for the IRS service-center keypunchers. It is good practice to leave those zero lines blank on the final return, but on worksheets it can be very handy to write in the zero to indicate that the information called for on that specific line has been considered.

Considering Lines 7–23, we can see that Line 7 (wages, salaries, and tips) represents probably the most important income source not discussed already in this text. Any wages or salaries received as an employee should be written here. The amount that belongs on Line 7 is gross wages, *not* net take-home pay. Each employee should have received a W-2 from his employer soon after the end of the tax year. Married couples filing joint returns should add the wages of both spouses, but *not* any children's wages, and enter the total.

If a Schedule B worksheet was completed even though it was not necessary to file it, interest and dividend totals are transferred to Lines 8 and 9a of Form 1040.

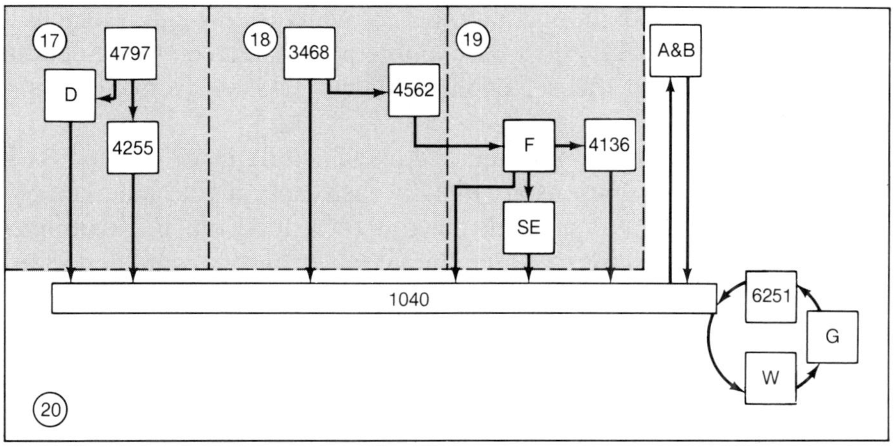

FIGURE 20–5. Farm and ranch tax reporting process.

The first $100 of dividends ($200 for married filing jointly) may be excluded on Line 9b. If a single individual received $300 in dividends, the full $100 exclusion could be claimed and only $200 would be written on Line 9c. However, if $75 was received in dividends the entire amount could be excluded. But Line 9c can *never* fall below zero.

Refunds of state and local income taxes received during the tax year, as a result of a previous year's tax return, must be written on Line 10 of Form 1040 *only if* state and local income taxes were used as an itemized deduction on Schedule A of last year's return. The reason for this line is that on any Schedule A (Line 6), the amount to be entered is the amount withheld or paid during the tax year. The federal tax return is always completed before the state return is begun. Consequently, the state or local income tax actually owed may be less than the amount withheld and thus less than the amount claimed as a deduction. Since Schedule A deductions are usually not the correct tax, any refund must be included in income to compensate. To see how this line actually works, let's consider three examples.

1. Ed Employee claimed $3,000 in state and local income tax on his 1985 Schedule A, since that amount was withheld from his 1985 wages by his state Department of Taxation and Revenue. After completing his state return he found his tax bill to be only $2,500; consequently the state owned him a refund of $500. Rather than recompute his federal return, he is allowed to report the $500 as income on his 1986 return.
2. Tom Taxpayer also had $3,000 in state and local income withheld in 1985 and received a refund a $500, since his 1985 tax was also only $2,500. Tom, however, does not need to report the $500 as income in 1986 since he did *not* itemize deductions on Schedule A in 1985.
3. Ron Rancher qualifies as a farmer and paid $4,000 state income taxes during 1985 as calculated on his 1984 return. He can claim $4,000 on his 1985 Schedule A. But since Ron did not hold a wage-paying job, he received no refund and thus has no entry to report on Form 1040, Line 10.

Any *alimony* (*not* child-support payments) received through a formal divorce decree is written on Line 11. *Business income* is transferred from Schedule C to Line 12 of Form 1040 and the net capital gains or (losses) from Schedule D are written on Line 13.

Capital gains distributions not included on Schedule D are written on Line 14 of Form 1040. These distributions are usually the result of actions taken by others on your behalf rather than as a result of your own decision. For example, some mutual funds and partnerships in which the taxpayer has invested generate capital gains that are passed to their investors. These capital gains do not belong on Schedule D and are reported on Form 1040. This decision between reporting on Schedule D or directly on Form 1040 is a difficult one and should be approached with caution.

Lines 15–21 provide for supplemental gains (business capital gains through IRC Section 1231) from Form 4797, taxable pensions and retirement account withdrawals, rent and royalty income from Schedule E, farm income from Schedule F, any unemployment compensation payments and certain social security benefits.

Adjustments to Income

The Internal Revenue Code allows several specific items to reduce income outside of the framework of itemized deductions of Schedule A. Lines 24 through 31 provide the opportunity to enter these *adjustments*. If specific requirements are met, individuals who move from one location to another to start a new job may reduce their income by claiming moving expenses on Form 3903. The net result from that form is transferred to Line 24 of Form 1040. Some expenses incurred by an employee that are necessary to his job may reduce income. Fill out IRS Form 2106 and transfer the net result to Line 25.

A relatively recent national policy has attempted to encourage individuals, both employers and self-employed persons, to build a personal pension or *retirement account*[4] rather than to rely solely on Social Security distributions for their retirement years. Contributions to qualified plans are excluded from taxable income, but these accounts cannot be withdrawn before retirement age without penalty.

Alimony paid may also reduce income if specific conditions are met.

Schedule W

Since the enactment of the Economic Recovery Tax Act of 1981, married couples may appeal to Schedule W to reduce their taxable incomes. Schedule W applies when both spouses earn income from wage or salary employment or from a business. Its intent is to correct what was once known as the "marriage tax," which penalized two-earner families.

Ten percent of the lower gross income of the two spouses may be excluded by using the simple Schedule W (Figure 20–6). But the maximum exclusion is $3,000. Assume Toni and Travis Taxpayer both work. Toni earned $40,000 in 1985 and

[4] IRAs, Schedule W adjustments, and income averaging were discontinued beginning in 1987 by the Tax Reform Act of 1986.

Schedule W (Form 1040)
Department of the Treasury
Internal Revenue Service (X)

Deduction for a Married Couple When Both Work

▶ For Paperwork Reduction Act Notice, see Form 1040 Instructions.
▶ Attach to Form 1040.

OMB No. 1545-0074
1985
20

Names as shown on Form 1040 | Your social security number

Step 1 Figure your earned income

		(a) You			(b) Your spouse	
1 Wages, salaries, tips, etc., from Form 1040, line 7. (Do not include nondisability pensions or annuities.)	1			1		
2 Net profit or (loss) from self-employment (from Schedules C and F (Form 1040), Schedule K-1 (Form 1065), and any other earned income)	2			2		
3 Add lines 1 and 2. This is your total earned income	3			3		

Step 2 Figure your qualified earned income

4 Adjustments from Form 1040, lines 25, 26, 27, and any repayment of sub-pay included on line 31. (See instructions below.)	4			4		
5 Subtract line 4 from line 3. This is your qualified earned income. If the amount in column (a) or (b) is zero (-0-) or less, stop here. You may not take this deduction	5			5		

Step 3 Figure your deduction

6 Compare the amounts in columns (a) and (b) of line 5 above. Write the smaller amount here. (Write either amount if 5(a) and 5(b) are the same.) ***Do not write more than $30,000***	6		
7 Percentage used to figure the deduction (10%)	7	x .10	
8 Multiply the amount on line 6 by the percentage on line 7. This is the amount of your deduction. Write the answer here and on Form 1040, line 30 ▶	8		

Instructions

Complete this schedule and attach it to your Form 1040 if you take the deduction for a married couple when both work. You may take the deduction if both you and your spouse:

- work and have qualified earned income, and
- file a joint return, and
- do not file **Form 2555** to exclude income or to exclude or deduct certain housing costs, and
- do not file **Form 4563** to exclude income.

There are three steps to follow in figuring the deduction on Schedule W.

Step 1 (lines 1, 2, and 3).—Figure earned income separately for yourself and your spouse.

Step 2 (lines 4 and 5).—Figure qualified earned income separately for yourself and your spouse by subtracting certain adjustments from earned income.

Step 3 (lines 6, 7, and 8).—Figure the deduction based on the **smaller** of:

- the qualified earned income entered in column (a) or (b) of line 5, whichever is less, **OR**
- $30,000.

Earned income.—This is generally income you receive for services you provide. It includes wages, salaries, tips, commissions, certain disability income, sub-pay, etc. (from Form 1040, line 7). It also includes income earned from self-employment (from Schedules C and F of Form 1040 and Schedule K-1 of Form 1065), and net earnings and gains (other than capital gains) from the disposition, transfer, or licensing of property that you created. Earned income does not include interest, dividends, social security or tier 1 railroad retirement benefits, IRA distributions, unemployment compensation, deferred compensation, or nontaxable income. It also does not include any amount your spouse paid you.

Caution: *Do not consider community property laws in figuring your earned income.*

Qualified earned income.—This is the amount on which the deduction is based. Figure it by subtracting certain adjustments from earned income.

These adjustments (and the related lines on Form 1040) are:

- Employee business expenses (from line 25).
- IRA deduction (from line 26).
- Keogh retirement plan deduction (from line 27).
- Repayment of supplemental unemployment benefits (sub-pay) included in the total on line 31. See the instructions on repayment of sub-pay on page 12 of the Form 1040 Instructions.

Enter the total of any adjustments that apply to your or your spouse's earned income in the appropriate column of line 4.

Example.—You earned a salary of $20,000 and had $3,000 of employee business expenses (Form 1040, line 25). Your spouse earned $17,000 and had an IRA deduction of $1,000 (Form 1040, line 26). Your qualified earned income is $17,000 ($20,000 minus $3,000) and your spouse's is $16,000 ($17,000 minus $1,000). Because your spouse's qualified earned income is less than yours, the deduction is figured on your spouse's income. Therefore, the deduction is $1,600 ($16,000 x .10).

Schedule W (Form 1040) 1985

91

FIGURE 20–6. IRS Schedule W.

Travis earned \$32,000. From Schedule W we can see that \$3,000 (10 percent of the maximum \$30,000 on Line 6) can be excluded and would be written on Form 1040, Line 30. If Travis had earned \$28,000, their maximum exclusion would be 10 percent of \$28,000 or \$2,800.

Adjusted Gross Income

Although only a simple subtraction on Form 1040 of Line 23 (Total Income) minus Line 31 (Total Adjustments), *Adjusted Gross Income* on Line 32 is a very important taxation benchmark. Adjusted Gross Income (*AGI*) represents the net income that will be subject to taxation on the back side of Form 1040.

At this point, it is generally best to return to Schedule A and compute the medical and dental expense deduction. Also, Line 32 should be carefully transferred to Line 33 (Figure 20–1(b)).

Tax Computation

Once AGI has been computed, we can now turn to the final adjustments and calculate the tax bill due on the net income received. AGI is first reduced by excess itemized deductions from Line 26 of Schedule A *if* the taxpayer chose to itemize (that is, if Schedule A Line 24 is larger than Line 25). If deductions are not itemized, the taxpayer may still claim 50 percent of his charitable contributions. Thus the maximum contributions claim is 50 percent of AGI.

After Line 35 is obtained, the next step is to calculate personal exemptions, which gives a \$1,040 reduction for each exemption claimed on Line 6f.[5] The result on Line 37 is then *Taxable Income*.

Using taxable income, we can finally calculate tax. There are two sets of tax calculation procedures: Tax Tables and Tax Rate Schedules. The tables (Figure 20–7) must be used if they are applicable, that is, if Line 37 taxable income is \$50,000 or less. If taxable income is greater than \$50,000, the schedules (Figure 20–8) must be used.

Assume that Ed and Ethel Earner calculated their taxable income (married filing jointly) to be \$4,977. By looking into the Tax Table, they find \$4,977 bracketed between \$4,950 and \$5,000. Consequently their total tax is \$158.

Assume that Bob and Betty Barber (also married filing jointly) earned \$74,000 taxable income. Since \$74,000 exceeds the \$50,000 limit of the Tax Table, they must turn to Tax Rate Schedule Y to compute their tax. Their \$74,000 income is greater than \$62,450 but less than \$89,090, the fourth line from the bottom of Schedule Y. Thus, their tax is \$20,639:

$$\$15,788 + 42\% \text{ of the amount over } \$62,450$$

or

$$\$15,788 + 0.42\,(\$74,000 - \$62,450) =$$
$$\$15,788 + \$4,851 = \$20,639$$

[5]The indexed personal exemptions amount for 1986 is \$1,080.

Every taxpayer also has the option to consult Schedule G—Income Averaging when computing taxes. We will discuss Schedule G in a later section of this chapter.

Credits and Other Taxes

Once the total regular income tax liability has been calculated and entered on Line 38, we can consider the tax credits and other types of taxes as listed on Form 1040. For all credits except the political contributions credit on Line 44, a supporting form or schedule must be completed and attached. The primary tax credits that apply to farmers and ranchers include the investment tax credit from Form 3468[6] (Line 48) and the partial credit for political contributions (up to $50 for half the political contributions made). Credits are added together on Line 49 but cannot exceed the total tax liability on Line 40.

Other taxes that must be entered on Lines 51–55 include the self-employment tax, which is transferred to Line 51 from Schedule SE, and the recapture of investment credit, which is transferred to Line 53 from Form 4255.

Another tax that often has an impact on farmers and ranchers is the Alternative Minimum Tax. That tax will be discussed later in this chapter.

Payments, Refunds, and Balance Owed

Line 56 shows the total net tax liability for this filing year. If no taxes had been withheld or paid during the year, this would be the tax that should be paid to the IRS. If some taxes have already been paid, however, we need to find through Lines 57–68 the amount to be paid or refunded. Typical tax payments during the year include withholding from wages earned (Line 57), estimated payments (Line 58) and the fuel tax credit (Line 62). Withholding amounts can be found on the W-2 supplied by your employer; the fuel tax was already calculated on Form 4136.

If total payments on Line 64 exceed total tax liability (Line 56) then the individual has paid to much and can claim a refund of the difference on Line 66. If tax liability exceeds payments, then the taxpayer still owes the difference (Line 68).

Filing the Tax Return

Hard as it may be to believe, the tax return package is now complete! All we need to do now is to copy our worksheets to final forms and put all of the forms together. The final return does not need to be typed, but all letters and numbers should be printed in indelible ink and must be clearly legible. Don't make the IRS guess at this point. Double check and triple check all numbers and especially mathematical computations before rewriting onto the final forms.

[6]At this point, Form 3468 must be finished as indicated in Chapter 18.

1985 Tax Table

Your zero bracket amount has been built into the Tax Table.

Based on Taxable Income

For persons with taxable incomes of less than $50,000.

Example: Mr. and Mrs. Brown are filing a joint return. Their taxable income on line 37 of Form 1040 is $25,325. First, they find the $25,300-25,350 income line. Next, they find the column for married filing jointly and read down the column. The amount shown where the income line and filing status column meet is $3,545. This is the tax amount they must write on line 38 of their return.

At least	But less than	Single	Married filing jointly *	Married filing separately	Head of a household
		Your tax is—			
25,200	25,250	4,513	3,523	5,614	4,177
25,250	25,300	4,528	3,534	5,633	4,191
→ 25,300	25,350	4,543	(3,545)	5,652	4,205
25,350	25,400	4,558	3,556	5,671	4,219

If line 37 (taxable income) is—		And you are—			
At least	But less than	Single	Married filing jointly *	Married filing separately	Head of a household
		Your tax is—			
$0	$1,775	$0	$0	$0	$0
1,775	1,800	0	0	2	0
1,800	1,825	0	0	5	0
1,825	1,850	0	0	7	0
1,850	1,875	0	0	10	0
1,875	1,900	0	0	13	0
1,900	1,925	0	0	16	0
1,925	1,950	0	0	18	0
1,950	1,975	0	0	21	0
1,975	2,000	0	0	24	0
2,000					
2,000	2,025	0	0	27	0
2,025	2,050	0	0	29	0
2,050	2,075	0	0	32	0
2,075	2,100	0	0	35	0
2,100	2,125	0	0	38	0
2,125	2,150	0	0	40	0
2,150	2,175	0	0	43	0
2,175	2,200	0	0	46	0
2,200	2,225	0	0	49	0
2,225	2,250	0	0	51	0
2,250	2,275	0	0	54	0
2,275	2,300	0	0	57	0
2,300	2,325	0	0	60	0
2,325	2,350	0	0	62	0
2,350	2,375	0	0	65	0
2,375	2,400	0	0	68	0

If line 37 (taxable income) is—		And you are—			
At least	But less than	Single	Married filing jointly *	Married filing separately	Head of a household
		Your tax is—			
2,400	2,425	2	0	71	2
2,425	2,450	5	0	73	5
2,450	2,475	8	0	76	8
2,475	2,500	11	0	79	11
2,500	2,525	13	0	82	13
2,525	2,550	16	0	84	16
2,550	2,575	19	0	87	19
2,575	2,600	22	0	90	22
2,600	2,625	24	0	93	24
2,625	2,650	27	0	95	27
2,650	2,675	30	0	98	30
2,675	2,700	33	0	101	33
2,700	2,725	35	0	104	35
2,725	2,750	38	0	106	38
2,750	2,775	41	0	109	41
2,775	2,800	44	0	112	44
2,800	2,825	46	0	115	46
2,825	2,850	49	0	117	49
2,850	2,875	52	0	120	52
2,875	2,900	55	0	123	55
2,900	2,925	57	0	126	57
2,925	2,950	60	0	129	60
2,950	2,975	63	0	132	63
2,975	3,000	66	0	135	66
3,000					
3,000	3,050	70	0	140	70
3,050	3,100	75	0	146	75
3,100	3,150	81	0	152	81
3,150	3,200	86	0	158	86
3,200	3,250	92	0	164	92
3,250	3,300	97	0	170	97
3,300	3,350	103	0	176	103
3,350	3,400	108	0	182	108

If line 37 (taxable income) is—		And you are—			
At least	But less than	Single	Married filing jointly *	Married filing separately	Head of a household
		Your tax is—			
3,400	3,450	114	0	188	114
3,450	3,500	119	0	194	119
3,500	3,550	125	0	200	125
3,550	3,600	131	4	206	130
3,600	3,650	137	9	212	136
3,650	3,700	143	15	218	141
3,700	3,750	149	20	224	147
3,750	3,800	155	26	230	152
3,800	3,850	161	31	236	158
3,850	3,900	167	37	242	163
3,900	3,950	173	42	248	169
3,950	4,000	179	48	254	174
4,000					
4,000	4,050	185	53	261	180
4,050	4,100	191	59	268	185
4,100	4,150	197	64	275	191
4,150	4,200	203	70	282	196
4,200	4,250	209	75	289	202
4,250	4,300	215	81	296	207
4,300	4,350	221	86	303	213
4,350	4,400	227	92	310	218
4,400	4,450	233	97	317	224
4,450	4,500	239	103	324	229
4,500	4,550	245	108	331	235
4,550	4,600	251	114	338	240
4,600	4,650	258	119	345	246
4,650	4,700	265	125	352	252
4,700	4,750	272	130	359	258
4,750	4,800	279	136	366	264
4,800	4,850	286	141	373	270
4,850	4,900	293	147	380	276
4,900	4,950	300	152	387	282
4,950	5,000	307	158	394	288

* This column must also be used by a qualifying widow(er).

Continued on next page

FIGURE 20–7(a) and (b). IRS 1985 tax table.

1985 Tax Table—*Continued*

If line 37 (taxable income) is— At least	But less than	And you are— Single	Married filing jointly *	Married filing sepa-rately	Head of a house-hold
		Your tax is—			
5,000					
5,000	**5,050**	314	163	401	294
5,050	**5,100**	321	169	408	300
5,100	**5,150**	328	174	415	306
5,150	**5,200**	335	180	422	312
5,200	**5,250**	342	185	429	318
5,250	**5,300**	349	191	436	324
5,300	**5,350**	356	196	443	330
5,350	**5,400**	363	202	450	336
5,400	**5,450**	370	207	457	342
5,450	**5,500**	377	213	464	348
5,500	**5,550**	384	218	471	354
5,550	**5,600**	391	224	478	360
5,600	**5,650**	398	229	485	366
5,650	**5,700**	405	235	492	372
5,700	**5,750**	412	240	499	378
5,750	**5,800**	419	246	506	384
5,800	**5,850**	426	252	513	390
5,850	**5,900**	433	258	520	396
5,900	**5,950**	440	264	527	402
5,950	**6,000**	447	270	534	408
6,000					
6,000	**6,050**	454	276	541	414
6,050	**6,100**	461	282	548	420
6,100	**6,150**	468	288	555	426
6,150	**6,200**	475	294	562	432
6,200	**6,250**	482	300	570	438
6,250	**6,300**	489	306	578	444
6,300	**6,350**	496	312	586	450
6,350	**6,400**	503	318	594	456
6,400	**6,450**	510	324	602	462
6,450	**6,500**	517	330	610	468
6,500	**6,550**	524	336	618	474
6,550	**6,600**	531	342	626	480
6,600	**6,650**	538	348	634	486
6,650	**6,700**	545	354	642	492
6,700	**6,750**	552	360	650	498
6,750	**6,800**	559	366	658	505
6,800	**6,850**	566	372	666	512
6,850	**6,900**	574	378	674	519
6,900	**6,950**	581	384	682	526
6,950	**7,000**	589	390	690	533
7,000					
7,000	**7,050**	596	396	698	540
7,050	**7,100**	604	402	706	547
7,100	**7,150**	611	408	714	554
7,150	**7,200**	619	414	722	561
7,200	**7,250**	626	420	730	568
7,250	**7,300**	634	426	738	575
7,300	**7,350**	641	432	746	582
7,350	**7,400**	649	438	754	589
7,400	**7,450**	656	444	762	596
7,450	**7,500**	664	450	770	603
7,500	**7,550**	671	456	778	610
7,550	**7,600**	679	462	786	617
7,600	**7,650**	686	468	794	624
7,650	**7,700**	694	474	802	631
7,700	**7,750**	701	480	810	638
7,750	**7,800**	709	486	818	645
7,800	**7,850**	716	492	826	652
7,850	**7,900**	724	498	834	659
7,900	**7,950**	731	505	842	666
7,950	**8,000**	739	512	850	673

If line 37 (taxable income) is— At least	But less than	And you are— Single	Married filing jointly *	Married filing sepa-rately	Head of a house-hold
		Your tax is—			
8,000					
8,000	**8,050**	746	519	858	680
8,050	**8,100**	754	526	866	687
8,100	**8,150**	761	533	874	694
8,150	**8,200**	769	540	882	701
8,200	**8,250**	776	547	890	708
8,250	**8,300**	784	554	898	715
8,300	**8,350**	791	561	906	722
8,350	**8,400**	799	568	915	729
8,400	**8,450**	806	575	924	736
8,450	**8,500**	814	582	933	743
8,500	**8,550**	821	589	942	750
8,550	**8,600**	829	596	951	757
8,600	**8,650**	836	603	960	764
8,650	**8,700**	844	610	969	771
8,700	**8,750**	851	617	978	778
8,750	**8,800**	859	624	987	785
8,800	**8,850**	866	631	996	792
8,850	**8,900**	874	638	1,005	799
8,900	**8,950**	882	645	1,014	806
8,950	**9,000**	890	652	1,023	813
9,000					
9,000	**9,050**	898	659	1,032	820
9,050	**9,100**	906	666	1,041	827
9,100	**9,150**	914	673	1,050	836
9,150	**9,200**	922	680	1,059	844
9,200	**9,250**	930	687	1,068	853
9,250	**9,300**	938	694	1,077	861
9,300	**9,350**	946	701	1,086	870
9,350	**9,400**	954	708	1,095	878
9,400	**9,450**	962	715	1,104	887
9,450	**9,500**	970	722	1,113	895
9,500	**9,550**	978	729	1,122	904
9,550	**9,600**	986	736	1,131	912
9,600	**9,650**	994	743	1,140	921
9,650	**9,700**	1,002	750	1,149	929
9,700	**9,750**	1,010	757	1,158	938
9,750	**9,800**	1,018	764	1,167	946
9,800	**9,850**	1,026	771	1,176	955
9,850	**9,900**	1,034	778	1,185	963
9,900	**9,950**	1,042	785	1,194	972
9,950	**10,000**	1,050	792	1,203	980
10,000					
10,000	**10,050**	1,058	799	1,212	989
10,050	**10,100**	1,066	806	1,221	997
10,100	**10,150**	1,074	813	1,230	1,006
10,150	**10,200**	1,082	820	1,239	1,014
10,200	**10,250**	1,090	827	1,248	1,023
10,250	**10,300**	1,098	834	1,257	1,031
10,300	**10,350**	1,106	841	1,266	1,040
10,350	**10,400**	1,114	848	1,275	1,048
10,400	**10,450**	1,122	855	1,284	1,057
10,450	**10,500**	1,130	862	1,293	1,065
10,500	**10,550**	1,138	869	1,302	1,074
10,550	**10,600**	1,146	876	1,313	1,082
10,600	**10,650**	1,154	883	1,324	1,091
10,650	**10,700**	1,162	890	1,335	1,099
10,700	**10,750**	1,170	897	1,346	1,108
10,750	**10,800**	1,178	904	1,357	1,116
10,800	**10,850**	1,186	911	1,368	1,125
10,850	**10,900**	1,194	918	1,379	1,133
10,900	**10,950**	1,202	925	1,390	1,142
10,950	**11,000**	1,210	932	1,401	1,150

If line 37 (taxable income) is— At least	But less than	And you are— Single	Married filing jointly *	Married filing sepa-rately	Head of a house-hold
		Your tax is—			
11,000					
11,000	**11,050**	1,218	939	1,412	1,159
11,050	**11,100**	1,226	946	1,423	1,167
11,100	**11,150**	1,234	953	1,434	1,176
11,150	**11,200**	1,242	960	1,445	1,184
11,200	**11,250**	1,250	967	1,456	1,193
11,250	**11,300**	1,259	974	1,467	1,201
11,300	**11,350**	1,268	981	1,478	1,210
11,350	**11,400**	1,277	988	1,489	1,218
11,400	**11,450**	1,286	995	1,500	1,227
11,450	**11,500**	1,295	1,002	1,511	1,235
11,500	**11,550**	1,304	1,009	1,522	1,244
11,550	**11,600**	1,313	1,016	1,533	1,252
11,600	**11,650**	1,322	1,023	1,544	1,261
11,650	**11,700**	1,331	1,030	1,555	1,269
11,700	**11,750**	1,340	1,037	1,566	1,278
11,750	**11,800**	1,349	1,044	1,577	1,286
11,800	**11,850**	1,358	1,051	1,588	1,295
11,850	**11,900**	1,367	1,058	1,599	1,303
11,900	**11,950**	1,376	1,065	1,610	1,312
11,950	**12,000**	1,385	1,072	1,621	1,320
12,000					
12,000	**12,050**	1,394	1,079	1,632	1,329
12,050	**12,100**	1,403	1,086	1,643	1,337
12,100	**12,150**	1,412	1,093	1,654	1,346
12,150	**12,200**	1,421	1,100	1,665	1,354
12,200	**12,250**	1,430	1,107	1,676	1,363
12,250	**12,300**	1,439	1,114	1,687	1,371
12,300	**12,350**	1,448	1,121	1,698	1,380
12,350	**12,400**	1,457	1,128	1,709	1,389
12,400	**12,450**	1,466	1,135	1,720	1,398
12,450	**12,500**	1,475	1,143	1,731	1,407
12,500	**12,550**	1,484	1,151	1,742	1,416
12,550	**12,600**	1,493	1,159	1,753	1,425
12,600	**12,650**	1,502	1,167	1,764	1,434
12,650	**12,700**	1,511	1,175	1,775	1,443
12,700	**12,750**	1,520	1,183	1,786	1,452
12,750	**12,800**	1,529	1,191	1,797	1,461
12,800	**12,850**	1,538	1,199	1,809	1,470
12,850	**12,900**	1,547	1,207	1,822	1,479
12,900	**12,950**	1,556	1,215	1,834	1,488
12,950	**13,000**	1,565	1,223	1,847	1,497
13,000					
13,000	**13,050**	1,574	1,231	1,859	1,506
13,050	**13,100**	1,583	1,239	1,872	1,515
13,100	**13,150**	1,592	1,247	1,884	1,524
13,150	**13,200**	1,601	1,255	1,897	1,533
13,200	**13,250**	1,610	1,263	1,909	1,542
13,250	**13,300**	1,619	1,271	1,922	1,551
13,300	**13,350**	1,628	1,279	1,934	1,560
13,350	**13,400**	1,637	1,287	1,947	1,569
13,400	**13,450**	1,646	1,295	1,959	1,578
13,450	**13,500**	1,656	1,303	1,972	1,587
13,500	**13,550**	1,666	1,311	1,984	1,596
13,550	**13,600**	1,676	1,319	1,997	1,605
13,600	**13,650**	1,686	1,327	2,009	1,614
13,650	**13,700**	1,696	1,335	2,022	1,623
13,700	**13,750**	1,706	1,343	2,034	1,632
13,750	**13,800**	1,716	1,351	2,047	1,641
13,800	**13,850**	1,726	1,359	2,059	1,650
13,850	**13,900**	1,736	1,367	2,072	1,659
13,900	**13,950**	1,746	1,375	2,084	1,668
13,950	**14,000**	1,756	1,383	2,097	1,677

* This column must also be used by a qualifying widow(er).

Continued on next page

1985 Tax Rate Schedules

Your zero bracket amount has been built into these Tax Rate Schedules.

Caution: You must use the Tax Table instead of these Tax Rate Schedules if your taxable income is less than $50,000 unless you use **Schedule G,** Income Averaging, to figure your tax. In that case, even if your taxable income is less than $50,000, use the rate schedules on this page to figure your tax.

Schedule X

Single Taxpayers

Use this Schedule if you checked **Filing Status Box 1** on Form 1040—

If the amount on Form 1040, line 37 is: *Over—*	*But not over—*	Enter on Form 1040, line 38	*of the amount over—*
$0	$2,390	—0—	
2,390	3,540	.......11%	**$2,390**
3,540	4,580	**$126.50 + 12%**	**3,540**
4,580	6,760	**251.30 + 14%**	**4,580**
6,760	8,850	**556.50 + 15%**	**6,760**
8,850	11,240	**870.00 + 16%**	**8,850**
11,240	13,430	**1,252.40 + 18%**	**11,240**
13,430	15,610	**1,646.60 + 20%**	**13,430**
15,610	18,940	**2,082.60 + 23%**	**15,610**
18,940	24,460	**2,848.50 + 26%**	**18,940**
24,460	29,970	**4,283.70 + 30%**	**24,460**
29,970	35,490	**5,936.70 + 34%**	**29,970**
35,490	43,190	**7,813.50 + 38%**	**35,490**
43,190	57,550	**10,739.50 + 42%**	**43,190**
57,550	85,130	**16,770.70 + 48%**	**57,550**
85,130	....	**30,009.10 + 50%**	**85,130**

Schedule Z

Unmarried Heads of Household

(including certain married persons who live apart—see page 5 of the instructions)

Use this schedule if you checked **Filing Status Box 4** on Form 1040—

If the amount on Form 1040, line 37 is: *Over—*	*But not over—*	Enter on Form 1040, line 38	*of the amount over—*
$0	$2,390	—0—	
2,390	4,580	.......11%	**$2,390**
4,580	6,760	**$240.90 + 12%**	**4,580**
6,760	9,050	**502.50 + 14%**	**6,760**
9,050	12,280	**823.10 + 17%**	**9,050**
12,280	15,610	**1,372.20 + 18%**	**12,280**
15,610	18,940	**1,971.60 + 20%**	**15,610**
18,940	24,460	**2,637.60 + 24%**	**18,940**
24,460	29,970	**3,962.40 + 28%**	**24,460**
29,970	35,490	**5,505.20 + 32%**	**29,970**
35,490	46,520	**7,271.60 + 35%**	**35,490**
46,520	63,070	**11,132.10 + 42%**	**46,520**
63,070	85,130	**18,083.10 + 45%**	**63,070**
85,130	112,720	**28,010.10 + 48%**	**85,130**
112,720	....	**41,253.30 + 50%**	**112,720**

Schedule Y

Married Taxpayers and Qualifying Widows and Widowers

Married Filing Joint Returns and Qualifying Widows and Widowers

Use this schedule if you checked **Filing Status Box 2 or 5** on Form 1040—

If the amount on Form 1040, line 37 is: *Over—*	*But not over—*	Enter on Form 1040, line 38	*of the amount over—*
$0	$3,540	—0—	
3,540	5,720	.......11%	**$3,540**
5,720	7,910	**$239.80 + 12%**	**5,720**
7,910	12,390	**502.60 + 14%**	**7,910**
12,390	16,650	**1,129.80 + 16%**	**12,390**
16,650	21,020	**1,811.40 + 18%**	**16,650**
21,020	25,600	**2,598.00 + 22%**	**21,020**
25,600	31,120	**3,605.60 + 25%**	**25,600**
31,120	36,630	**4,985.60 + 28%**	**31,120**
36,630	47,670	**6,528.40 + 33%**	**36,630**
47,670	62,450	**10,171.60 + 38%**	**47,670**
62,450	89,090	**15,788.00 + 42%**	**62,450**
89,090	113,860	**26,976.80 + 45%**	**89,090**
113,860	169,020	**38,123.30 + 49%**	**113,860**
169,020	....	**65,151.70 + 50%**	**169,020**

Married Filing Separate Returns

Use this schedule if you checked **Filing Status Box 3** on Form 1040—

If the amount on Form 1040, line 37 is: *Over—*	*But not over—*	Enter on Form 1040, line 38	*of the amount over—*
$0	$1,770	—0—	
1,770	2,860	.......11%	**$1,770**
2,860	3,955	**$119.90 + 12%**	**2,860**
3,955	6,195	**251.30 + 14%**	**3,955**
6,195	8,325	**564.90 + 16%**	**6,195**
8,325	10,510	**905.70 + 18%**	**8,325**
10,510	12,800	**1,299.00 + 22%**	**10,510**
12,800	15,560	**1,802.80 + 25%**	**12,800**
15,560	18,315	**2,492.80 + 28%**	**15,560**
18,315	23,835	**3,264.20 + 33%**	**18,315**
23,835	31,225	**5,085.80 + 38%**	**23,835**
31,225	44,545	**7,894.00 + 42%**	**31,225**
44,545	56,930	**13,488.40 + 45%**	**44,545**
56,930	84,510	**19,061.65 + 49%**	**56,930**
84,510	....	**32,575.85 + 50%**	**84,510**

Page 43

(63)

FIGURE 20–8. IRS 1985 tax rate schedules.

Do not rewrite any zeroes from worksheets unless the form instructions specifically call for a zero. Also, do not use minus signs to designate a negative number or a loss. Use parentheses instead. For example, if a loss of $1,000 is incurred, write the loss at (1,000) rather than − 1,000.

Finally, assemble all relevant forms and schedules as shown in Figure 20–9. Begin the package with Form 1040. Follow Form 1040 with all schedules arranged in alphabetical order and end with all forms in numerical order. Attach any W-2 statements from employers to the front and *staple* (do *not* use paper clips) on the left-hand side about the middle of the page. This system is preferred since it allows easy access to the schedule or form number at the upper left-hand corner and makes an orderly, permanently attached package. A check for payment of taxes due should be stapled on top of this package. Make sure the check can be detached without removing the staples that hold the rest of the package.

Form 1040 Example

Table 20–1 lists a series of example transactions which are entered directly on one of the four IRS forms or schedules considered so far in this chapter. These transactions should be analyzed and entered so as to complete the tax return example for Frank and Fran Farmer. Use the general sales tax tables and the Tax Tables or Tax Rate Schedules presented in Figures 20–3, 20–7, and 20–8 to calculate taxes due, and refer to the completed examples in Chapters 17, 18, and 19 to carry forward work already accomplished.

Analysis of these Form 1040 and Schedules A and B transactions is really quite straightforward, so it will be approached in a different manner than the transactions analyses in earlier chapters. Tables 20–2 through 20–6 list the lines of Schedules A and B, Form 1040, and Schedule W that should be filled in for Frank and Fran Farmer. The tables also show the source or calculation of the number entered. Figures 20–10 through 20–13 present the filled-in forms.

The completion process pattern outlined in this chapter should be followed to minimize paper shuffling. At this stage there is some unavoidable paper shuffling anyway, but a systematic approach can reduce the confusion. A quick outline follows:

- Step 1. Locate all necessary information and complete *Schedule A* (except those items needing AGI).
- Step 2. Locate all necessary information and complete *Schedule B*.
- Step 3. Locate all necessary information and complete *Form 1040,* Lines 1 through 29.
- Step 4. Complete *Schedule W*.
- Step 5. Compute *AGI* (Form 1040, Line 32).
- Step 6. Finish *Schedule A*.
- Step 7. Complete *Form 1040,* Lines 33 through 40.
- Step 8. Compute tax credit limitation, *Form 3468,* Lines 15–20.
- Step 9. Locate all necessary information and finish *Form 1040*.
- Step 10. Sign the return and write a check in the amount of Line 68, Form 1040.

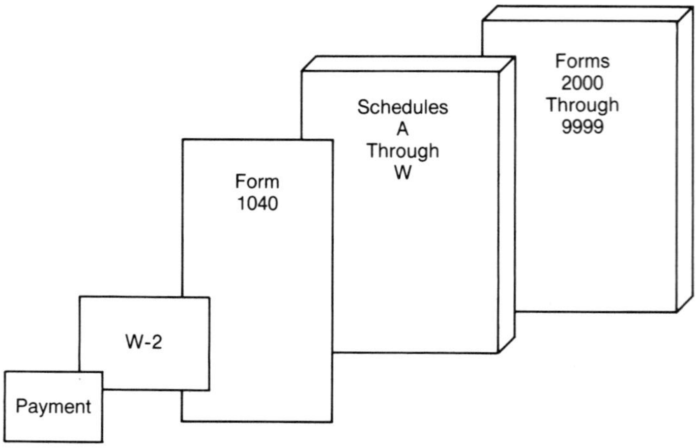

FIGURE 20–9(a). Tax return package, exploded view.

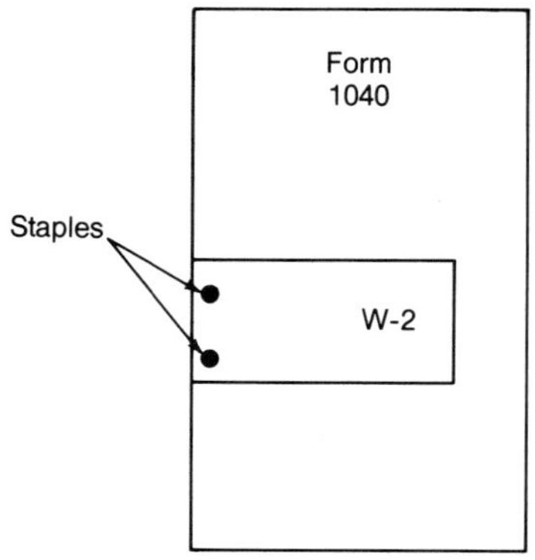

FIGURE 20–9(b). Tax return package, front view.

The net result of Frank and Fran Farmer's tax return is that they owe the IRS $34,845. Their next step is to make a photocopy of the final return for their records, package the forms, write the check, and send the package to IRS. Unfortunately, the process is not quite complete even yet. They must next turn to their State tax return. In most states that process is much easier than the federal return.

Additional Income Tax Forms

Earlier in this chapter, two additional tax forms were mentioned as ones that should be considered by farmers and ranchers. Those two forms, Schedule G and Form 6251 (Figures 20–14 and 20–15) often fall into a farm or ranch return, but for different reasons. Schedule G, *Income Averaging,* provides a mechanism to reduce taxes for individuals with highly volatile taxable incomes. Form 6251, *Alternative Minimum Tax,* is a difficult form that can add tax if the taxpayer has taken many of the credits and other *tax preference* items. Unfortunately, Schedule G is optional but Form 6251 *must* be filed if it applies.

TABLE 20–1 Example Form 1040 and Schedules A and B Transactions

Transaction	Description	Amount
1a	Fran's wages earned as a school principal	$28,000
1b	Federal income tax withheld	2,800
1c	State income tax withheld	600
2a	Frank's wages earned as a part-time welder	5,000
2b	Federal income tax withheld	500
2c	State income tax withheld	500
3	Medicine and drug bills	300
4	Doctor, dentist, hospital bills	1,250
5	Medical insurance bills	1,500
6	State income taxes paid with last year's return	3,500
7	Real estate taxes on personal home	400
8	Interest paid on home mortgage	1,202
9	Personal credit card interest charges	108
10	General cash charitable contributions	1,200
11	Interest earned from First National Bank	575
12	Dividends received on International Trucking Stock	100
13	Dividends received on IBM stock	400
General Background Data		
1	Frank Farmer (123-45-6789) and Fran Farmer (234-56-7890)	
2	Children: Fred, Age 16 and Faye, Age 14	
3	Address: Star Route 2, Box 18 Anytown, AB 88061	
4	Employer ID number: 12-3456789	

In addition to these two forms, we should also take a quick look at Form 1040X, the Amended Tax Return (Figure 20–16). If an error is found in a tax return that has already been filed, Form 1040X should be completed and sent to the IRS to correct the error.

Schedule G

Income averaging provisions of the IRC give an allowance to individuals with highly volatile incomes that may help reduce the tax impact of a big jump in income. Since farmers and ranchers typically see wide income swings due to weather patterns, pest problems, volatile output prices, and other reasons, the income average provisions can sometimes help save quite a few tax dollars. If past tax returns have been filed in an organized farm or ranch filing system, Schedule G (Figure 20–14) is very easy to complete.

Schedule G calls only for one figure from each tax return of the previous three years: taxable income. Line sources are clearly specified for the various tax years. The schedule also calls for several computations, but basically all that is needed is the current year's tax rate schedules (see Figure 20–8).

The basic indication for income averaging is that the current year's taxable income must be at least 40 percent greater than the average taxable income from the previous three years. So, if the last year or two were fairly poor income years

TABLE 20–2 Schedule A Analysis

Line Number	Item	Source or Calculation	Amount
1	Medicine and drugs	Table 20–1, Transaction 3	$ 300
2a	Doctors, etc.	Table 20–1, Transactions 4 and 5	2,750
3	Medical total	300 + 2,750	3,050
4	Medical limit	5% of Form 1040, Line 33: AGI 135,086 × 0.05 (To be calculated later)	6,754
5	Deductible medical	3,050 − 6,754 < 0	0
6	State and local income tax	Table 20–1, Transactions 1c, 2c, and 6: 600 + 500 + 3,500	4,600
7	Real estate tax	Table 20–1, Transaction 7	400
8a	General sales tax	Use Form 1040, Line 33: AGI and Figure 20–4. Income is over $100,000 limit. For family of four, sales tax allowance is 372 + 12(19). 12 units of $5,000 between $40,000 and $100,000.	600
10	Deductible taxes	4,600 + 400 + 600	5,600
11a	Home mortgage interest	Table 20–1, Transaction 8	1,202
12	Credit card charges	Table 20–1, Transaction 9	108
14	Deductible interest	1,202 + 108	1,310
15a	Cash contributions	Table 20–1, Transaction 10	1,200
18	Deductible contributions	1,200	1,200
24	Total itemized deductions	5,600 + 1,310 + 1,200	8,110
25	Standard deduction	Married filing jointly, Filing status 2	3,540
26	Excess itemized deductions	8,110 − 3,540	4,570

and this year a large taxable income was recognized, income averaging just may help. It will seldom, if ever, help in a down year, but since income averaging is extremely hard to predict, a schedule worksheet should be completed each year. If it helps, file it; if it doesn't help, don't include it in the final tax return package.

Form 6251

During the 1970s there was a great deal of concern raised over the tax preferences or advantages allowed to certain taxpayers. The concern primarily centered

TABLE 20–3 Schedule B Analysis

Line Number	Item	Source or Calculation	Amount
2	Interest income	Table 20–1, Transaction 11	$575
3	Total interest	$575	575
4	Dividend income	Table 20–1, Transaction 12	100
4	Dividend income	Table 20–1, Transaction 13	400
5	Total standard dividends	$100 + 400	500
10	Total dividends	$500 (Line 5)	500

TABLE 20–4 Form 1040 (Front Page) Analysis

Line Number	Item	Source or Calculation	Amount
1–16	Information	Table 20–1, General background data	—
7	Wages	Table 20–1, Transactions 1a and 2a, 28,000 + 5,000	$33,000
8	Interest income	Schedule B, Line 3	575
9a	Dividend income	Schedule B, Line 10	500
9b	Dividend exclusion	Form 1040 instructions	200
9c	Taxable dividends	Line 9a − 9b: 500 − 200	300
13	Capital gain	Schedule D, Line 23 (Figure 17–7(b))	5,860
15	Supplemental gains	Form 4797, Line 17(b)(2) (Figure 17–5)	27,078
19	Farm income	Schedule F, Line 39 (Figure 19–5)	71,073
23	Total income	Sum of Lines 7, 8, 9c, 13, 15, & 19 33,000 + 575 + 300 + 5,860 + 27,078 + 71,073	137,886
30	Schedule W exclusion	Schedule W, Line 8 (Figure 20–14)	2,800
31	Total adjustments	2,800	2,800
32	AGI[a]	Line 22 − Line 31: 137,886 − 2,800	135,086

[a]Carry this number to: (a) Form 1040, Line 33 (Figure 20–14)
(b) Schedule A, Line 4 analysis (Table 20–2)

on the nature of these preferences: they all seemed to benefit wealthy individuals, or at least wealthy individuals could take advantage of these items when low-income individuals could not. To respond to this concern, Congress created a minimum tax mechanism and later an alternative minimum tax mechanism. The minimum tax was repealed in 1981 but the alternative minimum tax remains.

The basic underlying concept of the alternative minimum tax is to add adjusted gross income (AGI) to most of the personal deductions claimed on Schedule A and other tax preference items such as the dividend exclusion and capital gains exclusion, among others. This sum is called the alternative minimum taxable income (AMTI). The alternative minimum tax then requires that an individual's tax bill be

TABLE 20–5 Schedule W Analysis

Line Number	Item	Source or Calculation	Amount
1a	Frank's wages	Table 20–1, Transaction 2a	$ 5,000
1b	Fran's wages	Table 20–1, Transaction 1a	28,000
2a	Farm income	Schedule F, Line 39 (Figure 19–5) Allocated entirely to Frank	71,073
3a	Frank's earned income	Line 1 + Line 2: 5,000 + 71,073	76,073
3b	Fran's earned income	Line 1: 28,000	28,000
5a	Frank's qualified income	Line 3a	76,073
5b	Fran's qualified income	Line 3b	28,000
6	Comparison	Lesser of 5(a) or 5(b): 76,073 ≷ 28,000	28,000
8	Exclusion	10% of comparison: 0.10 × 28,000	2,800

TABLE 20–6 Form 1040 (Back Page) Analysis

Line Number	Item	Source or Calculation	Amount
33	AGI	Form 1040, Line 32 (Figure 20–12)	$135,086
34a	Excess itemized deductions	Schedule A, Line 28 (Figure 20–10)	4,570
34	AGI minus excess deductions	Line 33 − Line 34a: 135,086 − 4,570	130,516
36	Exemption	Form 1040, Line 6f times $1,040: 4 × 1,040	4,160
37	Taxable income	Line 35 − Line 36: 130,516 − 4,000	126,356
38	Income tax	Tax Rate Schedule Y (Figure 20–8) 126,356 falls between 113,860 and 169,020. Therefore, tax is 38,123 plus 49% of the amount over 113,860: 38,123 + 0.49 (126,356 − 113,860) = 38,123 + 0.49 (12,496) = 38,123 + 6,123 =	44,246
40&46	Total income tax	Line 38 (Lines 7 through 22 of Form 3468 should be completed at this point)	44,246
48	Investment credit	Form 3468, Line 20 (Figure 18–5)	8,460
49	Total credits	Line 48	8,460
50	Balance	Line 46 − Line 49: 44,246 − 8,460	35,786
51	Self-employment tax	Schedule SE, Line 14 (Figure 19–6)	4,083
53	ITC recapture	Form 4255, Line 16 (Figure 17–6)	400
56	Total tax	Line 50 + Line 51 + Line 53: 35,786 + 4,083 + 400	40,269
57	Federal tax withheld	Table 20–1, Transactions 1b and 1c: 2,800 + 500	3,300
62	Fuel tax credit	Form 4136, Line 9 (Figure 19–7)	2,124
64	Total payments	Line 57 + Line 62: 3,300 + 2,124	5,424
68	Amount owed	Line 56 − Line 64: 40,269 − 5,424	34,845

at least 20 percent of the difference between AMTI and $30,000 ($40,000 for a married couple filing jointly). The net effect is that if a large amount of income was sheltered from taxation through these tax preferences, the alternative minimum tax will be activated to ensure that at least some tax is paid.

Form 6251 (Figure 20–16) details all of the items that must be added to AGI to compute AMTI. To complete this form, the taxpayer will need his Form 1040, Schedule A, Schedule D, and Form 4562 at the very least.

In good ordinary income years, farmers and ranchers will probably not be affected by the alternative minimum tax. But, in years in which capital sales make up a large proportion of taxable income, the alternative minimum tax is *very* likely to be a big factor. To be safe, Form 6251 should be completed each year in worksheet form.

Form 1040X

Mistakes in filling out a tax return are inevitable, especially since the time pressure is great on tax preparers. A well-organized records system and prelimi-

SCHEDULES A&B (Form 1040)
Department of the Treasury
Internal Revenue Service (X)

Schedule A—Itemized Deductions

(Schedule B is on back)

▶ Attach to Form 1040. ▶ See Instructions for Schedules A and B (Form 1040).

OMB No. 1545-0074
1985
07

Name(s) as shown on Form 1040: Frank and Fran Farmer
Your social security number: 123 45 6789

Medical and Dental Expenses (Do not include expenses reimbursed or paid by others.) (See Instructions on page 19.)	1 Prescription medicines and drugs; and insulin	1	300		
	2 a Doctors, dentists, nurses, hospitals, insurance premiums you paid for medical and dental care, etc.	2a	2,750		
	b Transportation and lodging	2b			
	c Other (list—include hearing aids, dentures, eyeglasses, etc.) ▶	2c			
	3 Add lines 1 through 2c, and write the total here	3	3,050		
	4 Multiply the amount on Form 1040, line 33, by 5% (.05)	4	6,754		
	5 Subtract line 4 from line 3. If zero or less, write -0-. **Total** medical and dental ▶			5	0
Taxes You Paid (See Instructions on page 20.)	6 State and local income taxes	6	4,600		
	7 Real estate taxes	7	400		
	8 a General sales tax (see sales tax tables in instruction booklet)	8a	600		
	b General sales tax on motor vehicles	8b			
	9 Other taxes (list—include personal property taxes) ▶	9			
	10 Add the amounts on lines 6 through 9. Write the total here. **Total** taxes ▶			10	5,600
Interest You Paid (See Instructions on page 20.)	11 a Home mortgage interest you paid to financial institutions	11a	1,202		
	b Home mortgage interest you paid to individuals (show that person's name and address) ▶	11b			
	12 Total credit card and charge account interest you paid	12	108		
	13 Other interest you paid (list) ▶	13			
	14 Add the amounts on lines 11a through 13. Write the total here. **Total** interest ▶			14	1,310
Contributions You Made (See Instructions on page 21.)	15 a Cash contributions. (If you gave $3,000 or more to any one organization, report those contributions on line 15b.)	15a	1,200		
	b Cash contributions totaling $3,000 or more to any one organization. (Show to whom you gave and how much you gave.) ▶	15b			
	16 Other than cash. (You must attach Form 8283 if over $500.)	16			
	17 Carryover from prior year	17			
	18 Add the amounts on lines 15a through 17. Write the total here. **Total** contributions ▶			18	1,200
Casualty and Theft Losses	19 Total casualty or theft loss(es). (You must attach Form 4684 or similar statement.) (See page 21 of Instructions.) ▶			19	
Miscellaneous Deductions (See Instructions on page 21.)	20 Union and professional dues	20			
	21 Tax return preparation fee	21			
	22 Other (list type and amount) ▶	22			
	23 Add the amounts on lines 20 through 22. Write the total here. **Total** miscellaneous ▶			23	
Summary of Itemized Deductions (See Instructions on page 22.)	24 Add the amounts on lines 5, 10, 14, 18, 19, and 23. Write your answer here.			24	8,110
	25 If you checked Form 1040 { Filing Status box 2 or 5, write $3,540; Filing Status box 1 or 4, write $2,390; Filing Status box 3, write $1,770 }			25	3,540
	26 Subtract line 25 from line 24. Write your answer here and on Form 1040, line 34a. (If line 25 is more than line 24, see the Instructions for line 26 on page 22.) ▶			26	4,570

For Paperwork Reduction Act Notice, see Form 1040 Instructions. Schedule A (Form 1040) 1985

(73)

FIGURE 20–10. Example of filled-in IRS Schedule A.

Schedules A&B (Form 1040) 1985 **Schedule B—Interest and Dividend Income** 08 OMB No. 1545-0074 Page 2

Name(s) as shown on Form 1040 (Do not enter name and social security number if shown on other side.) — Frank and Fran Farmer

Your social security number: 123 45 6789

Part I Interest Income

(See Instructions on pages 8 and 22.)

Also complete Part III.

If you received more than $400 in interest income, you must complete Part I and list ALL interest received. If you received interest as a nominee for another, or you received or paid accrued interest on securities transferred between interest payment dates, see page 22.

Interest income		Amount
1 Interest income from seller-financed mortgages. (See Instructions and show name of payer.) ▶	1	
2 Other interest income (list name of payer) ▶ First National Bank	2	575
3 Add the amounts on lines 1 and 2. Write the total here and on Form 1040, line 8 ▶	3	575

Part II Dividend Income

(See Instructions on pages 8 and 22.)

Also complete Part III.

If you received more than $400 in gross dividends and other distributions on stock, or you are electing to exclude qualified reinvested dividends from a public utility, complete Part II. If you received dividends as a nominee for another, see page 23.

Dividend income				Amount
4 Dividend income (list name of payer—include on line 4 capital gain distributions, nontaxable distributions, etc.) ▶ International Trucking			4	100
IBM				400
5 Add the amounts on line 4. Write the total here			5	500
6 Capital gain distributions. Enter here and on line 15, Schedule D.*	6			
7 Nontaxable distributions. (See Schedule D Instructions for adjustment to basis.)	7			
8 Exclusion of qualified reinvested dividends from a public utility. (See page 23 of Instructions.)	8			
9 Add the amounts on lines 6, 7, and 8. Write the total here			9	
10 Subtract line 9 from line 5. Write the result here and on Form 1040, line 9a ▶			10	500

**If you received capital gain distributions for the year and you do not need Schedule D to report any other gains or losses, do not file that schedule. Instead, enter 40% of your capital gain distributions on Form 1040, line 14.*

Part III Foreign Accounts and Foreign Trusts

(See Instructions on page 23.)

If you received more than $400 of interest or dividends, OR if you had a foreign account or were a grantor of, or a transferor to, a foreign trust, you must answer both questions in Part III.

	Yes	No
11 At any time during the tax year, did you have an interest in or a signature or other authority over a financial account in a foreign country (such as a bank account, securities account, or other financial account)? (See page 23 of the Instructions for exceptions and filing requirements for Form TD F 90-22.1.)		X
If "Yes," write the name of the foreign country ▶		
12 Were you the grantor of, or transferor to, a foreign trust which existed during the current tax year, whether or not you have any beneficial interest in it? If "Yes," you may have to file Forms 3520, 3520-A, or 926.		X

For Paperwork Reduction Act Notice, see Form 1040 Instructions. **Schedule B (Form 1040) 1985**

(74)

FIGURE 20–11. Example of filled-in IRS Schedule B.

Form 1040 — Department of the Treasury—Internal Revenue Service — U.S. Individual Income Tax Return 1985 (X)

For the year January 1-December 31, 1985, or other tax year beginning , 1985, ending , 19 . OMB No. 1545-0074

Use IRS label. Otherwise, please print or type.		
Your first name and initial (if joint return, also give spouse's name and initial) Last name: Frank and Fran Farmer		Your social security number: 123 45 6789
Present home address (number and street, including apartment number, or rural route): Star Route 2, Box 18		Spouse's social security number: 234 56 7890
City, town or post office, state, and ZIP code: Anytown, AB 88061	Your occupation: Farmer; Spouse's occupation: School Principal	

Presidential Election Campaign ▶ Do you want $1 to go to this fund? . . . Yes / No
If joint return, does your spouse want $1 to go to this fund? . . Yes / No
Note: *Checking "Yes" will not change your tax or reduce your refund.*

For Privacy Act and Paperwork Reduction Act Notice, see Instructions.

Filing Status

Check only one box.

1 ☐ Single
2 ☒ Married filing joint return (even if only one had income)
3 ☐ Married filing separate return. Enter spouse's social security no. above and full name here. ______
4 ☐ Head of household (with qualifying person). (See page 5 of Instructions.) If the qualifying person is your unmarried child but not your dependent, write child's name here. ______
5 ☐ Qualifying widow(er) with dependent child (year spouse died ▶ 19). (See page 6 of Instructions.)

Exemptions

Always check the box labeled Yourself. Check other boxes if they apply.

6a ☒ Yourself ☐ 65 or over ☐ Blind
b ☒ Spouse ☐ 65 or over ☐ Blind — Enter number of boxes checked on 6a and b ▶ 2

c First names of your dependent children who lived with you: Fred, Faye — Enter number of children listed on 6c ▶ 2

d First names of your dependent children who did not live with you (see page 6). (If pre-1985 agreement, check here ▶ ☐ .) — Enter number of children listed on 6d ▶

e Other dependents:

(1) Name	(2) Relationship	(3) Number of months lived in your home	(4) Did dependent have income of $1,040 or more?	(5) Did you provide more than one-half of dependent's support?

Enter number of other dependents ▶

f Total number of exemptions claimed (also complete line 36). . . . Add numbers entered in boxes above ▶ 4

Income

Please attach Copy B of your Forms W-2, W-2G, and W-2P here.

If you do not have a W-2, see page 4 of Instructions.

Please attach check or money order here.

Line	Description		Amount
7	Wages, salaries, tips, etc. (Attach Form(s) W-2.)	7	33,000
8	Interest income *(also attach Schedule B if over $400)*	8	575
9a	Dividends *(also attach Schedule B if over $400)* 500, 9b Exclusion 200		
c	Subtract line 9b from line 9a and enter the result	9c	300
10	Taxable refunds of state and local income taxes, if any, from the worksheet on page 9 of Instructions.	10	
11	Alimony received	11	
12	Business income or (loss) *(attach Schedule C)*	12	
13	Capital gain or (loss) *(attach Schedule D)*	13	5,860
14	40% of capital gain distributions not reported on line 13 (see page 9 of Instructions)	14	
15	Other gains or (losses) *(attach Form 4797)*	15	27,078
16	Fully taxable pensions, IRA distributions, and annuities not reported on line 17 (see page 9).	16	
17a	Other pensions and annuities, including rollovers. Total received 17a		
b	Taxable amount, if any, from the worksheet on page 10 of Instructions	17b	
18	Rents, royalties, partnerships, estates, trusts, etc. *(attach Schedule E)*	18	
19	Farm income or (loss) *(attach Schedule F)*	19	71,073
20a	Unemployment compensation (insurance). Total received 20a		
b	Taxable amount, if any, from the worksheet on page 10 of Instructions	20b	
21a	Social security benefits (see page 10). Total received 21a		
b	Taxable amount, if any, from worksheet on page 11. { Tax-exempt interest }	21b	
22	Other income (list type and amount—see page 11 of Instructions)	22	
23	Add lines 7 through 22. This is your **total income** ▶	23	137,886

Adjustments to Income

(See Instructions on page 11.)

24	Moving expense *(attach Form 3903 or 3903F)*	24			
25	Employee business expenses *(attach Form 2106)*	25			
26	IRA deduction, from the worksheet on page 12	26			
27	Keogh retirement plan deduction	27			
28	Penalty on early withdrawal of savings	28			
29	Alimony paid (recipient's last name ______ and social security no. ______)	29			
30	Deduction for a married couple when both work *(attach Schedule W)*	30	2,800		
31	Add lines 24 through 30. These are your **total adjustments** ▶			31	

Adjusted Gross Income

32 Subtract line 31 from line 23. This is your **adjusted gross income.** *If this line is less than $11,000 and a child lived with you, see "Earned Income Credit" (line 59) on page 16 of Instructions. If you want IRS to figure your tax, see page 13 of Instructions* ▶ 32 135,086

(19)

FIGURE 20–12(a). Example of filled-in IRS Form 1040, page 1.

Form 1040 (1985) Page 2

Tax Computation (See Instructions on page 13.)	33	Amount from line 32 (adjusted gross income)			33	135,086
	34a	If you itemize, attach Schedule A (Form 1040) and enter the amount from Schedule A, line 26. **Caution:** If you have unearned income and can be claimed as a dependent on your parents' return, check here ▶ ☐ and see page 13 of Instructions. Also see page 13 if you are married filing a separate return and your spouse itemizes deductions, or you are a dual-status alien.			34a	4,570
	b	If you do not itemize but you made charitable contributions, enter your cash contributions here. (If you gave $3,000 or more to any one organization, see page 14.)	34b			
	c	Enter your noncash contributions *(you must attach Form 8283 if over $500)*	34c			
	d	Add lines 34b and 34c. Enter the total	34d			
	e	Divide the amount on line 34d by 2. Enter the result here			34e	
	35	Subtract line 34a or line 34e, whichever applies, from line 33			35	130,516
	36	Multiply $1,040 by the total number of exemptions claimed on line 6f (see page 14)			36	4,160
	37	**Taxable income.** Subtract line 36 from line 35. Enter the result (but not less than zero)			37	126,356
	38	Enter tax here. Check if from ☐ Tax Table, ☑ Tax Rate Schedule X, Y, or Z, or ☐ Schedule G			38	44,246
	39	Additional taxes. (See page 14 of Instructions.) Enter here and check if from ☐ Form 4970, ☐ Form 4972, or ☐ Form 5544			39	
	40	Add lines 38 and 39. Enter the total ▶			40	44,246
Credits (See Instructions on page 14.)	41	Credit for child and dependent care expenses *(attach Form 2441)*	41			
	42	Credit for the elderly and the permanently and totally disabled *(attach Schedule R)*	42			
	43	Residential energy credit *(attach Form 5695)*	43			
	44	Partial credit for political contributions for which you have receipts	44			
	45	Add lines 41 through 44. These are your total personal credits			45	
	46	Subtract line 45 from line 40. Enter the result (but not less than zero)			46	
	47	Foreign tax credit *(attach Form 1116)*	47			
	48	General business credit. Check if from ☐ Form 3800, ☑ Form 3468, ☐ Form 5884, ☐ Form 6478	48	8,460		
	49	Add lines 47 and 48. These are your total business and other credits			49	8,460
	50	Subtract line 49 from line 46. Enter the result (but not less than zero) ▶			50	35,786
Other Taxes (Including Advance EIC Payments)	51	Self-employment tax *(attach Schedule SE)*			51	4,083
	52	Alternative minimum tax *(attach Form 6251)*			52	
	53	Tax from recapture of investment credit *(attach Form 4255)*			53	400
	54	Social security tax on tip income not reported to employer *(attach Form 4137)*			54	
	55	Tax on an IRA *(attach Form 5329)*			55	
	56	Add lines 50 through 55. This is your **total tax** ▶			56	40,269
Payments Attach Forms W-2, W-2G, and W-2P to front.	57	Federal income tax withheld	57	3,300		
	58	1985 estimated tax payments and amount applied from 1984 return	58			
	59	Earned income credit (see page 16)	59			
	60	Amount paid with Form 4868	60			
	61	Excess social security tax and RRTA tax withheld (two or more employers)	61			
	62	Credit for Federal tax on gasoline and special fuels *(attach Form 4136)*	62	2,124		
	63	Regulated Investment Company credit *(attach Form 2439)*	63			
	64	Add lines 57 through 63. These are your **total payments** ▶			64	5,424
Refund or Amount You Owe	65	If line 64 is larger than line 56, enter amount **OVERPAID** ▶			65	
	66	Amount of line 65 to be **REFUNDED TO YOU** ▶			66	
	67	Amount of line 65 to be applied to your 1986 estimated tax ▶	67			
	68	If line 56 is larger than line 64, enter **AMOUNT YOU OWE.** Attach check or money order for full amount payable to "Internal Revenue Service." Write your social security number and "1985 Form 1040" on it. ▶ Check ▶ ☐ if Form 2210 (2210F) is attached. See page 17. **Penalty:** $			68	34,845

Please Sign Here

Under penalties of perjury, I declare that I have examined this return and accompanying schedules and statements, and to the best of my knowledge and belief, they are true, correct, and complete. Declaration of preparer (other than taxpayer) is based on all information of which preparer has any knowledge.

Your signature	Date	Spouse's signature (if filing jointly, BOTH must sign)
Frank Farmer	2-20-85	Fran Farmer

Paid Preparer's Use Only

Preparer's signature ▶	Date	Check if self-employed ☐	Preparer's social security no.
Firm's name (or yours, if self-employed) and address ▶		E.I. No.	
		ZIP code	

(20)

FIGURE 20–12(b). Example of filled-in IRS Form 1040, page 2.

Schedule W
(Form 1040)
Department of the Treasury
Internal Revenue Service (X)

Deduction for a Married Couple When Both Work

▶ For Paperwork Reduction Act Notice, see Form 1040 Instructions.
▶ Attach to Form 1040.

OMB No. 1545-0074
1985
20

Names as shown on Form 1040: Frank and Fran Farmer
Your social security number: 123 : 45 : 6789

Step 1 Figure your earned income

		(a) You		(b) Your spouse
1 Wages, salaries, tips, etc., from Form 1040, line 7. (Do not include nondisability pensions or annuities.)	1	5,000	1	28,000
2 Net profit or (loss) from self-employment (from Schedules C and F (Form 1040), Schedule K-1 (Form 1065), and any other earned income)	2	71,073	2	
3 Add lines 1 and 2. This is your total earned income	3	76,073	3	28,000

Step 2 Figure your qualified earned income

4 Adjustments from Form 1040, lines 25, 26, 27, and any repayment of sub-pay included on line 31. (See instructions below.)	4		4	
5 Subtract line 4 from line 3. This is your qualified earned income. If the amount in column (a) or (b) is zero (-0-) or less, stop here. You may not take this deduction	5	76,073	5	28,000

Step 3 Figure your deduction

6 Compare the amounts in columns (a) and (b) of line 5 above. Write the smaller amount here. (Write either amount if 5(a) and 5(b) are the same.) ***Do not write more than $30,000***	6	28,000
7 Percentage used to figure the deduction (10%)	7	x .10
8 Multiply the amount on line 6 by the percentage on line 7. This is the amount of your deduction. Write the answer here and on Form 1040, line 30 ▶	8	2,800

Instructions

Complete this schedule and attach it to your Form 1040 if you take the deduction for a married couple when both work. You may take the deduction if both you and your spouse:

- work and have qualified earned income, and
- file a joint return, and
- do not file **Form 2555** to exclude income or to exclude or deduct certain housing costs, and
- do not file **Form 4563** to exclude income.

There are three steps to follow in figuring the deduction on Schedule W.

Step 1 (lines 1, 2, and 3).—Figure earned income separately for yourself and your spouse.

Step 2 (lines 4 and 5).—Figure qualified earned income separately for yourself and your spouse by subtracting certain adjustments from earned income.

Step 3 (lines 6, 7, and 8).—Figure the deduction based on the **smaller** of:

- the qualified earned income entered in column (a) or (b) of line 5, whichever is less, **OR**
- $30,000.

Earned income.—This is generally income you receive for services you provide. It includes wages, salaries, tips, commissions, certain disability income, sub-pay, etc. (from Form 1040, line 7). It also includes income earned from self-employment (from Schedules C and F of Form 1040 and Schedule K-1 of Form 1065), and net earnings and gains (other than capital gains) from the disposition, transfer, or licensing of property that you created. Earned income does not include interest, dividends, social security or tier 1 railroad retirement benefits, IRA distributions, unemployment compensation, deferred compensation, or nontaxable income. It also does not include any amount your spouse paid you.

Caution: *Do not consider community property laws in figuring your earned income.*

Qualified earned income.—This is the amount on which the deduction is based. Figure it by subtracting certain adjustments from earned income.

These adjustments (and the related lines on Form 1040) are:

- Employee business expenses (from line 25).
- IRA deduction (from line 26).
- Keogh retirement plan deduction (from line 27).
- Repayment of supplemental unemployment benefits (sub-pay) included in the total on line 31. See the instructions on repayment of sub-pay on page 12 of the Form 1040 Instructions.

Enter the total of any adjustments that apply to your or your spouse's earned income in the appropriate column of line 4.

Example.—You earned a salary of $20,000 and had $3,000 of employee business expenses (Form 1040, line 25). Your spouse earned $17,000 and had an IRA deduction of $1,000 (Form 1040, line 26). Your qualified earned income is $17,000 ($20,000 minus $3,000) and your spouse's is $16,000 ($17,000 minus $1,000). Because your spouse's qualified earned income is less than yours, the deduction is figured on your spouse's income. Therefore, the deduction is $1,600 ($16,000 x .10).

Schedule W (Form 1040) 1985

(91)

FIGURE 20–13. Example of filled-in IRS Schedule W.

Schedule G (Form 1040)
Department of the Treasury
Internal Revenue Service (L)

Income Averaging

▶ See instructions on back. ▶ Attach to Form 1040.

OMB No. 1545-0074
1985
16

Name(s) as shown on Form 1040 | Your social security number

Step 1 Add your income from 1982—1984

1982	1	Fill in the amount from your 1982 Form 1040 (line 37), Form 1040A (line 16), or Form 1040EZ (line 7). If less than zero, enter zero	1	
1983	2	Fill in the amount from your 1983 Form 1040 (line 37), Form 1040A (line 19), or Form 1040EZ (line 7). If less than zero, enter zero	2	
1984	3	Fill in the amount from your 1984 Form 1040 (line 37), Form 1040A (line 19), or Form 1040EZ (line 7). If less than zero, enter zero	3	
Total	4	Fill in all income less deductions earned outside of the U.S. or within U.S. possessions and excluded for 1982 through 1984 (include housing exclusion)	4	
	5	Add lines 1 through 4	5	

Step 2 Figure your averageable income

6	Divide the amount on line 5 by three (3)	6	
7	Multiply the amount on line 6 by 140% (1.4)	7	
8	Fill in your taxable income for 1985 from Form 1040, line 37	8	
9	If you received a premature or excessive distribution subject to a penalty under section 72, see instructions	9	
10	Subtract line 9 from line 8	10	
11	If you live in a community property state and are filing a separate return, see instructions	11	
12	Subtract line 11 from line 10. If less than zero, enter zero	12	
13	Write in the amount from line 7 above	13	
14	Subtract line 13 from line 12. This is your averageable income. If this line is $3,000 or less, do not complete the rest of this form	14	

Step 3 Figure your tax

15	Multiply the amount on line 14 by 25% (.25)			15	
16	Write in the amount from line 7 above			16	
17	Add lines 15 and 16			17	
18	Write in the amount from line 11 above			18	
19	Add lines 17 and 18			19	
20	Tax on amount on line 19 (from Tax Rate Schedule X, Y, or Z)			20	
21	Tax on amount on line 17 (from Tax Rate Schedule X, Y, or Z)	21			
22	Tax on amount on line 16 (from Tax Rate Schedule X, Y, or Z)	22			
23	Subtract line 22 from line 21	23			
24	Multiply the amount on line 23 by 3 *If you have no entry on line 9, skip lines 25 through 27 and go to line 28.*			24	
25	Tax on amount on line 8 (from Tax Rate Schedule X, Y, or Z)	25			
26	Tax on amount on line 10 (from Tax Rate Schedule X, Y, or Z)	26			
27	Subtract line 26 from line 25			27	
28	Add lines 20, 24, and 27. Write the result here. If it is less than the tax figured from the tax rate schedules or tax table, also write the result on Form 1040, line 38, and check the Schedule G box			28	

For Paperwork Reduction Act Notice, see Form 1040 instructions. | **Schedule G (Form 1040) 1985**

(25)

FIGURE 20–14. IRS Schedule G.

Form **6251** Department of the Treasury Internal Revenue Service

Alternative Minimum Tax Computation

▶ Attach to Forms 1040, 1040NR, 1041 or 990-T (Trust).

OMB No. 1545-0227 **1985** 32

Name(s) as shown on tax return	Identifying number

Line	Description	Box	Amount	Box	Amount
1	Adjusted gross income from Form 1040, line 33 or Form 1040NR, line 32 (see instructions)			1	
2	Deductions (see instructions):				
a (1)	Medical and dental expense from Schedule A, line 5	2a(1)			
(2)	Multiply Form 1040, line 33 by 5% (.05)	2a(2)			
(3)	Subtract line 2a(2) from line 2a(1). (If zero or less, enter zero)	2a(3)			
b	Contributions from Schedule A, line 18	2b			
c	Casualty and theft losses from Schedule A, line 19	2c			
d	Qualified interest on property used as a residence from Schedule A, line 11	2d			
e (1)	Interest, other than line 2d above, from Schedule A, line 14	2e(1)			
(2)	Net investment income (If zero or less, enter zero)	2e(2)			
(3)	Enter the smaller of line 2e(1) or line 2e(2)	2e(3)			
f	Gambling losses to the extent of gambling winnings from Schedule A, line 22	2f			
g	Estate tax allowable under section 691(c) from Schedule A	2g			
h	Estates and trusts only: Charitable deduction and income distribution deduction	2h			
i	Add lines 2a(3), b, c, d, e(3), f, g, and h			2i	
3	Subtract line 2i from line 1			3	
4	Tax preference items:				
a	Dividend exclusion	4a			
b	60% capital gain deduction	4b			
c	Accelerated depreciation on nonrecovery real property or 15 (or 18)-year real property	4c			
d	Accelerated depreciation on leased personal property or leased recovery property other than 15 (or 18)-year real property	4d			
e	Amortization of certified pollution control facilities	4e			
f	Mining exploration and development costs	4f			
g	Circulation and research and experimental expenditures	4g			
h	Reserves for losses on bad debts of financial institutions	4h			
i	Depletion	4i			
j	Incentive stock options	4j			
k	Intangible drilling costs	4k			
l	Add lines 4a through 4k			4l	
5	Alternative minimum taxable income (add lines 3 and 4(l)) (short period returns, see instructions)			5	
6	Enter: $40,000, if married filing joint return or Qualifying widow(er); $30,000, if single or head of household; $20,000, if married filing separate return or estate or trust			6	
7	Subtract line 6 from line 5. If zero or less, do not complete the rest of this form			7	
8	Enter 20% of line 7			8	
9	Amount from Form 1040, line 50, or Form 1040NR, line 51. (Do not include Form 1040, line 39, or Form 1040NR, line 41.) (estates and trusts, see instructions)			9	
10	Subtract line 9 from line 8. If zero or less, enter zero			10	
11	Foreign tax credit			11	
12	Alternative minimum tax (subtract line 11 from line 10). Enter on Form 1040, line 52, Form 1040NR, line 52, Form 1041, line 31, or Form 990-T, page 1, line 14			12	

Instructions

(Section References are to the Internal Revenue Code)

Paperwork Reduction Act Notice.— We ask for this information to carry out the Internal Revenue laws of the United States. We need it to ensure that taxpayers are complying with these laws and to allow us to figure and collect the right amount of tax. You are required to give us this information.

Who Must File.—File this form if : (a) You are liable for the alternative minimum tax; **or** (b) you have one or more tax preference items on lines 4c through 4k; **or** (c) you have an amount on line 2e(3), and line 2e(2) includes income other than interest and dividend income.

Individuals, estates or trusts may be liable if their adjusted gross income plus tax preference items listed on line 4 total more than line 6.

For more information, see **Publication 909,** Alternative Minimum Tax.

Minimum Tax Deferred From Earlier Year(s).—If a net operating loss carryover from an earlier year(s) reduces taxable income for 1985, and the net operating loss giving rise to the carryover resulted in the deferral of minimum tax in that earlier year(s), all or part of the deferred minimum tax may be includible as tax liability for 1985. Figure the deferred minimum tax in the worksheet in Publication 909 and enter it on Form 1040, line 52, or Form 1041, line 31. Write "Deferred Minimum Tax."

Partners, Beneficiaries, etc.—If you are a:

(1) Partner or shareholder of an S corporation, take into account separately your distributive share of items of income and deductions that enter into the computation of tax preference items.

(2) Beneficiary of an estate or trust, see section 58(c).

(3) Participant in a common trust fund, see section 58(e).

(4) Shareholder or holder of beneficial interest in a regulated investment company or a real estate investment trust, see section 58(f).

Carryback and Carryover of Unused Credits.—It may be necessary to figure the carryback or carryover of certain unused credits. See section 55(c)(3).

Note: If you have an earned income credit, you must reduce that credit by any alternative minimum tax.

(Continued on back) 275

Form **6251** (1985)

FIGURE 20–15. IRS Form 6251.

Form **1040X** (Rev. October 1985) (L)

Department of the Treasury—Internal Revenue Service

Amended U.S. Individual Income Tax Return

OMB No. 1545-0091
Expires 4-30-88

This return is for calendar year ▶ 19 , OR fiscal year ended ▶ , 19 .

Please print or type

Your first name and initial (if joint return, also give spouse's name and initial) Last name	**Your social security number**
Present home address (number and street, including apartment number, or rural route)	**Spouse's social security number**
City, town or post office, state, and ZIP code	Telephone number (optional) ( )

Enter below name and address as shown on original return (if same as above, write "Same"). If changing from separate to joint return, enter names and addresses used on original returns. (**Note:** *You cannot change from joint to separate returns after the due date has passed.)*

a Service center where original return was filed

b Has original return for the year being changed been audited? . ☐ Yes ☐ No
If "No," have you been notified that it will be? . . . . ☐ Yes ☐ No
If "Yes," identify IRS office

c Are you amending your return to include any item (loss, credit, deduction, other tax benefit, or income) relating to a tax shelter required to be registered? . . . . . . ☐ Yes ☐ No
If "Yes," you **MUST** attach **Form 8271,** Investor Reporting of Tax Shelter Registration Number.

d Filing status claimed. (**Note:** *You cannot change from joint to separate returns after the due date has passed.)*
On original return . ▶☐ Single ☐ Married filing joint return ☐ Married filing separate return ☐ Head of household ☐ Qualifying widow(er)
On this return . . ▶☐ Single ☐ Married filing joint return ☐ Married filing separate return ☐ Head of household ☐ Qualifying widow(er)

Income and Deductions		**A.** As originally reported or as adjusted (see instructions)	**B.** Net change—Increase or (Decrease)—explain on page 2	**C.** Correct amount
1 Total income (see instructions)	1			
2 Adjustments to income (see instructions)	2			
3 Adjusted gross income (subtract line 2 from line 1)	3			
4 Deductions (see instructions)	4			
5 Subtract line 4 from line 3	5			
6 Exemptions (see instructions)	6			
7 Taxable income (subtract line 6 from line 5)	7			
Tax Liability 8 Tax (see instructions). (Method used in col. C ________)	8			
9 Credits (see instructions)	9			
10 Subtract line 9 from line 8. Enter the result, but not less than zero	10			
11 Other taxes (such as self-employment tax, alternative minimum tax)	11			
12 Total tax liability (add line 10 and line 11)	12			
Payments 13 Federal income tax withheld and excess FICA and RRTA tax withheld	13			
14 Estimated tax payments	14			
15 Earned income credit	15			
16 Credits for Federal tax on gasoline and special fuels, regulated investment company, etc.	16			

17 Amount paid with Form 4868, Form 2688, or Form 2350 (application for extension of time to file)	17	
18 Amount paid with original return, plus additional tax paid after it was filed	18	
19 Total of lines 13 through 18, column C	19	
Refund or Amount You Owe		
20 Overpayment, if any, as shown on original return (or as previously adjusted by IRS)	20	
21 Subtract line 20 from line 19 (see instructions)	21	
22 **AMOUNT YOU OWE.** If line 12, col. C, is more than line 21, enter difference. Please pay in full with this return	22	
23 **REFUND** to be received. If line 12, column C, is less than line 21, enter difference.	23	

Please Sign Here Under penalties of perjury, I declare that I have filed an original return and that I have examined this amended return, including accompanying schedules and statements, and to the best of my knowledge and belief this amended return is true, correct, and complete. Declaration of preparer (other than taxpayer) is based on all information of which the preparer has any knowledge.

▶ Your signature | Date ▶ Spouse's signature (if filing jointly, BOTH must sign)

Paid Preparer's Use Only Preparer's signature ▶ | Date | Check if self-employed ☐ | Preparer's social security no.
Firm's name (or yours, if self-employed) and address ▶ | E.I. No. | ZIP code

For Paperwork Reduction Act Notice, see page 1 of separate Instructions.

BE SURE TO COMPLETE PAGE 2

(71)

FIGURE 20–16(a). IRS Form 1040X (front).

Form 1040X (Rev. 10-85) Page 2

Part I **Exemptions (see Form 1040 or Form 1040A Instructions)**
If claiming more exemptions, complete lines 1—9.
If claiming fewer exemptions, complete lines 1—6.

		A. Number originally reported	B. Net change	C. Correct number
1 Exemptions—yourself and spouse, 65 or over, blind	1			
2 Your dependent children who lived with you	2			
3 If amending your 1985 return, your dependent children who did not live with you	3			
4 Other dependents	4			
5 Total exemptions (add lines 1 through 4)	5			
6 Multiply $1,040 ($1,000, for tax years beginning before 1985) by the total number of exemptions claimed on line 5. Enter this amount here and on page 1, line 6	6			

7 First names of your dependent children who lived with you and were not claimed on original return: **Enter number ▶** ☐

8 If amending your 1985 return, first names of your dependent children who did not live with you and were not claimed on original return (see instructions). (If pre-1985 agreement, check here ☐ .) **Enter number ▶** ☐

9 Other dependents not claimed on original return:

(a) Name	(b) Relationship	(c) Number of months lived in your home	(d) Did dependent have income of $1,040 ($1,000, for tax years beginning before 1985) or more?	(e) Did you provide more than one-half of dependent's support?

Enter number ▶ ☐

Part II **Explanation of Changes to Income, Deductions, and Credits**

Enter the line number from page 1 for each item you are changing and give the reason for each change. Attach all supporting forms and schedules for items changed. Be sure to include your name and social security no. on any attachments.

If the change pertains to a net operating loss carryback, a general business credit carryback, or a research credit carryback, attach the schedule or form that shows the year in which the loss or credit occurred. See the instructions. Also, check here . . . . . . ▶ ☐

Part III **Presidential Election Campaign Fund**
Checking below will not increase your tax or reduce your refund.

If you did not previously want to have $1 go to the fund, but now want to . . . . . Check here ▶ ☐
If joint return and if spouse did not previously want to have $1 go to the fund, but now wants to . . . . . Check here ▶ ☐

(72)

FIGURE 20–16(b). IRS Form 1040X (back).

nary calculations can cut these mistakes significantly. But in some years an error just might be made. Farmers and ranchers should review their tax return a few months after it is filed, when the pressure is off, to guarantee accuracy. If an error is found, Form 1040X should be located.

Form 1040X (Figure 20–17[a] and [6]) is concerned *only* with changes to a Form 1040 that has *already been filed*. Generally, the net results should be entered on Form 1040X and the details should be provided on the applicable form. Although it may be overkill, send copies of original forms as well as corrected forms to minimize questions at the IRS Service Center. Any documentation that will support the claim will greatly increase chances of a successful amendment.

Assume a farmer discovers that he did not include a $5,000 interest payment to his lender when he completed his original return. He should then recompute Schedule F to increase the interest deduction by $5,000. And he should also compute Form 1040X, paying specific attention to Lines 1–5 and Lines 8–12. These two forms must be filed. But, a photocopy of the original Schedule F, photocopies of *all* interest statements, and a short note explaining the action, filed with Form 1040X, will definitely assist the IRS in processing the claim.

The taxpayer's tendency may be to file Form 1040X if the error discovered worked to his favor and to ignore amending the return if the error will force an additional payment. But it is generally in the best interest of the taxpayer to tell the IRS about the mistake before the IRS tells him.

Summary

The federal tax filing process is nearly complete for most individuals when Form 1040 is in hand. Form 1040 is a summary or umbrella form that wraps together all of the detailed and appendix-type supplementary material from the supporting forms and schedules. Before Form 1040 can be completed, the personal nonbusiness supporting schedules (Schedules A and B and Schedule W) must be considered. The remaining tax computation/adjustment forms (Schedule G and Form 6251) must also be considered.

Schedule A provides an opportunity to reduce taxable income through the IRC-allowable itemized personal deductions. A careful review of federal tax code sections relating to which deductions are allowable is crucial. The allowable deductions fall into six broad categories (medical and dental, taxes, interest, charitable contributions, casualty and theft losses, and miscellaneous) but each category and each deduction has a specific set of rules.

Schedule B is used to list sources and amounts of interest and dividend income but is not required for amounts under $400 each. Schedule W is a relatively new form that allows a deduction of 10 percent of the lower of two spouses' incomes when both are employed.

Quite a bit of paper shuffling is required when completing Form 1040, but the best way to minimize confusion and frustration is to take it line by line. The first three blocks of the form detail personal information (e.g., name, address, occupations, social security numbers, filing status, and number of exemptions). The fourth main block lists all sources of income by type, usually drawing on other forms and schedules for the documentation (for instance, Schedules C, D, E, and F and Form 4797) to justify the botton-line answer written on Form 1040. The last major block on the front side details adjustments to income (such as moving expense deduction, IRA contributions, and the Schedule W deduction). The last line on the front side is the computation of AGI: Adjusted Gross Income.

AGI is used to finish Schedule A computations and is also carried to Form 6251—Alternative Minimum Tax—to evaluate the effect of tax preference items. Then Schedule A results are brought back to the first block on the back side of Form 1040 to calculate taxable income. Taxable income is the second big step (after AGI) since this figure must be carried to the Tax Tables, Tax Rate Schedules, or Schedule G to determine the amount of tax due. Other major blocks on the back side of Form 1040 allow for tax-reducing credits to be listed (again justified by supporting forms such as Form 3468), other taxes due (such as the self-employment tax, alternative minimum tax, and recapture of ITC from Forms 4797, 6251, and 4255), payments made during the year and reconciliation of overpayment or underpayment.

The complete package can then be assembled in the order of: check, W-2 and 1099 forms, Form 1040, schedules in alphabetic order, and forms in numerical order. If an error is discovered after the package is filed, a Form 1040X can be completed and filed, as long as it is submitted within 3 years of the date when the erroneous return was filed.

CHAPTER 21

Tax Withholding and Estimated Tax

Because of the increased reliance of the U.S. Treasury on the income tax as its primary source of tax revenue, to enhance federal government cash flow and to insure accurate compliance with tax laws the Current Tax Payment Act was passed in 1943 to require taxpayers to pay taxes due throughout the calendar year. This amendment essentially changed the U.S. income tax from a once-each-year tax to a pay-as-you-go tax. Consequently taxpayers are required to pay taxes as they receive income during the year. There are two methods for taxpayers to meet this requirement: tax withholding and the estimated tax.

Tax withholding applies to employees. Employers are required to withhold income tax from the employee's gross wages or salary before the payroll check is issued. The amount withheld is an estimate based upon the wages paid for services performed over a specific time period. (An example of how this is accomplished will be provided later in this chapter). This estimate is affected by such items as the amount of wages (the higher the salary, the higher the withholding tax), the number of exemptions and the gross amount of itemized personal deductions.

Estimated taxes apply to self-employed persons and corporations and consequently to all businesses. The business is required to estimate its income for the tax year during the first 3 months of that year. Once income for the entire year is estimated, an estimated tax is calculated in much the same manner as the tax was figured on the actual tax return of the last chapter. This estimated tax is then paid in four equal, quarterly installments throughout the tax year. The final return reports any over or underpayment.

Because of special circumstances, such as the inability to accurately predict yields and prices 9 to 12 months in advance and a very strong political lobby, farmers and ranchers are exempted from many of the withholding and tax estimate rules that apply to other businesses and individuals. However, a strict definition of a farmer by the IRS must be met. This definition was spelled out in the introduction to this part of the textbook, but before continuing on, a review of that definition is necessary. Many individuals whom we would normally call farmers or ranchers have been surprised that under IRS rules they are *not* so classified. Misinterpretation or disregard for these rules can result in heavy fines.

A summary of the IRS definition of a farmer can be quickly stated: two-thirds or more of gross income for last year or two-thirds or more of estimated gross income for this year must be from farming or ranching for a taxpayer to be classified by the IRS as a farmer. *Gross income* includes farm or ranch income, wages from off-farm employment, wages received as a farm employee (including those from a farm or ranch corporation or partnership), interest and dividend income, and capital gains income (Schedule D and Form 4797), among others. To meet the two-thirds gross income test, compare gross income from farming (Schedule F, Line 12, Figure 19–2 plus Form 4797 gains from the sale of breeding livestock) to Total Income (usually Form 1040, Line 23, but also consider gross income from any Schedule C).

The final issue to be covered in this chapter is a string of informational forms, all of which fall under the heading of Form 1099. These forms must be used by many farmers and ranchers to report payments to independent contractors and other individuals. Farmers and ranchers will also likely receive some 1099s from financial institutions or from other individuals as a result of payments received. At the time a Form 1099 is filled out, one copy is given to the payee and one copy is sent to the IRS.

To begin this discussion of tax withholding and estimated tax, let's start with employee tax withholding. Farmers and ranchers are likely to find themselves dealing with tax withholding both from the side of being an employee of someone else and from the side of employing another person to work on the farm or ranch.

Tax Withholding

The tax withholding process begins with several definitions and requirements and progresses through quite a few IRS forms and procedures. Before any of the process is begun, an agricultural employer should obtain several IRS publications and copies of relevant forms. Most important of the publications and forms are IRS Publication 15, Circular E, *Employer's Tax Guide*; Publication 51, Circular A, *Agricultural Employer's Tax Guide*; and Publication 505, *Tax Withholding and Estimated Tax*; also obtain copies of Forms W-2 (included in Circular A), W-3, W-4, 943 and 943A.

The following discussion of the tax withholding process will be approached in a chronological order. We shall begin with the definitions and requirements to determine who must be concerned with tax withholding and then follow through on the mechanics.

Employers or Farmworkers

The IRS has written a specific definition of an employer of farmworkers.

In general, you are an employer of farmworkers if you employ workers who:

> Raise or harvest agricultural or horticultural products on a farm or ranch;
> Care for your farm and equipment when most of the care is done on a farm;
> Handle, process or package any agricultural or horticultural commodity if you produced over half of the commodity;
> Do work related to cotton ginning, turpentine or gum resin products; or
> Do housework in your private home if it is on a farm that is operated for profit. IRS Publication 51, Circular A.

The specific definition seems somewhat excessive and generally confusing, but like almost all IRS regulations, the definition must be precise to avoid problems of determining exactly who is obligated. The definition covers all field workers, ranch hands, farmers, domestic help, and such persons as milkers and feeders. Note that the definition, however, does *not* cover retail nurseries or independent businessmen such as veterinarians and mechanics actually employed by some other firm. The IRS definition also calls crew leaders employers of farmworkers. Crew leaders provide and pay groups of farmworkers to perform specific services for a farmer. This type of arrangement is common in some areas of the United States and involves primarily field harvesting, hoeing, and custom combine crews. While the distinction may seem picky, it is necessary to establish who is responsible to withhold and deposit taxes. If a farmer assembles a crew to harvest his onions, he is then the employer of the crew members and is responsible for any withholding taxes. On the other hand, if a farmer contracts with a crew leader to assemble the harvesting crew and if the farmer pays the crew leader who in turn pays the crew members, then the crew leader is actually the employer. The farmer may have to provide a Form 1099 with his payment to the crew leader, but the farmer does not handle any withholding for the individual crew members.

The primary purpose for the definition of an employer of a farmworker is that employers of farmworkers are *not* required by the IRS to withhold income taxes on wages paid to farm employees. But a voluntary agreement between employer and employee to withhold income taxes is possible and may be sought by the employee. If this voluntary agreement is made, the employee must complete an IRS Form W-4. After the form is completed and given to the employer, the employer should begin withholding the correct amount of tax. Let's first look at Form W-4 and then turn to determining the correct tax.

IRS Form W-4

IRS Form W-4 is a relatively simple form that should be filled out by an employee. The form, shown in Figure 21–1, details basic information about the employee such as name, address, Social Security number, marital status, and number of withholding allowances claimed. The process for determining number of allowances is somewhat similar to the process used in the information block on Form 1040 for exemptions, but the W-4 process is used *only* to determine allowances for withholding. This figure *cannot* be used directly on Form 1040. The allowance claim process can be followed through the worksheet attached to Form W-4 (Figure 21–2).

Form **W-4** (Rev. January 1986)

Department of the Treasury—Internal Revenue Service

Employee's Withholding Allowance Certificate

OMB No. 1545-C010
Expires: 11-30-87

1 Type or print your full name

2 Your social security number

Home address (number and street or rural route)

City or town, state, and ZIP code

3 Marital Status: ☐ Single ☐ Married ☐ Married, but withhold at higher Single rate
Note: If married, but legally separated, or spouse is a nonresident alien, check the Single box.

4 Total number of allowances you are claiming (from line F of the worksheet on page 2) . . . .

5 Additional amount, if any, you want deducted from each pay . . . . $

6 I claim exemption from withholding because (see instructions and check boxes below that apply):

a ☐ Last year I did not owe any Federal income tax and had a right to a full refund of **ALL** income tax withheld, **AND**

b ☐ This year I do not expect to owe any Federal income tax and expect to have a right to a full refund of **ALL** income tax withheld. If both a and b apply, enter the year effective and "EXEMPT" here . . . ▶ Year 19

c If you entered "EXEMPT" on line 6b, are you a full-time student? . . . . ☐**Yes** ☐**No**

Under penalties of perjury, I certify that I am entitled to the number of withholding allowances claimed on this certificate, or if claiming exemption from withholding, that I am entitled to claim the exempt status.

Employee's signature ▶ **Date ▶** , 19

7 Employer's name and address **(Employer: Complete 7, 8, and 9 only if sending to IRS)**

8 Office code

9 Employer identification number

Detach along this line. Give the top part of this form to employer; keep the lower part for your records.

Changes You Should Note.—Each withholding allowance has increased to $1,080 and the income tax brackets have expanded to reflect indexing. Therefore, income tax withholding has decreased. If you claim withholding allowances for deductions and credits, you should check to be sure that enough tax is being withheld.

Privacy Act and Paperwork Reduction Act Notice.—If you do not give your employer a completed Form W-4, you will be treated as a single person with no withholding allowances as required by law. We ask for this information to carry out the Internal Revenue laws of the United States. We may give the information to the Department of Justice for civil or criminal litigation and to the states and the District of Columbia for use in administering their tax laws.

Purpose.—You must complete Form W-4 so that your employer can withhold Federal income tax from your pay. Your Form W-4 will remain in effect until you change it. However, if you entered "EXEMPT" on line 6b, your W-4 will remain in effect only until February 15th of the next year. By correctly completing this form, the amount of tax withheld from your wages will closely match your tax liability.

If you got a large refund last year, you may be having too much tax withheld. If so, you may want to increase the number of your allowances on line 4 by claiming any other allowances you are entitled to. The kinds of allowances, and how to figure them, are explained later in detail.

If you owed a large amount of tax last year, you may not be having enough tax withheld. If so, you can claim fewer allowances on line 4, or have an additional amount withheld on line 5, or both.

If the number of withholding allowances you are entitled to claim decreases to less than you are now claiming, you must file a new W-4 with your employer within 10 days.

Publication 505, Tax Withholding and Estimated Tax, contains more information on withholding. You can get it from most IRS offices.

For more information about dependents, deductions, and tax credits you may qualify for, see the Form 1040 Instructions.

Penalty.—You may be fined $500 if you file, with no reasonable basis, a W-4 that results in less tax being withheld than is properly allowable. In addition, criminal penalties apply for wilfully supplying false or fraudulent information or failing to supply information requiring an increase in withholding.

Worksheet Instructions

A worksheet is provided on page 2. These instructions should help you to correctly fill out both the worksheet and the form.

Line A. Personal Allowances.—You can claim the following personal allowances: one for yourself, one if you are 65 or older, and one if you are blind.

If you are married and your spouse either does not work or is not claiming his or her allowances on a separate W-4, you may also claim the following allowances: one for your spouse, one if your spouse is 65 or older, and one if your spouse is blind.

Line B. Special Withholding Allowance.—Claim the special withholding allowance if you are single and have one job; you are married, have one job, and your spouse does not work; or the total wages earned on other jobs by you or your spouse (or both) is 10% or less of the combined total wages. Use this special withholding allowance only to figure your withholding. Do not claim it when you file your return.

Line C. Allowances for Dependents.—You may claim one allowance for each dependent you will be able to claim on your Federal income tax return.

Note 1: *If you are not claiming any deductions or credits or income averaging, skip D and E, add lines A, B, and C, enter the total on line F and carry the total over to line 4 of Form W-4.*

Note 2: *Before you claim allowances under D and E, total your nonwage taxable income (interest, dividends, self-employment income, etc.) and subtract this amount from estimated deductions you would otherwise enter in D1. If your nonwage income is greater than the amount of estimated deductions, you cannot claim any allowances under D. Moreover, you should take one-third of the excess (nonwage income over estimated deductions) and add this to the appropriate "A" value in Table 1 if determining allowances under E.*

Line D. Allowances for Estimated Deductions.— If you expect to itemize deductions or claim adjustments to income, you may be entitled to additional withholding allowances. See Schedule A (Form 1040) for deductions you can itemize and see the Adjustments to Income section on page 1 of Form 1040 for a list of adjustments. Do not include reimbursed moving expenses or IRA contributions made by your employer unless income tax has been withheld on them by your employer. If no tax has been withheld on them and you claim additional allowances for them, you will be underwithheld. For details, see Publication 505.

Once you have determined these deductions, enter the total on line D1 of the worksheet on page 2 and figure the number of withholding allowances for them.

Line E. Allowances for Tax Credits.—If you expect to take credits like those shown on the 1985 Form 1040, such as child care, use the table on page 2 to figure the number of additional allowances you can claim. You may estimate these credits. Include the earned income credit only if you are not receiving advance payment of it. Also, if you expect to income average, include the amount of the reduction in tax because of averaging and any excess social security tax withheld when using the table.

Line F.—Add lines A through E and enter the total on line F and on line 4 of your Form W-4.

Form W-4 Instructions

Identifying Information.—Fill in the identifying information in boxes 1 and 2. If you are married and want tax withheld at the regular rate for married persons, check "Married" in box 3. If you are married and want tax withheld at the higher Single rate (because both you and your spouse work, for example), check "Married, but withhold at higher Single rate" in box 3.

Line 4. Total Number of Allowances.—Use the worksheet on page 2 to figure your allowances. Enter the total on line 4.

If you are single and hold more than one job, you may not claim the same allowances with more than one employer at the same time. If you are married and both you and your spouse are employed, you may not both claim the same allowances with both of your employers at the same time. To have the highest amount of tax withheld, claim "0" allowances on line 4.

Line 5. Additional Amount You Want Deducted.—If you are not having enough tax withheld from your pay, you may ask your employer to withhold more by filling in an additional amount on line 5. Often, married couples, both of whom are working, and persons with two or more jobs need to have additional tax withheld. You may also need to have additional tax withheld because you have income other than wages, such as interest and dividends, capital gains, rents, alimony received, taxable social

Form **W-4** (Rev. 1-86)

1

FIGURE 21–1. IRS Form W-4.

security benefits, etc. Estimate the amount you will be underwithheld and divide that amount by the number of pay periods in the year. Enter the additional amount you want withheld each pay period on line 5.

Line 6. Exemption from Withholding.—You can claim exemption from withholding only if last year you did not owe any Federal income tax and had a right to a refund of all income tax withheld, **and** this year you do not expect to owe any Federal income tax and expect to have a right to a refund of all income tax withheld. If you qualify, check boxes 6a and b, write the year exempt status is effective and "EXEMPT" on line 6b, and answer "Yes" or "No" to the question on line 6c.

If you want to claim exemption from withholding next year, you must file a new W-4 with your employer on or before February 15 of next year. If you are not having Federal income tax withheld this year, but expect to have a tax liability next year, the law requires you to give your employer a new W-4 by December 1 of this year. If you are covered by social security, your employer must withhold social security tax.

Note: *Your employer must send to IRS any W-4 claiming more than 14 withholding allowances or claiming exemption from withholding if the wages are expected to usually exceed $200 a week. The employer is to complete boxes 7, 8, and 9 only on copies of the W-4 sent to IRS.*

Table 1—For Figuring Your Withholding Allowances for Estimated Tax Credits and Income Averaging (Line E)

Estimated Salaries and Wages from All Sources	Single Employees (A)	Single Employees (B)	Head of Household Employees (A)	Head of Household Employees (B)	Married Employees (When Spouse not Employed) (A)	Married Employees (When Spouse not Employed) (B)	Married Employees (When Both Spouses Are Employed) (A)	Married Employees (When Both Spouses Are Employed) (B)
Under $15,000	$ 90	$170	$ 0	$170	$ 50	$130	$ 0	$130
15,000-25,000	120	260	0	270	60	190	310	190
25,001-35,000	170	330	0	330	150	240	760	210
35,001-45,000	240	400	0	400	150	310	1,280	270
45,001-55,000	540	400	0	400	180	360	2,090	310
55,001-65,000	1,060	400	0	400	290	400	2,810	360
Over 65,000	2,050	400	550	400	460	400	3,490	400

Worksheet to Figure Your Withholding Allowances To Be Entered on Line 4 of Form W-4

A Personal allowances . . . . ▶ **A**

B Special withholding allowance (not to exceed 1 allowance—see instructions on page 1) . . . . ▶ **B**

C Allowances for dependents . . . . ▶ **C**

If you are not claiming any deductions or credits or income averaging, skip lines D and E.

D Allowances for estimated deductions:

1 Enter the total amount of your estimated itemized deductions, alimony payments, qualified retirement contributions including IRA and Keogh (H.R. 10) plans, deduction for a married couple when both work, business losses including net operating loss carryovers, moving expenses, employee business expenses, penalty on early withdrawal of savings, and charitable contributions for nonitemizers for the year . . . . . . ▶ **1** $

2 If you do not plan to itemize deductions, enter $540 on line D2. If you plan to itemize, find your total estimated salaries and wages amount in the left column of the table below. (Include salaries and wages of both spouses.) Read across to the right and find the amount from the column that applies to you. Enter that amount on line D2. . . . . . ▶ **2** $

Estimated salaries and wages from all sources:	Single and Head of Household Employees (only one job)	Married Employees (one spouse working and one job only)	Employees with more than one job or Married Employees with both spouses working [1]
Under $20,000	$3,000	$4,100	35% of estimated salaries and wages
20,000-40,000	3,000	4,200	23% of estimated salaries and wages
40,001-60,000	8% of estimated salaries and wages	4,200	19% of estimated salaries and wages
Over $60,000	10% of estimated salaries and wages	7% of estimated salaries and wages	18% of estimated salaries and wages

3 Subtract line D2 from line D1 (But not less than zero) . . . . . . ▶ **3** $

4 Divide the amount on line D3 by $1,080 (increase any fraction to the next whole number). Enter here . . . ▶ **D**

E Allowances for tax credits and income averaging: use Table 1 above for figuring withholding allowances

1 Enter tax credits, excess social security tax withheld, and tax reduction from income averaging . $

2 Enter the column (A) amount from Table 1 for your salary range and filing status (single, etc.). However, enter 0 if you claim 1 or more allowances on line D4 . . . . . . $

3 Subtract line 2 from line 1 (If zero or less, do not complete lines 4 and 5) . . . . . $

4 Find the column (B) amount from Table 1 for your salary range and filing status . . . . $

5 Divide line 3 by line 4. Increase any fraction to the next whole number. This is the maximum number of withholding allowances for tax credits and income averaging. Enter here . . . . . . ▶ **E**

Example: A taxpayer who expects to file a Federal income tax return as a single person estimates annual wages of $12,000 and tax credits of $650. The $12,000 falls in the wage bracket of under $15,000. The value in column (A) is $90. Subtracting this from the estimated credits of $650 leaves $560. The value in column (B) is $170. Dividing $560 by $170 gives 3.3. Since any fraction is increased to the next whole number, show 4 on line E.

F Total (add lines A through E). Enter total here and on line 4 of Form W-4 . . . . . . ▶ **F**

[1] If you earn 10% or less of your total wages from other jobs or one spouse earns 10% or less of the couple's combined total wages, you can use the "Single and Head of Household Employees (only one job)" or "Married Employees (one spouse working and one job only)" table, in section D 2.

☆ U.S. GOVERNMENT PRINTING OFFICE: 1985-O-463-048

FIGURE 21–2. IRS Form W-4 worksheets.

Let's take an example and determine the maximum number of allowances which can be claimed. Walt Worker expects to earn $18,000 in 1986; his wife Wilma is not employed. They further estimate that they will generate $4,000 in itemized deductions and will put $2,000 into an Individual Retirement Account. They will not have any capital or business losses or any tax credits. Walt and Wilma have two children under age 16 who live at home.

Turning to the worksheet in Figure 21–2, Walt and Wilma can claim two personal allowances on Line A and two allowances on Line C for their children. They may claim the special withholding allowance on Line B since Wilma is not employed (see the instructions in Figure 21–1). On Line D1, they may write $6,000, which is the sum of $4,000 itemized deductions plus $2,000 IRA contributions. Since gross wages are expected to be $18,000, the amount to be entered on Line D2 is $4,100. Line D3 will then be $6,000 − $4,100 = $1,900. Dividing by $1,080 yields 1.76, which should be rounded to 2 and entered on Line D4. Line E is blank since no tax credits are expected. Consequently Line F is the sum of Lines A, B, C, and D, or 7. Walt and Wilma may claim 7 withholding allowances on Line 4 of Form W–4.

Social Security Withholding

Unlike income tax withholding, Social Security tax withholding on employee wages is *not* optional. The Social Security tax on employees *must* be withheld if:

1. Cash wages paid total $150 or more duing a year, or
2. The employee worked for 20 or more days during the year.

These two tests apply separately to each individual employed and do not account for any wages the employer earned from another employer.

The primary exception to Social Security withholding is that Social Security taxes *cannot* be withheld from wages earned by a spouse or by children under age 21.

The tax rate for 1986 is 7.15 percent of all cash wages paid. That amount should be deducted from the employee's paycheck. There is an additional 7.15 percent tax that represents the employer's contribution. So, in effect, the Social Security tax is 14.3 percent of wages paid, half paid by the employee and half by the employer.

Calculating Withholding Taxes

IRS Publication 15, Circular E, *Employer's Tax Guide,* presents a series of tables and schedules to tell the employer how much tax should be withheld from an employee's paycheck. The easiest method for most farmers and ranchers will be to use a table such as the example shown in Figure 21–3. But, if the farmer or rancher owns a computer and would like to program the withholding calculation, the tax rate schedules (such as those in Figure 21–4) may be used.

Let's once again determine the withholding taxes for Walt and Wilma Worker.

MARRIED Persons—MONTHLY Payroll Period

(For Wages Paid After December 1985)

And the wages are—		And the number of withholding allowances claimed is—										
At least	But less than	0	1	2	3	4	5	6	7	8	9	10
		The amount of income tax to be withheld shall be—										
$0	$220	$0	$0	$0	$0	$0	$0	$0	$0	$0	$0	$0
220	224	1	0	0	0	0	0	0	0	0	0	0
224	228	1	0	0	0	0	0	0	0	0	0	0
228	232	2	0	0	0	0	0	0	0	0	0	0
232	236	2	0	0	0	0	0	0	0	0	0	0
236	240	2	0	0	0	0	0	0	0	0	0	0
240	248	3	0	0	0	0	0	0	0	0	0	0
248	256	4	0	0	0	0	0	0	0	0	0	0
256	264	5	0	0	0	0	0	0	0	0	0	0
264	272	6	0	0	0	0	0	0	0	0	0	0
272	280	7	0	0	0	0	0	0	0	0	0	0
280	288	7	0	0	0	0	0	0	0	0	0	0
288	296	8	0	0	0	0	0	0	0	0	0	0
296	304	9	0	0	0	0	0	0	0	0	0	0
304	312	10	0	0	0	0	0	0	0	0	0	0
312	320	11	1	0	0	0	0	0	0	0	0	0
320	328	12	2	0	0	0	0	0	0	0	0	0
328	336	13	3	0	0	0	0	0	0	0	0	0
336	344	14	4	0	0	0	0	0	0	0	0	0
344	352	15	5	0	0	0	0	0	0	0	0	0
352	360	15	6	0	0	0	0	0	0	0	0	0
360	368	16	6	0	0	0	0	0	0	0	0	0
368	376	17	7	0	0	0	0	0	0	0	0	0
376	384	18	8	0	0	0	0	0	0	0	0	0
384	392	19	9	0	0	0	0	0	0	0	0	0
392	400	20	10	0	0	0	0	0	0	0	0	0
400	420	21	11	2	0	0	0	0	0	0	0	0
420	440	24	14	4	0	0	0	0	0	0	0	0
440	460	26	16	6	0	0	0	0	0	0	0	0
460	480	29	18	8	0	0	0	0	0	0	0	0
480	500	31	20	10	0	0	0	0	0	0	0	0
500	520	33	23	13	3	0	0	0	0	0	0	0
520	540	36	25	15	5	0	0	0	0	0	0	0
540	560	38	27	17	7	0	0	0	0	0	0	0
560	580	41	30	19	9	0	0	0	0	0	0	0
580	600	43	32	21	11	2	0	0	0	0	0	0
600	640	47	36	25	15	5	0	0	0	0	0	0
640	680	53	41	30	19	9	0	0	0	0	0	0
680	720	58	46	35	24	14	4	0	0	0	0	0
720	760	64	51	39	29	18	8	0	0	0	0	0
760	800	70	57	44	33	23	13	3	0	0	0	0
800	840	75	63	50	38	27	17	7	0	0	0	0
840	880	81	68	56	43	32	21	11	2	0	0	0
880	920	86	74	61	49	37	26	16	6	0	0	0
920	960	92	79	67	54	42	31	20	10	0	0	0
960	1,000	98	85	72	60	47	36	25	15	5	0	0
1,000	1,040	104	91	78	65	53	41	30	19	9	0	0
1,040	1,080	110	96	84	71	58	46	35	24	14	4	0
1,080	1,120	117	102	89	77	64	51	39	29	18	8	0
1,120	1,160	123	109	95	82	70	57	44	33	23	13	3
1,160	1,200	130	115	101	88	75	63	50	38	27	17	7
1,200	1,240	136	122	107	93	81	68	56	43	32	21	11
1,240	1,280	142	128	114	99	86	74	61	49	37	26	16
1,280	1,320	149	134	120	106	92	79	67	54	42	31	20
1,320	1,360	155	141	126	112	98	85	72	60	47	36	25
1,360	1,400	162	147	133	118	104	91	78	65	53	41	30
1,400	1,440	169	154	139	125	110	96	84	71	58	46	35
1,440	1,480	177	160	146	131	117	102	89	77	64	51	39
1,480	1,520	184	168	152	138	123	109	95	82	70	57	44
1,520	1,560	191	175	159	144	130	115	101	88	75	63	50
1,560	1,600	198	182	166	150	136	122	107	93	81	68	56
1,600	1,640	205	189	173	157	142	128	114	99	86	74	61
1,640	1,680	213	196	180	164	149	134	120	106	92	79	67
1,680	1,720	220	204	187	171	155	141	126	112	98	85	72
1,720	1,760	227	211	195	178	162	147	133	118	104	91	78
1,760	1,800	236	218	202	186	169	154	139	125	110	96	84
1,800	1,840	245	225	209	193	177	160	146	131	117	102	89
1,840	1,880	254	234	216	200	184	168	152	138	123	109	95
1,880	1,920	263	243	223	207	191	175	159	144	130	115	101
1,920	1,960	271	252	232	214	198	182	166	150	136	122	107
1,960	2,000	280	260	241	222	205	189	173	157	142	128	114
2,000	2,040	289	269	249	230	213	196	180	164	149	134	120
2,040	2,080	298	278	258	238	220	204	187	171	155	141	126
2,080	2,120	307	287	267	247	227	211	195	178	162	147	133
2,120	2,160	316	296	276	256	236	218	202	186	169	154	139

Page 36

(Continued on next page)

FIGURE 21–3. Tables for calculating withholding tax.

Source: IRS Publication 15, Circular E, *Employee's Tax Guide.*

Tables for Percentage Method of Withholding
(For Wages Paid After December 1985)

TABLE 1—If the Payroll Period With Respect to an Employee is Weekly

(a) SINGLE person—including head of household:

If the amount of wages is: / **The amount of income tax to be withheld shall be:**

Not over $28 0

Over—	But not over—		of excess over—
$28	—$87	12%	—$28
$87	—$192	$7.08 plus 15%	—$87
$192	—$302	$22.83 plus 19%	—$192
$302	—$457	$43.73 plus 25%	—$302
$457	—$577	$82.48 plus 30%	—$457
$577	—$687	$118.48 plus 34%	—$577
$687		$155.88 plus 37%	—$687

(b) MARRIED person—

If the amount of wages is: / **The amount of income tax to be withheld shall be:**

Not over $50 0

Over—	But not over—		of excess over—
$50	—$199	12%	—$50
$199	—$398	$17.88 plus 17%	—$199
$398	—$490	$51.71 plus 22%	—$398
$490	—$600	$71.95 plus 25%	—$490
$600	—$710	$99.45 plus 28%	—$600
$710	—$930	$130.25 plus 33%	—$710
$930		$202.85 plus 37%	—$930

TABLE 2—If the Payroll Period With Respect to an Employee is Biweekly

(a) SINGLE person—including head of household:

If the amount of wages is: / **The amount of income tax to be withheld shall be:**

Not over $57 0

Over—	But not over—		of excess over—
$57	—$173	12%	—$57
$173	—$385	$13.92 plus 15%	—$173
$385	—$605	$45.72 plus 19%	—$385
$605	—$913	$87.52 plus 25%	—$605
$913	—$1,154	$164.52 plus 30%	—$913
$1,154	—$1,373	$236.82 plus 34%	—$1,154
$1,373		$311.28 plus 37%	—$1,373

(b) MARRIED person—

If the amount of wages is: / **The amount of income tax to be withheld shall be:**

Not over $100 0

Over—	But not over—		of excess over—
$100	—$398	12%	—$100
$398	—$795	$35.76 plus 17%	—$398
$795	—$980	$103.25 plus 22%	—$795
$980	—$1,199	$143.95 plus 25%	—$980
$1,199	—$1,419	$198.70 plus 28%	—$1,199
$1,419	—$1,859	$260.30 plus 33%	—$1,419
$1,859		$405.50 plus 37%	—$1,859

TABLE 3—If the Payroll Period With Respect to an Employee is Semimonthly

(a) SINGLE person—including head of household:

If the amount of wages is: / **The amount of income tax to be withheld shall be:**

Not over $62 0

Over—	But not over—		of excess over—
$62	—$188	12%	—$62
$188	—$417	$15.12 plus 15%	—$188
$417	—$655	$49.47 plus 19%	—$417
$655	—$989	$94.69 plus 25%	—$655
$989	—$1,250	$178.19 plus 30%	—$989
$1,250	—$1,488	$256.49 plus 34%	—$1,250
$1,488		$337.41 plus 37%	—$1,488

(b) MARRIED person—

If the amount of wages is: / **The amount of income tax to be withheld shall be:**

Not over $108 0

Over—	But not over—		of excess over—
$108	—$431	12%	—$108
$431	—$861	$38.76 plus 17%	—$431
$861	—$1,061	$111.86 plus 22%	—$861
$1,061	—$1,299	$155.86 plus 25%	—$1,061
$1,299	—$1,538	$215.36 plus 28%	—$1,299
$1,538	—$2,014	$282.28 plus 33%	—$1,538
$2,014		$439.36 plus 37%	—$2,014

TABLE 4—If the Payroll Period With Respect to an Employee is Monthly

(a) SINGLE person—including head of household:

If the amount of wages is: / **The amount of income tax to be withheld shall be:**

Not over $123 0

Over—	But not over—		of excess over—
$123	—$375	12%	—$123
$375	—$833	$30.24 plus 15%	—$375
$833	—$1,310	$98.94 plus 19%	—$833
$1,310	—$1,978	$189.57 plus 25%	—$1,310
$1,978	—$2,500	$356.57 plus 30%	—$1,978
$2,500	—$2,976	$513.17 plus 34%	—$2,500
$2,976		$675.01 plus 37%	—$2,976

(b) MARRIED person—

If the amount of wages is: / **The amount of income tax to be withheld shall be:**

Not over $217 0

Over—	But not over—		of excess over—
$217	—$863	12%	—$217
$863	—$1,723	$77.52 plus 17%	—$863
$1,723	—$2,123	$223.72 plus 22%	—$1,723
$2,123	—$2,598	$311.72 plus 25%	—$2,123
$2,598	—$3,075	$430.47 plus 28%	—$2,598
$3,075	—$4,028	$564.03 plus 33%	—$3,075
$4,028		$878.52 plus 37%	—$4,028

Page 20

FIGURE 21–4. Tables for percentage method of withholding income taxes.

Source: IRS Publication 15, Circular E, *Employee's Tax Guide.*

Recall that Walt's wages were expected to be $18,000 in 1986 ($1,500 per month). Consulting Figure 21–3, Walt's employer should withhold $82 from his monthly paycheck. Walt's $1,500 monthly gross falls between $1,480 and $1,520 and he claims seven withholding allowances. Furthermore, Walt's employer should withhold $107.25 in Social Security taxes from his paycheck each month.

So, for each month of 1986, Walt's net paycheck is $1,310.75:

$1,500.00	gross wages
−82.00	income tax withholding
−107.25	Social Security tax
$1,310.75	net wages

Don't forget that Walt's employer must match Walt's $107.25 Social Security contribution.

Payment of Withholding Taxes

Eventually the income and Social Security taxes withheld by an employer must be paid. Generally, these taxes are deposited with an IRS-authorized financial institution, which includes any bank with National or N.A. in its name and many other banks, savings and loan associations, and credit unions. When these payments are due depends upon the total amount of taxes withheld from all employees during specific time periods.

A summary of the tax deposit rules is shown in Figure 21–5. Taxes are deposited with a financial institution by using one of the coupons supplied by IRS (Figure 21–6). The excerpt from IRS Publication 51, Circular A in Box 21–1 provides further explanation of these rules and a few examples.

IRS Form 943

The tax withholding year is closed with IRS Form 943 (Figure 21–7). This form summarizes all of the wages paid during the year to all employees (Line 2), FICA (Social Security) and income taxes withheld during the year (Lines 3 and 6), and taxes deposited (Line 10). Form 943, which must be filed by January 31 following the year for which the tax return is filed, consequently computes the amount of taxes still due after the close of the tax year.[1] The amount of taxes reported on Line 10 as being deposited during the year must be broken down by month on the bottom half of Form 943.

Forms W-2 and W-3

The final step of the tax withholding process is the completion and distribution of Forms W-2 and W-3. Form W-2 (Figures 21–8) is a multiple-copy form which

[1] If *tax* liability is $3,000 or more during any month, Form 943A should also be completed. Although used by some farmers and ranchers, this tax liability is generated by a gross payroll in the neighborhood of $15,000 *per month*. See IRS Publication 51, Circular A.

payments, or file tax returns. Such an employer may be required to deposit taxes in a special trust account for the U.S. Government and file monthly tax returns on **Form 941-M**, Employer's Monthly Federal Tax Return. Employers are required to file monthly when written notification is received from the District Director. Penalties are provided for failure to deposit taxes and to make payments.

The amount of taxes determines the frequency of deposits. You owe these taxes when you pay the wages, not when your payroll period ends. The rules and examples below explain how often to deposit taxes.

Note: *If any date shown falls on a Saturday, Sunday, or legal holiday, use the next regular workday for depositing taxes.*

Depositing Withheld Income and Social Security Taxes.—

Rule 1—Less than $500 at end of quarter.—If at the end of the calendar quarter your total undeposited taxes for the quarter are less than $500, you do not have to deposit the taxes. You may pay the taxes to IRS with Form 941 (or 941E), or you may deposit them by the end of the next month.

Rule 2—Less than $500 at the end of any month.—If at the end of any month your total undeposited taxes are less than $500, you do not have to make a deposit. You may carry the taxes over to the following month within the quarter. (See Example A.)

Rule 3—$500 or more but less than $3,000 at the end of any month.—If at the end of any month your total undeposited taxes are $500 or more but less than $3,000, you must deposit the taxes within 15 days after the end of the month.

Exception: If this occurs at the end of a month in which you made a deposit of $3,000 or more, see Examples D and E.

Rule 4—$3,000 or more at the end of any eighth-monthly period.—Each month is divided into eight deposit periods that end on the 3rd, 7th, 11th, 15th, 19th, 22nd, 25th, and last day of the month. If at the end of any eighth-monthly period your total undeposited taxes are $3,000 or more, deposit the taxes within 3 banking days after the end of the eighth-monthly period. (See Examples B, C, D, and E.) Do not count as banking days local holidays observed by authorized financial institutions, Saturdays, Sundays, and legal holidays.

You will be considered to meet Rule 4 if:

- You deposit at least 95% of the tax liability for the eighth-monthly period within 3 banking days after the end of the period; and
- You deposit any underpayment as follows:

(a) If the eighth-monthly period is in the first or second month of the quarter, you deposit the underpayment with the first deposit that is required to be made after the 15th of the following month.

(b) If the eighth-monthly period is in the last month of the quarter, you deposit any underpayment of $500 or more by the due date of the return. (Any underpayment less than $500 can be paid with Form 941 (or 941E) as explained in rule 1.)

Exception: If this is the first time you are required to make a deposit within 3 banking days after the end of an eighth-monthly period, you may deposit the taxes by the 15th of the next month (instead of within 3 banking days after the eighth-monthly period) if you meet all of the following conditions:

- You were not required to deposit taxes under rule 4 during the 4 quarters preceding the current quarter.
- You were not required to deposit taxes under rule 4 during earlier months of this quarter.
- Your total undeposited taxes at the end of any eighth-monthly period during this month are less than $10,000.

Example A.—The taxes on wages paid in October are $450, and the taxes on wages paid in November are $550. No deposit is required for October (because of rule 2), but add the $450 to the $550 for November and deposit the total ($1,000) by December 15 (as required by rule 3).

Example B.—The taxes on wages paid from the 1st through the 3rd of the month are $3,500. Deposit these taxes within 3 banking days after the 3rd of the month (as required by rule 4).

Example C.—The taxes on wages paid from the 4th through the 7th of a month are $2,500 and the taxes on wages paid from the 8th through the 11th are $2,000. A separate deposit is not required for the $2,500, but add it to the $2,000 and deposit the total ($4,500) within 3 banking days after the 11th of the month (as required by rule 4).

Example D.—The taxes on wages paid from the 23rd through the 25th of January are $3,500. Deposit these taxes within 3 banking days after the 25th of the month (as required by rule 4). The taxes on wages paid from the 26th through the end of January are $2,500. Since a deposit was already made for an eighth-monthly period during the month, a separate deposit is not required for the $2,500 (because of the exception to rule 3). Carry the $2,500 over and add it to the taxes on wages paid in February. Then follow rules 3 and 4 to determine when the next deposit is required. (However, if this occurs in the last month of a quarter, deposit any balance due of $500 or more but less than $3,000 by the end of the next month.)

Example E.—Wages are paid on Friday for the prior week. In the first quarter of 1986, paydays are on January 3, 10, 17, 24, and 31; February 7, 14, 21, and 28; and March 7, 14, 21, and 28. Taxes are $1,050 for each payday.

Enter $1,050 on lines A, C, E, G, H, J, L, N, P, R, T, V, and X of Form 941. Enter $5,250 on Total line I; $4,200 on lines II and III; and $13,650 on line IV.

You must make a deposit within 3 banking days of January 19, February 7, February 28, and March 22 because undeposited taxes reached at least $3,000 for the eighth-monthly periods ending on those dates.

The undeposited taxes at the end of January of $2,100 (for paydays on the 24th and 31st) may be carried over to February because of the exception to rule 3 as explained in Example D. At the end of February, there is no carryover since the undeposited taxes are $3,150 and must be deposited within 3 banking days. At the end of March, the undeposited taxes of $1,050 (for the payday on the 28th) must be deposited by April 30 (because of the exception to rule 3 for the last month of a quarter as explained in Example D).

Depositing Backup Withholding.—For tax deposit purposes, you can either combine backup withholding with other taxes reported on Form 941 or 941E and deposit the combined total, or you can treat backup withholding as a separate tax and deposit it separately following the same deposit rules used for social security and withheld income taxes.

If you treat backup withholding as a separate tax, show the amounts for deposit purposes on **Schedule A (Form 941)**, Record of Federal Backup Withholding Tax Liability, and when depositing this tax, mark the "Sch. A" entry box on the deposit coupon. Schedule A (Form 941) must be attached to Form 941 or 941E.

State and local government employers should deposit withheld income tax only. Do not deposit contributions payable under a section 218 Federal-State social security coverage agreement. Send any social security contributions and earnings reports to the appropriate state agency.

Summary of Deposit Rules for Social Security Taxes and Withheld Income Tax

Deposit Rule	Deposit Due
(1) If at the end of the quarter your total undeposited taxes for the quarter are less than $500:	(1) No deposit is required. You may pay the taxes to IRS with Form 941 (or 941E), or you may deposit them by the due date of the return.
(2) If at the end of any month your total undeposited taxes are less than $500:	(2) No deposit is required. You may carry the taxes over to the following month.
(3) If at the end of any month your total undeposited taxes are $500 or more but less than $3,000:	(3) Within 15 days after the end of the month. (No deposit is required if you made a deposit for an eighth-monthly period during the month under rule 4. However, if this occurs in the last month of the quarter, deposit any balance due by the due date of the return.)
(4) If at the end of any eighth-monthly period (the 3rd, 7th, 11th, 15th, 19th, 22nd, 25th, and last day of each month) your total undeposited taxes are $3,000 or more:	(4) Within 3 banking days after the end of the eighth-monthly period.

Depositing Federal Unemployment (FUTA) Taxes.—

For deposit purposes, figure FUTA tax quarterly. Deposit any amount due by the last day of the first month after the quarter ends.

Determine whether you must deposit tax for any of the first 3 quarters in a year. Figure the total tax by multiplying by .008 that part of the first $7,000 of each employee's annual wages that you paid during the quarter.

If this amount (plus any amount not yet deposited for any earlier quarter) is more than $100, deposit it during the first month after the quarter. But if it is $100 or less, you do not have to deposit it. Just add it to the amount for deposit the next quarter.

If the tax reportable on Form 940 (including any credit reductions under Code section 3302(c)), minus amounts deposited for the year, is more than $100, deposit all of the tax by January 31. If your tax for the year (minus deposits) is $100 or less, you may either deposit it or pay it with Form 940 by January 31.

Page 8

FIGURE 21–5. Deposit rules for tax withholdings.

Source: IRS Publication 51, Circular A, *Agricultural Employer's Tax Guide.*

AMOUNT OF DEPOSIT (Do NOT type; please print.)

DOLLARS	CENTS

TAX YEAR MONTH →

EMPLOYER IDENTIFICATION NUMBER →

Darken only one TYPE OF TAX		Darken only one TAX PERIOD
941	Sch. A	1st Quarter
940	1120	2nd Quarter
943	990-T	3rd Quarter
720	990-C	4th Quarter
CT-1	1042	

BANK NAME/ DATE STAMP

Name __________

Address __________

City __________

State __________ ZIP __________

IRS USE ONLY

66

Telephone number ( ) __________

FOR BANK USE IN MICR ENCODING

Federal Tax Deposit Coupon
Form **8109-B** (Rev. 11-85)

↑ SEPARATE ALONG THIS LINE AND SUBMIT TO DEPOSITARY WITH PAYMENT ↑

OMB NO. 1545-0257

IMPORTANT

Read instructions carefully before completing Form 8109-B Federal tax deposit coupons.

Note: The entries other than in the name and address section and the telephone number are processed by optical scanning equipment and must be completed by hand in the manner specified. The name and address section and the telephone number may be completed other than by hand. Do not use photocopies to make your deposits.

Paperwork Reduction Act Notice.—We ask for this information to carry out the Internal Revenue laws of the United States. We need it to ensure that taxpayers are complying with these laws and to allow us to figure and collect the right amount of tax. You are required to give us this information.

Purpose of Form.—Use Form 8109-B deposit coupons to make tax deposits **only** in the following two situations: (1) You have reordered your preprinted deposit coupons **(Form 8109)** but you have not yet received them; or (2) you are a new entity and have already been assigned an employer identification number (EIN), but have not yet received your initial supply of preprinted deposit coupons (Form 8109). **Note:** *You should get your reordered coupons (or your initial supply) within 5-6 weeks of the reorder (or receipt of your EIN). If you do not, please contact your local IRS office.*

If you have applied for an EIN, have not received it, and a deposit must be made, make the deposit with your Internal Revenue Service Center. Make it payable to IRS and show on it your name (as shown on **Form SS-4**, Application for Employer Identification Number), address, kind of tax, period covered, and date you applied for an EIN. Do not use Form 8109-B in this situation.

How To Complete Form 8109-B.—Enter your name exactly as shown on your return or other IRS correspondence, address, and employer identification number in the spaces provided. If you are required to file Form 1120, 990-C, or 990-T, enter the month in which your tax year ends in the "TAX YEAR MONTH" boxes. For example, if your tax year ends in January, enter 01; if it ends in June, enter 06; if it ends in December, enter 12. Please make your entries for employer identification number and tax year month (if applicable) in the manner specified in *Amount of Deposit* below. Darken one box each for "Type of Tax" and "Tax Period" as explained below.

Amount of Deposit.—

Enter the amount of the deposit in the space provided.

DO enter amount legibly, forming the characters as shown below:

1	2	3	4	5	6	7	8	9	0

DO handprint money amounts without using dollar signs, commas, decimal points, or leading zeros. The commas and decimal point are already shown in the entry area.

For example, a deposit of $7,635.22 would be entered like this:

DOLLARS								CENTS	
				7	6	3	5	2	2

If the deposit is for whole dollars only, enter "00" in the "CENTS" boxes.

Types of Tax.—

Form 941	—	Withheld Income and Social Security (FICA) Taxes. (Includes Form 941 series of returns.)
Form 940	—	Federal Unemployment (FUTA) Tax. (Includes Form 940PR.)
Form 943	—	Agricultural Withheld Income and Social Security (FICA) Taxes. (Includes Form 943PR.)
Form 720	—	Excise Tax.
Form CT-1	—	Railroad Retirement Tax.
Schedule A	—	Backup Withholding. (Reported on Forms 941 and 941E.)
Form 1120	—	Corporation Income Tax. (Includes Form 1120 series of returns, except Form 1120-FSC.)
Form 990-T	—	Exempt Organization Business Income Tax.
Form 990-C	—	Farmers' Cooperative Association Income Tax.
Form 1042	—	Withholding at Source.

How To Determine the Proper Tax Period.—

(a) Payroll Taxes and Withholding:
(Forms 941, Schedule A, 940, 943, 1042, and CT-1)

If your liability was incurred during:

- January—March, darken the 1st quarter box
- April—June, darken the 2nd quarter box
- July—September, darken the 3rd quarter box
- October—December, darken the 4th quarter box

Note: *If the liability was incurred during one quarter and deposited in another quarter, darken the box for the quarter in which the tax liability was incurred. For example: If the liability was incurred in March and deposited in April, darken the 1st quarter box.*

(b) Excise Taxes (Form 720):

Generally, follow the instructions listed under (a) above. For exceptions, see the Form 720 instructions.

(Continued on back of page.)

Department of the Treasury
Internal Revenue Service

*U.S. GOVERNMENT PRINTING OFFICE: 1985—496-431

Form 8109-B (Rev. 11-85)

FIGURE 21–6. Federal Tax Deposit coupon, IRS Form 8109-B.

Box 21–1

From IRS Publication 51, Circular A

1. Less than $500 at end of December.—If at the end of December your total undeposited taxes for the year are less then $500, you do not have to deposit the taxes. You may pay the taxes to IRS with Form 943, or you may deposit them by January 31.

2. Less than $500 at the end of any month except December.—If at the end of any month except December your total undeposited taxes are less than $500, you do not have to make a deposit. Carry the taxes over to the following month. (See Example A.)

3. $500 or more but less than $3,000 at the end of any month.—If at the end of any month your total undeposited taxes are $500 or more but less than $3,000, you must deposit the taxes within 15 days after the end of the month. Exception: If your undeposited taxes are $500 or more but less than $3,000 at the end of a month during which you already made a deposit of $3,000 or more in an earlier period of the month (because of rule 4 below), you do not have to deposit them. Carry them over into the next month instead. (See Example D.)

4. $3,000 or more at the end of any eighth-monthly period.—Each month is divided into eight deposit periods that end on the 3rd, 7th, 11th, 15th, 19th, 22nd, 25th, and last day of the month. If at the end of any eight-monthly period your total undeposited taxes are $3,000 or more, deposit the taxes within 3 banking days after the end of the eighth-monthly period. (See Examples B, C, D, and E.) Do not count as banking days local holidays observed by authorized financial institutions, Saturdays, Sundays, and legal holidays.

You will be considered to meet rule 4 if:

You deposit at least 95% of the tax liability for the eighth-monthly period within 3 banking days after the end of the period, and

You deposit any underpayment as follows:

(a) If the eighth-monthly period is in a month other than December, you deposit the underpayment with the first deposit that is required to be made after the 15th of the following month.

(b) If the eighth-monthly period is in December, you deposit any underpayment of $500 or more by January 31. (Any underpayment less than $500 can be paid with Form 943 as explained in rule 1.)

Example A.—The taxes on wages paid in October are $450, and the taxes on wages paid in November are $550. No deposit is required for October (because of rule 2), but add the $450 to the $550 for November and deposit the total ($1,000) by December 15 (as required by rule 3).

Example B.—The taxes on wages paid from the 1st through the 3rd of a month are $3,500. Deposit these taxes within 3 banking days after the 3rd of the month (as required by rule 4).

Example C.—The taxes on wages paid from the 4th through the 7th of a month are $2,500 and the taxes on wages paid from the 8th through the 11th are $2,000 and deposit the total ($4,500) within 3 banking days after the 11th of the month (as required by rule 4).

Example D.—The taxes on wages paid from the 23rd through the 25th of a month are $3,500. Deposit these taxes within 3 banking days after the 25th of the month (as required by rule 4). The taxes on wages paid from the 26th through the end of the month are $2,500. Since a deposit was already made for an eighth-monthly period during the month, a separate deposit is not required for the $2,500 (because of the exception to rule 3). Carry the $2,500 over and add it to the taxes on wages paid in the next month. Then follow rules 1 through 4 to determine when the next deposit is required. (However, if this occurs in December, deposit any balance due of $500 or more but less than $3,000 by January 31.)

Example E.—Wages are paid on Monday for the prior week. In the first quarter of 1986 paydays are on January 3, 10, 17, 24, and 31; February 7, 14, 21, and 28; and March 7, 14, 21, and 28. Taxes are $1,050 for each payday.

Use Form 943A and enter $1,050 on lines A, C, E, G, H, J, L, N, P, R, T, V, and X. Enter $5,250 on Total line I; $4,200 on lines II and III; and $13,650 on line IV.

You must make a deposit within 3 banking days of January 19, February 7, February 28, and March 22 because undeposited taxes reached at least $3,000 for the eighth-monthly periods ending on those dates.

The undeposited taxes at the end of January of $2,100 (for paydays on the 24th and 31st) may be carried over to February because of the exception to rule 3 as explained in Example D. At the end of February, there is no carryover since the undeposited taxes are $3,150 and must be deposited within 3 banking days. At the end of March, the undeposited taxes of $1,050 (for the payday of the 28th) may be carried over to April because of the exception to rule 3 as explained in Example D. However, if this occurs at the end of December, the undeposited taxes of $1,050 must be deposited by January 31 (because of the exception to rule 3 for December as explained in Example D.)

Form **943** Department of the Treasury Internal Revenue Service

Employer's Annual Tax Return for Agricultural Employees
(For Social Security Taxes and Withheld Income Tax)

OMB No. 1545-0035
1985

T	
FF	
FD	
FP	
I	
T	

Your name, address, employer identification number, and calendar year of return. (If not correct, please change.)

Name (as distinguished from trade name) — Calendar Year 1985
Trade name, if any — Employer identification number
Address and ZIP code

If address is different from prior return, check here ▶ ☐

If you are not liable for returns in the future, write "FINAL RETURN" here ▶

1a	Number of agricultural employees employed in the pay period that includes March 12, 1985 ▶	1a	
b	If you are a subsidiary corporation AND your parent corporation files a consolidated Form 1120, enter parent corporation's employer identification number (EIN) ▶	1b	
2	Taxable cash wages paid during the year (see instructions on page 2)	2	
3	Social security taxes (multiply line 2 by 14.1%)	3	
4	Adjustment of social security taxes (attach statement—see instructions on page 2)	4	
5	Social security taxes as adjusted	5	
6	Federal income tax withheld (see instructions on page 2)	6	
7	Total taxes (add lines 5 and 6)	7	
8	Advance earned income credit (EIC) payments, if any (see instructions on page 4)	8	
9	Net taxes (subtract line 8 from line 7)	9	
10	Total amount deposited for 1985, including any overpayment from 1984, as shown in your records	10	
11	Undeposited taxes due (subtract line 10 from line 9—this should be less than $500). Pay to Internal Revenue Service ▶	11	

12 If line 10 is more than line 9, enter overpayment here ▶ $ and check if to be: ☐ Applied to next return, or ☐ Refunded.

Check if you are a first-time 3-banking-day depositor (see deposit rules on page 4) ▶ ☐

Record of Federal Tax Liability

(See deposit rules on page 4)

If your tax liability for any month in 1985 is $3,000 or more, do not use the schedule below. Instead, attach **Form 943A**, Agricultural Employer's Record of Federal Tax Liability, or a statement showing your tax liability for each eighth-monthly period during which you had a payday. Copies of Form 943A are in the **1985 Circular A**, Agricultural Employer's Tax Guide, and additional copies can be obtained at IRS offices.

Deposit period ending		Tax liability for month
A January 31	A	
B February 28	B	
C March 31	C	
D April 30	D	
E May 31	E	
F June 30	F	
G July 31	G	
H August 31	H	
I September 30	I	
J October 31	J	
K November 30	K	
L December 31	L	
M Total liability for year (add lines A through L)	M	

Instructions

If your taxes for the year (line 9) are less than $500, you do not have to complete this record. You may pay the taxes with Form 943 or deposit them by January 31, 1986.

If your taxes for any month are $500 or more but less than $3,000, show your tax liability next to the month. (If your tax liability for any month is $3,000 or more, attach Form 943A or a statement showing your tax liability for each eighth-monthly period during which you had a payday. See rule (4) under "Deposit Rules" on page 4.)

The total tax liability for the year (line M) should equal net taxes (line 9).

Under penalties of perjury, I declare that I have examined this return, including accompanying schedules and statements, and to the best of my knowledge and belief it is true, correct, and complete.

Date ▶ **Signature ▶** **Title (Owner, etc.) ▶**

Please file this form with your Internal Revenue Service Center (see instructions on "Where to File").

For Paperwork Reduction Act Notice, see page 2. Form **943** (1985)

FIGURE 21–7. IRS Form 943.

1 Control number	22222	For Paperwork Reduction Act Notice, see back of Copy D. OMB No. 1545-0008	For Official Use Only
2 Employer's name, address, and ZIP code		3 Employer's identification number	4 Employer's State number
		5 Statutory employee ☐ Deceased ☐ Legal rep. ☐ 942 emp. ☐ Subtotal ☐ Void ☐	
		6 Allocated tips	7 Advance EIC payment
8 Employee's social security number	9 Federal income tax withheld	10 Wages, tips, other compensation	11 Social security tax withheld
12 Employee's name (first, middle, last)		13 Social security wages	14 Social security tips
		16 *	
		17 State income tax / 18 State wages, tips, etc.	19 Name of State
15 Employee's address and ZIP code		20 Local income tax / 21 Local wages, tips, etc.	22 Name of locality

Form **W-2 Wage and Tax Statement 1985**

Copy A For Social Security Administration
* See Instructions for Forms W-2 and W-2P

Department of the Treasury
Internal Revenue Service

Do NOT CUT or Separate Forms on This Page

1 Control number	22222	For Paperwork Reduction Act Notice, see back of Copy D. OMB No. 1545-0008	For Official Use Only
2 Employer's name, address, and ZIP code		3 Employer's identification number	4 Employer's State number
		5 Statutory employee ☐ Deceased ☐ Legal rep. ☐ 942 emp. ☐ Subtotal ☐ Void ☐	
		6 Allocated tips	7 Advance EIC payment
8 Employee's social security number	9 Federal income tax withheld	10 Wages, tips, other compensation	11 Social security tax withheld
12 Employee's name (first, middle, last)		13 Social security wages	14 Social security tips
		16 *	
		17 State income tax / 18 State wages, tips, etc.	19 Name of State
15 Employee's address and ZIP code		20 Local income tax / 21 Local wages, tips, etc.	22 Name of locality

Form **W-2 Wage and Tax Statement 1985**

Copy A For Social Security Administration
* See Instructions for Forms W-2 and W-2P

Department of the Treasury
Internal Revenue Service

Do NOT CUT or Separate Forms on This Page

1 Control number	22222	For Paperwork Reduction Act Notice, see back of Copy D. OMB No. 1545-0008	For Official Use Only
2 Employer's name, address, and ZIP code		3 Employer's identification number	4 Employer's State number
		5 Statutory employee ☐ Deceased ☐ Legal rep. ☐ 942 emp. ☐ Subtotal ☐ Void ☐	
		6 Allocated tips	7 Advance EIC payment.
8 Employee's social security number	9 Federal income tax withheld	10 Wages, tips, other compensation	11 Social security tax withheld
12 Employee's name (first, middle, last)		13 Social security wages	14 Social security tips
		16 *	
		17 State income tax / 18 State wages, tips, etc.	19 Name of State
15 Employee's address and ZIP code		20 Local income tax / 21 Local wages, tips, etc.	22 Name of locality

Form **W-2 Wage and Tax Statement 1985**

Copy A For Social Security Administration
* See Instructions for Forms W-2 and W-2P

Department of the Treasury
Internal Revenue Service

FIGURE 21–8. Social Security Administration copy of IRS Form W-2.

summarizes the gross wages paid, FICA taxes withheld, federal income taxes withheld, and state income taxes withheld for *each* employee. The cover form, Copy A is filed not with the IRS, but with the Social Security Administration (SSA). The last copy, Copy D, is maintained by the employer. All remaining copies must be given to the employee. Copies of the W-2 statement must be given to employees no later than January 31 and to the SSA by February 29.

The copies of W-2s filed with SSA must be accompanied by Form W-3 (Figure 21–9). Form W-3 simply adds all of the W-2s and reports total wages and total taxes withheld. Form W-3 actually just restates the information provided already on Form 943. At first glance that seems repetitive, but remember that Form 943 is sent to IRS and Form W-3 is sent to SSA.

Estimated Tax

The pay-as-you-go tax system works fairly smoothly for employees (although it obviously generates quite a bit of paperwork for the employer), but self-employed individuals have no employer to pay into the system for them. This distinction, however, doesn't necessarily mean that the IRS does not want their money.

Self-employed people must look to the tax estimate as their alternative to withholding. Although it was stated earlier in this chapter that farmers and ranchers are exempt from many of the estimate rules, we shall take a quick look at the estimate reporting process and then specify the rules that apply to farmers and ranchers.

IRS Form 1040-ES

Self-employed persons and others who have substantial income from sources other than wages subject to withholding should obtain a copy of Form 1040-ES at the *beginning* of the tax year. After reviewing the specific instructions on Form 1040-ES, page 1 (Figure 21–10[a]), the taxpayer should estimate taxable income from all sources (AGI). This process is oversimplified, at least in appearance, by Form 1040-ES. A projected income statement and/or cash flow budget (see Chapters 14 and 16) should be developed before this tax form is attempted. These background forms are necessary prior to working through Form 1040-ES because the tax worksheet shown in Figure 21–10(c) begins with AGI.

Once AGI has been estimated, the remainder of the estimated tax worksheet can be completed fairly easily. Current year tax rates (Figure 21–10[d]) are always provided in the Form 1040-ES package.

The estimated tax determined on page 3, Line 16 is payable in four equal quarterly installments, due on April 15, June 15 and September 15 of the tax year and January of the following year.[2] Each of these payments must be accompanied

[2]These dates assume a calendar-year records system. If some other fiscal year is used by the business for filing taxes, due dates are the 15th day of the 4th, 6th, and 9th months of the fiscal year and the 1st month of the following fiscal year.

Please do not staple

1 Control number	33333	For Official Use Only OMB No. 1545-0008

Kind of Payer and Tax Statements Transmitted ☐	2 941/941E ☐ Military ☐ 943 ☐ CT-1 ☐ 942 ☐ Medicare Fed. emp. ☐	3 W-2 ☐ W-2P ☐	4	5 Number of statements attached

6 Allocated tips	7 Advance EIC payments	8
9 Federal income tax withheld	10 Wages, tips, and other compensation	11 Social security tax withheld
12 Employer's State number	13 Social security wages	14 Social security tips

Place Label Here

15 Employer's identification number	16 Establishment number
17 Employer's name	18 Gross annuity, pension, etc. (Form W-2P)
	20 Taxable amount (Form W-2P)
19 Employer's address and ZIP code (If available, place label over boxes 15, 17, and 19.)	21 Income tax withheld by third-party payer

Under penalties of perjury, I declare that I have examined this return and accompanying documents, and to the best of my knowledge and belief they are true, correct, and complete. In the case of documents without recipients' identifying numbers, I have complied with the requirements of the law in attempting to secure such numbers from the recipients.

Signature ▶ Title ▶ Date ▶

Form **W-3 Transmittal of Income and Tax Statements 1985** Department of the Treasury Internal Revenue Service

Please return this entire page with the accompanying Forms W-2 or W-2P to the Social Security Administration address for your State as listed below.

If your legal residence, principal place of business, office or agency is located in ▼	Use this address ▼
Alaska, Arizona, California, Colorado, Hawaii, Idaho, Iowa, Minnesota, Missouri, Montana, Nebraska, Nevada, North Dakota, Oregon, South Dakota, Utah, Washington, Wisconsin, Wyoming	Social Security Administration Salinas Data Operations Center Salinas, CA 93911
Alabama, Arkansas, Florida, Georgia, Illinois, Kansas, Louisiana, Mississippi, New Mexico, Oklahoma, South Carolina, Tennessee, Texas	Social Security Administration Albuquerque Data Operations Center Albuquerque, NM 87180
Connecticut, Delaware, District of Columbia, Indiana, Kentucky, Maine, Maryland, Massachusetts, Michigan, New Hampshire, New Jersey, New York, North Carolina, Ohio, Pennsylvania, Rhode Island, Vermont, Virginia, West Virginia	Social Security Administration Wilkes-Barre Data Operations Center Wilkes-Barre, PA 18769
If you have no legal residence or principal place of business in any State	Social Security Administration Wilkes-Barre Data Operations Center Wilkes-Barre, PA 18769

Paperwork Reduction Act Notice.—We ask for this information to carry out the Internal Revenue laws of the United States. We need it to ensure that taxpayers are complying with these laws and to allow us to figure and collect the right amount of tax. You are required to give us this information.

FIGURE 21–9. IRS Form W-3.

Form **1040-ES**

Department of the Treasury
Internal Revenue Service

Estimated Tax for Individuals

▶ **This form is primarily for first-time filers of estimated tax.**

OMB No. 1545-0087

1986

Instructions

Paperwork Reduction Act Notice.—We ask for this information to carry out the Internal Revenue laws of the United States. We need it to ensure that taxpayers are complying with these laws and to allow us to figure and collect the right amount of tax. You are required to give us this information.

Purpose of Form.—Use this form to figure and pay your estimated tax.

Estimated tax is the amount of tax you expect to owe for the year after subtracting the amount of tax you expect to have withheld and the amount of any credits you plan to take.

Generally, you do not have to pay estimated tax if your 1986 income tax return will show (1) a tax refund, or (2) a tax balance due of less than $500. Also, you do not have to pay estimated tax if you were a U.S. citizen or resident and you had no tax liability for the full 12-month preceding tax year.

This form is primarily for first-time filers. After your first payment-voucher is received in the Internal Revenue Service Center, IRS will mail you a 1040-ES package. Your name, address, and social security number will be preprinted on the vouchers. You should use these vouchers in making the **remaining** payments of estimated tax for the year. Using the preprinted vouchers will speed processing, reduce the chance of error, and help save processing costs.

This form can also be used if you did not receive a 1040-ES package, or if you lost it. Complete the appropriate payment-voucher and mail it with your payment to your Internal Revenue Service Center (see page 2 for the address).

A. Who Must Make Estimated Tax Payments

The rules below are for U.S. citizens or residents and for res dents of Puerto Rjco, Virgin Islands, Guam, or American Samoa. (If you are a nonresident alien, use **Form 1040-ES (NR).**)

Generally, you must pay estimated tax if you expect to owe, after subtracting your withholding and crecits, at least $500 in tax for 1986; and you expect your withholding to be less than:

1. 80% of the tax shown on your 1986 tax return, or

2. 100% of the tax shown on your 1985 tax return (assuming the return covered all 12 months).

Note: *If you must make estimated tax payments and receive salaries and wages, you may not be having enough tax withheld during the year. To avoid making estimated tax payments, consider asking your employer to take more tax out of your earnings. To do this, file a new* **Form W-4,** *Employee's Withholding Allowance Certificate, with your employer and make sure you will not owe $500 or more in tax.*

B. How To Figure Your Estimated Tax

Use the Estimated Tax Worksheet on page 3, the 1986 Tax Rate Schedules on page 4, and your 1985 tax return as a guide for figuring your estimated tax.

Note: *You may not make joint estimated tax payments if: (1) either you or your spouse is a nonresident alien, (2) you are separated under a decree of divorce or separate maintenance, or (3) you have different tax years.*

Most of the items on the worksheet are self-explanatory. However, the instructions below provide additional information for filling out certain lines.

Caution: Generally, you are required to itemize your deductions if:

- you have unearned income of $1,080 or more and can be claimed as a dependent on your parent's return;
- you are married filing a separate return and your spouse itemizes deductions;
- you file **Form 4563,** Exclusion of Income From Sources in United States Possessions; **OR**
- you are a dual-status alien.

For more information, see the 1985 Instructions for **Form 1040.** If you **MUST** itemize and line 2b of the Estimated Tax Worksheet is more than line 2a, subtract 2a from 2b. Add this amount to line 1 of the worksheet and enter the total on line 3. Disregard the instructions for lines 2c, 2d, and 3 on the worksheet.

The following tax law changes for 1986 may affect your 1986 estimated tax.

1. Deduction for charitable contributions. The amount you can deduct as a charitable contribution if you do not itemize deductions has been increased. You can deduct all of your allowable contributions to qualified organizations, subject to limitations based on your adjusted gross income.

2. Tax indexing. The Tax Rate Schedules and the personal exemption deduction have been adjusted to correspond to the increase in the Consumer Price Index. Also, these adjustments affect the Tax Table, the zero bracket amount, and the return filing requirements.

For more details, see **Publication 505,** Tax Withholding and Estimated Tax, and **Publication 553,** Highlights of 1985 Tax Changes.

Line 7—Additional taxes.—Enter on line 7 any additional taxes from:

- **Form 4970,** Tax on Accumulation Distribution of Trusts;
- **Form 4972,** Special 10-Year Averaging Method; **OR**
- **Form 5544,** Multiple Recipient Special 10-Year Averaging Method.

Line 9—Credits.—For details on credits you may take, see Form 1040, lines 41 through 49, and the related instructions. Also, include any overpaid windfall profit tax attributable to amounts withheld for oil removal in 1986.

Line 11—Self-employment tax.—If you and your spouse make joint estimated tax payments and both have self-employment income, figure the estimated self-employment tax separately. Enter the total amount on line 11.

Line 12—Other taxes.—Enter on line 12 any other taxes from:

- Recapture of investment credit;
- Premature distribution from an IRA;
- Section 72 penalty taxes;
- Advance earned income credit payments;
- Alternative minimum tax; **AND**
- Excise tax on golden parachute payment.

C. How To Use the Payment-Voucher

Each payment-voucher has the date when the voucher is due for calendar year taxpayers. Please use the correct voucher.

(1) Enter your name, address, and social security number in the space provided on the payment-voucher. If you are filing a joint payment-voucher, your spouse's name and social security number should be included on the voucher. If you file a joint payment-voucher and have different last names, please separate them with an "and." For example, "John Brown and Mary Smith."

(2) Enter the net amount of your payment on line 1 of the payment-voucher. If you paid too much tax on your 1985 Form 1040, you may have chosen to apply the overpayment to your estimated tax for 1986. If so, you may apply all or part of the overpayment to any voucher. Send the payment-voucher to IRS **ONLY** when you are making a payment.

(3) Tear off the voucher at the perforation.

(4) Attach, but do not staple, your check or money order to the payment-voucher. Make check or money order payable to Internal Revenue Service. Please write your social security number and "**1986 Form 1040-ES**" on your check or money order. **Do not** include any balance due on your Form 1040 with your check for estimated tax. Please fill in the Record of Estimated Tax Payments on page 2 so you will have a record of your payments.

(5) Mail your payment-voucher to the Internal Revenue Service Center for the place where you live. Use the address for your state shown on page 2.

D. When To Pay Your Estimated Tax

The general rule is that you must make your first estimated tax payment by April 15, 1986. You may either pay all of your estimated tax at that time or pay in four equal amounts that are due by April 15, 1986; June 16, 1986; September 15, 1986; and January 15, 1987. Exceptions to the general rule are listed below.

(1) Other payment dates.—In some cases, such as a change in income, you may have to make your first estimated tax payment after April 15, 1986. The payment dates are as follows:

If the requirement is met after:	Payment date is:
• March 31 and before June 1	June 16, 1986
• May 31 and before Sept. 1	Sept. 15, 1986
• August 31	Jan. 15, 1987

Note: *You may use the "Amended Estimated Tax Schedule" on page 2 to figure your amended estimated tax.*

If the first payment you are required to make is due:

- June 16, 1986, enter ½;
- September 15, 1986, enter ¾;
- January 15, 1987, enter all;

of line 16 (less any 1985 overpayment applied to this installment) on line 17 of the worksheet and on line 1 of the payment-voucher.

If you file your 1986 Form 1040 by February 2, 1987, and pay the entire balance due, then you do not have to make the payment which would otherwise be due January 15, 1987.

(Continued on page 2)

FIGURE 21–10(a). IRS Form 1040-ES, page 1.

(2) Farmers and fishermen.—If at least two-thirds of your gross income for 1985 or 1986 is from farming or fishing, you may do one of the following:

- Pay all of your estimated tax by January 15, 1987; **OR**
- File Form 1040 for 1986 by March 2, 1987, and pay the total tax due. In this case, you do not need to make estimated tax payments for 1986.

(3) Fiscal year.—If your return is on a fiscal year basis, your due dates are the 15th day of the 4th, 6th, and 9th months of your fiscal year and the 1st month of the following fiscal year. If any date falls on a Saturday, Sunday, or legal holiday, use the next regular workday.

E. Penalty for Not Paying Enough Estimated Tax

You may be charged a penalty for not paying enough estimated tax, or for not making the payments on time in the required amount (even if you have an overpayment on your tax return). The penalty does not apply if each required payment is timely and the total tax paid:

- is at least 80% (66⅔% for farmers and fishermen) of the amount of income tax (including alternative minimum tax) and self-employment tax due as shown on your return for 1986, or 80% of the tax due if no return was filed;
- is 100% of the tax shown on your return for 1985 (you must have filed a return for 1985 and it must have been for a full 12-month year); **OR**
- is 80% of the tax figured by annualizing the taxable income, alternative minimum taxable income, and adjusted self-employment income received for the months ending before the due date of the installment. For more information, see Publication 505.

Note: *The penalty may be waived by IRS under certain conditions, such as casualty, disaster, or other unusal circumstances. See Publication 505 for details.*

Caution: *You may be required to make payments of past due amounts to avoid further penalty. You may have to make these payments if you do not make your estimated tax payments on time, or if you did not pay the correct amount for a previous payment date.*

Example: On June 1, 1986, you find out that you should have made an estimated tax payment for April 15. You should immediately fill out the payment-voucher due April 15, 1986, and send in the required amount (¼ × 1986 estimated tax).

If you changed your name because of marriage, divorce, etc., and you made estimated tax payments using your old name, you should attach a brief statement to the front of your 1986 Form 1040. In it explain all the estimated tax payments you and your spouse made during 1986, the name of the Service Center where you made the payments, and the name(s) and social security number(s) under which you made payments.

F. Where To File Your Payment-Voucher

Mail your payment-voucher to the Internal Revenue Service Center for the place where you live.

If you are located in: ▼	Use this address: ▼
New Jersey, New York–New York City, and counties of Nassau, Rockland, Suffolk, and Westchester	Holtsville, NY 00501
New York (all other counties), Connecticut, Maine, Massachusetts, Minnesota, New Hampshire, Rhode Island, Vermont	Andover, MA 05501
District of Columbia, Delaware, Maryland, Pennsylvania	Philadelphia, PA 19255
Alabama, Florida, Georgia, Mississippi, South Carolina	Atlanta, GA 31101
Kentucky, West Virginia, Michigan, Ohio	Cincinnati, OH 45999
Kansas, Louisiana, New Mexico, Oklahoma, Texas	Austin, TX 73301
Alaska, Arizona, California (counties of: Alpine, Amador, Butte, Calaveras, Colusa, Contra Costa, Del Norte, El Dorado, Glenn, Humboldt, Lake, Lassen, Marin, Mendocino, Modoc, Napa, Nevada, Placer, Plumas, Sacramento, San Joaquin, Shasta, Sierra, Siskiyou, Solano, Sonoma, Sutter, Tehama, Trinity, Yolo, and Yuba), Colorado, Idaho, Montana, Nebraska, Nevada, North Dakota, Oregon, South Dakota, Utah, Washington, Wyoming	Ogden, UT 84201
California (all other counties), Hawaii	Fresno, CA 93888
Illinois, Iowa, Missouri, Wisconsin	Kansas City, MO 64999
Arkansas, Indiana, North Carolina, Tennessee, Virginia	Memphis, TN 37501
American Samoa	Philadelphia, PA 19255
Guam	Commissioner of Revenue and Taxation Agana, GU 96910
Puerto Rico (or if excluding income under section 933) Virgin Islands: Non-permanent residents	Philadelphia, PA 19255
Virgin Islands: Permanent residents	V.I. Bureau of Internal Revenue P.O. Box 3186 St. Thomas, VI 00801
A.P.O. or F.P.O. address of:	Miami—Atlanta, GA 31101 New York—Holtsville, NY 00501 San Francisco—Fresno, CA 93888 Seattle—Ogden, UT 84201
Foreign country: U.S. citizens and those filing Form 2555 or Form 4563, even if you have an A.P.O. or F.P.O. address	Philadelphia, PA 19255

Record of Estimated Tax Payments

Payment number	(a) Date	(b) Amount	(c) 1985 overpayment credit applied	(d) Total amount paid and credited (add (b) and (c))
1				
2				
3				
4				
Total . . . . ▶				

Note: If you are not required to make the estimated tax payment due April 15, 1986, at this time, you may have to make a payment by a later date. See Instruction D(1).

Amended Estimated Tax Schedule (Use if your estimated tax changes during the year)

1 Amended estimated tax		1	
2 Less: **a** Amount of 1985 overpayment chosen for credit to 1986 estimated tax and applied to date	2a		
b Estimated tax payments to date.	2b		
c Total of lines 2a and b		2c	
3 Unpaid balance (subtract line 2c from line 1)		3	
4 Amount to be paid		4	

Page 2

(38)

FIGURE 21–10(b). IRS Form 1040-ES, page 2.

1986 Estimated Tax Worksheet (Keep for Your Records—Do Not Send to Internal Revenue Service)

1	Enter amount of Adjusted Gross Income you expect in 1986	1	
2 a	If you plan to itemize deductions, enter the estimated total of your deductions. If you must itemize deductions, see Instruction B, page 1. (If you do not plan to itemize deductions, skip lines 2a–2c and go on to line 2d.)	2a	
b	Enter { $3,670 if married filing a joint return (or qualifying widow(er)); $2,480 if single (or head of household); $1,835 if married filing a separate return }	2b	
c	Subtract line 2b from line 2a (if zero or less, enter zero)	2c	
d	If you do not itemize deductions, enter your allowable deduction, if any, for charitable contributions (see Instruction B.1 on page 1)	2d	
3	Subtract line 2c or 2d, whichever applies, from line 1	3	
4	Exemptions (multiply $1,080 times number of personal exemptions)	4	
5	Subtract line 4 from line 3	5	
6	Tax. (Figure your tax on line 5 by using Tax Rate Schedule X, Y, or Z in these instructions. DO NOT use the Tax Table or Tax Rate Schedule X, Y, or Z in the 1985 Form 1040 Instructions.)	6	
7	Enter any additional taxes (see line 7 Instructions)	7	
8	Add lines 6 and 7	8	
9	Credits (see line 9 Instructions)	9	
10	Subtract line 9 from line 8	10	
11	Self-employment tax. Estimate of 1986 self-employment income $____________________; if $42,000 or more, enter $5,166.00; **if less, multiply the amount by .123** (see line 11 Instructions for additional information)	11	
12	Other taxes (see line 12 Instructions)	12	
13 a	Total. Add lines 10 through 12	13a	
b	Earned income credit and credit from **Form 4136**	13b	
c	Total. Subtract line 13b from line 13a	13c	
14	Required annual payment. Enter amount of line 13c that you are required to pay	14	
	Caution: Generally, if you do not prepay at least 80% (66⅔% for farmers and fishermen) of your 1986 tax liability or 100% of the tax shown on your 1985 return, whichever is smaller, you may be subject to a penalty for not paying enough estimated tax (see Instruction E). To avoid this, make sure your estimate is as accurate as possible. If you are unsure of your estimate, you may want to pay more than 80% (or 66⅔%) of the amount you have shown on line 13c. For more information, get Publication 505.		
15	Income tax withheld and estimated to be withheld (including income tax withholding on pensions, annuities, certain deferred income, etc.) during 1986	15	
16	Balance (subtract line 15 from line 14). (**Note:** If line 13c less line 15 is less than $500, you are not required to make estimated tax payments.) If you are applying an overpayment from 1985 to 1986 estimated tax, see Instruction C(2), page 1	16	
17	If the first payment you are required to make is due April 15, 1986, enter ¼ of line 16 (less any 1985 overpayment that you are applying to this installment) here and on line 1 of your payment-voucher(s). You may round off cents to the nearest whole dollar	17	

Tear off here

Form **1040-ES** | **1986 Payment-Voucher**

Department of the Treasury
Internal Revenue Service

OMB No. 1545-0087

Return this voucher with check or money order payable to the Internal Revenue Service.
Please do not send cash or staple your payment to this voucher.

(Calendar year—Due Jan. 15, 1987)

1 Amount of payment $____________________	Please type or print	Your social security number	Spouse's number, if joint payment
2 Fiscal year filers enter year ending __________ (month and year)		First name and middle initial (of both spouses if joint payment)	Last name
		Address (Number and street)	
File only if you are making a payment of estimated tax.		City, state, and ZIP code	

For Paperwork Reduction Act Notice, see instructions on page 1.

Page 3

(39)

FIGURE 21–10(c). IRS Form 1040-ES, page 3.

1986 Tax Rate Schedules

Caution: Do not use these Tax Rate Schedules to figure your 1985 taxes. Use only to figure your 1986 estimated taxes.

SCHEDULE X—Single Taxpayers

If line 5 is: Over—	but not over—	The tax is:	of the amount over—
$0	$2,480	—0—	
2,480	3,670	-------- 11%	$2,480
3,670	4,750	$130.90 + 12%	3,670
4,750	7,010	260.50 + 14%	4,750
7,010	9,170	576.90 + 15%	7,010
9,170	11,650	900.90 + 16%	9,170
11,650	13,920	1,297.70 + 18%	11,650
13,920	16,190	1,706.30 + 20%	13,920
16,190	19,640	2,160.30 + 23%	16,190
19,640	25,360	2,953.80 + 26%	19,640
25,360	31,080	4,441.00 + 30%	25,360
31,080	36,800	6,157.00 + 34%	31,080
36,800	44,780	8,101.80 + 38%	36,800
44,780	59,670	11,134.20 + 42%	44,780
59,670	88,270	17,388.00 + 48%	59,670
88,270	--------	31,116.00 + 50%	88,270

SCHEDULE Z—Heads of Household

If line 5 is: Over—	but not over—	The tax is:	of the amount over—
$0	$2,480	—0—	
2,480	4,750	-------- 11%	$2,480
4,750	7,010	$249.70 + 12%	4,750
7,010	9,390	520.90 + 14%	7,010
9,390	12,730	854.10 + 17%	9,390
12,730	16,190	1,421.90 + 18%	12,730
16,190	19,640	2,044.70 + 20%	16,190
19,640	25,360	2,734.70 + 24%	19,640
25,360	31,080	4,107.50 + 28%	25,360
31,080	36,800	5,709.10 + 32%	31,080
36,800	48,240	7,539.50 + 35%	36,800
48,240	65,390	11,543.50 + 42%	48,240
65,390	88,270	18,746.50 + 45%	65,390
88,270	116,870	29,042.50 + 48%	88,270
116,870	--------	42,770.50 + 50%	116,870

SCHEDULE Y—Married Taxpayers and Qualifying Widows and Widowers

Married Filing Joint Returns and Qualifying Widows and Widowers

If line 5 is: Over—	but not over—	The tax is:	of the amount over—
$0	$3,670	—0—	
3,670	5,940	-------- 11%	$3,670
5,940	8,200	$249.70 + 12%	5,940
8,200	12,840	520.90 + 14%	8,200
12,840	17,270	1,170.50 + 16%	12,840
17,270	21,800	1,879.30 + 18%	17,270
21,800	26,550	2,694.70 + 22%	21,800
26,550	32,270	3,739.70 + 25%	26,550
32,270	37,980	5,169.70 + 28%	32,270
37,980	49,420	6,768.50 + 33%	37,980
49,420	64,750	10,543.70 + 38%	49,420
64,750	92,370	16,369.10 + 42%	64,750
92,370	118,050	27,969.50 + 45%	92,370
118,050	175,250	39,525.50 + 49%	118,050
175,250	--------	67,553.50 + 50%	175,250

Married Filing Separate Returns

If line 5 is: Over—	but not over—	The tax is:	of the amount over—
$0	$1,835	—0—	
1,835	2,970	-------- 11%	$1,835
2,970	4,100	$124.85 + 12%	2,970
4,100	6,420	260.45 + 14%	4,100
6,420	8,635	585.25 + 16%	6,420
8,635	10,900	939.65 + 18%	8,635
10,900	13,275	1,347.35 + 22%	10,900
13,275	16,135	1,869.85 + 25%	13,275
16,135	18,990	2,584.85 + 28%	16,135
18,990	24,710	3,384.25 + 33%	18,990
24,710	32,375	5,271.85 + 38%	24,710
32,375	46,185	8,184.55 + 42%	32,375
46,185	59,025	13,984.75 + 45%	46,185
59,025	87,625	19,762.75 + 49%	59,025
87,625	--------	33,776.75 + 50%	87,625

Page 4

(40)

FIGURE 21–10(d). IRS Form 1040-ES, page 4.

Form **1040-ES** | **1986 Payment-Voucher**
Department of the Treasury
Internal Revenue Service

OMB No. 1545-0087

Return this voucher with check or money order payable to the Internal Revenue Service.
Please do not send cash or staple your payment to this voucher.

(Calendar year—Due Sept. 15, 1986)

	Please type or print		
1 Amount of payment $ ____________		Your social security number	Spouse's number, if joint payment
2 Fiscal year filers enter year ending ____________ (month and year)		First name and middle initial (of both spouses if joint payment)	Last name
		Address (Number and street)	
File only if you are making a payment of estimated tax.		City, state, and ZIP code	

For Paperwork Reduction Act Notice, see instructions on page 1.

Tear off here

Form **1040-ES** | **1986 Payment-Voucher**
Department of the Treasury
Internal Revenue Service

OMB No. 1545-0087

Return this voucher with check or money order payable to the Internal Revenue Service.
Please do not send cash or staple your payment to this voucher.

(Calendar year—Due June 16, 1986)

	Please type or print		
1 Amount of payment $ ____________		Your social security number	Spouse's number, if joint payment
2 Fiscal year filers enter year ending ____________ (month and year)		First name and middle initial (of both spouses if joint payment)	Last name
		Address (Number and street)	
File only if you are making a payment of estimated tax.		City, state, and ZIP code	

For Paperwork Reduction Act Notice, see instructions on page 1.

Tear off here

Form **1040-ES** | **1986 Payment-Voucher**
Department of the Treasury
Internal Revenue Service

OMB No. 1545-0087

Return this voucher with check or money order payable to the Internal Revenue Service.
Please do not send cash or staple your payment to this voucher.

(Calendar year—Due April 15, 1986)

	Please type or print		
1 Amount of payment $ ____________		Your social security number	Spouse's number, if joint payment
2 Fiscal year filers enter year ending ____________ (month and year)		First name and middle initial (of both spouses if joint payment)	Last name
		Address (Number and street)	
File only if you are making a payment of estimated tax.		City, state, and ZIP code	

For Paperwork Reduction Act Notice, see instructions on page 1.

Page 5
(41)

FIGURE 21–10(e). IRS Form 1040-ES, page 5.

by the payment voucher shown in Figure 21–10(e) and should be recorded on the payment record shown in Figure 21–10(b).

The temptation of inexperienced taxpayers at this point is to avoid many of the difficult estimation procedures and just write a fictitious number for taxes due. While this process may meet the letter of the law and let the taxpayer think he is getting away with something, that feeling will not last too long. Note the specific caution written between Lines 14 and 15 of the estimated tax worksheet of Figure 21–10(c). If the estimated tax paid in during the year is less than 80 percent of the actual tax liability determined on Form 1040, a penalty will be assessed by IRS. On the other hand, if the tax estimate is overstated and overpaid, the taxpayer has lost the use of the overpaid cash during the year. That amounts to an interest-free loan to the IRS! Consequently, the tax estimate should be as accurate as possible.

An exact estimate is not likely, so any difference is either paid or reclaimed through the normal Form 1040 return process.

Special Rules for Farmers and Ranchers

As noted several times by now, qualifying farmers and ranchers do not need to complete a Form 1040-ES tax estimate. However, the same process of estimating taxable income through a projected income statement and a cash flow budget should be followed to determine if the strict IRS definition of a farmer will be met. If a farmer or rancher actually meets the definition, he can then appeal to the special rules and avoid Form 1040-ES throughout the tax year.

But after the tax year closes, a farmer must once again examine Form 1040-ES and consider his options. Two alternatives exist: (1) ignore Form 1040-ES altogether and file the completed Form 1040 package by March 1; or (2) file a Form 1040-ES and pay the estimated tax by January 15 and then file the complete Form 1040 package by April 15. The two alternatives should be considered as a choice which must be made by the taxpayer and his tax return preparer. Filing the Form 1040 package by March 1 reduces the paperwork load since it avoids Form 1040-ES altogether. It also allows the farmer to hold onto his tax money for the additional 6 weeks between January 15 and March 1. Being able to delay payment for 45 days on a large tax amount due can save quite a bit of interest. On the other hand, filing the Form 1040-ES allows an additional 6 weeks between March 1 and April 15 to get all records in order and complete Form 1040.

IRS Form 2210F

Do not imagine that a small amount of tax can be paid with Form 1040-ES, so that the additional time between March 1 and April 15 can be used without the loss of any major cash resources. The same penalty for underpayment of estimated tax described earlier applies here as well. In fact, if the estimated tax is underpaid at all, one more IRS form, Form 2210F, must be filed.

Form 2210F, Figure 21–11, is relatively easy to complete, but as can be seen in Part II, a penalty is assessed on the amount underpaid. That underpayment penalty is assessed beginning January 15 and can quickly become a significant

Form **2210F**

Department of the Treasury
Internal Revenue Service

Underpayment of Estimated Tax by Farmers and Fishermen

▶ **Attach to Form 1040**

OMB No. 1545-0140

1985

57

Name(s) as shown on Form 1040 | **Social security number**

Part I **Figuring Your Underpayment**

1	1985 tax after credits (from Form 1040, line 50)		1
2	Other taxes (see instructions)		2
3	Add lines 1 and 2		3
4	Earned income credit	4	
5	Credit for federal tax on gasoline and special fuels	5	
6	Credit for overpaid windfall profit tax attributable to amounts withheld	6	
7	Add lines 4, 5, and 6		7
8	Subtract line 7 from line 3		8
9	Multiply line 8 by 66⅔% (.6667)	9	
10	Withholding taxes from 1985 Form 1040, lines 57 and 61		10
11	Subtract line 10 from line 8. If result is less than $500, do not complete rest of form		11
12	Enter your 1984 tax (see instructions)		12
13	Enter the smaller of line 9 or line 12		13
14	Amounts withheld during 1985 and amounts paid or credited by January 15, 1986		14
15	Underpayment. (Subtract line 14 from line 13. If zero or less, you do not owe a penalty.)		15

Part II **Figuring the Penalty**

16	Enter the date line 15 was paid or April 15, 1986, whichever is earlier	16
17	Number of days after January 15, 1986, to and including the date on line 16	17
18	Penalty. $\frac{\text{Number of days on line 17}}{365}$ x 10% x underpayment on line 15. Check the box below line 68 on Form 1040 and show this amount in the space provided. If you owe tax, add penalty to tax and show total on line 68. If you are due a refund, subtract penalty from overpayment on line 65.	18

Instructions

Paperwork Reduction Act Notice.—We ask for this information to carry out the Internal Revenue laws of the United States. We need it to ensure that taxpayers are complying with these laws and to allow us to figure and collect the right amount of tax. You are required to give us this information.

Changes You Should Note.—New rules apply for determining whether you owe a penalty for underpaying your estimated tax. You may owe a penalty if, for 1985, you did not pay at least 66⅔% of your tax liability for the current year or 100% of your tax liability for the preceding year, whichever is less.

Who Should Use This Form.—If at least two-thirds of your 1984 or 1985 gross income is from farming or fishing, use this form to see if you must pay a penalty for underpaying your estimated tax. (Otherwise, use Form 2210.) Use Part I to see if you paid enough estimated tax. If not, use Part II to figure the penalty.

However, if you file your return and pay the tax due by March 3, 1986, do not file this form. You do not have to pay a penalty.

For a definition of gross income from farming and fishing, and more information, see **Publication 505**, Tax Withholding and Estimated Tax.

Note: *You will not have to pay a penalty or complete this form if you had no tax liability for 1984, you were a U.S. citizen or resident for all of 1984, and your 1984 tax return was (or would have been had you been required to file) for a full 12 months.*

Line-by-Line Instructions

Line 2—Other taxes.—Enter on this line the following:

- Self-employment tax (Form 1040, line 51);
- Alternative minimum tax (Form 1040, line 52);
- Tax from recapture of investment credit (Form 1040, line 53);
- Tax on premature distributions from an IRA (Form 5329, Part II, only);
- Section 72 penalty taxes;
- Excise tax on golden parachute payments; and
- Advance earned income credit payments.

Line 11.—If this line is less than $500, you do not owe a penalty and need not attach this form to your tax return.

Line 12—1984 tax.—Figure your 1984 tax by using your taxes and credits from your 1984 tax return. Use the same type taxes and credits as shown on lines 1 through 8 of this form.

If you did not file a return for 1984 or if your 1984 tax year was for less than 12 months, do not complete this line. Instead, enter the amount from line 9 on line 13. However, see the "note" under the instructions for Who Should Use This Form, above.

Line 14.—Enter your tax withheld and estimated tax payments from Form 1040, lines 57, 58, and 61.

Line 15.—If your underpayment was due to a casualty or other unusual circumstance, or because in 1984 or 1985 you retired after age 62 or became disabled and your underpayment was due to reasonable cause, explain in the margin of Part II and do not complete the rest of form.

Form **2210F** (1985)

(181)

FIGURE 21–11. IRS Form 2210F.

amount if a sloppy estimating job was done or if taxable income was grossly underestimated in an attempt to avoid the unavoidable. A farmer or rancher who files his complete return by March 1 without having filed a Form 1040-ES can ignore Form 2210-F.

IRS Form 1099

One final confusing tax issue has arisen in the last decade: the series of informational returns called Form 1099 by IRS. The purpose of the Form 1099 group is to inform IRS of large payments made to nonemployees. These forms are essentially a substitute for or alternative to W-2s given to employees.

Farmers and ranchers will commonly receive a few Form 1099s for their services or for interest, dividend and mutual fund income. When received, the process is easy to follow: report them within the Form 1040 package as income received. But filling them out for others is quite a bit more difficult. Examples of just a few of the current Form 1099s are shown in Figures 21–12 through 21–15.

Common to each Form 1099 are the information blocks for the payer's name and address, the recipient's name and address, and the amount paid. However, there is not one magic figure which triggers when a Form 1099 should be filled out. A summary of rules for the major Form 1099s is shown in Figures 21–16(a) and (b).

The most common information return that will be filled out by a farmer or rancher is Form 1099-MISC (Figure 21–14). This form should be completed for every nonemployee, subcontractor or independent business person (such as the labor crew leader, a veterinarian, or a custom combiner) who was paid more than $600 by the farmer or rancher throughout the year. The individual who performed

7979 ☐ VOID ☐ CORRECTED For Official Use Only

Type or machine print PAYER'S name, street address, city, state, and ZIP code

1a Date of sale

1b CUSIP No.

OMB No. 1545-0715

1985

Statement for Recipients of

Proceeds from Broker and Barter Exchange Transactions

2 Stocks, bonds, etc.

Reported to IRS ☐ Gross proceeds ☐ Gross proceeds less commissions and option premiums

PAYER'S Federal identification number

RECIPIENT'S identification number

3 Bartering

4 Federal income tax withheld

Copy A For Internal Revenue Service Center

For Paperwork Reduction Act Notice and instructions for completing this form, see Instructions for Form 1099 Series, 1098, 5498, and 1096.

Type or machine print RECIPIENT'S name (first, middle, last)

Street address

City, state, and ZIP code

Account number (optional)

5 Description

Regulated Futures Contracts

6 Profit or (loss) realized in 1985

7 Unrealized profit or (loss) on open contracts—12/31/84

8 Unrealized profit or (loss) on open contracts—12/31/85

9 Aggregate profit or (loss)

Form **1099-B**

Do NOT Cut or Separate Forms on This Page

Department of the Treasury - Internal Revenue Service

FIGURE 21–12. IRS Form 1099-B.

9292 ☐ VOID ☐ CORRECTED For Official Use Only

Type or machine print PAYER'S name, street address, city, state, and ZIP code			OMB No.1545-0112 **1985** Statement for Recipients of
PAYER'S Federal identification number	RECIPIENT'S identification number	1 Earnings from savings and loan associations, credit unions, bank deposits, bearer certificates of deposit, etc.	
Type or machine print RECIPIENT'S name (first, middle, last)		2 Amount of forfeiture	3 U.S. Savings Bonds, etc.
Street address		4 Federal income tax withheld	
City, state, and ZIP code		5 Foreign tax paid (if eligible for foreign tax credit)	6 Foreign country or U.S. possession
Account number (optional)			

Interest Income

Copy A For Internal Revenue Service Center

For Paperwork Reduction Act Notice and instructions for completing this form, see Instructions for Form 1099 Series, 1098, 5498, and 1096.

Form **1099-INT**

Do NOT Cut or Separate Forms on This Page

Department of the Treasury - Internal Revenue Service

FIGURE 21–13. IRS Form 1099-INT.

9595 ☐ VOID ☐ CORRECTED For Official Use Only

Type or machine print PAYER'S name, street address, city, state, and ZIP code		1 Rents	OMB No.1545-0115 **1985** Statement for Recipients of
		2 Royalties	
PAYER'S Federal identification number	RECIPIENT'S identification number	3 Prizes and awards	4 Federal income tax withheld
Type or machine print RECIPIENT'S name (first, middle, last)		5 Fishing boat proceeds	6 Medical and health care payments
Street address		7 Nonemployee compensation	8 Substitute payments in lieu of dividends or interest
City, state, and ZIP code		9 Payer made direct sales of $5,000 or more of consumer products to a buyer (recipient) for resale ▶☐	
Account number (optional)			

Miscellaneous Income

Copy A For Internal Revenue Service Center

For Paperwork Reduction Act Notice and instructions for completing this form, see Instructions for Form 1099 Series, 1098, 5498, and 1096.

Form **1099-MISC**

Do NOT Cut or Separate Forms on This Page

Department of the Treasury - Internal Revenue Service

FIGURE 21–14. IRS Form 1099-MISC.

9191 ☐ VOID ☐ CORRECTED For Official Use Only

Type or machine print PAYER'S name, street address, city, state, and ZIP code	1 Gross dividends and other distributions on stock	OMB No.1545-0110 **1985** Statement for Recipients of	**Dividends and Distributions**
	2 Dividends qualifying for exclusion		
PAYER'S Federal identification number / RECIPIENT'S identification number	3 Dividends not qualifying for exclusion	4 Federal income tax withheld	**Copy A For Internal Revenue Service Center** For Paperwork Reduction Act Notice and instructions for completing this form, see Instructions for Form 1099 Series, 1098, 5498, and 1096.
Type or machine print RECIPIENT'S name (first, middle, last)	5 Capital gain distributions	6 Nontaxable distributions (if determinable)	
Street address	7 Foreign tax paid	8 Foreign country or U.S. possession	
City, state, and ZIP code	**Liquidation Distributions**		
	9 Cash	10 Noncash (Fair market value)	
Account number (optional)			

OCR Form
Do NOT Photocopy

Form **1099-DIV** **Do NOT Cut or Separate Forms on This Page** Department of the Treasury - Internal Revenue Service

FIGURE 21–15. IRS Form 1099-DIV.

the nonemployee services must supply his Social Security number or Taxpayer Identification number. The contractor must comply by completing IRS Form W-9 (Figure 21–17). The payer collects those forms and must file them with IRS. A very nice modification to IRS Form W-9 was made by Mike Hardin at Oklahoma State University. His modification, shown in Figure 21–18, can be used as a written contract for the work to be performed. If the contractor refuses to provide his identification number on Form W-9, the payer must withhold 20 percent of the total amount owed for deposit with the IRS or be subject to a fine himself.

One copy of each Form 1099 should be provided to the payee after the close of the tax year and one copy should be retained by the payer for his own records. A third copy should be sent to the IRS. All of the Form 1099s issued should be gathered and bundled with a Form 1096. A separate Form 1096 should be used for *each* type of 1099.

Form 1096 (Figure 21–19) is also a simple form even if it is a nuisance. Simply fill in the payer's name, address, and identification number and then mark *one* of the Form 1099 boxes.

Summary

All employed persons are subject to income tax and Social Security tax withholding, a system whereby the employer collects an estimated amount of tax from

Form 1099-INT

File Form 1099-INT, Statement for Recipients of Interest Income, for each person (1) to whom you paid amounts reportable in Box 1 of at least $10 (except for the $600 limit described in the instructions below in Box 1), (2) for whom you withheld and paid foreign tax on interest if the recipient can claim credit for the tax on his or her income tax return, or (3) for each person from whom you withheld any Federal income tax under the backup withholding rules regardless of the amount of the payment.

Give a copy of the official form, or a substitute form containing substantially similar language, to each recipient. If you have furnished Forms 1099-INT to a recipient for amounts received during the year on a transactional basis, for example, window transactions, do not include these same amounts in a 1099-INT furnished to the same recipient for other payments during the year.

Only report interest payments made in the course of your trade or business (including Federal, state, and local government agencies and activities deemed nonprofit), or for which you were a nominee/middleman, or from which you withheld Federal income tax or foreign tax.

Note: *Do not report tax-exempt or tax-deferred interest, such as interest on municipal bonds or interest that is earned but not distributed from an individual retirement arrangement (IRA).*

An exempt-interest dividend from a regulated investment company retains its tax-exempt status and is not reported on Form 1099-DIV or 1099-INT.

Report in:

Box 1.—Amounts, whether or not designated as interest, that are paid or credited to any person's account by savings and loan associations, mutual savings banks not having capital stock represented by shares, building and loan associations, cooperative banks, homestead associations, credit unions, or similar organizations. Include interest on bank deposits, accumulated dividends or delayed death benefits paid by a life insurance company, indebtedness (including bonds, debentures, notes and certificates other than those of the U.S. Treasury) issued in registered form or of a type offered to the public, or from which you withheld Federal income tax or foreign tax. Also include payments of interest made in the course of your trade or business not meeting these criteria if they total $600 or more for any person.

Also show original issue discount on short-term obligations of one year or less and interest on all bearer certificates of deposit.

Do not include in Box 1 interest on tax-free covenant bonds, which is reportable on Form 1042S or dividends from money market funds, which is reportable on Form 1099-DIV. Also, do not include any description in Box 1. If you wish to show a description, use the blank box above Box 1.

Box 2.—Amounts of interest or principal forfeited due to premature withdrawal of time deposits and deductible by the recipient from gross income. The amount in Box 1 should not be reduced by the forfeiture. For detailed instructions for determining the amount of forfeiture deductible by the depositor, see Rev. Ruls. 75-20, 1975-1 C.B. 29 and 75-21, 1975-1 C.B. 367.

Box 3.—Amounts of interest on U.S. Savings Bonds, Treasury bills, Treasury notes, and Treasury bonds. Do not include this amount in Box 1.

Box 4.—Amounts for backup withholding. For example, persons who have not furnished their taxpayer identification number to you in the manner required become subject to withholding at a 20% rate on payments required to be reported in Box 1 (which may be reduced by the amount reported in Box 2) on this form.

Box 5.—Foreign tax withheld and paid on interest. Report this amount in U.S. dollars.

Box 6.—The name of the foreign country or U.S. possession to which the withheld tax applies.

Form 1099-MISC

File Form 1099-MISC, Statement for Recipients of Miscellaneous Income, for each person, other than corporations (Boxes 6 and 8 must be reported to corporations), to whom you have paid at least $600 (unless a different amount is indicated below) in rents, royalties, and other payments reportable on the form, or from whom you have withheld any Federal income tax under the backup withholding rules regardless of the amount of the payment. Give a copy to each recipient. Report only payments made in the course of your trade or business (including Federal, state, or local government agencies and activities deemed nonprofit), for which you were a nominee/middleman, or from which you withheld Federal income tax or foreign tax. If you are a broker, you must also report payments received as described in Box 8.

Report in:

Box 1.—Amounts paid to recipients for all types of rents, such as real estate rentals paid for office space, machine rentals (for example, hiring a bulldozer to level your parking lot), and pasture rentals (for example, farmers paying for the use of grazing land). If the machine rental is a part of a contract that includes both the use of the machine and the operator, the rental should be prorated between the rent of the machine (reported in Box 1) and the operator's charge (reported as nonemployee compensation in Box 7).

Box 2.—Gross royalty payments before reduction for severance and other taxes which may have been withheld and paid. Include in this box royalties paid by a publisher directly to an author or literary agent or paid by a literary agent to an author. Do not include surface royalties. They should be reported in Box 1.

Box 3.—Prizes and awards that are not for services rendered. Prizes or awards for services rendered by employees should be reported on Form W-2. Prizes and awards for services rendered by nonemployees, such as an award for the top commission salesperson, are reported in Box 7. Include the fair market value of merchandise won on game shows.

Box 4.—Amounts for backup withholding. For example, persons who have not furnished their taxpayer identification number to you become subject to withholding at a 20% rate on payments required to be reported in Boxes 1, 2 (net of severance taxes), 3, 5 (to the extent paid in cash), 6, 7, and 8 on this form.

Box 5.—The share of all proceeds from the sale of a catch or the fair market value of a distribution in-kind to each crewmember of fishing boats with normally fewer than 10 crewmembers.

Box 6.—Payments by medical and health care insurers to each physician or other supplier or provider of services under health, accident, and sickness insurance programs. In the case of a corporation, list the corporation as the recipient rather than the individual providing the services. (See Rev. Ruls. 69-595, 1969-2 C.B. 242, and 70-608, 1970-2 C.B. 286.)

The exemption for issuing Form 1099-MISC to a corporation does not apply to corporations, including professional corporations, that provide medical or health care services.

Box 7.—Nonemployee compensation. Include fees, commissions, prizes, awards, or other forms of compensation for services rendered for your trade or business by an individual who is not your employee. Also include expenses incurred for the use of an entertainment facility that you treat as compensation to a nonemployee.

Examples of payments to be reported in Box 7 are:

(1) Attorney's and accountant's fees for professional services.

(2) Fees paid by one professional to another, such as fee-splitting or referral fees.

(3) Payments by attorneys to witnesses or experts in legal adjudication.

(4) Insurance company payments to an auto repair shop under a repair contract indicating an amount for labor and another amount for parts.

(5) Commissions paid to nonemployee salespersons, subject to repayment but not repaid during the calendar year.

(6) A fee paid to a nonemployee and travel reimbursement for which the nonemployee did not account to the payer, if the fee and reimbursement aggregate at least $600.

(7) Payments to nonemployee entertainers for services.

Note: *In order to qualify for relief under section 530 of the Revenue Act of 1978 as extended by Pub. L. 97-248, employers may file Form 1099-MISC. Additional requirements for relief are discussed in* ***Publication 15****, Employer's Tax Guide, under "Who Are Employees?"*

(205) **Page 7**

FIGURE 21–16(a). IRS Summary of rules of major 1099s.

Guide to Information Returns

FORM NUMBER	TITLE	WHAT TO REPORT	AMOUNTS TO REPORT	DUE DATE	
				TO IRS	TO RECIPIENT
1042S	Income Subject to Withholding Under Chapter 3, Internal Revenue Code	Payments subject to withholding under Chapter 3 of the Code, including interest, dividends, royalties, pensions and annuities, and compensation for personal services.	All amounts	March 15	March 15
1098	Mortgage Interest Statement	Mortgage interest you received in the course of your trade or business from individuals.	$600 or more	February 28	January 31
1099-A	Information Return for Acquisition or Abandonment of Secured Property	Information about the acquisition or abandonment of property that is security for a debt for which you are the lender.	All amounts	February 28	January 31
1099-B	Statement for Recipients of Proceeds from Broker and Barter Exchange Transactions	Sales or redemptions of securities, futures transactions, commodities, and bartering exchange transactions.	All amounts	February 28	January 31
1099-DIV	Statement for Recipients of Dividends and Distributions	Distributions, such as dividends, capital gains, or nontaxable distributions, that were paid on stock, and distributions in liquidation of $600 or more.	$10 or more, except $600 or more for liquidations	February 28	January 31
1099-G	Statement for Recipients of Certain Government Payments	Unemployment compensation, state and local income tax refunds, agricultural payments, taxable grants, and discharge of indebtedness.	$10 for unemployment and tax refunds; $600 for all others	February 28	January 31
1099-INT	Statement for Recipients of Interest Income	Interest payments not including interest on an IRA, SEP, or DEC.	$10 or more ($600 in some cases)	February 28	January 31
1099-MISC	Statement for Recipients of Miscellaneous Income (Also, use this form to report the occurrence of direct sales of $5,000 or more of consumer goods for resale.)	Rent or royalty payments; prizes and awards that are not for services, such as winnings on TV or radio shows. Payments to crewmembers by owners or operators of fishing boats. Report payments of proceeds from sale of catch. Payments to a physician, physicians' corporation or other supplier of health and medical services. Issued mainly by medical assistance programs or health and accident insurance plans. Payments for services performed for a trade or business by people not treated as its employees. Examples: fees to subcontractors or directors, expenses incurred for use of an entertainment facility treated as compensation to a nonemployee and golden parachute payments. Substitute dividend and interest payments reportable by brokers.	$600 or more All payments $600 or more $600 or more $10 or more	February 28	January 31
1099-OID	Statement for Recipients of Original Issue Discount	Original issue discount.	$10 or more	February 28	January 31
1099-PATR	Statement for Recipients (Patrons) of Taxable Distributions Received from Cooperatives	Distributions from cooperatives to their patrons.	$10 or more	February 28	January 31
1099-R	Statement for Recipients of Total Distributions from Profit-Sharing, Retirement Plans, Individual Retirement Arrangements, Etc.	Distributions from retirement or profit-sharing plans or from individual retirement arrangements (IRA's). Use Form 1099-R only if the distribution closed the payee's account.	All amounts	February 28	January 31
5498	Individual Retirement Arrangement Information	Contributions (including rollover contributions) to an individual retirement arrangement (IRA), simplified employee pension (SEP), or a plan that accepts qualified deductible voluntary employee contributions (DEC's).	All amounts	May 31	May 31 (IRA, SEP) January 31 (DEC)

Page 11
(209)

FIGURE 21–16(b). Guide to Form 1099 Information Returns.

Source: IRS Publication 15, Circular E, *Employer's Tax Guide.*

Form **W-9**
(Rev. July 1984)
Department of the Treasury
Internal Revenue Service

Payer's Request for Taxpayer Identification Number and Certification

Give This Form to the Payer, Middleman, Broker, or Barter Exchange

Please print or type

Name as shown on account (If joint account, list first and circle the name of the person or entity whose number you enter in Part I below.)

Address

City, State, and ZIP code

List account number(s) here ▶

Part I **Taxpayer Identification Number—For All Accounts**

Enter your taxpayer identification number in the appropriate box. For most individuals, this is your social security number. If you do not have a number, see *How to Obtain a TIN.*

Note: *If the account is in more than one name, see the chart on page 2 for guidelines on which number to give the payer.*

Social security number

OR

Employer identification number

Part II **For Payees Exempt From Backup Withholding (See Instructions)**

Certification.—Under penalties of perjury, I certify that:

(1) The number shown on this form is my correct Taxpayer Identification Number (or I am waiting for a number to be issued to me), **and**

(2) I am not subject to backup withholding either because I have not been notified by the Internal Revenue Service (IRS) that I am subject to backup withholding as a result of a failure to report all interest or dividends, or the IRS has notified me that I am no longer subject to backup withholding.

Certification Instructions.—You must cross out item (2) above if you have been notified by IRS that you are subject to backup withholding because of underreporting interest or dividends on your tax return. However, if after being notified by IRS that you were subject to backup withholding you received another notification from IRS that you are no longer subject to backup withholding, do not cross out item (2). (Also see *Certification* under *Specific Instructions*.)

Please Sign Here **Signature ▶** **Date ▶**

Instructions

(Section references are to the Internal Revenue Code.)

Purpose of Form

Complete this form and give it to the payer of interest, dividends, and certain other payments (including broker and barter exchange transactions) so that you will not be subject to the 20% backup withholding that became effective January 1, 1984.

Use this form to report and certify your taxpayer identification number (TIN) to the payer, to certify that you are not subject to backup withholding because of underreporting interest and dividends on your tax return, and to claim exemption from backup withholding if you are an exempt payee.

If you do not complete this form properly and return it to the payer, the payer may be required to withhold 20% of payments made to you.

Note: *If a payer gives you a form other than a W-9 to request your TIN, you must use the payer's form.*

What Is Backup Withholding

The Interest and Dividend Tax Compliance Act of 1983 requires payers to withhold and pay to IRS 20% of payments of interest, dividends, and certain other payments under certain conditions. This is called "backup withholding." If you give the payer your correct TIN, certify your TIN when required, and report all your taxable interest and dividends on your tax return, your payments will not be subject to backup withholding.

Payments you receive will be subject to backup withholding if:

(1) You do not furnish your TIN to the payer, or

(2) IRS notifies the payer that you furnished an incorrect TIN, or

(3) You are notified by IRS that you are subject to backup withholding because you failed to report all your interest and dividends on your tax return (for interest and dividend accounts only), or

(4) You fail to certify to the payer that you are not subject to backup withholding under (3) above (for interest and dividend accounts opened after 1983 only), or

(5) You fail to certify your TIN. This applies only to interest, dividend, broker, or barter exchange accounts opened after 1983, or broker accounts considered inactive in 1983.

For other payments, you are subject to backup withholding only if (1) or (2) above applies.

Certain payees and payments are exempt from backup withholding and information reporting. See *Payees and Payments Exempt from Backup Withholding,* on this page, and *Exempt Payees and Payments* under *Specific Instructions,* on page 2, if you are an exempt payee.

How to Obtain a TIN

If you do not have a TIN, you should apply for one immediately. To apply for the number obtain **Form SS-5,** Application for a Social Security Number Card (for individuals), or **Form SS-4,** Application for Employer Identification Number (for businesses and all other entities), at your local office of the Social Security Administration or the Internal Revenue Service. Complete and file the appropriate form according to its instructions.

If you do not have a TIN, write "Applied For" in the space for the TIN in Part I, sign and date the form, and give it to the payer. You will then have 60 days to obtain a TIN and furnish it to the payer. During the 60-day period, the payments you receive will not be subject to the 20% backup withholding. However, if the payer does not receive your TIN from you within 60 days, backup withholding will begin and continue until you furnish your TIN to the payer.

Note: *Writing "Applied For" on the form means that you have already applied for a TIN,* ***OR*** *that you intend to apply for one in the near future.*

As soon as you receive your TIN, complete another Form W-9, include your new TIN, sign and date the form, and give it to the payer.

Payees and Payments Exempt from Backup Withholding

The following lists payees that are exempt from backup withholding and information reporting. For interest and dividends, all listed payees are exempt. For broker transactions, payees listed in (1) through (13), and a person registered under the Investment Advisers Act of 1940 who regularly acts as a broker are exempt. Payments subject to reporting under sections 6041 and 6041A are generally exempt from backup withholding only if paid to payees described in items (1) through (6), except that a corporation that provides medical and health care services or bills and collects payments for such services is not exempt from backup withholding or information reporting. Only payees described in items (2) through (6) are exempt from backup withholding for barter exchange transactions, patronage dividends, and payments by certain fishing boat operators.

Form **W-9** (Rev. 7-84)

(17)

FIGURE 21–17. IRS Form W-9.

Form **W-9**
(October 1983)
Department of the Treasury
Internal Revenue Service

Payer's Request for Taxpayer Identification Number

Please print or type

Name as shown on account (if joint account, also give joint owner's name)

Address

City, State, and ZIP code

List account number(s) here (See Instructions) ▶ ....

PART I.—Taxpayer Identification Number

Enter the taxpayer identification number in the appropriate box. For most individual taxpayers, this is the social security number.
Note: *If the account is in more than one name, see the chart on page 2 for guidelines on which number to give the payer.*

Social security number

OR

Employer identification number

PART II.—Backup Withholding On Accounts Opened After 12/31/83

Check the box if you are NOT subject to backup withholding under the provisions of section 3406(a)(1)(C) of the Internal Revenue Code . . . . ▶ ☐
(See **Highlight** below.)

Certification.—Under the penalties of perjury, I certify that the information provided on this form is true, correct, and complete.

Signature ▶ **Date ▶**

**Failure to provide correct taxpayer identification number will require the payer to withhold 20% backup withholding from payments.

INDEPENDENT CONTRACTOR SERVICE AGREEMENT

Payer: Name

Address

City, State, and ZIP Code

Contractor: Name

Permanent Address

Information City, State, and ZIP Code

Service to be provided: ____________________

Amount and timing of payments: ____________________

FIGURE 21–18. Modified IRS Form W-9 and contractor service agreement.

Source: Mike L. Hardin, Oklahoma State University.

DO NOT STAPLE 6969 ☐ CORRECTED

Form **1096** Department of the Treasury Internal Revenue Service	**Annual Summary and Transmittal of U.S. Information Returns**	OMB No. 1545-0108 **1986**

Type or machine print FILER'S name Street address City, state, and ZIP code	Enter in Box 1 or 2 below the identification number you used as the filer on the attached information returns. Do not fill in both Boxes 1 and 2.	
	1 Employer identification number	3 Total number of documents
	2 Social security number	4 Federal income tax withheld

Check only one box below to indicate the type of forms attached.

W-2G 32	1098 81	1099-A 80	1099-B 79	1099-DIV 91	1099-G 86	1099-INT 92	1099-MISC 95	1099-OID 96	1099-PATR 97	1099-R 98	5498 28
☐	☐	☐	☐	☐	☐	☐	☐	☐	☐	☐	☐

Under penalties of perjury, I declare that I have examined this return and accompanying documents and, to the best of my knowledge and belief, they are true, correct, and complete. In the case of documents without recipients' identification numbers, I have complied with the requirements of the law in attempting to secure such numbers from the recipients.

Signature ▶ .... Title ▶ .... Date ▶ ....

Please return this entire page to the Internal Revenue Service. Photocopies are NOT acceptable.

Paperwork Reduction Act Notice

We ask for this information to carry out the Internal Revenue laws of the United States. We need it to ensure that taxpayers are complying with these laws and to allow us to figure and collect the right amount of tax. You are required to give us this information.

Instructions

Changes You Should Note.—If you have received a preprinted label from IRS, place it in the name and address area of the form using the brackets as indicators. Make any necessary corrections on the label.

Do not mark in the boxes at the bottom of this page. The boxes are for use by Internal Revenue Service only.

For 1986 and later years, transmit Forms W-2G and 1099-R to IRS with Form 1096.

The box for the number of forms without TINs has been eliminated. In Box 4, enter the total Federal income tax withheld as shown on Forms 1099 or W-2G filed with this Form 1096.

Purpose of Form.—Use this form to transmit Forms W-2G, 1098, 1099, and 5498 to the Internal Revenue Service.

Completing Form 1096.—If you are not using a preprinted label, enter the filer's name, address, and taxpayer identification number (TIN) in the spaces provided on the form. A filer includes a payer, a recipient of mortgage interest payments, a broker, a barter exchange, a trustee or issuer of an individual retirement arrangement (including an IRA, SEP, or DEC), and a lender who acquires an interest in secured property or who has reason to know that the property has been abandoned. Individuals not in a trade or business should enter their social security number in Box 2; sole proprietors and all others should enter their employer identification number in Box 1. However, sole proprietors who are not required to have an employer identification number should enter their social security number in Box 2.

Group the forms by form number and submit each group with a separate Form 1096. For example, if you must file both Forms 1099-DIV and Forms 1099-INT, complete one Form 1096 to transmit your Forms 1099-DIV and another Form 1096 to transmit your Forms 1099-INT.

In Box 3, enter the number of forms you are transmitting with this Form 1096. Do not include blank or voided forms in your total. Enter the number of correctly completed forms, not the number of pages, being transmitted. For example, if you send one page of three-to-a-page Forms 5498 with a Form 1096, and you have correctly completed two Forms 5498 on that page, enter 2 in Box 3 of Form 1096. Check the appropriate box to indicate the type of form you are transmitting.

If you are transmitting forms that include TINs and others with no TINs, you may submit them with one Form 1096, as long as they are all the same form number. Show the total number of documents being transmitted in Box 3.

If you are filing a Form 1096 for corrected information returns, mark over the "X" in the CORRECTED box at the top of this form.

For more information about filing, see the separate Instructions for Forms 1099, 1098, 5498, 1096, and W-2G.

☐ ☐☐☐☐☐☐ **For Official Use Only**

★U.S.GPO:1985-0-463-157 ★E.I.#430814328 Form **1096** (1986)

FIGURE 21–19. IRS Form 1096.

each paycheck before wages are paid. These tax collections are then paid to the IRS. Self-employed persons must estimate their tax bill quarterly and pay the estimated amount. Because it is impossible to estimate income accurately 6 to 9 months in advance, farmers do not have to file quarterly estimates. They must, however, file an estimate by January 15 or, if they choose not to file that estimate, their return is due March 1.

Nonemployee payments (such as payments to independent contractors) are reported to IRS at the end of the year on Form 1099s just as total wages paid to employees are reported on Form W-2.

PART V

Business Management Analysis

We are finally at a point of reflection in this discussion of the farm and ranch records process. It is time to look back on what we have done so far to see what kind of insight about a particular business can be gained from the long, tedious record-keeping and reporting process. Much of what has been accomplished so far was started because of the needs imposed by other people. A good businessman, however, will realize that what he started for others can become a major part of helping himself.

Although a primary purpose of the cash transactions records discussed in Chapter 5 and the asset records of Chapter 7 was to provide the basis for the tax return of the chapters in Part IV, an astute businessman can use those same concepts to control cash flow, the lifeblood of a successful business. He can also begin to use this information to maneuver his income stream and purchasing patterns to minimize the pain of the income and self-employment taxes.

The inventory records, asset inventories, and financial inventories discussed in Chapters 6 through 8 and the cash transactions records were developed primarily to prepare the three major financial statements described in Part III. Although a primary reason for developing financial statements was to apply for credit, the good businessman should realize that there is a great amount of information built into these statements. Lenders are not really sadists, they look for specific information about the business applicant in requiring financial statements. A far-sighted businessman will look for the same information and maybe a bit more to help himself.

Business problem solving, financial control, and forward planning are by far the most interesting and challenging topics in farm and ranch business management. Although it would seem only fair to now dwell on these more interesting concepts, especially after going through the least interesting and most tedious records topics, they are lengthy, time consuming, and many are well beyond the scope of this book. Although these three topics will not be covered here in detail, it is very fortunate that there are several excellent references available for further study in farm and ranch business analysis.

We shall now turn our attention toward analyzing the records which have been developed throughout the first twenty-one chapters of this book. We shall try to learn how to extract as much pertinent, important information as possible from these records. Consequently, we will not attempt to duplicate the excellent coverage of farm and ranch financial control and forward planning found in such books as Kay's *Farm Management,* Penson, Klinefelter, and Lins' *Farm Investment and Financial Analysis,* or several other books currently available.[1] The chapters in this final part of the textbook will concentrate on using the information now available.

Analysis typically takes several paths. The first begins with *financial ratio analysis,* primarily because the information is easily gathered from financial statements, is well understood by financial advisors and lenders, and is very important in understanding the business as a whole. A second path of analysis concentrates on comparing financial ratios and other information from one year to similar information from previous years. This type of analysis, called *trend analysis* or *horizontal analysis,* can be a very effective way to detect strengths in the growth of a business and weaknesses before they become serious or unsolvable. A third path of analysis considers relationships among various factors within a business, usually income statement items. Because these items are generally arranged in a columnar format, this type of analysis is often referred to as *vertical analysis.* A fourth path that our analysis process can take is to compute data from one farm or ranch and compare it to data gathered from similar farms or ranches. *Comparative analysis* can be a very fruitful exercise but in agriculture the comparisons are often difficult, if not impossible, to find.

While it may not be very clear at this point, none of these types of analysis is exclusive; they often use the same basic information and often overlap. There is in fact, another way to view the business analysis process. That view is that the four types of analysis mentioned above (ratio, horizontal, vertical, and comparative) are really just methods to begin analyzing either the whole business or a specific enterprise (Figures V–1 and V–2).

The first two chapters of this Part will approach business analysis in the latter fashion. We will define a series of ratios and other parameters in Chapter 22 and begin the process of actually using that information in Chapter 23.

Chapter 24 covers a topic somewhat unrelated to the other two chapters in this part of the book. Its primary emphasis is on how to go about putting a records system into place to meet individualized needs. The earlier chapters of this book have provided a broad range of records to choose from. Not all are necessary for every farm

[1]Although not a complete list, the following books comprise a comprehensive yet relatively inexpensive addition set for a good farm and ranch business library:

Ronald D. Kay, *Farm Management: Planning, Control, and Implementation,* 2nd ed. (New York: McGraw-Hill Book Company, 1986)

Peter H. Calkins, and Dennis D. DiPietre, *Farm Business Management: Successful Decisions in a Changing Environment* (New York: Macmillan Publishing Co., Inc., 1983)

John B. Penson, Jr., Danny A. Klinefelter, and David A. Lins, *Farm Investment and Financial Analysis* (Englewood Cliffs, NJ: Prentice-Hall, Inc., 1982)

John B. Penson, Jr., and Clair J. Nixon, *Understanding Financial Statements in Agriculture* (College Station, TX: Agri-Information Corporation, 1983)

John B, Penson, Jr., and Clair J. Nixon, *How to Analyze Financial Statements in Agriculture* (College Station, TX: Agri-Information Corporation, 1985)

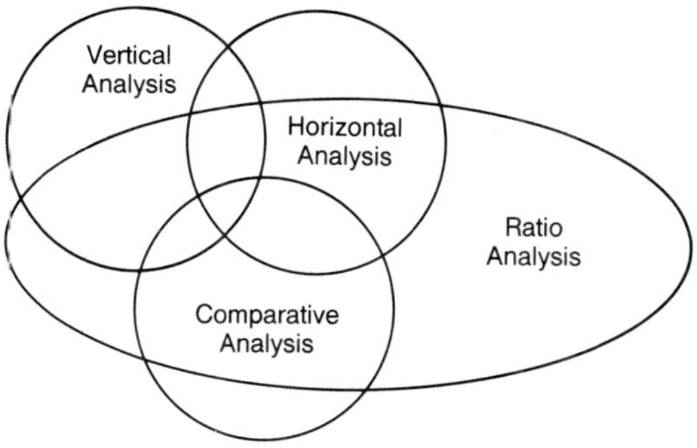

FIGURE V–1. Methods of business analysis.

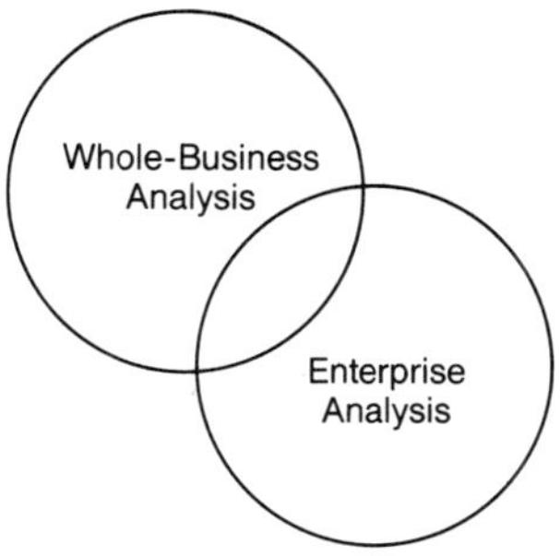

FIGURE V–2. Types of business analysis.

or ranch business. The most applicable parts need to be selected to fit individual circumstances. Furthermore, a way to make these records operational must be found: should a commercial package be used, should a self-designed system with no more structure than a series of accounting pads be used, or should a computer be purchased?

CHAPTER 22

Measuring Business Performance

The impetus to begin farm or ranch business analysis usually starts with one of three major elements: (1) a desire on the part of the farmer or rancher to generate more earnings from his business; (2) a statement or formulation of a particular goal by the farmer or rancher concerning a continuing problem such as insufficient funds to meet family living needs, poor earnings, or eroding equity; or (3) the statement of a problem noted or caused by an external source, such as a loan payment delinquency, a large tax bill due which will severely hamper operating plans, or a credit application rejection.

Most farmers and ranchers will at one time or another begin to feel paranoid or persecuted by the problems which are native to a farm or ranch business: low or volatile prices, impossibly high taxes, drought, excessive rain, pests, and a host of others. The tendency of many is to blame others for their problems to such an extent that they feel that nothing can be done. In fact, in a recent survey of Washington state farmers, in reply to a question as to what was the primary problem facing agriculture, none of the top ten responses related to items under the farmer's control.

Despite the feeling that so many of the factors influencing farm and ranch earnings are beyond the control of the individual farmer or rancher, there are many things an astute businessman can do to ensure success or improve his business and personal situation. By taking charge of events and shouldering responsibility for decisions rather than continually blaming others or external events, a farmer or rancher can help determine his own destinty. Although luck is definitely helpful, industrial psychologists will generally tell us that the most successful businessmen are those individuals who take charge of their own affairs, make sound decisions based on the best information available, and shoulder the responsibiity for their decisions.

A good decision, as distinguished from a lucky decision or break, is by its nature based on sound information. The best information is right at the fingertips of the decision maker: in his records system. To make it work, about all that he or she needs is a set of tools to pull the information apart and put it back together again in a logical fashion. The businessman must first discover what he has in his business and then must objectively analyze it to find strong points and weak points. The final step is to reassemble the parts into a plan for the future. The road map of the plan is his set of written long-term and short-term goals. The vehicle to achieve those goals is a set of strategies based upon sound economic planning.[1] But the beginning is always a thorough knowledge and analysis of the current operation, which can only be accomplished through a fundamentally strong set of farm or ranch records.

Performance Analysis Measures

The best way to begin analyzing a business is to perform a series of mechanical calculations built from existing financial statements and other farm and ranch records. These calculations can usually be done quickly and easily and can be laid out in a logical fashion for later thought. In other words, first lay out the most useful measures of business performance. After the measures are in hand, attention can be turned to finding patterns and identifying problems.

Probably the most useful place to begin these calculations is with a set of measures called financial ratios. These ratios can be calculated directly from information contained in the financial statements. For ease of definition, a long list of financial and physical performance measures is presented in Table 22–1. These measures can be roughly divided into six categories:

1. Liquidity
2. Solvency
3. Profitability
4. Financial (economic) efficiency
5. Physical efficiency
6. Marketing efficiency

Many of the measures listed apply to the business as a whole and many apply to a specific enterprise. We shall stress these uses in Chapter 23.

Each one of the six categories requires some explanation, but as noted in the introduction to this chapter, it is beyond the scope of this text to provide all of the financial background each of these measures deserves. Consequently, these descriptions should be viewed as an introduction to further studies of agricultural financial analysis, such as that provided by Penson and Lins.[2]

[1] See Ronald D. Kay, *Farm Management: Planning, Control, and Implementation,* 2nd ed. (New York: McGraw-Hill Book Company, 1986).

[2] See John B. Penson, Jr., and David A. Lins, *Agricultural Finance: An Introduction to Micro and Macro Concepts* (Englewood Cliffs, New Jersey: Prentice-Hall, Inc., 1980).

TABLE 22–1 Financial and Production Performance Measures

Category and Ratio	Definition
Liquidity	
Current ratio	Current assets ÷ current liabilities
Quick ratio	Liquid assets ÷ current liabilities
Debt structure ratio	Current liabilities ÷ total liabilities
Solvency	
Leverage ratio	Total liabilities ÷ net worth
Net capital ratio	Total assets ÷ total liabilities
Equity-value ratio	Net worth ÷ total assets
Current-debt-to-worth ratio	Current liabilities ÷ net worth
Profitability	
Rate of return on total capital	(Return to total capital ÷ total assets) × 100%
Rate of return on equity	(Return to equity capital ÷ net worth) × 100%
Profit-to-asset ratio	Net farm or ranch income ÷ total assets
Profit-to-equity ratio	Net farm or ranch incomes ÷ net worth
Economic Efficiency	
Turnover ratio	Value of farm or ranch production ÷ total productive assets
Gross ratio	Total farm or ranch expenses ÷ value of farm or ranch production
Profit-to-gross income ratio	Net income ÷ value of farm or ranch production
(enterprise ratios)	
Crop yield per dollar of cost	Per-acre yield ÷ cost of production
Livestock production per dollar of cost	Livestock production ÷ cost of production
Total costs per acre	Total costs for each enterprise ÷ acreage in that enterprise
Total costs per dollar of receipts	Total costs ÷ receipts
Total costs per unit	Total costs ÷ units of production (cwt., bu., head)
Variable costs per acre	Variable costs ÷ acreage
Variable costs per unit	Variable costs ÷ units of production
Fixed costs per acre	Fixed costs ÷ acreage
Fixed costs per unit	Fixed costs ÷ units of production
Labor costs per acre	Labor costs ÷ acreage
Labor costs per unit	Labor costs ÷ units of production
Power and machinery costs per acre	Power and machinery costs ÷ acreage
Power and machinery costs per unit	Power and machinery costs ÷ units of production
Feed costs per unit	Feed costs ÷ units of production
Total investment per acre or unit of production	Total assets ÷ total acreage or total units of livestock
Building investment per unit	Building assets ÷ units
Machinery investment per unit	Machinery assets ÷ units
Family earnings per dollar invested	Net family withdrawals ÷ total assets
Physical Efficiency	
Crop yield per acre	
Livestock yield per head or unit	
Eggs per hen	
Calves per cow (percentage calf crop)	
Pigs per litter	
Pigs raised per sow	
Pounds of milk per cow	
Marketing Efficiency	
Price received rate	(Price received ÷ average price) × 100%

Liquidity

The financial concept of *liquidity* refers to financial soundness of a business in a general sense, but more specifically, liquidity refers to the ease with which a business can meet its debt obligations as they come due or to meet unexpected cash requirements. Assets are often ranked according to their liquidity, that is, how quickly and easily they can be converted into cash. Since debt obligations must be met in cash, the liquidity of an asset is an important concept. In fact, the classified balance sheet, as discussed in Chapter 13, attempts to rank or classify assets into one of three categories (current, intermediate, and long-term) according to their liquidity. In a bit broader sense, liquidity of the business is defined as the ability to convert assets to cash to meet current obligations. The underlying concern is that the business should be able to meet its current obligations with current assets, or by maintaining reserves of unused credit but without being forced to sell its productive assets, such as machinery, breeding livestock, or land.

Several measures of liquidity are available, each of which highlights a slightly different concern. The most common measure is the *current ratio,* but many financial experts argue that the *quick ratio* is more relevant. Another, slightly different measure is the *debt structure ratio.*

Current Ratio

The current ratio is simply the ratio of current assets to current liabilities:

$$\text{Current Ratio} = \frac{\text{Current Assets}}{\text{Current Liabilities}}$$

Current assets are those that turn over during a normal year's operation, whereas current liabilities are those that will come due within one year. Examples of current assets include cash, inventories of crops and livestock to be sold, feed, supplies, and prepaid expenses. Current liabilities include short-term notes, principal on intermediate and long-term notes due within one year, and accrued interest.

A ratio of 1:1 is considered to be the lowest that the current ratio should ever be. If it falls below 1:1, current liabilities exceed current assets and the farmer or rancher *must* rely upon income or further borrowing at the proper moment to meet current obligations as well as living requirements. Most lenders feel that a 2:1 ratio or higher is safe. While lenders are usually a bit too conservative in their rules of thumb, a 2:1 current ratio is probably a reasonable goal for most farmers and ranchers. A higher ratio may mean the owner is retaining too many assets in cash or near-cash forms. Those assets might be better used to purchase intermediate or long-term productive assets to enhance profitability.

Quick Ratio

The *quick ratio,* often called the *acid-test ratio,* defines liquid assets more narrowly than the current ratio.

$$\text{Quick Ratio} = \frac{\text{Liquid Assets}}{\text{Current Liabilities}}$$

where Liquid Assets = Cash + Accounts Receivable + Marketable Securities

The definition of *liquid assets* specifically excludes inventories of crops, livestock, and supplies, prepaid expenses, and cash investment in growing crops. The purpose behind excluding these items for a measure of liquidity is that inventories of supplies, prepaid expenses, and cash investment in growing crops usually cannot be converted to cash (sold) to meet debt obligations. Furthermore, sales of inventories such as crops and livestock may need to be delayed to achieve market price expectations.

A quick ratio value of 1:1 or greater is considered to be relatively sound, as with the current ratio. Nevertheless a few lenders would feel a 2:1 value is necessary.

Debt Structure Ratio

The *debt structure ratio,* or *current debt ratio,* measures current liabilities relative to total liabilities.

$$\text{Debt Structure Ratio} = \frac{\text{Current Liabilities}}{\text{Total Liabilities}}$$

The debt structure ratio indicates the portion of total debt which must be repaid within one year.[3] A ratio of 1:4 suggests that 25% of total debt must be repaid within one year. The higher the ratio, the greater will be the drain on cash reserves and the more the risk of being unable to repay current liabilities on time. A high ratio of current to total liabilities may indicate a need to refinance some short-term

A great deal of care must be used in the interpretation of the debt structure ratio. In fact, it probably should never be used without one of the other ratios to put it in perspective. An example of an erroneous conclusion might be an individual who owns $1,000,000 in assets but only owes $1,000 in total liabilities. Even if his debt structure ratio is 1:1 (which by itself would indicate a high degree of risk), his financial position is obviously quite strong. On the other hand if current assets just equal current liabilities and the debt structure ratio is very acceptable (such as 1:10), financial difficulties probably lie not too far ahead.

Solvency

Solvency is a financial concept that relates to the overall financial strength or equity position of the business owner. In the extreme, an insolvent business is technically bankrupt; in other words, liabilities exceed assets. Solvency can be defined as the ability to convert all assets to cash, to use that cash to retire all

[3]It is once again important to remember that principal payments on longer-term liabilities due within one year are properly classified as current liabilities.

debt, and to have some cash left over. Since this conversion will not be done often, solvency really describes the long-run financial strength of the business.

The most common measure of solvency is the *leverage ratio*. Another way of expressing the same concept is the *net capital ratio*. Although both of these ratios may be thought to be unnecessary since a business could be called solvent if assets exceed liabilities, leaving net worth positive, the ratios put the picture of solvency in relative terms. For example, a farmer with a net worth of $10,000 is much safer if his liabilities total $15,000 than he is if liabilities total $1,500,000. Other solvency measures include the *current debt-to-worth ratio* and the *equity-value ratio*.

Leverage Ratio

The leverage ratio measures liabilities relative to net worth.

$$\text{Leverage Ratio} = \frac{\text{Total Liabilities}}{\text{Net Worth}}$$

The leverage ratio is a very important measure of long-term financial strength because it measures the relative contribution of the businessman's own capital to borrowed (another individual's) capital. While it is an absolute requirement that the leverage ratio be *positive* (a negative value means net worth is negative), a value of 1:1 is considered to be a minimum by many. A value of 1:1 means creditors actually own half of the business. A value of 2:1 means creditors have $2 invested for every dollar invested by the business owner. Thus, a value larger than 1:1 implies a greater risk exposure, a lower equity cushion if the business would happen to fail. Furthermore, a softening of land values, typical in the early 1980s, can cause solvency positions to weaken rapidly if care is not taken to interpret risk exposure through the leverage ratio.

Net Capital Ratio

The *net capital ratio* measures virtually the same concept as the leverage ratio since there is a direct mathematical relationship between the two.

$$\text{Net Capital Ratio} = \frac{\text{Total Assets (TA)}}{\text{Total Liabilities (TL)}}$$

The net capital ratio must be greater than 1:1 since a value such as 1:2 would imply that liabilities exceed assets and thus net worth is zero. The net capital ratio, like the leverage ratio, measures the ability of the business to repay debts if forced to liquidate. A net capital ratio of 2:1 is equivalent to a leverage ratio value of 1:1.[4]

[4]The relationship between the net capital ratio and the leverage ration can be seen easily:

$$LR = \frac{TL}{NW} = \frac{TL}{TA - TL} = \frac{TL/TL}{TA/TL - TL/TL} = \frac{1}{TA/TL - 1}$$

$$\text{Since: } NCR = \frac{TA}{TL},\ LR = \frac{1}{NCR - 1}$$

Equity/Value Ratio

The equity/value ratio once again restates the information contained in the leverage ratio in a different manner.

$$\text{Equity/Value Ratio} = \frac{\text{Net Worth}}{\text{Total Assets}}$$

The equity-value ratio provides a quick reference point for the percentage of the business owned and paid for by its operator. A value of 1:2 implies that the businessman owns 50% of his business free and clear and creditors have a 50% claim on his assets. This 1:2 value is equivalent to a leverage ratio of 1:1.[5]

Current-Debt-to-Worth Ratio

The current-debt-to-worth ratio measures a different concept than the other solvency ratios. It is actually more closely aligned with the debt structure ratio than it is with the other solvency ratios.

$$\text{Current-Debt-to-Worth Ratio} = \frac{\text{Current Liabilities}}{\text{Net Worth}}$$

This ratio is a less severe measure of solvency in that it recognizes that all liabilities are not created equal, in the sense that although a long-term liability is real and must eventually be retired, there is quite a bit of time before the debt actually becomes due.

A high current-debt-to-worth ratio suggests more about the need to restructure debt than about insolvency or eventual bankruptcy. But it can be a definitive early indicator of impending problems.

Profitability

Many commonly used measures have been developed to provide a sense of how efficiently the business is using its productive assets to generate income. Unlike the liquidity and solvency ratios, *profitability measures* use income statement data as well as balance sheet information. Profitability measures either determine total amounts of income or relate some measure of net income to the value of assets or equity investment. The relationships are usually expressed in percentage terms, such as "a 4% rate of return on equity capital," and are often compared to market interest rates on other nonfarm or ranch investments.

[5] The relationship between the equity-value ratio and the leverage ratio can also be seen clearly:

$$LR = \frac{TL}{NW} = \frac{TA - NW}{NW} = \frac{TA}{NW} - 1$$

$$\text{Since } EVR = \frac{NW}{TA},\ LR = \frac{1}{EVR} - 1$$

The most common profitability measures are net farm income, return to operator labor and management, *rate of return on total capital, rate of return on equity capital,* and *profit ratios.*

Net Farm Income

As defined in Chapter 14, *net farm* or *ranch income* measures the accrual-based income for the farming or ranching operation for a given period of time, usually one year. Unfortunately different users define *net farm income* differently, but the most consistent and defensible definition is that net farm income is return to operator labor, management, equity (or net worth) and risk. Although more lengthy, this definition is more specific as it clearly details which resources have not yet had a charge computed and allocated. In other words, when net farm income is calculated, a return to operator labor (such as wages for work performed), management, equity, or risk (pure economic profit) has not yet been charged and subtracted from gross returns.

Measurements of net farm income are definitely necessary and useful as they provide an indication of the level of profitability. But absolute values such as a dollar amount of income can be difficult to interpret. A $20,000 net income for a 100-acre grain farm may well be very good whereas a $20,000 net income for a 10,000-acre fruit and vegetable operation is likely to be unacceptable. In other words, the absolute dollar amounts must be interpreted relative to size of operation, expectations, and needs of the owners.

Return to Operator Labor and Management

A measure of the return to operator labor and management can provide an estimate of the salary earned by the operator through his personal input of time, planning, and other managerial tasks. It specifically excludes the contribution of his capital to profit (defined as net farm or ranch income). Capital, whether invested in a farm, ranch, or other investment opportunity should earn a return of its own.

Usually an assumed interest rate is multiplied by owner's equity and that amount is subtracted from net farm income to obtain return to operator labor and management. This return measure, like net farm income, is an absolute dollar amount rather than a percentage and must be interpreted carefully. Usually it should be compared to the size of operation to judge adequacy.

Rate of Return on Total Capital

The key to proper calculation of the *rate of return on total capital* lies in proper definition of the return figure needed in the numerator of the rate equation.

$$\text{Rate of Return on Total Capital} = \frac{\text{Return to Total Capital}}{\text{Total Assets}} \times 100\%$$

The return to total capital is defined as net farm or ranch income (from the income statement) plus interest paid on farm or ranch loans minus the value of unpaid operator and family labor and management. Interest paid on business loans is added back to net farm or ranch income since it represents the return to borrowed capital. (The interest the borrower pays is the income of the lender and the return to his loaned money.) The value of unpaid labor and management is often difficult to estimate, but it must be deducted from net farm or ranch income because that amount is a return to labor and management, essentially an equivalent of wages or salaries, rather than a return or income amount generated by asset investment.

Typically, rate of return on total capital in agriculture will be significantly less than rates of return or interest rates common to other industries. A 2 to 3 percent rate of return on farm and ranch capital is not uncommon. These low return figures can cause a great concern about the overall financial strength of U.S. agriculture, but we must be careful not to compare apples to oranges. It is often stated that farmers and ranchers are cash poor and land rich or that they live poor and die rich. Much of the great discrepancy between agricultural rates of return and industrial rates can be explained by analyzing the expectations of farmers and ranchers with respect to future appreciation in asset values. Because they pay more for farm and ranch land than can be justified on a strict rate-of-return basis, it must be true that farmers and ranchers are willing to trade current income for future security and expected capital gain profits, or that they want to hold land as a hedge against expected inflation. This trend of reasoning partially explains why farmers and ranchers will not sell their land to reinvest the proceeds in savings accounts even though the farm and ranch rates of return on capital are usually much lower.

Rate of Return on Equity Capital

The rate of return on equity capital is defined and interpreted in much the same way as the rate of return on total capital.

$$\text{Rate of Return on Equity Capital} = \frac{\text{Return to Equity Capital}}{\text{Net Worth}} \times 100\%$$

The same problem exists here with the rate of return to equity capital as in the previous section: the definition of *return to equity capital.* Actually the calculation is easy if return to total capital has already been computed. Return to equity capital is defined as net farm or ranch income minus the value of unpaid operator and family labor and management. Or it can be defined as return to total capital minus interest paid on farm or ranch loans. The arithmetic manipulations of interest paid can be very confusing. Interest paid represents returns to borrowed capital and we are now concerned only with returns to equity capital.

Since rates of return on agricultural assets are typically lower than other return rates, including the interest rate on borrowed capital, the rate of return on equity capital will usually be less than the rate of return on total capital. This rate of return provides a somewhat clearer picture of farm and ranch profitability than the rate of return on total capital since it excludes the lender's returns and capital.

Profit Ratios

Profit ratios are sometimes used in place of the two previous rates of return, primarily because they are easier to compute. Profit ratios simply use net income in place of the return figures computed above.

$$\text{Profit-to-Asset Ratio} = \frac{\text{Net Farm or Ranch Income}}{\text{Total Assets}}$$

$$\text{Profit-to-Equity Ratio} = \frac{\text{Net Farm or Ranch Income}}{\text{Net Worth}}$$

While it may seem nice to view overall profitability of the business with respect to income generated by all resources, for the reasons detailed above, these profit ratios mix returns to the various factors of production and thus generate a number that is difficult to interpret.

Economic Efficiency

A fourth category of performance measures allows analysis of how efficiently the business is spending its income dollars or generating income dollars. Given that income is always hard to come by and that the cash to buy new assets is always difficult to generate, it will be useful when analyzing business problems to have a few measures of how efficiently the business is creating and using income. There are several economic efficiency measures, but three of the most commonly used measures include the turnover ratio, the gross ratio, and the profit-to-gross income ratio. In addition to these three whole-business efficiency ratios, there are a whole host of efficiency ratios that can be applied to individual crop or livestock enterprises.

Turnover Ratio

The turnover ratio provides a measure of how intensively or how efficiently the productive assets available to a farmer or rancher are used in generating income. This ratio is quite different from the profitability ratios (and the liquidity and solvency ratios as well) since it will pull the value of rented assets into the calculations.

$$\text{Turnover Ratio} = \frac{\text{Value of Farm or Ranch Production}}{\text{Total Productive Assets}}$$

Value of farm production can be taken directly from the income statement of Chapter 14. Value of farm production is a measure of gross income, except that it deducts purchases of feed and feeder livestock from sales and inventory changes. Total productive assets include total assets from the balance sheet plus the value of all rented assets.

Like the profitability ratios, a higher turnover ratio is preferred since it implies greater efficiency, which should translate to higher profitability. Interpretation is difficult, though, because the link between higher turnover efficiency and higher profitability is not automatic. Efficient use of income also plays an important part in higher profitability.

Gross Ratio

The gross ratio helps to analyze the link between greater efficiency of resource use and higher profitability as it directly measures how efficiently income dollars are spent. The numerator, total farm or ranch expenses, once again comes directly from the income statement.

$$\text{Gross Ratio} = \frac{\text{Total Farm or Ranch Expenses}}{\text{Value of Farm or Ranch Production}}$$

Expenditure efficiency is enhanced by low gross ratio values. When coupled with a high turnover ratio, a low gross ratio implies good profitability and efficiency of resource use to generate net income.

Profit-to-Gross-Income Ratio

The profit-to-gross-income ratio turns the gross ratio around to provide a quick reference to profitability.

$$\text{Profit-to-Gross-Income Ratio} = \frac{\text{Net Income}}{\text{Value of Farm or Ranch Production}}$$

Interpretation is similar, but opposite in direction, to the gross ratio.

Enterprise Ratios

The list of enterprise efficiency ratios shown in Table 22–1 is not a complete list of items that should concern farmers and ranchers about individual enterprises. But this list should provide an idea of the types of items producers should look at. Their interpretations are relatively simple—lower cost ratios and higher income ratios imply greater profitability.

Physical Efficiency

Unlike the performance measures described previously in this chapter, the physical efficiency measures cannot be taken from any financial statement, with the possible exception of the cash flow budget schedules described in Chapter 15. Each of these concepts deals solely with physical production, output per unit of input, and is not concerned with dollars expended.

Too many agricultural economists downplay the importance of a strong set of physical efficiency ratios. Although intepretation can be misleading, these measures are extremely useful and necessary to a complete whole-farm or whole-ranch business analysis.

While it is true that a good technical producer cannot always produce himself out of financial difficulty, it is also true that all the business management skills in the world will not help an individual who has produced nothing to sell.

The economics of achieving maximum crop and livestock yields is often very shaky. Generally, a profit-maximizing level of production is *less* than maximum physical production. As a result, great caution must be exercised when interpreting and analyzing physical performance measures. Sound economic principles and a healthy dose of common sense must be applied to determine the best levels for physical efficiency measures. These principles are the primary concern of farm and ranch business management courses and textbooks such as Ronald Kay's *Farm Management: Planning, Control, and Implementation,* 2nd ed. (New York: McGraw-Hill, 1986).

Marketing Efficiency

Oftentimes a thorough business analysis will not turn up a definitive answer from the measures calculated in the first five performance measure categories. The indications of low profitability may be there, but a specific cause may not appear. Many times it is necessary to look directly at the prices received and prices paid by farmers and ranchers in relation to what they could have received or paid to find the answer to a low-profitability problem.

Marketing efficiency measures are less developed than the measures in our previous five categories. Simply comparing prices received and paid by an individual to averages can seem deceptively unsophisticated, but it is nonetheless very effective.

Marketing is not a subject to be taken lightly. It is often the maker or breaker of profitable farms and ranches.

Frank Farmer Example

To get a better idea of how all of these ratios and other business performance measures, look at Frank Farmer's 12/31/85 Balance Sheet and 1985 Income Statement (Figures 22–1 and 22–2). Before you begin to calculate ratios and rates of return, take a few minutes to study the general conditions of Frank's operation as well as some of the details. After a little bit of thought, calculate some of the performance measures. The calculations for some of the whole-business ratios are shown in Table 22–2. Refer to the definitions throughout this chapter and in the summary in Table 22–1.

Financial Ratio Interpretation and Analysis

Using, interpreting, and analyzing all of the measures described in this chapter and laid out in Table 22–1 is a challenging and difficult process. No two farms or

B1

B1

ASSETS

Current Assets	Modified Cost	Market Value
Cash (on hand) or in checking	$ 102,000	$ 102,000
Savings accounts & time certificates	43,000	43,000
Hedging account equity		
Marketable bonds & securities (Sch. 1)	10,000	15,000
Notes & accounts receivable (good)		
Livestock & poultry to be sold:		

Kind	No.	Avg. Wt.	$/Unit	Value
CALVES	40	500#	$63	12,600

	Modified Cost	Market Value
Livestock & poultry total	12,600	12,600

Crops and feed:

Item	Quant.	Unit	$/Unit	Value
BH				
COTTON	150	BALES	290	43,500
HAY	100	T	80	8,000

	Modified Cost	Market Value
Crops and feed total	51,500	51,500

Cash investment in growing crops:

Crop	Acres	$/Ac.	Value

	Modified Cost	Market Value
Supplies (Sch. 2)	4,000	4,000
Prepaid expenses	5,000	5,000
Other		
TOTAL CURRENT ASSETS	$ 228,100	$ 233,100
Intermediate Assets		
Notes & accounts receivable (good)	$	$
Machinery, equipment, trucks (Sch. 9):		203,000
Cost or Basis $		
Less accumulated dep $	150,423	
Breeding stock (Sch. 4)	25,000	25,000
Retirement accounts (IRA, HR-10)		
Cash value of life insurance (Sch. 7)	6,000	6,000
Securities not readily mktable (Sch. 1)		
Personal & recreational vehicles (Sch. 3)	7,000	7,000
Household goods & personal effects	4,000	4,000
Other		
TOTAL INTERMEDIATE ASSETS	$ 192,423	$ 245,000
Fixed Assets		
Contracts & notes receivable	$	$
Farm real estate (Sch. 5):		1,200,000
Cost or Basis $		
Less accumulated dep $	600,000	
Non-farm real estate (Sch. 6)		
Other		
TOTAL FIXED ASSETS	$ 600,000	$ 1,200,000
TOTAL ASSETS	$ 1,020,523	$ 1,678,100

LIABILITIES AND NET WORTH

Current Liabilities	Modified Cost	Market Value
Accounts payable: Labor $ ___ Feed & seed $ ___ Repairs ___ Fert. & chem ___ Storage ___ Fuel & oil ___ Other ___ Mach. hire ___	$ 8,000	$ 8,000
Medical & other personal		
Notes payable within 12 months (to whom, maturity, purpose)		
FIRST NATIONAL	75,000	75,000
Amount delinq. on notes payable		
Principal portion of longer term debts due within 12 months (Sch. 8):		
Intermediate	10,000	10,000
Long term	75,000	75,000
Principal delinq. (interm. & long term)		
Estimated accrued interest on:		
Accounts $ ___ Notes $ ___		
Intermediate liabilities $ ___		
Long term liabilities $ ___	14,265	14,265
Estimated accrued tax liability:		
Property	3,400	3,400
Real estate		
Employer payroll withholdings	5,120	5,120
Income & Soc. Sec.	35,677	35,677
Interest delinq. (interm. & long term)		
Accrued rents & lease payments		
Other (including relatives)		
Contingent income tax liability:		
Current assets		23,718
Marketable securities		1,000
TOTAL CURRENT LIABILITIES	$ 226,462	$ 251,180
Intermediate Liabilities (Sch. 8)		
(Principal due beyond 12 months)		
Notes payable	$ 20,000	$ 20,000
Sales contracts		
Life insurance policy loans		
Other		
Contingent income tax liability:		
Machinery		26,289
Breeding stock		5,000
Securities not readily mktable		
Contingent income tax liability & interest penalty on retirement accts.		
TOTAL INTERMEDIATE LIABILITIES	$ 20,000	$ 51,289
Long Term Liabilities (Sch. 8)		
(Principal due beyond 12 months)		
Mortgage on farm real estate	$ 350,000	$ 350,000
Land contracts		
Mortgage on non-farm real estate		
Other		
Contingent capital gains tax liability:		
Real estate		120,000
TOTAL LONG TERM LIABILITIES	$ 350,000	$ 470,000
TOTAL LIABILITIES	$ 596,462	$ 772,469
NET WORTH	$ 424,061	$ 905,631
TOTAL LIABILITIES AND NET WORTH	$ 1,020,523	$ 1,678,100

Working capital (CA÷CL) ________ Debt to worth ratio (TL÷NW) ________

FIGURE 22–1. Frank Farmer balance sheet, December 31, 1985.

1 USDA-FmHA
Form FmHA 1930-3
(5-12-83)

INCOME STATEMENT

FORM APPROVED
OMB No. 0575-0061
Expires 4/86

☒ Actual
☐ Projected

For 12 Month Period
Ending December 31, 1985

Name Frank Farmer

Address Star Route 2, Box 18 Anytown, AZ 88061

REVENUE				
Crops and feed:				
cash sales	$ 243,330 (1a)			
inventory change (Sch. 1)	(±) + 14,000 (1b)	$ 257,330 (1)		
Livestock & poultry sold:				
cash sales	$ 112,074 (2a)			
inventory change (Sch. 1)	(±) − 10,000 (2b)	102,074 (2)		
Breeding stock:				
cash sales	$ 750 (3a)			
inventory change (Sch. 1)	(±) + 2,000 (3b)	2,750 (3)		
Products: livestock and poultry — cash sales		(4)		
Custom work: cash		(5)		
Government payments and patronage dividends		4,000 (6)		
Income from hedging transactions (Sch. 2)	(±)	5,450 (7)		
Other		(8)		
Adjustment in notes and accounts receivable (Sch. 1)	(±)	(9)		
Gross revenue (Add lines 1 thru 9)	$	371,604 (10)		
Less livestock & poultry purchases (feeder & breeding)	(−)	55,000 (11)		
Less feed purchased	(−)	(12)		
VALUE OF FARM PRODUCTION			$	316,604 (a)
EXPENSES				
Cash operating expenses (Sch. 3)	$	121,318 (13)		
Expense adjustment (unused assets) (Sch. 4)	(±)	+ 2,000 (14)		
Expense adjustment (unpaid items) (Sch. 5)	(±)	(15)		
Depreciation:				
Machinery and equipment	$ (16a)			
Fixed farm improvements	$ (16b)	60,167 (16)		
Total operating expenses (Add lines 13 thru 16)				183,485 (b)
Income from farm operations			(a−b) $	133,119 (c)
Less interest expense (Sch. 6)			(−)	63,431
Gain or loss on disposal of farm assets (Sch. 7)			(±)	+ 39,220
NET FARM INCOME			$	108,908 (d)
NON-FARM INCOME				
Operator's wage off farm (net of expenses incurred)	$	5,000 (17)		
Spouse's wage off farm (net of expenses incurred)		28,000 (18)		
Interest and dividends (including int. on retirement accts.)		1,075 (19)		
Gain or loss on sale of non-farm assets (Sch. 8)	(±)	+ 230 (20)		
Net income — other farms		(21)		
Net income — non-farm real estate		(22)		
Other		(23)		
NET NON-FARM INCOME (Add lines 17 thru 23)			$	34,305 (e)
Income before income taxes and extraordinary items			(d+e) $	143,213 (f)
Provision for income and Soc. Sec. taxes (all income) (Sch. 9)				38,677 (g)
Income before extraordinary items			(f−g) $	104,536 (h)
Extraordinary items (explain)				(i)
NET INCOME			$	104,536

The information requested on this form is being collected in order (1) for FmHA to make loan and/or servicing decisions and (2) to provide management assistance to borrowers. The FmHA retains the form in the borrower's file for future reference. The information collected on this form is required in order for a borrower to obtain the benefits of a farmer program loan from FmHA.

FIGURE 22–2. Frank Farmer income statement, 1985.

TABLE 22-2 Selected Financial and Production Performance Measures for Frank Farmer

Category and Ratio	Calculation: Modified Cost	Calculation: Market Value	Source
Liquidity			
Current ratio	$\frac{228,100}{226,462} = 1.01$	$\frac{233,100}{251,180} = 0.93$	Balance sheet
Quick ratio[a]	$\frac{155,000}{226,462} = 0.68$	$\frac{160,000}{251,180} = 0.64$	Balance sheet
Debt structure ratio	$\frac{226,462}{596,462} = 0.38$	$\frac{251,180}{772,469} = 0.33$	Balance sheet
Solvency			
Leverage ratio	$\frac{596,462}{424,061} = 1.41$	$\frac{772,469}{905,631} = 0.85$	Balance sheet
Net capital ratio	$\frac{1,020,523}{596,462} = 1.71$	$\frac{1,678,100}{772,469} = 2.17$	Balance sheet
Equity-value ratio	$\frac{424,061}{1,020,523} = 0.42$	$\frac{905,631}{1,678,100} = 0.54$	Balance sheet
Current debt-to-worth ratio	$\frac{226,462}{424,061} = 0.53$	$\frac{251,180}{905,631} = 0.28$	
Profitability			
Net farm income	108,908		Income statement
Return to operator labor & management[b]	108,908 − 0.10 (424,061) = 66,502	108,908 − 0.10 (905,631) = 18,345	Income statement
Rate of return on total capital[c]	$\frac{142,339}{1,020,523} \times 100\% = 13.95\%$	$\frac{142,339}{1,678,100} \times 100\% = 8.48\%$	IS/BS
Rate of return on equity capital	$\frac{78,908}{424,061} \times 100\% = 18.61\%$	$\frac{78,908}{905,631} \times 100\% = 8.71\%$	IS/BS
Profit-to-asset	$\frac{108,908}{1,020,523} = 0.107$	$\frac{108,908}{1,678,100} = 0.065$	IS/BS
Profit-to-equity ratio	$\frac{108,908}{424,061} = 0.257$	$\frac{108,908}{905,631} = 0.120$	IS/BS
Economic Efficiency			
Turnover ratio[d]	$\frac{316,604}{1,220,523} = 0.259$	$\frac{316,604}{1,878,100} = 0.169$	Income statement
Gross ratio	$\frac{246,916}{316,604} = 0.780$		Income statement
Profit-to-gross income ratio	$\frac{108,908}{316,604} = 0.344$		Income statement

[a] Excludes livestock and poultry to be sold, crops and feed, supplies and prepaid expenses.
[b] Assumes opportunity cost of capital is 10%.
[c] Assumes value of unpaid operator and family labor and management is $30,000.
[d] Assumes $200,000 value for rented cropland.

ranches are alike, so none is directly comparable to another. The size of operation, debt structure, debt load, type of assets controlled, age of operator, and many other factors enter to distort direct comparisons. As a result, analysis is closer to being an art than a science despite all of the fancy ratios and rates of return.

A framework to approach analysis is needed to begin the process of uncovering or even defining problems. We will try to provide a few alternative approaches to begin analysis in the next chapter.

Summary

A very common approach to analyzing the performance of any business, including farms and ranches, begins with the calculation of a series of well-accepted performance analysis measures. Each of these measures usually provides a clear view of only one small piece of the business, but when put together, a string of these measures can often point to problem areas in the business.

There are six major categories of financial and physical performance measures:

1. Liquidity
2. Solvency
3. Profitability
4. Financial (or economic) efficiency
5. Physical efficiency
6. Marketing efficiency

Liquidity refers to the ability to generate cash to meet current obligations. Liquidity ratios are calculated from numbers on the balance sheet. Examples of liquidity ratios include the current, quick, and debt-structure ratios. Solvency ratios are also calculated from balance sheet information, but solvency refers to the overall financial strength or equity position of the owner. Typical solvency ratios include the leverage and net capital ratios.

Profitability measures turn to the income statement to determine how well a business is performing by measuring the ability of the business to earn income from the productive assets available. Examples of profiability measures include rates of return on total capital and equity capital, profit ratios, and returns to various factors of production. Economic efficiency measures are very similar to profitability measures and also use the income statement as the primary information source. They measure efficiency in generating income. Typical efficiency ratios include the turnover and gross ratios as well as several enterprise ratios.

Physical and marketing efficiency measures usually require further details concerning individual enterprises rather than the aggregated whole-business reports such as the balance sheet and income statement. Physical efficiency measures can present a story of how efficiently the farmer or rancher is producing his commodities, whereas the marketing efficiency ratios can tell how efficiently he is pricing his commodities.

CHAPTER 23

Methods of Business Analysis

Troubleshooting farm and ranch business problems is an extremely difficult process primarily because although the symptom may be known and steps to cure the problem can be prescribed, diagnosing the specific problem or cause of the problem is not an easy process. Problems often have more than one cause and at times several apparently different problems can be traced to just one cause. Troubleshooting difficulties can be directly related to the steps in the scientific method of problem solving first suggested by John Dewey:

1. Clearly define the problem.
2. Gather objective facts and arrange them in relevant order.
3. Evaluate and analyze the data.
4. Draw conclusions or make decisions based upon the analysis.
5. Take appropriate action.

Step 1, recognition and definition of the problem, is the stumbling block. Steps 2 and 3 are relatively mechanical processes, whereas steps 4 and 5 truly represent the role of a manager; he must decide, act and bear responsibility for those actions.

Before we proceed to problem identification methods, let's put this troubleshooting difficulty in a clearer light. Osburn and Schneeberger[1] note that the three most common symptoms of farm and ranch earnings problems are:

1. Low return on investment
2. Low labor and management return
3. Cash flow problems

[1]Donald D. Osburn and Kenneth C. Schneeberger, *Modern Agricultural Management* (Reston, VA: Reston Publishing Company, 1983), pp. 134–149.

Suppose Ron Rancher comes to you for financial advice. Ron is having difficulty meeting his debt commitments and providing a decent living for his family at the same time. You are well acquainted with Ron and know him to be a serious individual who is working very hard to make a success of his 350-cow ranch. Now your responsibility is to help him figure out what is going wrong and suggest ways to correct those problems. Where do you begin?

Hopefully, it is clear from the first twenty-two chapters of this book that you are going to encourage Ron to get his records system into shape and prepare a complete set of accurate financial statements. Although Ron is frustrated with this suggestion because it means a lot of paper-and-pencil work, he is convinced. In fact, Ron sets his mind to the task, gets his records and financial statements in shape, and even calculates all of the financial ratios and performance measures defined in Chapter 22. Now he is doubly frustrated—he has gone through a long, difficult process of organizing a financial records system but the problem of meeting debt and family commitments still exists. What next, Advisor?

With the symptom and all of the records data back in your lap, it becomes your task to help evaluate the records. You will soon learn that the answers to his problems may well be found in his records, but those answers have a convenient way of hiding themselves. More numbers are not necessarily better for digging out the root cause of the problem, just as an additional shovel for a man who already has six does not help him get a tree stump out of the ground.

What you need is a way to begin evaluating the mass of information Ron has prepared. In essence, you are forced to proceed at this point with a still shaky definition of the problem, Step 1 of the decision process described by Dewey. Rummaging around in the records and statements will not get you to an answer rapidly unless you get lucky. Rather, you need to develop a logical, orderly procedure to process these records.

The Diagnostic Decision Tree

One of the best diagnostic tools available to any analyst is to follow, or develop on his own, a systematic process called a *decision tree*. The purpose of a decision tree, such as the one developed by Osburn and Schneeberger and shown in Figure 23–1, is to pose a series of yes/no questions at the beginning (or left side) and progress towards specific problems (at the right side). Although many of the possible solutions posed in this decision tree are beyond the scope of this book, many can be identified solely with the information already discussed. At nearly every node or juncture in the decision, one of the financial ratios or performance measures can be called upon to provide an answer to the specific question.

This highly stylized method of diagnosing farm and ranch business problems may be excessive for some individuals. Not many financial analysts use it either, but that lack of use should not be taken as an indictment of the procedure; rather, few analysts have been trained in *how* to use it. But even if analysts were comfortable with this process, more analytical methods are needed to answer its questions. These methods are necessary primarily because the raw numbers developed in Chapter 22 are difficult to interpret. Reference points are needed to judge the

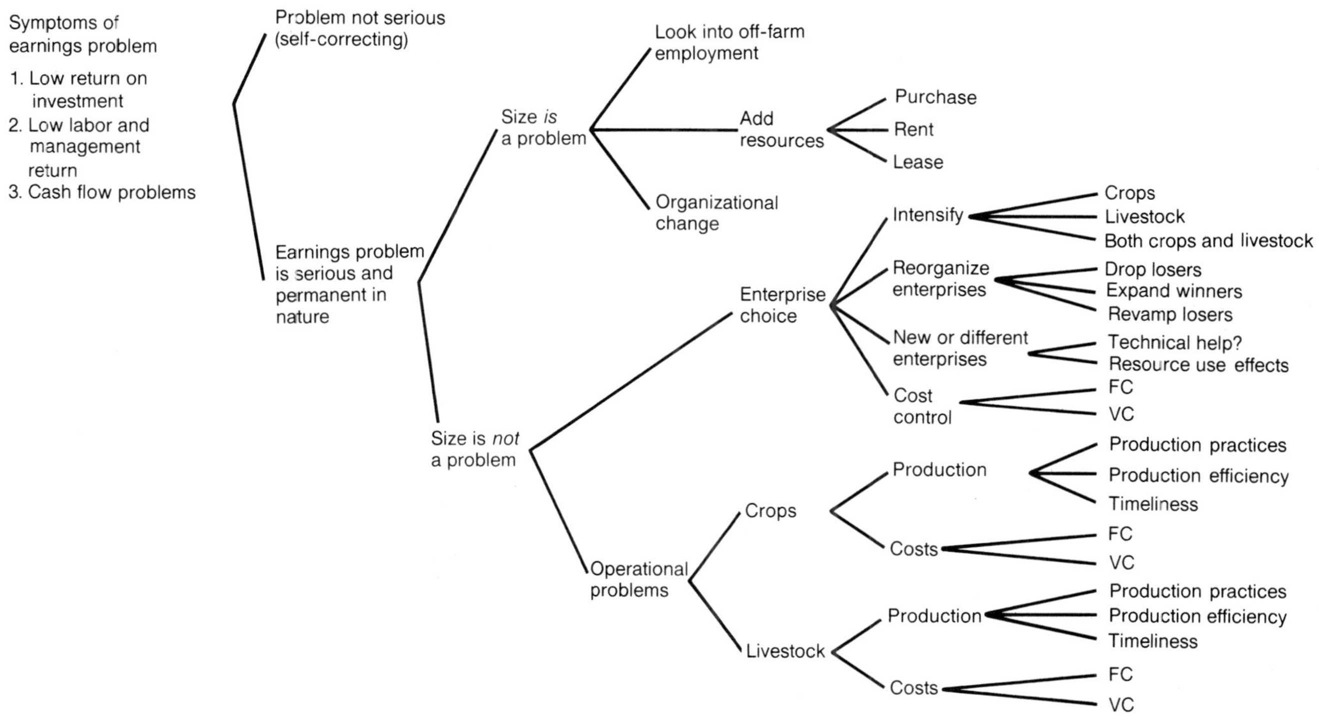

FIGURE 23–1. Diagnostic decision tree.

Source: Osburn and Schneeberger, p. 138.

relative strength or weakness of the ratio for a particular operation. These reference points can often be gained through horizontal, vertical, and comparative analysis methods.

Horizontal Analysis

Horizontal analysis is one of the most powerful tools available for diagnosing or even discovering farm and ranch business problems. The concept of horizontal analysis, often called *trend analysis,* is relatively simple. Several succeeding years' data are lined up side by side and trends of how particular measures have moved through time are assessed. For example, a trend in the current ratio of 2:1, 1:1, 1:3, and 1:5 over four years indicates clearly that the liquidity position is eroding. Though not yet a cause for great alarm, the last year's current ratio is approaching a slim margin. Debt restructuring might be needed or the ability to repay operating notes might be lost.

A convenient format to arrange data for horizontal analysis has been developed by the American Bankers' Association. That format is reproduced in Figure 23–2. It allows information from the last four years to be written side by side. The ratio analysis section can be expanded to include all or part of the measures defined in Chapter 22.

Horizontal analysis by itself is still a tricky process. It may help define the problem or it may just point out weaknesses, potential problems, or symptoms

FORM BA-2
(page 1 of 3)

COMPARATIVE ANALYSIS SHEET

Date ____________________

Name ____

Address ____

STATEMENT DATE:								
ASSETS *(line references are to Balance Sheet AB-2 or BB-2)*								
1. Current Assets (line 1)								
2. Intermediate Assets (line 2)								
3. Fixed Assets (line 3)								
4. TOTAL ASSETS (line 7)								
LIABILITIES AND NET WORTH *(line references are to Balance Sheet AB-2 or BB-2)*								
5. Current Liabilities (line 4)								
6. Intermediate Liabilities (line 5)								
7. Long-term Liabilities (line 6)								
8. TOTAL LIABILITIES (line 8)								
9. NET WORTH (line 9)								
INCOME STATEMENT *(line references are to Form AI-2 or BI-2)*								
10. Gross from sale of livestock purchased (line 1)								
11. Gross from sale of other livestock and livestock products (line 2)								
12. Gross from sale of crops (line 3)								
13. Other income from agriculture (line 4)								
14. Government payments (line 5)								
15. TOTAL CASH INCOME (line 6)								

FIGURE 23–2(a). ABA comparative analysis sheet, Form BA-2, page 1.

Source: Agricultural Bankers Division, *Agricultural Credit Analysis Handbook* (Washington, D.C.: American Bankers Association, 1975)

FORM BA-2
(page 2 of 3)

STATEMENT DATE:				
INCOME STATEMENT *(Continued)* *(line references are to Form AI-2 or BI-2)*				
16. Feeder livestock purchased (line 7)				
17. Livestock expenses (line 8)				
18. Crop expenses (line 9)				
19. Overhead and other expenses (line 10)				
20. Subtotal (line 11)				
21. TOTAL CASH OPERATING EXPENSES (line 12)				
PROFIT AND LOSS STATEMENT *(line references are to Form AI-2 or BI-2)*				
22. Net cash income from operation (line 13)				
23. Net change in inventory (line 14a)				
24. Net change in capital assets (line 14b)				
25. PROFIT (LOSS) FROM OPERATION (line 15)				
26. Other income (line 16)				
27. GROSS INCOME (line 17)				
28. Family living expenses, or gross salaries for principals, and income taxes (line 18)				
29. TOTAL INCOME (line 19)				

FIGURE 23–2(b). ABA comparative analysis sheet, Form BA-2, page 2.

without pinpointing the specific *cause* of the problem. Used in conjunction with other analysis methods and/or the decision tree, however, it can be extremely useful.

Comparative Analysis

Comparative analysis is a second very powerful analytical tool that can be used to discover or define farm and ranch problems. The basis of this method is to

FORM BA-2
(page 3 of 3)

STATEMENT DATE:				
LOAN/REPAYMENT SUMMARY				
Short term:				
amount outstanding				
amount paid				
promptness/delay				
Intermediate term:				
amount outstanding				
amount paid				
promptness/delay				
Long term:				
amount outstanding				
amount paid				
promptness/delay				
RATIO ANALYSIS				

FIGURE 23–2(c). ABA comparative analysis sheet, Form BA-2, page 3.

compare the performance measures previously defined to the same measures from similar businesses. Although comparative analysis may not be sufficient to allow statements about how the case farm or ranch stacks up against the optimum, it can allow statements of how the case example is performing relative to other producers in similar situations.

An excellent example of comparative values from the Illinois Farm Business Farm Management Association is provided in Figures 23–3 and 23–4. Although these records are available only to members of that Association, the results are usually published as shown in Figures 23–5(a) and (b) and 23–6.

The major problem with comparative analysis is the availability of comparison data. Only a handful of states have a data bank anywhere close to Illinois FBFM's. While it may be possible to compare a farm in west central Indiana to an average for east central Illinois, it is quite useless to try to compare that average to an irrigated vegetable-cotton farm in southern New Mexico. These standards are hard to obtain and usually require an expensive superstructure system.

FARM management analysis program

1981 **FARM BUSINESS ANALYSIS**

Name: SAMPLE FARM
Type: GRAIN
Soil Rating: 86-100
No. In Average: 21

Acres: 0800-1999
Feed-% Fed Range: 1-40
Feed-% Fed Average: 7
Feed-% Fed My Farm: 3

County: 201
Farm: 127
Fieldman: 199

INCOME AND EXPENSES*	OPERATOR	LANDLORD	TOTAL	AV FARM OPR SHR.	AV FARM TOTAL
FARM INCOME					
CROPS, GOVERNMENT PAYMENTS	175986	108032			270766
LIVESTOCK, LIVESTOCK PRODUCTS	7185				32772
OTHER FARM RECEIPTS	9590	1000			13917
TOTAL CASH OPERATING INCOME	192761	109032	301793		317456
INVENTORY CHANGE	29974	4591			2358
ACCTS. RECEIVABLE (net change)	2947-				140-
FARM PRODUCTS USED	350	350			2029
PURCHASED FEED	3488-				11637-
PURCHASED LIVESTOCK					8559-
GROSS FARM RETURNS	216650	113973	330623		301506
FARM OPERATING EXPENSES					
FERTILIZER	28981	18770			37836
PESTICIDES	11729	6990			13860
SEED	9343	7081			12567
DRYING & STORAGE	5164	1631			5297
LIVESTOCK SUPPLIES & SERVICES	58				1136
POWER & EQUIPMENT	27130	617			33912
INTEREST	12681			31050	31310
HIRED LABOR	16441				11557
BUILDING REPAIR	3573				2269
INSURANCE	4708				6747
TAXES	3991	16000			18307
CASH RENT & SETTLEMENTS	1000	808			8015
MISCELLANEOUS	2101				2032
TOTAL CASH OPERATING EXPENSE	126900	51897	178797		184844
FARM PRODUCED INPUTS	350	350			1413
PREPAID EXPENSE (— if increase)	6809-	2303-			1983-
ACCOUNTS PAYABLE (+ if increase)					1147-
TOTAL OPERATING EXPENSE	120441	49944	170385		183128
INCOME BEFORE DEPRECIATION	96209	64029	160238	48993	118378
LESS DEPRECIATION	32189	1833	34022	31197	33714
FARM OPERATING INCOME	64020	62196	126216	17796	84663
GAIN OR LOSS ON MCHY & BLDG SOLD				187	231
NET FARM INCOME	64020	62196	126216	17984	84895
INTEREST ON EQUITY CAPITAL	50417	96512	146929	24135	119363
UNPAID FAMILY LABOR				154	154
OPERATOR(S) LABOR AND MGMT. INCOME	13603	34316-	20713-	6305-	34622-
UNPAID OPERATOR LABOR	12900		12900	17814	17814
MANAGEMENT RETURNS	703	34316-	33613-	24119-	52436-
LABOR AND MGMT. INCOME PER OPERATOR	13603			5440-	

LAND USE AND CROP ANALYSIS

CROP	ACRES	CROP YIELDS		LAND USE %		GRAIN PRICES RECEIVED		GRAIN PRICES PAID	
	MY FARM	MY FARM	AV. FARM	MY FARM	AV. FARM	OPER.	AV.	OPER.	AV.
CORN	590	147	145	61.4	48.8	3.02	2.95	3.46	2.92
SOYBEANS	341	45	42	35.5	44.8	7.28	7.23		
WHEAT			55		2.4		3.47		4.00
OATS			119		.2		2.17		2.15
DC-SOYBEANS			23		.8				
LEGUME GRASS	30	4	3	3.1	.7				
MISCS CROPS					3.1				
TOTAL	961			100.0	100.8				

12-81 20M *FOR DETAILED CASH FLOW SEE FINANCIAL SUMMARY PAGE. SEPTEMBER 02, 1982

FARM RETURNS AND COSTS			TOTAL FARM	PER TILLABLE ACRE MY FARM	AV FARM	DIFF
FARM RETURNS:						
CROP RETURNS			326188	339.43	297.96	41.47
LVSTK RETURN ABOVE FEED			3208-	3.34-	2.38	5.72-
CUSTOM WORK			1134	1.18	3.54	2.36-
OTHER FARM RECEIPTS			4701	4.89	2.47	2.42
GROSS VALUE OF FARM PRODUCTION			328815	342.16	306.36	35.80
FARM COSTS:						
SOIL FERTILITY			38601	40.17	37.29	2.88
PESTICIDES			18719	19.48	14.61	4.87
SEED			17124	17.82	14.57	3.25
CROP TOTAL			74444	77.47	66.47	11.00
UTILITIES			1381	1.44	2.18	.74-
MACHINERY REPAIRS			9679	10.07	12.37	2.30-
MACHINE HIRE			677	.70	3.00	2.30-
FUEL & OIL			14056	14.63	16.85	2.22-
AUTO-FARM SHARE			1374	1.43	.97	.46
MACHINERY DEPRECIATION			30191	31.42	28.95	2.47
POWER & EQUIPMENT TOTAL			57358	59.69	64.32	4.63-
DRYING AND STORAGE			4217	4.39	5.41	1.02-
BUILDING REPAIR			3573	3.72	2.37	1.35
BUILDING DEPRECIATION			3831	3.99	6.24	2.25-
BUILDING TOTAL			11621	12.09	14.02	1.93-
	Months	Av. Months				
LABOR UNPAID	12.0	16.7	12900	13.42	18.76	5.34-
LABOR PAID	17.0	10.4	16441	17.11	12.06	5.05
LABOR TOTAL	29.0	27.1	29341	30.53	30.82	.29-
LIVESTOCK SUPPLIES & SERVICES			58	.06	1.19	1.13-
INSURANCE			4708	4.90	7.04	2.14-
TAXES			19991	20.80	19.11	1.69
MISCELLANEOUS			5297	5.51	2.28	3.23
INTEREST CHG, LAND 2.8 %, OTHER 14.0%			159610	166.09	156.08	10.01
OTHER COSTS, TOTAL			189664	197.36	185.70	11.66
TOTAL NON-FEED COSTS			362428	377.14	361.33	15.81
GAIN OR LOSS ON MCHY & BLDG. SOLD					.24	.24-
MANAGEMENT RETURNS			33613-	34.98-	54.73-	19.75
PROD. PER $1. NON-FEED COST			.91		.85	
FARM PRODUCTION PER MAN YEAR			136061		129822	
FARM PROD. PER $1,000 CAPITAL			78		73	
LIVESTOCK VOLUME	CWT-BEEF		200		75	
VALUE OF FEED FED			9303	9.68	20.47	10.79-

% OF PRODUCTION	DOLLARS	MY %	AV %	LAND INFORMATION	MY	AV
TOTAL OPERATING EXPENSE	168577	51.2	59.6	TOTAL ACRES	1094	1023
DEPRECIATION	34022	10.3	11.4	TILLABLE ACRES	961	958
FARM OPERATING INCOME	126216	38.3	28.8	TOTAL CROP ACRES	961	965
GROSS VALUE OF PROD	328815	100.0	100.0	SOIL RATING	95	94

INVESTMENT	OPERATOR Beg.-Yr.	OPERATOR End-Yr.	LANDLORD Beg.-Yr.	LANDLORD End-Yr.	TOTAL FARM Beg. Yr.	TOTAL FARM End-Yr.
CROPS AND FEED	120864	151928	112524	117115	233388	269043
LIVESTOCK	29685	28595			29685	28595
MACHINERY	115080	116615			115080	116615
BUILDINGS	21006	19008	18102	16269	39108	35277
SOIL FERTILITY						
LAND VALUE (CURRENT)	801990	801990	2944152	2944152	3746142	3746142
TOTAL CAP INVESTMENT	1088625	1118136	3074778	3077536	4163403	4195672

RATE EARNED ON FARM INVESTMENT % LANDLORD 2.02 AV. 2.17 TOT FARM 3.01 AV. 2.43
INVESTMENT PER MAN MY FARM 1729463 AV. 1763921

Farm Management Ass'ns and Farm Analysis groups cooperating with University of Kentucky and University of Illinois, Departments of Agricultural Economics

FIGURE 23–3. Farm management analysis program, business analysis.

FARM management analysis program

1981 LIVESTOCK ANALYSIS
FARM BUSINESS ANALYSIS ILLINOIS

Name: SAMPLE FARM

County: 201
Farm: 127
Fieldman: 199

BEEF COW ENTERPRISE	MY FARM CALVES SOLD	MY FARM 21 CALVES FED	CALVES SOLD: AV FARM	CALVES SOLD: HIGH RETURN FARMS 2	AV CALVES FED OUT	AV ALL FARMS
NUMBER OF FARMS			195	64	185	449
NUMBER OF COWS IN HERD		30	45	49	48	45
ANIMAL UNITS IN HERD		56	62	68	81	70
TOTAL POUNDS PRODUCED		19980	24686	26774	38451	31133
TOTAL RETURNS		6095	11518	16127	18449	14759
VALUE OF FEED FED		9303	9848	8233	19798	14673
RETURNS PER $100 FEED FED		65	116	195	93	100
RETURNS ABOVE FEED/COW		106-	37	161	28-	1
POUNDS OF BEEF PER COW IN HERD		666	548	546	801	691
DEATH LOSS:						
POUNDS		5350	1049	875	1793	1415
PERCENT OF POUNDS PRODUCED		26.7	4.2	3.2	4.6	4.5
FEED COST/UNIT—1000 LBS MILK OR 100 LBS BEEF		46.56	39.89	30.74	51.48	47.13
PRICE RECEIVED FOR 100 LBS SOLD		59.97	59.65	65.36	58.65	59.54
LBS FEED PER UNIT OF MILK AND BEEF:						
GRAIN		185	100	48	347	251
PROTEIN AND MINERALS		18	25	23	37	34
TOTAL CONCENTRATES		204	125	72	384	286
HAY AND DRY ROUGHAGE		850	688	537	577	616
CORN SILAGE			243	134	431	366
OTHER SILAGE			64	74	130	109
PASTURE DAYS		39	37	33	28	31
PASTURE DAYS PER ANIMAL UNIT		139	150	131	133	141
HAY EQUIV. PER COW—TONS		5.9	4.6	3.8	5.8	5.3
COST PER 100LBS. OF CONCENTRATES		7.77	7.52	8.45	6.34	6.57

SHEEP ENTERPRISE	MY FARM	AV FARM
NUMBER OF FARMS		50
TOTAL POUNDS PRODUCED		6098
TOTAL RETURNS		2169
VALUE OF FEED FED		2555
RETURNS PER $100 FEED FED		84
PERCENT LAMB CROP		127
DEATH LOSS: PERCENT OF LBS PRODUCED		10.4
FEED COST PER 100 LBS PRODUCED		41.90
PRICE RECEIVED PER 100 LBS OF SHEEP & WOOL SOLD		51.80
POUNDS FEED PER 100 LBS PRODUCED:		
CONCENTRATES		285
HAY AND DRY ROUGHAGE		524
OTHER SILAGE		
CORN SILAGE		32
PASTURE DAYS		24
HAY EQUIV.		1133
COST PER 100LBS. OF CONCENTRATES		7.03
PRICES USED ON FEED RECORDS		
PASTURE DAY	.35	
CORN	2.98	
WHEAT	3.80	
OATS	2.00	
MILK	1.15	
SOYBEANS	7.00	
2 HI-BEEF ⅓ - RETURN/$100 FEED EXCEEDS		135
3 HI-DAIRY ⅓ - RETURN ABOVE FEED/COW EXCEED		1175
4 LO-DAIRY ⅓ - RETURN ABOVE FEED/COW BELOW		930

DAIRY CATTLE ENTERPRISE	MY FARM	AV FARM	NUMBER OF COWS: 40—80	NUMBER OF COWS: 80+	EFFICIENCY: HIGH 3	EFFICIENCY: LOW 4
NUMBER OF FARMS		316	182	72	103	110
NUMBER OF COWS		62.4	58.2	99.6	64.3	58.9
PERCENT OF COWS DRY		14.2	14.4	14.0	13.6	15.6
ANIMAL UNITS IN HERD		114	108	180	122	104
TOTAL CWT OF MILK PRODUCED		8483	7859	13770	9685	7160
POUNDS OF BEEF PRODUCED		38499	35953	61291	43979	31018
TOTAL RETURNS		128668	119279	209230	155019	99195
VALUE OF FEED FED		64044	59175	103272	66424	60086
RETURNS PER $100 FEED FED		200	201	202	233	165
RETURNS ABOVE FEED PER COW		1035	1032	1063	1377	663
POUNDS OF BEEF PER COW		616	617	615	683	526
POUNDS OF MILK PER COW		13594	13503	13825	15062	12156
POUNDS OF BUTTERFAT PER COW		503	499	511	557	449
DEATH LOSS: PERCENT OF POUNDS PRODUCED		9.3	9.4	8.9	6.8	13.6
FEED COST/UNIT—1000 LBS MILK—100 LBS BEEF		51.92	51.66	51.89	47.16	58.55
PRICE RECEIVED FOR: 100 LB MILK		13.19	13.13	13.33	13.26	12.94
100 LB BEEF		52.55	52.47	53.57	58.95	48.00
LBS OF FEED/UNITS OF MILK OR BEEF: GRAIN		344	343	346	321	372
PROTEIN AND MINERALS		87	81	95	80	94
TOTAL CONCENTRATES		431	425	442	401	466
HAY AND DRY ROUGHAGE		274	288	218	211	346
CORN SILAGE		648	626	704	529	726
OTHER SILAGE		460	466	513	475	493
PASTURE DAYS		1	1			3
PASTURE DAYS PER ANIMAL UNIT		18	11	8	10	32
HAY EQUIV. PER COW, TONS		7.2	7.1	6.9	6.8	7.7
COST PER 100LBS. OF CONCENTRATES		7.43	7.49	7.33	7.50	7.52

MISC. LIVESTOCK ENTERPRISE	MY FARM	AV FARM POULTRY
NUMBER OF LITTERS FARROWED		
NUMBER OF COWS		
AVERAGE NUMBER OF HENS		17378
TOTAL POUNDS PRODUCED		28062
TOTAL RETURNS		148941
VALUE OF FEED FED		123156
RETURNS PER $100 FEED FED		120
POUNDS OF BEEF PER COW: PORK/LITTER		
PIGS WEANED PER LITTER		
POUNDS OF MILK PER COW		
EGGS PRODUCED PER HEN		226
DEATH LOSS: PERCENT OF POUNDS PRODUCED		
AVERAGE WEIGHT PURCHASED		
PRICE PAID PER 100 LBS.		
AVERAGE WEIGHT SOLD		
FEED COST/UNIT		.35
PRICE RECEIVED FOR: 100 LB. MILK		
100 LB. BEEF, PORK		
DOZEN EGGS SOLD		.49
LBS. OF FEED/UNIT: GRAIN		
PROTEIN AND MINERALS		
TOTAL CONCENTRATES		3.8
HAY AND DRY ROUGHAGE		
CORN SILAGE		
OTHER SILAGE		
HAY EQUIV. PER COW, TONS		
COST PER 100 LBS. OF CONCENTRATES		9.25

SEPTEMBER 02, 1982

Farm Management Ass'ns and Farm Analysis groups cooperating with University of Kentucky and University of Illinois Departments of Agricultural Economics

FIGURE 23–4. Farm management analysis program, livestock analysis.

1983 LIVESTOCK ENTERPRISE STANDARDS
from
Illinois Farm Business Records

Hog Enterprise	All Farms	Sow and Litter Efficiency High[1]	Low[2]	Litters Farrowed 10-49	200+	Pig Producers
Number of Farms	819	265	271	164	203	23
Pounds of Pork Produced	266,427	351,658	169,264	55,324	575,122	100,657
Pounds of Pork/Litter	1,718	1,776	1,627	1,784	1,691	529
Total Returns	114,342	158,347	67,185	22,869	250,379	67,593
Value of Feed Fed	80,529	96,066	59,828	18,439	169,397	41,429
Returns per $100 Feed Fed	141	164	112	124	147	163
Returns Above Feed per Litter	218	314	70	142	238	137
Number of Litters Farrowed	155	198	104	31	340	190
Pigs Farrowed per Litter	9.25	9.34	8.94	9.13	9.39	9.34
Pigs Weaned per Litter	7.27	7.55	6.88	7.03	7.28	7.93
Litters Farrowed/Female Year	1.72	1.79	1.57	1.39	1.80	1.50
Pigs Weaned/Female Year	12.65	13.79	10.90	10.23	13.05	11.16
Death Loss: Numbers	66	66	55	12	154	57
Pounds	4,860	5,205	4,109	1,248	10,440	1,954
% of Pounds Produced	1.8	1.4	2.4	2.2	1.8	1.9
Weight per Hog Sold	237	235	240	249	233	50
Price Received per 100 Pounds	$ 46.27	$ 47.11	$ 44.96	$ 45.44	$ 46.69	$ 79.87
Total Return per 100 Pounds Produced	42.91	45.02	39.69	41.33	43.53	67.15
Feed Cost per 100 Pounds Produced	30.23	27.32	35.35	33.33	29.45	41.16
Pounds Feed per 100 Pounds Produced:						
Farm Grains	315	292	363	352	306	319
Commercial Feeds	83	78	92	79	85	151
Total Concentrates	398	370	455	432	391	471
Pasture Days						
Cost per 100 lbs. of Commercial Feeds	$ 15.51	$ 14.48	$ 16.71	$ 17.54	$ 14.85	$ 15.77
Cost per 100 lbs. of Concentrates	7.57	7.37	7.74	7.70	7.51	8.73

1/ *High 1/3 Hog Return Per $100 Feed Fed Exceeds 148.*
2/ *Low 1/3 Hog Return Per $100 Feed Fed is Below 127.*

Dairy Cattle Enterprise	All Farms	No. of Cows 40-80	80+	Efficiency High[3]	Low[4]
Number of Farms	256	144	61	87	86
Number of Cows	62.3	58.5	95.8	67.3	58.7
Percent of Cows Dry	13.6	13.5	13.7	13.2	14.4
Animals Units in Herd	117	108	183	131	105
Total Cwt. of Milk Produced	8,850	8,290	13,990	10,397	7,479
Pounds of Beef Produced	40,377	38,083	63,180	48,529	29,044
Total Returns	$125,184	$118,048	$198,321	$154,644	$ 99,427
Value of Feed Fed	70,007	65,016	109,450	75,306	65,757
Returns per $100 Feed Fed	178	181	181	205	151
Returns above Feed per Cow	885	906	927	1,178	573
Pounds of Beef per Cow	648	650	659	721	494
Pounds of Milk per Cow	14,205	14,170	14,603	15,448	12,741
Pounds of Butterfat per Cow	525	538	540	571	471
Death Loss: % of Pounds Produced	9.8	9.9	9.1	7.0	15.5
Feed Cost/Unit-1,000 lbs. Milk or 100 lbs. Beef	$ 54.32	$ 53.73	$ 53.89	$ 49.38	$ 63.32
Price Received for: 100 lbs. Milk	12.63	12.70	12.60	12.75	12.57
100 lbs. Beef	45.62	44.69	47.09	49.14	43.16
Lbs. of Feed/Units of Milk or Beef:					
Grain	328	339	308	302	367
Protein and Minerals	102	92	117	86	139
Total Concentrates	431	431	425	389	507
Hay and Dry Roughage	260	271	214	225	327
Corn Silage	696	684	692	624	830
Other Silage	405	391	458	405	365
Pasture Days	1	1	1		2
Pasture Days per Animal Unit	14	12	14	11	20
Hay Equiv. per Cow, Tons	7.2	7.2	7.2	7.2	7.2
Cost per 100 lbs. of Concentrates	7.71	7.57	7.85	7.85	7.56

3/ *High 1/3 Dairy Return Above Feed Per Cow Exceeds 980.*
4/ *Low 1/3 Dairy Return Above Feed Per Cow is Below 740.*

Enterprise smaller than the following sizes were not included in the analyses on the page: hogs--10 litters; beef and dairy herds--10 cows; feeder cattle--10,000 lbs. weight produced. When the value of feed fed was calculated, grains were priced at the average farm prices as follows: corn--$3.06; oats--$1.91; wheat--$3.33; Commercial feeds were priced at actual cost, hay and silage at farm values and pasture at $.35 per animal unit pasture day.

Prepared by
Illinois FBFM Association in Cooperation with
University of Illinois, Department of Agricultural Economics--Champaign-Urbana
March, 1984

FIGURE 23–5(a). Livestock Enterprise Comparative Standards, page 1.

Beef Cow Enterprise	Calves Sold Ave.	Calves Sold High Return[5/]	Ave. Calves Fed Out	All Farms
Number of Farms	210	68	183	458
Number of Cows in Herd	40	47	41	40
Animal Units in Herd	55	67	70	62
Total Pounds Produced	21,660	26,479	34,954	28,307
Total Returns	$11,915	$16,930	$19,851	$15,387
Value of Feed Fed	9,632	9,306	17,200	13,345
Returns per $100 Feed Fed	123	181	115	115
Returns Above Feed/Cow	57	162	64	51
Pounds of Beef per Cow in Herd	541	563	852	707
Death Loss: Pounds	1,349	1,148	1,907	1,605
% of Pounds Produced	6.2	4.3	5.4	5.6
Feed Cost/Unit-1,000 lbs. Milk or 100 lbs. Beef	$ 44.46	$ 35.14	$ 49.20	$ 47.14
Price Received for 100 lbs. Sold	55.33	59.47	57.23	56.25
Lbs. Feed per Unit of Milk and Beef:				
Grain	107	83	322	238
Protein and Minerals	30	31	45	39
Total Concentrates	138	115	368	277
Hay and Dry Roughage	674	506	537	607
Corn Silage	332	196	377	395
Other Silage	83	52	68	77
Pasture Days	39	32	26	29
Pasture Days per Animal Unit	155	128	130	135
Hay Equiv. per Cow, Tons	4.7	3.8	5.6	5.2
Cost per 100 lbs. of Concentrates	7.87	8.12	6.56	6.83

5/ High 1/3 Return Per $100 Feed Fed Exceeds 140.

Feeder Cattle Enterprise: Average Farm	Purchase Weight 0-500	Purchase Weight 500-749	Purchase Weight 750+	Items	Feeder Pig Enterprise: Average Farm
324	81	229	14	Number of Farms	164
133,591	79,058	152,073	146,818	Total Pounds Produced	150,957
$ 83,987	$ 49,685	$ 95,619	$ 92,180	Total Returns	$ 50,253
62,557	35,376	71,194	78,536	Value of Feed Fed	42,315
134	140	134	117	Returns per $100 Feed Fed	118
590	456	602	776	Average Weight Purchased	49
$ 60.00	$ 59.85	$ 59.96	$ 60.72	Price Paid per 100 Lbs.	$ 81.38
1,050	1,000	1,053	1,144	Average Weight Sold	232
$ 61.11	$ 60.85	$ 61.04	$ 62.71	Price Received per 100 Pounds	$ 47.00
62.87	62.85	62.88	62.79	Total Returns per 100 Pounds Produced	33.29
46.83	44.75	46.82	53.49	Feed Cost per 100 Pounds Produced	28.03
5	5	5	2	Death Loss: Numbers	45
2,972	2,625	3,177	1,629	Pounds	3,429
2.2	3.3	2.0	1.1	% of Pounds Produced	2.2
				Pounds Feed per 100 lbs. Produced:	
532	488	530	703	Grain	309
46	43	47	45	Protein and Minerals	71
578	532	577	748	Total Concentrates	381
50	86	46	27	Hay and Dry Roughage	-
160	209	148	197	Other Silage	-
733	607	774	446	Corn Silage	-
-	-	-	-	Cost per 100 lbs. of Commercial Feeds	$ 15.39
-	-	-	-	Cost per 100 lbs. of Concentrates	$ 7.33

Sheep Enterprise	Average Farm
Number of Farms	54
Total Pounds Produced	5,958
Total Returns	$2,235
Value of Feed Fed	2,848
Returns per $100 Feed Fed	78
Percent Lamb Crop	124
Death Loss: % of lbs. Prod.	9.2
Feed Cost per 100 lbs. Produced	$ 47.80
Price Recd. per 100 lbs. mutton and wool sold	$ 44.94
Pounds Feed per 100 lbs. Prod.:	
Concentrates	320
Hay and Dry Roughage	570
Other Silage	67
Corn Silage	33
Pasture Days	23
Hay Equiv.	1,187
Cost per 100 lbs. of Concentrates	7.49

Poultry Enterprise	200-1,999 Hens	Flocks 2,000+
Number of Farms	5	5
Pounds of Poultry Produced	1,219	9,672
Total Returns	$ 5,053	$144,021
Value of Feed Fed	4,522	101,440
Returns per $100 Feed Fed	112	142
Returns above Feed Fed per Hen	.85	3.09
Average Number of Hens	627	13,783
Eggs Produced per Hen	166	241
Percent Production	45	66
Price per Dozen Eggs Sold	.70	.60
Cost per 100 lbs. of Concentrates	8.01	8.06
Feed Units*	9,485	283,298
Feed Cost per Unit	.48	.36
Lbs. of Concentrates/Feed Unit	5.95	4.44

* *One dozen eggs or 1.5 pounds weight produced.*

FIGURE 23–5(b). Livestock enterprise comparative standards, page 2.

1982

FARM TYPE ***	Hi GRAIN	Lo GRAIN	Hi GRAIN	Lo GRAIN	Hi GRAIN	Lo GRAIN	Hi GRAIN	Lo GRAIN	Hi GRAIN	Lo GRAIN	Hi GRAIN	Lo GRAIN
SOIL RATING -RANGE	86-100	86-100	86-100	86-100	86-100	86-100	86-100	86-100	86-100	86-100	86-100	86-100
FEED FED VOLUME %	0	0	0	0	0	0	0	0	0	0	0	0
RANGE IN SIZE -TOTAL ACRES	260-339	260-339	340-499	340-499	500-649	500-649	650-799	650-799	800-949	800-949	950-1199	950-1199
NO. OF FARMS	20	20	63	63	47	47	26	26	16	16	11	11
TOTAL ACRES	307	308	430	419	556	565	716	728	861	864	1059	1034
TILLABLE ACRES	300	299	418	403	539	543	701	697	838	843	1018	991
SOIL RATING	94	94	94	94	94	93	94	94	95	95	91	92
TOTAL MONTHS LABOR	12.3	13.2	13.1	14.2	14.8	15.9	16.6	20.0	17.1	20.4	24.9	22.6
CROP RETURNS	313.83	293.92	333.73	289.80	333.66	284.15	336.87	296.75	344.55	299.50	381.66	313.58
LVSTK RETURNS ABOVE FEED	.00	.00	.01	.02-	.04-	.01	.04-	.00	.00	.05	.00	.00-
CUSTOM WORK	5.02	3.18	4.49	6.23	3.18	3.29	1.44	1.38	2.83	3.03	3.82	3.55
OTHER FARM RECEIPTS	3.55	3.30	4.75	6.69	5.58	5.73	3.83	3.06	5.58	2.54	3.29	4.41
GROSS VALUE OF FARM PROD. / TA. ***	322.40	300.39	342.97	302.70	342.38	293.19	342.10	301.19	352.96	305.12	388.77	321.55
SOIL FERTILITY	31.38	41.59	36.49	38.32	33.93	40.91	37.74	38.54	33.45	41.48	38.62	38.23
PESTICIDES	15.78	17.72	15.40	18.40	16.95	18.03	14.52	19.84	16.06	15.46	16.22	23.55
SEED	14.58	17.01	13.81	15.54	13.59	16.85	13.32	16.44	12.67	17.13	13.50	15.86
CROP TOTAL *	61.74	76.32	65.69	72.27	64.47	75.78	65.58	74.81	62.18	74.07	68.34	77.64
UTILITIES	2.73	3.81	2.31	2.85	2.27	2.56	2.20	2.47	1.81	2.20	1.89	1.91
MACHINERY REPAIRS	9.44	14.01	10.16	14.90	10.96	13.26	9.69	12.32	11.53	15.49	9.97	12.97
MACHINERY HIRE	2.89	4.59	2.68	5.63	2.54	3.73	2.77	3.21	3.95	2.79	4.05	3.53
FUEL & OIL	12.17	15.27	12.67	14.92	13.87	14.45	11.64	14.73	10.51	14.79	14.74	13.98
AUTO-FARM SHARE	2.28	3.54	1.34	1.98	1.49	1.45	1.05	1.06	.76	1.49	.71	1.14
MACHINERY DEPRECIATION	22.84	37.01	28.50	38.28	27.10	36.76	26.30	33.68	30.94	31.65	22.88	38.69
POWER & EQUIPMENT TOTAL *	52.35	78.23	57.67	78.56	58.22	72.21	53.66	67.46	59.49	68.41	54.24	72.21
DRYING AND STORAGE	5.72	10.16	8.34	10.35	8.66	8.62	10.53	8.79	7.55	10.66	6.40	9.38
BUILDING REPAIR	1.06	2.35	.80	2.28	.94	2.02	.67	1.55	.54	1.15	.98	1.40
BUILDING DEPRECIATION	2.99	8.73	4.85	9.80	4.59	7.75	4.35	6.64	5.33	4.57	4.06	4.66
BUILDING TOTAL *	9.77	21.25	14.00	22.43	14.19	18.38	15.55	16.98	13.41	16.38	11.43	15.44
LABOR UNPAID	39.80	41.64	29.68	30.75	24.53	24.53	18.95	19.56	16.38	18.81	16.99	13.06
HIRED LABOR	3.81	8.27	3.70	7.25	4.46	6.08	6.04	11.95	4.98	8.04	9.58	12.11
LABOR TOTAL *	43.61	49.91	33.37	38.00	28.99	30.61	24.99	31.51	21.35	26.85	26.58	25.17
LVSTK SUPPLIES & SERVICES	.18	.09	.19	.22	.10	.06	.07	.17	.22	.04	.13	.03
TAXES	21.58	21.89	21.19	22.75	21.10	22.17	19.46	22.52	21.75	21.15	19.78	20.18
INSURANCE & MISCELLANEOUS	6.97	11.15	7.02	9.23	6.04	7.44	5.94	7.25	5.85	7.63	6.70	6.79
INTEREST ON CAPITAL	131.14	154.13	140.41	151.31	137.68	147.71	137.43	148.37	142.14	143.88	133.12	146.14
TOTAL NON-FEED COSTS / T.A. ***	327.34	412.97	339.55	394.77	330.80	374.38	322.67	369.09	326.40	358.40	320.32	363.61
GAIN OR LOSS ON MCHY & BLDG SOLD	.24	.18	2.92	1.19	1.89	2.16	4.15	.86	1.30	.33	.07	.45
MANAGEMENT RETURNS / T.A. ***	4.70-	112.40-	6.34	90.87-	13.47	79.03-	23.58	67.04-	27.87	52.95-	68.52	41.61-
NET FARM INCOME OPR SHARE	19138	5238	24725	7943	28806	2668	29157	7712	43938	1628	49640	479
LABOR AND MGMT. INCOME PER OPR ***	11867	16297-	15111	10782-	19590	16413-	22708	12673-	25168	16260-	25657	13467-
PRODUCTION / $1.00 NON-FEED COST	.98	.73	1.01	.77	1.04	.78	1.06	.82	1.08	.85	1.21	.88
PRODUCTION / MAN	94538	81734	130838	103240	149674	120166	173666	125849	207521	151165	191048	169189
RATE EARNED-%	3.43	1.08	3.89	1.56	4.05	1.82	4.29	2.11	4.44	2.39	5.55	2.75
% OF FARM PRODUCTION-COE	45.3	63.5	47.8	62.3	48.1	61.9	49.7	63.2	45.4	62.2	42.9	62.4
-DEPR.	8.0	15.4	9.7	15.8	9.2	15.1	8.9	13.3	10.2	11.8	6.9	13.4
-OPR INCOME	46.6	20.9	42.3	21.7	42.5	22.8	41.3	23.4	44.2	25.8	50.1	24.1
% OF LAND IN- CORN & CORN SILAGE	47.4	56.9	48.5	53.0	51.5	51.6	52.9	51.4	46.7	51.7	49.5	53.0
SOYBEANS	47.6	42.0	45.1	42.5	43.0	44.8	44.1	46.1	47.7	46.8	42.4	44.6
WHEAT	.9	.4	1.0	1.1	.3	.3	1.2	.5	.1	.7	.2	1.0
OATS	.0	.0	.0	.0	.1	.1	.0	.2	.0	.0	.0	.0
DIVERTED ACRES	.4	.2	.7	.4	.3	.6	.5	.3	.6	.5	.8	1.4
ALL HAY-PASTURE	.1	.5	.1	.3	.4	.4	.1	.3	.2	.2	.2	.0
CORN YIELD	156	146	161	145	159	147	160	151	160	150	166	153
SOYBEAN YIELD	47	43	47	45	47	44	47	45	49	46	50	48
WHEAT YIELD	54	59	65	59	62	46	58	53	74	74	66	39
PRICE RECEIVED-CORN	2.39	2.51	2.53	2.43	2.56	2.41	2.56	2.56	2.58	2.48	2.46	2.58
PRICE RECEIVED-SOYBEANS	5.90	6.02	6.04	6.00	5.98	5.97	5.96	5.97	6.10	6.01	6.14	6.13
CAPITAL INVESTMENT / MAN	1081441	1052736	1438662	1318171	1632522	1550867	1904425	1610287	2253298	1887780	1786069	2000599
CAPITAL PURCHASES - OPR SHARE	9688	7023	11671	13634	11628	18725	13962	20466	29975	15834	26691	30684
VALUE OF FEED FED / TA.	0	0	0	0	0	0	0	0	0	0	0	0
CWT OF PORK PRODUCED	0	0	0	0	0	0	1	0	0	1	0	0
CWT OF BEEF PRODUCED	0	0	0	0	0	0	0	2	0	1	0	1
AVERAGE NO. OF DAIRY COWS	0	0	0	0	0	0	0	0	0	0	0	0

FIGURE 23–6. Profitability group comparison example.

A couple of twists can be applied to comparative analysis. With the consultation of lenders and others familiar with financial aspects of farming and ranching in a specific area, a set of performance measure standards might be estimated. The most useful comparison might be to compare actual performance statistics against objective standards (from systems such as Illinois FBFM or from expert estimates) *and* against goals established by the producer himself. Many states also publish crop and livestock enterprise budgets for commodities important to their area. These "typical-farm" budgets can also be used as guideposts for comparative analysis.

Vertical Analysis

Vertical analysis deserves a few comments even though it is really more of a subset of comparative analysis than a stand-alone method. The distinctive feature of vertical analysis is its contrast to horizontal analysis. Vertical analysis is primarily used for income statement items and concentrates on data from a particular year rather than over a series of years. A typical example of vertical analysis is to compare fuel expenditures, fertilizer expenditures, and labor expenditures individually to total expenditures. If fertilizer expenditures amount to 20 percent of total expenditures while the area average is 30 percent, the case farmer may not be following normal practices in his area *or* he may be much more efficient than the standard.

Usually vertical analysis is encompassed by the calculation of the performance measures of Chapter 22 and requires comparison data.

Summary

As mentioned several times already, all of the analytical processes available are usually needed to discover and identify specific farm or ranch business problems. The overall process requires patience, but more importantly, it requires a sharp mind to peer into each performance measure, to judge that measure against theory and field comparisons, and to develop a consistent hypothesis or pattern. Eventually that pattern will emerge to assist the analyst in prescribing a cure.

Although there are several different tools and methods that can be used to process financial analysis information, they all have one thing in common: they follow a scientific method of problem solving. The steps in that problem-solving method involve:

1. Problem definition
2. Information collection
3. Data analysis
4. Decision making
5. An action response

Following this method, a farmer or rancher must state his problem and he must also make a commitment to collect the relevant financial information about his own business through his record system. He must also evaluate the information in a logical process, through horizontal analysis (usually over a period of years), through comparative analysis (comparing his situation to theory, to accepted norms or guidelines, or to averages to similar operations), or through vertical analysis (evaluations of data within a year).

CHAPTER 24

Choice of a Record System

Although a complete farm and ranch records system has been described in this text, all of the parts may not be necessary for each farmer or rancher and some may not fit his particular operation. Given a familiarity with a thorough and complete financial records package, each farmer and rancher must analyze and decide what fits his needs.

System Selection

Many university and commercial record books are available and several excellent personal computer packages are now available at relatively low cost. And of course, the option to design a system built upon parts of other systems or built upon standard accountant's analysis forms is always possible. Accountants, other bookkeepers, and supervised systems may fit particular circumstances best. In choosing a system based upon all of the various elements described throughout this text, probably the best evaluation mechanism is to (1) review Chapter 4, (2) obtain or look at each available prepared hand-computed package, (3) consider a computer purchase and investigate the adaptability of commercial programs to your needs, (4) discuss your needs with other farmers and ranchers, lenders, accountants, County Extension Agents, and Extension farm and ranch business specialists, and (5) evaluate each alternative with respect to cost, ease of use, and adaptability. Do not necessarily select a fancy system if it will not be used on a day-to-day continual basis. Make sure that the information recorded or caculated is worth the effort and the cost.

Figure 24–1 presents an excellent checklist for evaluation of a computer system. However, almost all of the same questions should be asked about all potential systems including prepared, self-developed, and accountant systems.

SURVEY LIST FOR A FARM RECORDS SYSTEM

General

A good farm financial records system depends on records which are complete, are accurate, and can be effectively used. The records can be used as a (1) service tool, (2) diagnostic tool, (3) indicator of progress, and (4) forward planning tool. The selection of a particular system, however, is a process which depends on a great many variables. These include general characteristics of the vendor, required information, desired information, and the system's ability to accomodate various entries which may arise on the farm. This general overview section allows you to judge various characteristics related to both the vendor and the farm records system. Check either the yes or no column and make any appropriate comments.

Item	Yes	No	Comments
1 Reliable and reputable vendor readily accessible to user			
2 Vendor updates system to accommodate changing tax laws			
3 Vendor updates system to meet changing needs of users			
4 Vendor provides general training in the use of the system and interpretation of the results			
5 System provides user opportunity to expand capabilities			
6 System provides for cash and/or accrual accounting			
7 System provides for single and/or double entry			
8 System allows for user-specified fiscal year			
9 System provides for different types of business arrangements			
10 System provides for the separation of business and personal income and expenses			
11 System provides clear, concise user instructions which are well indexed			
12 System provides trouble-shooting guide to any errors due to data entry or incorrect commands			
13 System writes checks			
14 System has printer or other back-up method to retain data in case of computer failure			
15 System is acceptable to your accountant or tax practitioner for tax filing			

Required Information

This required information represents output from the farm records systems which is needed to meet minimum income tax needs. A user needs to have information pertaining to income from (1) purchased items held for sale, (2) capital items, and (3) ordinary income. Other information concerning income entries is also required to complete income tax requirements. Expense information about ordinary expenses, purchased items held for sale, and capital items along with other considerations completes the income tax requirements. Cash accuracy checks are necessary if one is to trust the numbers received as output from the farm records system.

Item	Yes	No	Comments
A. Income			
1. Items purchased "held for sale"	XX	XX	XXX
a. Tax information			
b. Cash flow information			

Item	Yes	No	Comments
2. Capital items which a. Do qualify for capital gains treatment b. Do not qualify for capital gains treatments	XX	XX	XXX
3. Ordinary a. Cash b. Noncash c. C.C.C. loans/sales d. Hedging e. Other items relevant to your situation	XX	XX	XXX
4. Other considerations a. Trades and exchanges b. Barter c. Casualty losses d. Credits e. Reimbursements, refunds, and so forth f. Exclusion from income g. Nontaxable income h. Other items relevant to your situation	XX	XX	XXX
B. Expenses			
1. Ordinary a. Operating b. "Held for sale" items	XX	XX	XXX
2. Capital a. Purchases b. Depreciation or cost recovery	XX	XX	XXX
3. Other considerations a. Barter b. Soil and water conservation c. Land clearing d. Timber e. Sand, gravel, minerals, and so forth f. Nondeductible items g. Payroll reports h. Provides 1099 Misc., Int., NEC, information i. Other items relevant to your situation	XX	XX	XXX
C. Other possible records 1. Social Security 2. F.U.T.A. 3. OSHA 4. Workmen's Compensation 5. Retirement plans 6. Other	XX	XX	XXX
D. Cash accuracy check			

Desired Information

Additional desired information may not be required for income tax purposes, but it is necessary if you are going to complete the farm business analysis. This desired information includes the "Big Three" basic farm financial records; accuracy checks, enterprise and other accounts; and measures of business performance. In order to complete the analysis, you may want to measure a particular year's results against your budgeted estimates and the trend of measures appropriate for your business. Consistent asset valuation and techniques of deriving financial measures of performance are essential.

Item	Yes	No	Comments
A. Net Worth Statement			
1. Assets a. Current; inventory records b. Intermediate; capital asset records c. Long term; capital asset records	XX	XX	XXX

Item	Yes	No	Comments
2. Liabilities	XX	XX	XXX
a. Current; loan accounts			
b. Intermediate; loan accounts			
c. Long term; loan accounts			
3. Valuation method used for:	XX	XX	XXX
a. Business management			
b. Income tax			
c. Credit			
d. Estate			
4. Personal and business separated			
B. Farm Earnings Statement			
1. Cash Operating Statement	XX	XX	XXX
a. Operating Income			
b. Operating Expenses			
2. Inventory Adjustments			
C. Cash Flow Report			
1. In flows	XX	XX	XXX
2. Out flows			
D. Accuracy Checks			
1. Cash flow	XX	XX	XXX
2. Profitability/Net Worth			
3. Liabilities			
E. Enterprise Accounts	XX	XX	XXX
1. Crop	XX	XX	XXX
a. Financial			
b. Production			
2. Livestock	XX	XX	XXX
a. Financial			
b. Production			
3. Other accounts	XX	XX	XXX
a. Labor			
b. Marketing			
c. Seed			
d. Experimental			
e. Individual Machine			
f. Family			
F. Business Analysis	XX	XX	XXX
1. Profitability measures	XX	XX	XXX
a. Causes			
b. Effects			
2. Solvency	XX	XX	XXX
a. Causes			
b. Effects			
3. Liquidity/Cash Flow	XX	XX	XXX
a. Causes			
b. Effects			

FIGURE 24–1. Checklist for evaluation of records systems.

Source: R. A. Leuning and R. M. Klemme. *Selecting a Farm Business Records System.* University of Wisconsin, Managing the Farm Publication, Vol. XIII, No. 4, July 1982.

Specialized Production Records

One last consideration must influence the final decision of a financial records system. That consideration is whether the system chosen will fit in with a specialized production records system such as those used by the Dairy Herd Improvement Associations or livestock breed associations. Although those systems do not usually stress financial records, much of the information needed for financial records is already in those production records. Avoiding duplication will help to ensure continued use of both records systems.

Filing in the Farm or Ranch Office

It is obvious by now that a lot of paper will be used by a complete farm or ranch records system. The record books, receipts, cancelled checks, computer printouts and data disks, financial statements, and tax returns must all be maintained in easily accessible locations for later analysis or use. Dedicating one room of a house or service building is almost imperative as is investment in office furniture and file storage.

Organizing files is a problem all to itself, but a quick, excellent reference to farm and ranch office filing is available. That reference, *Put Paperwork in its Place* by Jorgensen, Anderson, and Geiken, can greatly simplify the crush of paperwork that is inevitable in modern farm and ranch business management.[1]

[1]Ann Jorgensen, Julianne Anderson, and Janice Geiken, *Put Paperwork in its Place* (Des Moines, Iowa: Wallace-Homestead Book Company, 1983).

Glossary

A* = *L* + *NW Assets = Liabilities + Net Worth, the fundamental accounting equation.

ACRS The accelerated cost recovery system.

Accelerated cost recovery system (ACRS) A depreciation system adopted under ERTA 81 that standardized many depreciation variables including useful life, salvage value and depreciation rate.

Account payable The amount owed on open account for goods and services to creditors.

Account statement A record of transactions during a specified period concerning open account balances and charges.

Accounting The system of recording, reporting, and verifying assets, liabilities, income and expenses.

Accounting entity A separate individual, family, or business activity that requires a separate set of accounting records.

Accounting system An accounting and office procedure for the accumulation and reporting of financial data.

Accounts Separate records of goods and services purchased, used or sold.

Accounts receivable The amount owed to the business on open account for goods and services rendered to debtors.

Accrual accounting A system of accounting that recognizes income and expenses when they are effectively earned or incurred, rather than when cash changes hands.

Accrued liability A liability that has been incurred but not yet paid. For example, if mortgage interest is to be paid annually on February 1, 11 months of interest would have accrued by January 1, although the cash payment is not yet due.

Accumulated depreciation The total of all depreciation claimed on an asset from the time it was purchased to the present.

Acid-test ratio Same as the quick ratio; the ratio of quick or liquid assets to current liabilities.

Adjusted basis Original basis minus accumulated depreciation. Also called book value, however, the proper term according to IRS is adjusted basis.

Adjusted gross income Total income reported for income tax purposes minus specific adjustments for employee business expenses, moving expenses, two-wage household adjustments and others.

Agricultural Stabilization and Conservation Service A USDA agency responsible for administering price support programs.

Alimony Payments made by an individual to an ex-spouse. These payments usually reflect the input of one spouse in helping the other to achieve income.

Alternative minimum tax A tax process that requires high-income taxpayers to pay some income tax by recognizing tax preference items.

Amended tax return A process that allows taxpayers to correct errors in returns previously filed.

Amortization The reduction of a debt by regular payments (including interest and principal) sufficient to pay off a loan by maturity.

Amortization table A table consisting of various interest rates and time periods showing periodic payments required to amortize a loan.

Amortized loan A loan in which the amount borrowed is repaid in installments over the term of the loan.

Annual percentage rate (APR) Often referred to as APR, it reflects the interest and noninterest costs associated with a loan expressed as an annual percentage. APR may differ substantially from the contractural interest rate stated on the note.

Annuity A series of payments of a fixed amount covering a number of years.

Appraisal The process of estimating value of assets (generally real property).

Appreciation The long-term increase in value of an asset due to inflation, other economic conditions or an increase in demand for that asset. If, for example, land or corporate stock is valued above what was paid for it, the asset is said to have appreciated.

Asset Any item that is owned. Full value of land or of stored grain is classified as an asset, regardless of whether there is a mortgage or lien on that property.

Asset management The act of planning for asset use to yield returns.

Audit A check of business record books or tax returns for accuracy and completeness.

Average costs A concept related to the production function and related cost curves, representing average cost per unit of output. Average cost is defined as total costs divided by output, which is total variable costs plus total fixed costs divided by output.

Average physical product A concept related to the production function representing the average amount of output produced by a given amount of variable input, i.e., output divided by input.

Balance sheet A statement of all items of value owned (assets)—the claims against those assets, including debts owed (liabilities)—and the owner's net worth or equity at a specific point in time. The balance sheet is also called a net worth statement.

Balloon payment The final lump sum payment on a loan whose payments were calculated on a long-term amortization schedule whereas the loan was intended to be repaid in a few years.

Bankruptcy An economic and legal concept that arises when a business can no longer function because it has no financial resources with which to continue.

Barter income The products or services provided by another individual or business for which cash is not paid but rather for which other products or services are traded.

Basis A term used in income tax preparation and valuation of assets. The basis of an item generally refers to the value of cash and other assets paid for the item when originally purchased.

Beginning cash balance The amount of cash on hand at the beginning of a fiscal period.

Beginning inventory The amounts of products on hand at the beginning of a fiscal period.

Beneficiary The person who receives benefits, usually refers to the person who receives life insurance benefits upon the death of the policy holder.

Bond A debt obligation that usually bears interest and has a maturity that is generally at least 5 years. Interest payments are usually made at specific intervals to the bondholder and the principal is paid back at maturity.

Book value The cost of an item less its accumulated depreciation allowances.

Bookkeeping Keeping track of income and expenses for income tax, credit application, or management analysis purposes. Records provide the necessary basis for all forms of business analysis.

Boot The additional amount of money required to equalize a trade.

Budget A written estimate of future income and expenses covering a definite time period. Budgets can take several different forms, such as a partial budget, cash flow budget, a whole farm budget, or others, depending on the need or decision to be made.

Business analysis The process of determining the strengths and weaknesses of a business through careful scrutiny of records and financial statements.

Capital As used in capital assets, capital investments, capital improvements—to describe money (owned or borrowed) invested in anticipation of a return.

Capital assets Long-lived or productive assets; for example, tractors or breeding livestock.

Capital expenditures Payments made to purchase long-lived assets.

Capital expensing An income tax provision that allows extra depreciation to be claimed in the year of purchase.

Capital gains The differences between an asset's purchase price and its selling price, if positive. If negative, the difference is referred to as a capital loss.

Capital gains exclusion Prior to 1987, a certain amount (recently it was 60%) of gain on the sale of capital assets was excluded from taxation.

Cash A general reference to paper currency or coins.

In accounting terms it represents an account on the balance sheet including paper currency, coins, checks and bank balances.

Cash accounting method A system of recognizing income and expense items only at the time the actual cash is paid or received. It is also called cash basis.

Cash flow Actual cash coming into or going out of the business.

Cash flow budget A written estimate of projected cash income and outgo for a definite period. It is usually prepared on an annual period, by months, showing cash receipts, expenditures, and cash needed or surplus for each period.

Cash flow control The financial management process of monitoring and controlling cash flow and the cash position of the business.

Cash flow planning The financial management process of forecasting cash flows and of developing a cash flow budget.

Cash flow statement A cash flow budget.

Cash flow summary A condensed cash flow budget that shows the net cash position for one year without monthly or quarterly detail.

Cash journal A record book containing all cash receipts and expenditures.

Cash value of life insurance The cash value built through time by a whole-life insurance policy.

Casualty losses Losses of property due to theft, adverse weather, accident, or other catastrophe.

Certificate of Deposit (CD) A debt instrument of a bank that pays interest. Maturities range from a few weeks to several years. Denominations are as small as $100 and have no upper limit.

Check processing The action of a clearing bank involving the physical movement of the check from the receiver's bank to the writer's bank.

Check register A simple ledger to record the date, amount, and receiver of the check.

Closing and adjusting journal and ledger entries Entries in a double-entry bookkeeping system that summarize and/or close account balances.

Closing costs Expenses involved in transferring real estate. They include but are not limited to title searches, insurance, fees to file deeds, and mortgages and survey fees.

Collateral An asset that is pledged to secure a loan until the loan is repaid.

Commodity futures contract An agreement to deliver or accept delivery of a specific commodity with other provisions of the contract standardized by the individual exchange. Futures contracts can be sold or bought only through organized exchanges.

Communications satellites Satellites that are in geocentric orbits (approximately 23,300 miles above the earth's surface) such that they orbit the earth at the same speed that the earth rotates on its axis and thus appear to be stationary.

Comparative analysis A business analysis technique that involves comparing one business' records with another business or group of businesses.

Computer hardware The actual physical equipment, including the memory chips, keyboard, printer, and central processing unit (CPU).

Computer network A hookup involving several computers such that the individual computers can be used to communicate with other computers and/or such that the individual computers can be used simultaneously to gain speed and increased capacity.

Computer software The operations and commands that the computer executes. These are often thought of as programs that perform certain functions.

Consolidated financial statements Financial statements (balance sheets, income statements, cash flow statements) that combine business and personal items, including assets, income, and sources of funds.

Constructive receipt A tax term that describes proceeds from sales that may not yet be converted to cash or deposited in a bank account, but nonetheless are available. Example: a check received for sale of a commodity in late December, but not cashed until January.

Contingent liabilities A liability or debt incurred only if some specific event should occur, such as the sale of an asset or business.

Contingent tax liabilities A contingent liability that recognizes that income taxes would be due if a specific event should occur.

Contra account An account that offsets another account. For example, machinery depreciation is recorded in an account (called accumulated depreciation), which offsets the machinery account.

Contractual rate The rate agreed upon by each contracting party. Usually refers to the interest rate charged for loans or received on securities. May not be the annual percentage rate (APR).

Convertibility The ease with which an asset can be converted to cash (i.e., sold).

Cooperative stock A share of ownership of a corporation.

Coordinated Financial Statements (CFS) A comprehensive, systematic system of four financial statements developed specifically for agriculture by Thomas Frey and Danny Klinefelter.

Coordinated Financial Statements of Agriculture (CFS) A system of financial statements developed to meet the unique needs of U.S. farms and ranches.

Corporate distributions The net profit of a corporation that is allocated to the various debt instruments and stockholders.

Corporate stock A share of ownership of a corporation.

Corporation A business form composed of stockholders and treated as an artifical person.

Cost recovery A new term or system referring to what used to be called depreciation. Cost recovery began with ERTAM81 and ACRS.

Credit Money borrowed with the understanding that it will be repaid. Alternatively, credit may be referred to as the amount of money that can be borrowed, i.e., a credit reserve.

Credit application A series of questions usually concerning an individual's assets, liabilities, employment, and income. Used to render judgment about the ability of an individual to repay a loan.

Credit rating A judgment of an individual's ability to handle credit based on past performance, assets owned, repayment capacity, character, and other factors.

Credit risk The risk assumed by the lender that the borrower will not repay a loan.

Credit worthiness The proven ability of a borrower to use and repay borrowed funds.

Creditor One to whom a debt is owed—a lender.

Current asset An asset which is or may be converted to cash easily (such as stocks and bonds), one which will be sold within 1 year (such as grain in storage, feeder livestock, or a cull cow), or one with a useful life of less than 1 year.

Current debt-to-worth ratio A financial ratio that relates current liabilities to net worth.

Current liability A liability that will be paid off within 1 year, such as an operating loan, interest accrued on a long-term loan and due within 1 year, principal due within 1 year, or an accrued income tax liability.

Current market value The price or value of an asset in a given market during the current or present time period.

Current ratio A financial ratio that relates current assets to current liabilities.

Dairy Herd Improvement Associations Associations developed to assist dairymen with production records and milk testing.

Data processing The act of assembling raw information (primary data) into a more useable or condensed form usually involving a computer.

Debit An entry on the left side of an account causing an increase in assets or a decrease in liabilities and net worth.

Debt An obligation owed to another. Alternatively, debt refers to the total sum of money owed to others.

Debt financing The process of purchasing an asset with funds arranged through a loan. Usually involves a mortgage or lien on the purchased asset.

Debt loads The total amount of debt owed. The term is usually used in the context of how much debt a business can service and repay.

Debt repayment capacity The capacity that a business has to repay its debts out of current earnings.

Debt restructuring A financial process undertaken by a borrower and a lender to change the nature (not the amount) of debt. Usually involves lengthening the terms of loans.

Debt service records A ledger of payments made towards retiring a loan. Includes the amount of principal and interest and date of action.

Debt structure ratio A financial ratio relating current liabilities to total liabilities.

Decision making The process of determining a course of action.

Decision tree A business analysis tool that asks a series of questions in a logical order to help isolate or identify specific problems.

Declining-balance method A specific depreciation method once recognized by the IRS. The declining balance method allowed a faster depreciation (more depreciation in early years) than the straight-line method.

Deductible business expenses Legitimate business expenses that can be used to reduce taxable business income.

Deductible personal expenses Legitimate personal expenses that can be used to reduce taxable income.

Deduction An expense that is considered under tax law to be a legitimate reduction in income.

Depreciable assets Assets that qualify for depreciation. Depreciable assets must be business assets, be purchased, and have a definable useful life.

Depreciable basis The basis amount of an asset that qualifies for depreciation.

Depreciated value Book value of a depreciable asset.

Depreciation Annual accounting charge for the deterioration, obsolescence, or wear of an asset whose productive life is more than 1 year.

Depreciation recapture An IRS system that requires some depreciation claimed to be included in income when the asset is sold.

Depreciation schedule A record that shows the asset's purchase price, depreciable basis, and the history of depreciation charges against the asset.

Disposable income The amount of net income left after taxes.

Dividends Corporate profits distributed to shareholders.

Double-entry An accounting system that uses two entries, a debit and credit, for each transaction.

Dual valuation analysis The process of valuing assets on the basis of their original cost and on the basis of their current market value.

Earnings The value obtained from a profit and loss statement. Generally thought of as net profit after income taxes.

Economic Recovery Tax Act of 1981 (ERTA) The major tax law that introduced many changes to the Internal Revenue Code. One of its most important provisions was the adoption of ACRS.

Economic analysis A method used to determine if resources are allocated in a technical and market-efficient way.

Economic efficiency Use of resources in such a way as to maximize profits.

Economic useful life The length of time an asset can contribute earnings in excess of costs.

Ending cash balance The amount of cash on hand at the end of a fiscal period.

Ending inventory The amounts of products on hand at the end of a fiscal period.

Enterprise A particular activity of a business that can generally be separated from other activities.

Enterprise analysis An analysis of costs and returns specific to each crop and livestock commodity produced on the farm or ranch.

Enterprise budget A detailed listing of projected prices, yields, and costs associated with a specific crop or livestock commodity.

Enterprise ratios Financial ratios that relate to individual enterprises.

Entry A notation or recording of a transaction in a journal or ledger.

Equity The part of an asset or an entire business owned outright by the operator. For example, equity in a business is the value of the business less the amount owed. Sometimes expressed as percent of ownership.

Equity value ratio A financial ratio relating net worth to total assets.

ERTA Economic Recovery Tax Act of 1981.

Estate planning The process of planning for the transfer of property to heirs such that the goals and desires of the person making the transfer are met.

Estate tax A tax on the transfer of property to heirs.

Estimated tax The amount of tax that an individual or business projects to be due for the remaining portion of the tax year.

Excess itemized deductions The difference between total itemized deductions and the standard deduction.

Exemptions The number of people being claimed as dependents on a tax return. The IRS allows a specified amount of income to be excluded from taxation for each dependent or exemption.

Expenditures All items of cash outflow. Most expenditures are also business expenses, but others (such as the purchase of a machine or principal payments on a loan) are not.

Expenses Costs incurred in doing business, whether paid in cash or not. Many expenses are also expenditures, but others (such as depreciation) are not.

Face value The value stated on a security or insurance policy. Upon maturity or a stated condition the face value is the amount paid.

Family assets Assets owned by individuals, not by a business.

Family corporations Corporations formed by families to operate a family business.

Family liabilities Liabilities entered into to purchase family (nonbusiness) assets.

Family living budget A projection or plan of family living expenditures.

Farm Credit System (FCS) A farmer-owned cooperative that has become a major agricultural lender.

Farmers Home Administration (FmHA) A USDA agency that lends funds to farmers, ranchers, rural homeowners, and rural governments under strict eligibility rules. Usually FmHA is called the lender of last resort because its borrowers must not be able to qualify for commercial loans.

Farmers Tax Guide An IRS publication designed to assist taxpayers with farm and ranch income tax questions.

FICA Federal Insurance Contributions Act; a law passed in the 1930s that created the social insurance/retirement safety net now commonly called Social Security.

Filing status One of several broad categories [single,

married filing jointly, married filing separately, head of household, or qualifying widow(er)] used by IRS to classify taxpayers and determine the appropriate tax rates.

Financial ratio analysis A method of business analysis that emphasizes the use of financial ratios to measure business performance to discover strengths and weaknesses.

Financial ratios One of the several types of measurements used to indicate the position of a business. Ratios are often used to make comparisons with other businesses.

Financial statement Usually refers to a Balance Sheet, although it may include all financial statements such as the Income Statement and Cash Flow Budget.

Fixed assets Long-term assets.

Fixed costs Expenses that do not vary with level of production, such as depreciation and personal property taxes. For example, personal property taxes are the same on a tractor, regardless of whether it is used on one acre, 300 acres, or is not used at all.

Flow statement A financial statement that covers a period of time, thus showing a moving picture of the business. A flow statement is a direct contrast to a stock statement.

Foreclosure A legal action taken by a creditor to take possession of collateral in the event of a business' failure to make loan payments.

Fundamental accounting equation $A = L + NW$. All financial statements and all transactions under accrual accounting conform to this relationship. The equation must always remain in balance.

FUTA Federal Unemployment Tax Act; a law passed that requires certain employers to pay a tax based on employees' earnings into a fund that is used to compensate unemployed workers.

Futures market An organized market composed of nonprofit exchanges where futures contracts are bought and sold.

GAAP Generally Accepted Accounting Principles; a set of guidelines promulgated by the Fair Accounting Standards Board and adhered to by CPAs.

General journal The original book of entry for all transactions in an accrual accounting system.

Gross income Gross returns.

Gross margins A concept used in developing the Whole Farm Plan. Gross margins are equal to income over variable costs, i.e., gross receipts minus variable costs.

Gross ratio A financial ratio relating total farm or ranch expenses to the value of farm or ranch production.

Gross returns Total cash receipts from an enterprise, i.e., total yield times price.

Gross wages The total amount of wages paid to an employee without deductions for income taxes, social security taxes, and insurance.

Half-year convention Method of determining length of time an asset is owned in its first year. Under ACRS, all assets, regardless of actual purchase date, are assumed to have been purchased July 1, so each asset is credited with one-half year's depreciation for year of purchase.

Hedging Use of the futures markets to price commodities.

Historical (actual) financial statements Financial statements that describe the current or past position or activity of the business. Actual statements contrast to projected or pro forma statements.

Holding period The length of time an asset must be held before sale to qualify for the capital gains exclusion.

Horizontal analysis A business analysis technique that involves comparing one year's records with another year's or years' records; trend analysis.

Income Money or increased value of inventory earned. An accrual accounting concept, income reflects money earned from commodities produced or services rendered, regardless of whether that commodity was actually sold on the market or payment was received.

Income averaging An IRS method allowed prior to 1987 to allow several years' incomes to be averaged so as to avoid paying tax at high marginal tax rates.

Income statement Also called operating statement or profit and loss statement. An accounting of income and expenses for a period of time (usually 1 year) for the purpose of calculating net farm or ranch income or net loss for the period.

Income tax A tax based upon income of both individuals and corporations.

Income tax reporting The annual process of completing and filing an income tax return.

Inflation A continuous increase in the general level of prices, thus a decline in the purchasing power of money.

Interest A calculated rent paid for the use of someone else's money, or an opportunity cost on the use of owned capital.

Interest income Interest earned on savings accounts or other financial investments.

Interests on operating capital and equipment investment A calculated cost, or opportunity cost, on the use of capital in the business. For some farmers and ranchers, the interest cost might be a cash outlay whereas for others it might be an imputed cost.

Interest payable Interest that is owed or has been incurred but is not yet due.

Intermediate assets Assets that are less liquid than current assets or that have a useful life of more than 1 year but less than 10 or 15 years. Examples include trucks, breeding or dairy livestock, and machinery.

Intermediate liabilities Notes or mortgages held on intermediate assets.

Internal revenue code (IRC) The collection of legislative tax law; the basic tax law as passed by Congress.

Inventory control Managing inventory levels to meet business needs yet minimizing the costs of maintaining given levels of inventory.

Invoices A shipping form showing the items shipped, location, time, and price.

IRS The Internal Revenue Service; a U.S. Department of Treasury agency responsible for collection of federal taxes and for writing tax regulations.

ITC recapture An IRS system that requires some claimed investment tax credit to be repaid when the asset is sold.

Itemized deductions Personal deductions from taxable income allowed by IRS. Examples include medical expenses, taxes paid to other governments, interest paid, and charitable contributions.

Joint ownership Ownership that is legally in more than one person's name.

Joint returns Tax returns filed by a husband and wife on one form. A married couple may file their return jointly (as a couple) or separately (file two returns).

Joint ventures An association of individuals to carry out a business activity.

Journal The record book that has the first transaction of record. From the journal the entry will be transferred to the various ledger accounts.

Landlord The owner or lessor of land or other fixed asset.

Lease A contractual (verbal or written) obligation to use some asset belonging to another individual for a specified time period and for a specified fee. Only a fine distinction exists between a rental agreement and a lease.

Ledger A general term for the collection of various accounts that shows assets, liabilities, income, and expenses.

Legal description A set of surveying systems used to describe the location and boundaries of land.

Lessee The person who leases an asset from an owner (lessor).

Lessor The person who owns and leases an asset to an operator (lessee).

Leverage Use of borrowed funds as supplemental equity capital. The leverage ratio measures the degree to which borrowed capital exceeds net worth.

Leverage ratio A financial ratio relating liabilities to net worth.

Liability A debt or obligation expressed in terms of money.

Lien A creditor's claim against property. A mortgage is a lien against property.

Life insurance Insurance purchased by a policy holder that requires the insurance company to pay a specified amount to a beneficiary when the policy holder dies.

Like-kind property An exchange involving property that is considered for tax purposes to be similar and acceptable as a substitute.

Line of credit An agreed-upon amount, which may be borrowed in the course of normal operation. Interest is usually charged only on the amount actually used, and only over the period in which it is actually used.

Linear programming A mathematical technique that can be used to optimize the combination of enterprises on a farm or ranch. A sophisticated computer program to develop a whole farm or ranch plan.

Liquid assets Assets that can be quickly converted to cash.

Liquidity The ease and speed with which assets could be turned into cash. The term may also refer to a general business condition of having sufficient cash to meet obligations.

Loan application The package of documents submitted to a lender when requesting the extension of a loan.

Loan origination fees Charges against the receiver of a loan for obtaining the loan. Usually expressed as points.

Loan repayment capacity The ability, either through collateral or earnings, to repay a loan.

Long-term assets Assets with a useful life of more than 10 years. Examples include land and buildings.

Long-term liabilities Notes or mortgages held on long-term assets.

Loss carryovers An IRS system that allows excess tax losses accumulated in one year to be transferred or carried over to other years so as to offset income in those years.

Marginal costs A concept related to the production function and related cost curves, and represents the additional cost of one additional unit of output. Marginal cost is defined as a change in total cost divided by a related change in output.

Marginal physical product Marginal physical product is a concept related to the production function and represents the increase in output generated by the last unit of variable input applied, i.e., change in output produced by change in input used.

Marginal revenue Marginal revenue is a concept related to the production function. It is defined as the additional revenue generated by an additional unit of output, i.e., change in total revenue divided by change in output.

Marketing All activities involved in getting a product produced and distributed to consumers.

Marketing plan A plan of action for a business that involves goal setting and decisions about pricing, quality, quantity, timing, location, and methods of selling.

Microcomputers Powerful computers that are priced and sized to be attractive to individuals and small businesses.

Modified cash/accrual accounting system A hybrid system that retains the ease of cash systems but also maintains inventory records to allow year-end conversion to accrual financial statements.

Modified cost Book value.

Monitoring worksheet A worksheet used to monitor or compare cash flow as it occurs to that projected in the cash flow budget.

Mortgage A long-term lien held on a long-term asset.

Moving average Series of numbers that is averaged; then, based upon some criteria, a new number is added to the series and one removed. For a three-day average of prices, a new day is added and the oldest day removed so that the series always contains 3 numbers.

Mutual funds Financial pools created by mutual fund companies to pool small investors' funds in purchasing stocks or bonds.

Net capital ratio A financial ratio relating total assets to total liabilities.

Net farm or ranch income Returns to labor, management, capital, land, and risk, i.e., gross returns minus purchased inputs and fixed costs.

Net operating profit Gross returns minus total operation expenses.

Net worth Assets minus liabilities. The value of owner's equity, the amount that the business owner would have left over if all assets were sold and all liabilities repaid.

Nonbusiness assets Personal assets; assets not associated with the business.

Nondepreciable asset An asset that is not considered to be used up during a production process, such as land. Also, an asset such as a personal asset that does not qualify for depreciation in a business or tax sense.

Nonfarm or nonranch income Personal or other business income.

Nonrevolving line or a fixed amount loan A specified upper limit loan that can be taken. A nonrevolving line differs from a revolving line in that a revolving line limits the amount that can be outstanding at any one time, whereas a nonrevolving line limits the amount that can be borrowed.

Nontaxable exchange An exchange involving like-kind assets.

Note payable A loan owed to another business, such as a bank, or to another person.

Note receivable A loan that is owed to you or your business.

Off-farm or off-ranch income Income that is generated separate from the farm or ranch business.

Operating capital Operating expenses minus fixed costs. The amount of cash required for all purchased inputs (including labor, fuel, oil, and repairs) to produce a crop or livestock commodity, but excluding machinery, equipment, and land investments.

Operating expenses The total of all costs attributable to the production of a crop or livestock commodity, except interest.

Operating loan A loan obtained to cover normal operating (variable) expenses.

Opportunity cost Cost of using a resource in one enterprise when it could be used in alternative enterprises or investment opportunities, measured in terms of the return that would be obtained from using the resource in the alternative investment.

Optimal The best; usually refers to the combination of enterprises that maximizes profit.

Ordinary income Income earned from operations, wages, interest, rent, or other sources that is taxable; contrast with capital gains.

Outstanding loan balance The amount of principal remaining to be paid.

Overhead costs Fixed costs not directly associated with production, such as insurance, employee benefits, land taxes, and utilities. These costs, once agreed to, occur without regard to level of production, or whether production exists at all.

P & L statement Profit and loss statement; an income statement.

Par value The value of a stock or bond when issued.

Partial budgeting A management technique or planning procedure that lists items of receipts and expenses affected by a particular change in procedure or organization.

Partnership An association of individuals who agree to pool some items such as money, time, and talent and share in the profits or losses of a joint business.

Partnership distributions The share of profits or losses due each member of the partnership based upon each partner's percentage of ownership.

Perquisites Noncash benefits provided to employees.

Personal financial records Financial records or accounts maintained on personal or family concerns.

Physical efficiency use of resources in such a way as to maximize production.

Planning The process of applying forethought and forecasting to an operation—trying to anticipate scope, character, and results of such an operation.

Points A point equals 1% of some value, usually in reference to an original loan balance. Points are sometimes an additional cost of obtaining a loan.

Posting The process of transferring transactions from the journal to ledger accounts.

Premiums Expenditures or amounts paid to purchase insurance.

Prepaid expenses Expenditures paid before the good or service has been used or consumed.

Present value An estimate of today's value of money that will be received or spent at some future date. For example, if $1 is to be received in 2 years, and the inflation rate is 10%, the net present value of the $1 is only 83 cents.

Price volatility A measure of how prices move above and below the average price. Also called price variance.

Principal Face value of a debt instrument such as a loan or deposit on which interest is either paid or owed.

Pro forma An accounting term usually used in conjunction with the income statement which means the statement is a projection into the future or a budget.

Production function The physical relationship between the amounts of variable inputs applied to a crop or livestock commodity and the response of resulting output (typically yield or gain).

Profit Difference between the cost of production and the selling price for a product.

Profit maximization In economic terms, it involves producing an output that equates the marginal cost of producing one more unit with the marginal revenue.

Profit ratios Financial ratios used to measure profitability of the business.

Profit-to-asset ratio A financial ratio relating net income to total assets.

Profit-to-equity ratio A financial ratio relating net income to net worth.

Profit-to-gross income ratio A financial ratio relating net income to value of farm or ranch income.

Profitability The ability or capacity of the business to generate net income and cover costs.

Profitability measures Financial ratios and other calculations to measure profitability.

Property class Groups of assets.

Property tax A tax levied upon real property usually based on its value.

Pure economic profit Return to risk; the amount of earnings left over after all factors of production have been paid, including opportunity costs.

Quick ratio A financial ratio relating quick or liquid assets to current liabilities; the acid test ratio.

Rate of return on equity capital Return to owner's equity divided by equity, expressed as a percentage.

Rate of return on investment Net operating profit divided by the total machinery, equipment, and land investment. It measures profitability of assets in percentage terms.

Real estate Real property.

Real property Property that is land or permanently attached to land (e.g., buildings).

Recapture A system used by the IRS to require the repayment of tax credits and depreciation.

Receipt The term receipt can be used in two major ways: (1) a written record of a transaction or (2) money received. In cash accounting terminology, a receipt is any item of cash inflow, whether or not that money can be classified as income.

Recognized gain Gain or appreciation realized upon sale of an asset. The difference between sale price and adjusted basis.

Records association An association of individuals who pay an annual fee to receive advice and accounting services for their business. The associations are usually nonprofit corporations.

Records system The accounting/bookkeeping system chosen for a business. Could be cash, accrual, or a mixture of the two.

Refinanced An outstanding loan balance that is usually partially reduced and its maturity extended.

Rent A contractual (verbal or written) obligation to use some asset belonging to another individual for a specified time period and for a specified fee. Only a fine distinction exists between a rental agreement and a lease.

Rental income Income earned on property and paid to its owner.

Repayment capacity Ability of a business to produce receipts above expenditures. In other words, ability of a business to repay money borrowed for that business.

Resale value The expected price an item will command in the event it must be sold.

Residual claim Net worth; the value of assets minus the claims (or liabilities) of everyone except the owner.

Retirement accounts Financial accounts set up with banks or other financial institutions to provide funds for retirement.

Return to land and risk Net operating profit minus the interest charge on the use of machinery, equipment and operating capital.

Return to risk Return to land and risk minus an additional charge for land investment. Because the only residual is the return to risk, this concept is usually called pure profit because payments have been made to all other factors of production.

Revolving line of credit A type of operating loan.

Roll over To move from one time period to another to change maturity.

Rollover loans Short-term loans obtained with the anticipation the loan will be renewed at its due date.

S & L's Savings and loan institutions.

Sales contract A contract specifying terms of payment, usually involving an amortized loan between a buyer and seller of real estate.

Sales taxes Taxes collected on retail sales, usually expressed as a percent and collected by retailers for state and local governments.

Salvage value The expected price an item will command when its useful life is over. Usually scrap or slaughter value.

SEC Securities and Exchanges Commission; the federal agency that regulates publicly-held corporations and stock exchanges.

Securities Financial instruments such as stocks and bonds that have value.

Self-employment tax A tax levied on the income of self-employed individuals that represents their contribution to Social Security.

Share lease distributions The allocations made between the lessee and lessor as specified in the lease concerning certain expenses and income.

Short-term asset Current asset.

Single entry An accounting system involving only one transaction recording in a journal.

Social security tax A tax levied upon wages and self employment earnings to provide retirement and disability income to individuals.

Soil Conservation Service (SCS) A USDA agency responsible for technical assistance to farmers and ranchers for soil and water conservation.

Sole proprietorship The simplest form of business in which one individual owns and controls the business.

Solvency A term used to describe the long-term financial position of the business. A business is called solvent if the value of assets exceed liabilities and insolvent (or technically bankrupt) if liabilities exceed assets.

Speculation Activities involving buying and selling assets (often commodity futures contracts) in the anticipation of making a profit based on increase in asset value.

Standard deduction The amount IRS taxpayers are to deduct regardless of whether they can itemize deductions on Schedule A.

Statement of owner equity A short financial analysis that reconciles two balance sheets with the intervening.

Stock purchase requirements A stipulation that shares of ownership must be purchased in order to use the business. A requirement of Farm Credit Services and other cooperatives.

Stock statement A financial statement that presents financial conditions at a specific point in time, thus it is a snapshot, as opposed to a moving picture. A stock statement is a direct contrast to a flow statement.

Straight-line depreciation A specific depreciation method that allocates an equal amount of depreciation to each year of useful life.

Sum-of-years digits method A specific depreciation method once recognized by the IRS. SYD allowed a faster depreciation (more depreciation in early years) than the straight-line method.

Surety The assets (or collateral) pledged to back up a loan.

SYD Sum-of-years digits; a depreciation method once recognized by the IRS.

T-account A graphic accounting teaching device that aids in learning the relationships between debits and credits for various types of accounts.

Tangible asset An asset that has physical substance. Items such as goodwill, patents, and secret processes are considered intangibles.

Tax audit An audit of tax returns conducted by IRS to verify accuracy.

Tax credits Special provisions of tax law that reduce tax liability (not taxable income).

Tax liability The dollar amount of taxes due to all governments.

Tax planning The process of computing expected tax liability before the year is over so as to legally maneuver taxable income to offset taxable income.

Tax preference The so-called loophole; tax advantages provided to certain individuals or activities that reduce tax liability.

Tax withholding Taxes collected by an employer to the IRS or other governments, usually deducted from gross wages.

Taxable income The amount of income, after all deductions and adjustments, that is used to calculate income tax liability.

Tenant The renter or leasor of property. In agriculture, the typical tenant is the operator but not the owner of a farm or ranch.

Term life insurance Life insurance that remains in force only for a specified period of time and has no residual benefits. Unlike whole-life insurance, if premium payments on term life policies lapse, insurance is cancelled and policyholders have no cash value.

Time value of money A concept that reflects the fact that the present value of a given amount of money is greater if received now than if received in the future.

Total costs of production Variable costs plus fixed costs incurred in production.

Transaction Any financial event. Every transaction must be recorded in a general journal under an accrual accounting system. Only when cash changes hands is a transaction recognized under a cash accounting system.

Transactions journal A record book used to record all transactions of a business over some time period.

Trend analysis A business analysis technique that involves comparing one year's records with another year's or years' records; horizontal analysis.

Trial balance A bookkeeping/accounting process of checking to see that the net values of ledger accounts are in balance, i.e., the net value of asset accounts equals the net value of all liability and all net worth accounts.

Truth in Lending Act of 1968 A law that requires the terms of loans to be spelled out in a common format and to be prominently displayed. One of the most important components of TLA68 was the common use of APR.

Truth in Lending Simplification and Reform Act of 1980 A law that modified the Truth in Lending Act of 1968.

Turnover ratio A financial ratio relating the value of farm or ranch production to total productive assets.

USDA U.S. Department of Agriculture; a federal agency which oversees agricultural price support, education, conservation, research, and extension activities and food distributions programs.

Unemployment taxes Taxes levied on employers that create a compensation fund for unemployed workers.

Unpaid balance The portion of a loan or open account that has not been paid.

Useful life The number of years when an asset can provide its intended service.

Valuation guides Books published by various companies that provide estimates of the value of tractors, cars, trucks, and other machinery and equipment.

Value of farm or ranch production Gross income minus purchased livestock and purchased feed. This value represents the total income generated on this farm or ranch.

Variable costs of production Expenses that vary with the level of production, such as labor, fuel, oil and repairs, feed, fertilizer, and seed.

Variable interest rates Interest rates on borrowed money that vary from time to time, depending on the interest rates on major money markets or the prime interest rate.

Vertical analysis A business analysis technique that involves comparing components of the income statement to net income.

Vouchers A written record of a business transaction used as an internal control for the business.

Whole farm or ranch budget A Pro Forma (or Projected) Income Statement built from the enterprises, selected in the Whole Farm or Ranch Plan.

Whole farm or ranch plan A forward plan of farm/ranch enterprise organization for the coming year. An attempt to determine the most profitable combination of enterprises within the constraints of available fixed resources, rotations, conservation and financial determinants.

Whole-life insurance policies Life insurance that builds a residual or cash value. Policy terms differ greatly but usually include provisions for a predetermined premium that will not increase as the policyholder grows older.

Withdrawals Funds extracted from the business, usually for family living expenses, gifts, or distributions to owners.

Withholdings Deductions from paychecks for such items as social security and income taxes. Employers deduct these amounts, hold them in special funds and eventually deposit them with appropriate taxing authorities.

Word processors Computers designed primarily as electronic typewriters with memory and editing capability.

Working capital The amount of capital or current assets available for use in operating the business. Commonly figured as the amount by which current assets exceed current liabilities.

Zero bracket amount The standard deduction on income tax returns.

Zero-basis assets Assets that have no original basis (or purchase price). Typical examples include raised breeding livestock.

Index